Strategies for Sustainability

Series Editors
Lawrence Susskind
Ravi Jain

For further volumes:
http://www.springer.com/series/8584

Strategies for Sustainability

Aims and Scope
The series, will focus on "implementation strategies and responses" to environmental problems – at the local, national, and global levels. Our objective is to encourage policy proposals and prescriptive thinking on topics such as: the management of sustainability (i.e. environment-development trade-offs), pollution prevention, clean technologies, multilateral treaty-making, harmonization of environmental standards, the role of scientific analysis in decision-making, the implementation of public-private partnerships for resource management, regulatory enforcement, and approaches to meeting inter-generational obligations regarding the management of common resources. We will favour trans-disciplinary perspectives and analyses grounded in careful, comparative studies of practice, demonstrations, or policy reforms. We will not be interested in further documentation of problems, prescriptive pieces that are not grounded in practice, or environmental studies. Philosophically, we will adopt an open-minded pragmatism – "show us what works and why" – rather than a particular bias toward a theory of the liberal state (i.e. "command-and- control") or a theory of markets.

We invite Authors to submit manuscripts that:
Prescribe how to do better at incorporating concerns about sustainability into public policy and private action.

Document what has and has not worked in practice.

Describe what should be tried next to promote greater sustainability in natural resource management, energy production, housing design and development, industrial reorganization, infrastructure planning, land use, and business strategy.

Develop implementation strategies and examine the effectiveness of specific sustainability strategies. Focus on trans-disciplinary analyses grounded in careful, comparative studies of practice or policy reform.

Provide an approach "...to meeting the needs of the present without compromising the ability of future generations to meet their own needs," and do this in a way that balances the goal of economic development with due consideration for environmental protection, social progress, and individual rights.

The Series Editors welcome any comments and suggestions for future volumes

Mitsuhiko Kawakami • Zhen-jiang Shen
Jen-te Pai • Xiao-lu Gao • Ming Zhang
Editors

Spatial Planning and Sustainable Development

Approaches for Achieving Sustainable Urban Form in Asian Cities

Editors
Mitsuhiko Kawakami
Kanazawa University
Kanazawa, Ishikawa
Japan

Zhen-jiang Shen
Kanazawa University
Kanazawa, Ishikawa
Japan

Jen-te Pai
Chengchi University
Taipei, Taiwan

Xiao-lu Gao
Institute of Geographic Sciences and Natural Resources Research
Chinese Academy of Sciences
Beijing, China

Ming Zhang
School of Architecture
University of Texas at Austin
Austin, Texas, USA

ISBN 978-94-007-9871-7 ISBN 978-94-007-5922-0 (eBook)
DOI 10.1007/978-94-007-5922-0
Springer Dordrecht Heidelberg New York London

Printed on acid-free paper

Springer is part of Springer Science+Business Media (www.springer.com)

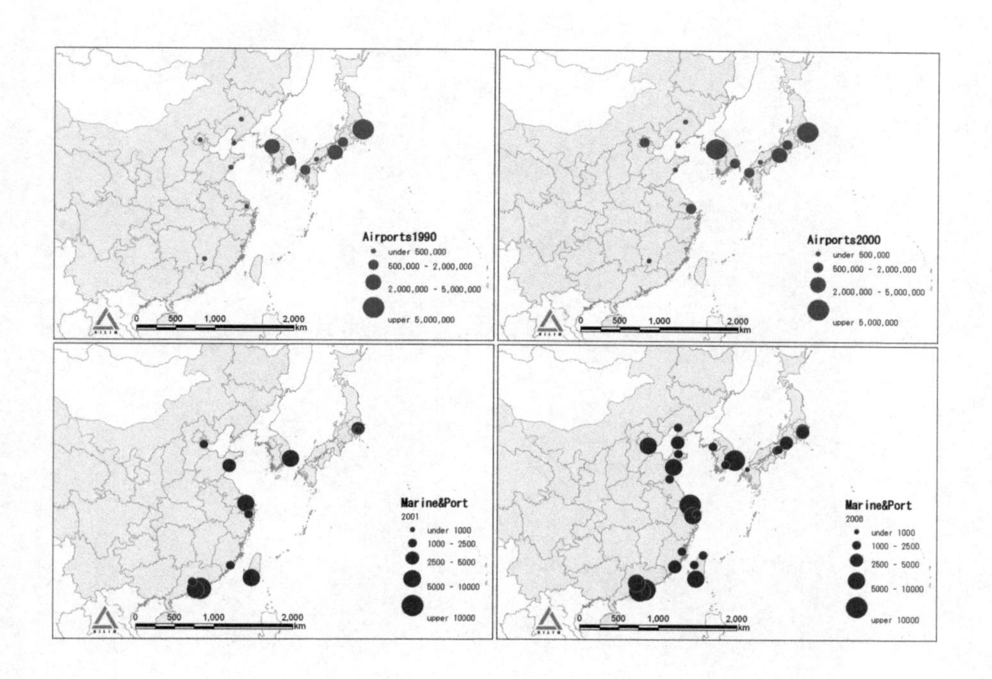

Airports1990
under 500,000
500,000 - 2,000,000
2,000,000 - 5,000,000
upper 5,000,000
Airports2000
under 500,000
500,000 - 2,000,000
2,000,000 - 5,000,000
upper 5,000,000
Marine&Port
2001
under 1000
1000 - 2500
2500 - 5000
5000 - 10000
upper 10000
Marine&Port
2008
under 1000
1000 - 2500
2500 - 5000
5000 - 10000
upper 10000
0 500 1,000 2,000 km

Preface on Behalf of the Editors

Researchers across the world are concerned with sustainable urban forms, and this field is particularly significant for policy planners aiming for sustainable and smart growth. This book investigates the impact of policy on sustainable urban forms through spatial planning implementation, which has been examined by analyzing Asian planning experiences from a multidisciplinary viewpoint that involves different professional planning in various fields, such as land use, transportation, geography, and environment.

Sustainable urban form represents the objectives of spatial planning and relevant policies. For example, the compact city, one of the concepts of sustainable urban form with a high density of urban settlements, has revitalized central city areas, and mixed land use has been widely accepted in Europe, North America, and Asia. Urban form is a result of the interactions between stakeholders using spatial planning and relevant planning policies and considering economic, social, and ecological aspects. Private investors are always competing for low cost and high profit; thus, the land demand of such private sectors as industries, shopping malls, and housing development projects tends to be located at the urban fringes, thereby resulting in urban sprawl. In this book, the authors argue that sustainable urban form is possible under effective urban policies in the process of spatial planning implementation. "Public policy" in this book refers broadly to government actions in planning and implementation.

Overall, this book attempts to provide insight on achieving sustainable urban form by focusing on planning practices at the city level and certain metropolitan areas in different Asian countries. Currently, some cities in developing countries are experiencing rapid urban growth, whereas many cities in developed countries are experiencing urban decline because of depopulation and an aging society. We can attempt to learn from both sides in order to achieve sustainable urban forms that employ a multidisciplinary approach, considering natural resources, aging societies and population transformations, housing developments, transportation and land use, and landscapes.

Although local governments have made many efforts to implement the compact city concept in many of the developed cities of Japan and South Korea, urban sprawl has substantially influenced city form. In order to find a sustainable urban form in the developing period, many developing Asian cities nowadays are learning from the experiences of their European counterparts. However, most of the cities in developing Asian countries, such as those in China, still pay more attention to economic development and physical planning, following the history of urban sprawl in European, US, or Asian developed cities. Therefore, conflicts are emerging between economic extension and compact urban areas in Asian cities. In such situations, we believe that effective planning policies are necessary for reaching a sustainable urban form.

Public planning policies are important in achieving sustainable urban form and controlling urban sprawl in both developing and developed countries. For cities experiencing urban growth, it is important for local governments to set demand allocation patterns, such as for industry, housing, and transportation. Meanwhile, setting the balance between social and ecological quality in the economic development process is an important task of local governments. For those cities undergoing urban decline, it is important for city governments to make effective decisions on public planning policies in order to prevent decreasing population, to improve urban regeneration, and to increase centrality. It is also important to introduce a new public transportation system for improving access to downtown areas. Moreover, cooperation between public actors and the private sector is important for using new advanced environmental technology to improve ecological functions in dense urban areas.

We have organized this book into five parts. The first two parts concern urbanization and sustainable society, and the following three parts deal with landscaping and ecological systems for sustainable development.

In the first two parts, we focus on planning issues regarding urbanization and de-urbanization. In Part I (Urbanization and Planning Approaches) and Part II (Housing and Transportation), policy measures in planning and design are taken as important tools to achieve sustainable urban form.

In Part I, we see that through decades of urban development, the local cities in developed Asian countries are now experiencing urban decline from large-scale development projects on the urban fringe—namely, urban sprawl. Spatial strategies for improving centrality and increasing population in urban areas are taken into account for preventing this trend. We have focused on planning issues in downtown areas with advanced depopulation and aging societies resulting from urban sprawl. We have also introduced some planning practices to decrease the negative influences of urban decline, such as urban regeneration by implementing appropriate design guidelines and developing urban facilities for aging societies. On the other hand, in developing countries where approaching urban development with economic growth leads to exploitation and use of natural resources in excess, public efforts for spatial planning are expected to encourage an environmentally friendly development in order to improve sustainable society.

In the next part, we focus on public policies regarding housing and transportation. Housing policies are challenged against a background of rapid urbanization. Many traditional dwellings have been abandoned in favor of flat roof houses; meanwhile, traditional culture is absent. This part tries to explore a sustainably oriented housing development, while keeping traditional society in the historical areas. On the other hand, we also suggest pursuing housing policies through enriching the methodology for predicting housing demand patterns.

Research has recently been carried out on strategies whereby public transportation system can provide a solution for traffic congestion in urban areas. In this part, we do not include these popular transportation topics. Rather, we present some unique and new ways to achieve a low-carbon transport society; for example, we are interested in representative technological innovation, such as personal mobility vehicles (PMV). We investigate the significant sociopsychological factors that can influence the acceptance of PMVs in society so that public policy may be formed on the application of PMVs. We also examine the advantages of bicycle transportation and conduct a city-wide evaluation of the walking accessibility and bus availability of urban facilities and public transit; walking and bicycle transportation are now considered as completely pollution-free methods.

Policy makers have now found that urban transportation energy as the main part of urban energy consumption has a strong relationship with urban form. We attempt to develop some tools for evaluating plan alternatives in terms of transportation energy consumption. A theoretical model is introduced by our colleagues in this stage, which is likely to be applied in the Beijing metropolitan area.

The remaining three parts consider environment and ecological issues: Part III (Green Design and Landscape), Part IV (Agricultural and Ecological Systems), and Part V (Urban Vulnerability).

Part III looks at local governments' support for the development of new green technologies, such as green curtains and green roof systems for improving the urban thermal environment and reducing CO2 emissions. Under such local environmental policies, some case studies show the benefits of green technologies for ecological functioning and urban landscape in dense urban areas of Japan. However, because environmental planning and urban planning are separated in most planning systems in Asian countries, even in developed countries, we attempt to argue that integration of environmental planning, including ecological vulnerability, with urban planning, is very important. Additionally, we describe a system framework for assessment and regulation of ecological security when implementing urban planning.

In Part IV, on agricultural and ecological systems, we show that rapid economic growth and urbanization have led to a series of resource and environmental problems. We discuss the existing agricultural status and environmental impact and propose new agricultural planning and policies whereby agriculture may not only exert its production functions but also fulfill landscape and ecological functions for making a more comfortable and sustainable living environment. In terms of spatial planning to deal with these issues, spatial indicators that reflect the patterns of land use and social patterns, such as land ownership, are very useful for achieving sustainable landscape management. We present some case studies in which geospatial techniques were used as new planning tools that played an important role in the spatial planning process.

Finally, the perspective switches from ecology back to urban systems. The part on urban vulnerability shows that, during urban or economic development with its consequent competition for land, vulnerability to urban systems floods, drought, and pollution has become a widespread concern. We recommend establishing a pragmatic overall index in order to increase the number of reference values for disaster assessments and disaster preventions based on spatial planning and relevant planning policies.

In the introduction to this book, we argue that public efforts are important in all case studies, from planning and design to policy-making. The key contribution of the book concerns the role of public actors in implementing spatial planning. The sustainable urban form is examined according to different scales—such as the human, urban structure, landscape and ecological structure, and global scales—which are related to planning and design issues at the national, regional, and urban levels.

School of Environmental Design,
Kanazawa University, Japan

Mitsuhiko Kawakami
and Zhen-jiang Shen

Acknowledgement

We would like to express our deep appreciation to all the authors, each of whom is from the International Community of Spatial Planning and Sustainable Development, for their outstanding contributions to this book. We are also indebted to the series editors of Strategies for Sustainability, Professors Lawrence Susskind and Ravi Jain, for their kind invitations and encouragement, and to Dr. Tamara Welschot and Mrs. Judith Terpos of Springer, for their kind editorial work and help in publishing within a book a diversity of topics pertaining to "Spatial planning and sustainable development: Approaches for Achieving Sustainable Urban Form in Asian Cities."

We are also deeply indebted to our colleagues Dr. Kyung-rock Ye and Dr. Ryosuke Ando, whose help and stimulating suggestions helped us in soliciting contributions from a variety of interdisciplinary fields. Special thanks go to Professor Qi-zhi Mao of Tsinghua University; Professor Yi Liu of the Institute of Geographic Sciences and Natural Resources Research (Chinese Academy of Sciences), Mainland China; and Professor Cheng-ming Feng of Jiaotong University, Professor Chian-yuan Lin of Taiwan University, Taiwan, for their efforts in supporting the International Community of Spatial Planning and Sustainable Development.

Editors

Mitsuhiko Kawakami
School of Environmental Design
Kanazawa University,
Kakuma Machi, Kanazawa City,
Japan, 920-1192
kawakami@t.kanazawa-u.ac.jp
Tel.: 0081-76-234-4914

Zhen-jiang Shen*
School of Environmental Design
Kanazawa University,
Kakuma Machi, Kanazawa City,
Japan, 920-1192
shenzhe@t.kanazawa-u.ac.jp
Tel.0081-76-234-4650

Jen-te Pai
School of Social Science
Chengchi University, 64, Sec. 2
Chinan Rd. Taipei, Taiwan
brianpai@nccu.edu.tw
Tel. 00886-2-29393091-51663

Xiao-lu Gao
Institute of Geographic Sciences and Natural Resources Research,
Chinese Academy of Sciences
11A, Datun, Chaoyang, Beijing, China
gaoxl@igsnrr.ac.cn
Tel. 0086-10-64889075

Ming Zhang
School of Architecture
University of Texas at Austin
Austin, TX 78712, U.S.A.
zhangm@mail.utexas.edu
Tel.001-512-471-0139

*Corresponding editor

About the Editors

Mitsuhiko Kawakami is a Professor at the School of Environmental Design, Kanazawa University, Kanazawa City, Japan. Dr. Kawakami serves as Chairperson of the Urban Planning Committee and the Land Use Committee of Ishikawa Prefecture, Advisor of Urban Planning for Kanazawa City, and Director of the Kanazawa Citizens' Research Organization. He also serves as the Commissioner of Urban Planning and Design at the Architectural Institute of Japan and a Councilor of the City Planning Institute of Japan. His research interests include land-use planning, housing planning, planning support systems, and environmental design. He is the main investigator of a number of historical conservation projects in Kanazawa City that support local policy decision-making on urban conservation. He also participated in the revision of Kanazawa's Master Plan and a diversity of urban projects. Recently, Dr. Kawakami was selected as the President of the nonprofit organization Kanazawa Traditional House (Machiya).

Zhen-jiang Shen is a Professor at the School of Environmental Design, Kanazawa University, Kanazawa City, Japan. His research interests include the urban policy of China, decision support systems for planning, and design through the use of GIS and VR. Recently, Dr. Shen has been collaborating with the Beijing Municipal Commission of Urban Planning for research on metropolitan growth simulation. He has also served as an Advisor of urban planning in local cities within Ishikawa Prefecture, Japan. In planning practice, he has also participated in a diversity of urban projects, including an early-stage historical conservation plan for Beijing, which won the second prize awarded by the Ministry of Construction, China, in 1987. In 2010, he won the Heritage Conservation Award, Region IV, UIA (International Union of Architects) for his work on historical landscape visualization for traditional temple-building preservation, in Kanazawa, Japan.

Jen-te Pai is an Associate Professor at the National Cheng Chi University in Taiwan and also serves as the Secretary-General of the Taiwan Institute of Urban Planning. He received a Ph.D. degree from National Taiwan University and worked as a government officer in the Ministry of Transportation and Communication.

His teaching and research areas include urban and regional planning, urban design, industrial cluster analysis, and disaster-prevention planning.

Xiao-lu Gao is a Professor at the Institute of Geographic Sciences and Natural Resources Research at the Chinese Academy of Sciences and Vice Director of the Key Laboratory of Regional Sustainable Development Modeling, Chinese Academy of Sciences. She holds a Bachelor's Degree in architecture and a Master's Degree in city planning and design, both from Tsinghua University, as well as a Ph.D. in urban engineering from the University of Tokyo. As an urban planner and geographer, her research interests include urban and regional development policies, urban and housing analysis, the evaluation of urban environments, and the construction of planning support systems. In recent years, she has published a number of research papers in a number of geographical and planning journals, including *Urban Studies, Landscape and Urban Planning, Land Use Policy, Journal of Geographical Sciences, Environment and Planning B and C, Housing Studies, and Habitat International*.

Ming Zhang is tenured Associate Professor and Graduate Advisor in the Community and Regional Planning Program at the University of Texas, Austin. He specializes in planning for land use–transportation integration, GIS applications, and planning in an international setting. Prior to joining UT, Austin, Dr. Zhang held several academic and professional positions, including tenure-track Assistant Professor in the Department of Landscape Architecture and Urban Planning, at Texas A&M University; Research Scientist at the Rockefeller Institute of Government in Albany, New York; and Lecturer and licensed Planner/Architect at the Huazhong (central China) University of Science and Technology, Wuhan, China. Dr. Zhang has published research papers in the *Journal of the American Planning Association*, *Journal of Planning Education and Research*, and *Urban Studies*.

Contents

1 Overview: Spatial Planning for Achieving Sustainable Urban Forms 1
Zhen-jiang Shen and Mitsuhiko Kawakami

Part I Urbanization and Sustainable Society: Section Urbanization and Planning Approach

2 The Possibility of Sharing Spatial Data and Research Cooperation Within East Asia Countries: For Sustainable and Balanced Regional Growth 15
Kyung-rock Ye

3 A Study on Classification of Downtown Areas Based on Small and Medium Cities in Korea 33
Bum-hyun Lee

4 Significance and Limitations of the Support Policy for Marginal Hamlets in the Strategy of Self-sustaining Regional Sphere Development 51
Ryohei Yamashita and Tomohiro Ichinose

5 Continuity of Relations Between Local Living Environments and the Elderly Moved to a Group Living 69
Tatsuya Nishino

6 The Use of Indicators to Assess Urban Regeneration Performance for Climate-Friendly Urban Development: The Case of Yokohama Minato Mirai 21 91
Osman Balaban

7 **Imagination and Practice of Collaborative Landscape, Ecological, and Cultural Planning in Taiwan: The Case of Taichung County and Changhua County** 117
Li-wei Liu and Pei-yin Ko

Part II Urbanization and Sustainable Society: Section Housing and Transportation

8 **Sustainable-Oriented Study on Conservation Planning of Cave-Dwelling Village Culture Landscape** 137
An-rong Dang, Yan Zhang, and Yang Chen

9 **Characteristic of Sustainable Location for Townhouse Development in Bangkok and Greater Metropolitan Area, Thailand** 155
Siwaporn Klimalai and Kiyoko Kanki

10 **Modeling Housing Demand Structure: An Example of Beijing** 173
Xiao-lu Gao

11 **The Role of the Knowledge Community and Transmission of Knowledge: A Case of Bicycle SMEs in Taiwan** 189
Jen-te Pai and Tai-shan Hu

12 **Acceptability of Personal Mobility Vehicles to Public in Japan: Results of Social Trial in Toyota City** 213
Ryosuke Ando, Ang Li, Yasuhide Nishihori, and Noriyasu Kachi

13 **Urban Form, Transportation Energy Consumption, and Environment Impact Integrated Simulation: A Multi-agent Model** 227
Ying Long, Qi-zhi Mao, and Zhen-jiang Shen

14 **Mapping Walking Accessibility, Bus Availability, and Car Dependence: A Case Study of Xiamen, China** 249
Hui Wang

Part III Landscape and Ecological System, Sustainable Development: Section Green Design and Landscape

15 **Effects of Green Curtains to Improve the Living Environment** 271
Masashi Kato, Tsukasa Iwata, Norimitsu Ishii, Kimihiro Hino, Junichiro Tsutsumi, Ryo Nakamatsu, Yoshitaka Nishime, Koji Miyagi, and Masakazu Suzuki

16 **A Comparison of Green Roof Systems with Conventional Roof for the Storm Water Runoff** 287
Sachiko Kikuchi and Hajime Koshimizu

17 Evaluation and Regulation of Ecological Security When Implementing Urban Planning: Review and Suggestions for Spatial Planning and Sustainable Development in China 305
Lin-yu Xu and Zhi-feng Yang

Part IV Landscape and Ecological System, Sustainable Development: Section Agriculture and Ecological System

18 An Investigation of Changes in the Urban Shadow of Beijing Metropolis Under Agricultural Structural Adjustment in China 325
Dai Wang, Yue-fang Si, Wen-zhong Zhang, and Wei Sun

19 The Spatial Planning of Agricultural Production in Beijing Toward Producing Comfortable and Beautiful Living Environment 339
Feng-rong Zhang and Hua-fu Zhao

20 Simplified Ecological Planning Method for Sustainable Landscape Management by *Humantope Index*: Patterns of Land-Use Continuity, Historical Land Use and Landownership 353
Misato Uehara

21 Land Cover Analysis with High-Resolution Multispectral Satellite Imagery and Its Application for the CO_2 Flux Estimation 381
Jung-Rack Kim, Shih-Yuan Lin, Eun-Mi Chang, In-Hee Lee, and He-Won Yun

Part V Landscape and Ecological System, Sustainable Development: Section Vulnerability of Urban System

22 Taiwan's Five Major Metropolitan Areas of Taiwan Vulnerability Assessment of Flood Disaster Comparison Study 401
Jen-te Pai

23 The Post-Disaster Reconstruction and Socioeconomic Vulnerabilities in the Historical Site of an Island City: A Case Study of a Fire Incident in Nan-Gan Township, Lien-Chiang County, Taiwan 417
Chi-tung Hung, Wen-yen Lin, and Ju-yin Cheng

24 Sustainable Communities in Hilly, Mountainous and Heavy Snowfall Areas 431
Asako Yuhara and Kyung-rock Ye

25 A Vulnerability Study from Water Perspective on the Largest City of China 443
Guang-wei Huang and Zhen-jiang Shen

Index 455

Contributors

Ryosuke Ando Research Department, Toyota Transportation Research Institute, Toyota, Aichi, Japan

Osman Balaban Institute of Advanced Studies, United Nations University, Yokohama, Japan

Eun-Mi Chang Ziin Consulting Inc., Jongno, Seoul, South Korea

Yang Chen School of Architecture, Tsinghua University, Haidian District, Beijing, China

Ju-yin Cheng Department of Urban Development, Taipei City Government, Xinyi District, Taipei City, Taiwan

An-rong Dang School of Architecture, Tsinghua University, Haidian District, Beijing, China

Xiao-lu Gao Key Laboratory of Regional Sustainable Development Modeling, Institute of Geographic Sciences and Natural Resources Research, Chinese Academy of Sciences, Chaoyang, Beijing, China

Kimihiro Hino Department of Housing and Urban Planning, Building Research Institute, Tsukuba City, Ibaraki Prefecture, Japan

Tai-shan Hu Department of Architecture and Urban Planning, Chung Hua University, Hsinchu, Taiwan

Guang-wei Huang Sophia University, Chiyoda-ku, Tokyo, Japan

Chi-tung Hung Department of Urban Planning and Disaster Planning, Ming Chuan University, Guishan Township, Taoyuan County, Taiwan

Tomohiro Ichinose Department of Environment and Information Studies, Keio University, Fujisawa, Kanagawa, Japan

Norimitsu Ishii Department of Housing and Urban Planning, Building Research Institute, Tsukuba City, Ibaraki Prefecture, Japan

Tsukasa Iwata Department of Housing and Urban Planning, Building Research Institute, Tsukuba City, Ibaraki Prefecture, Japan

Noriyasu Kachi Research Department, Toyota Transportation Research Institute, Aichi, Toyota, Japan

Kiyoko Kanki Graduate School of Engineering, Kyoto University, Nishikyo-ku, Kyoto, Japan

Masashi Kato Department of Housing and Urban Planning, Building Research Institute, Tsukuba City, Ibaraki Prefecture, Japan

Mitsuhiko Kawakami School of Environmental Design, Kanazawa University, Kakuma Machi, Kanazawa City, Japan

Sachiko Kikuchi Graduate School of Life Sciences, Tohoku University, Aoba-ku, Sendai, Japan

Jung-Rack Kim Department of Geoinformatics, University of Seoul, Dongdaemun-gu, Seoul, South Korea

Siwaporn Klimalai Graduate School of Engineering, Kyoto University, Nishikyo-ku, Kyoto, Japan

Pei-yin Ko Department of Urban Planning and Spatial Information, Feng Chia University, Seatwen, Taichung, Taiwan

Hajime Koshimizu School of Agriculture, Meiji University, Tamaku, Kawasaki, Japan

Bum-hyun Lee Korea Research Institute for Human Settlement, Dongan-gu, Anyang-si, Gyeonggi-do, South Korea

In-Hee Lee Department of Environment and Ecology Research, Chungnam Development Institute, Yonesuwon-gil, Gong-Ju, Chungcheonnamdo, South Korea

Ang Li Urban Transport Center, Ministry of Housing and Urban–Rural Development, China

Shih-Yuan Lin Department of Land Economics, National Chengchi University, Wenshan District, Taipei, Taiwan

Wen-yen Lin Department of Urban Planning and Disaster Planning, Ming Chuan University, Guishan Township, Taoyuan County, Taiwan

Li-wei Liu Department of Urban Planning and Spatial Information, Feng Chia University, Seatwen, Taichung, Taiwan

Ying Long Beijing Institute of City Planning, Beijing, China

Qi-zhi Mao School of Architecture, Tsinghua University, Haidian District, Beijing, China

Koji Miyagi Ocean Expo Research Center, Motobu-chom Okinawa Prefecture, Japan

Ryo Nakamatsu Department of Civil Engineering and Architecture, University of Ryukyus, Nishihara-cho, Okinawa Prefecture, Japan

Yasuhide Nishihori Chubu Branch, Chuo Fukken Consultants Co. Ltd, Nagoya, Japan

Yoshitaka Nishime Ocean Expo Research Center, Motobu-cho, Okinawa Prefecture, Japan

Tatsuya Nishino School of Environmental Design, Kanazawa University, Kakuma Machi, Kanazawa City, Japan

Jen-te Pai Department of Land Economics, National Chengchi University, Wenshan District, Taipei City, Taiwan

Zhen-jiang Shen School of Environmental Design, Kanazawa University, Kakuma Machi, Kanazawa City, Japan

Yue-fang Si Department of Economic Geography, Justus-Liebig-University Giessen, Giessen, Germany

Wei Sun Institute of Geographic Sciences and Natural Resources Research, Chinese Academy of Sciences, Chaoyang District, Beijing, China

Masakazu Suzuki Faculty of Arts and Design, University of Tsukuba, Tsukuba City, Ibaraki Prefecture, Japan

Junichiro Tsutsumi Department of Civil Engineering and Architecture, University of Ryukyus, Nishihara-cho, Okinawa Prefecture, Japan

Misato Uehara Department of Forest Science, Shinshu University, Nagano, Japan

Dai Wang Institute of Geographic Sciences and Natural Resources Research, Chinese Academy of Sciences, Chaoyang District, Beijing, China

Hui Wang Department of Urban Planning, Xiamen University, Xiamen, China

Cross-Straits Institute of Urban Planning at Xiamen University, Xiamen University, Xiamen, China

Lin-yu Xu State Key Joint Laboratory of Environment Simulation and Pollution Control, School of Environment, Beijing Normal University, Haidian District, Beijing, China

Ryohei Yamashita Tokyo University of Science, Yamazaki, Noda, Chiba, Japan

Zhi-feng Yang State Key Joint Laboratory of Environment Simulation and Pollution Control, School of Environment, Beijing Normal University, Haidian District, Beijing, China

Kyung-rock Ye Graduate School of Frontier Sciences, The University of Tokyo, Bunkyo District, Tokyo, Japan

Asako Yuhara National Institute for Land and Infrastructure Management, Tsukuba, Ibaraki, Japan

He-Won Yun Department of Geoinformatics, University of Seoul, Dongdaemun-gu, Seoul, South Korea

Feng-rong Zhang College of Resources and Environmental Sciences, China Agricultural University, Haidian District, Beijing, China

Wen-zhong Zhang Institute of Geographic Sciences and Natural Resources Research, Chinese Academy of Sciences, Chaoyang District, Beijing, China

Yan Zhang School of Architecture, Tsinghua University, Haidian District, Beijing, China

Hua-fu Zhao College of Land Science and Technology, China Geosciences University, Haidian District, Beijing, China

Chapter 1
Overview: Spatial Planning for Achieving Sustainable Urban Forms

Zhen-jiang Shen and Mitsuhiko Kawakami

Abstract In this chapter, as an introduction and summary of this book, we will summarize its key contributions regarding the role of public actors in the implementation of spatial planning. We will introduce each chapter, showing how it will contribute to the book's main conclusions.

Keywords Asian cities • Urbanization • Housing • Transportation • Agriculture • Ecological system • Urban vulnerability

1.1 Introduction

This book presents some case studies regarding spatial planning in pursuing sustainable urban form, based on existing concepts of sustainable urban form in the planning practice of various Asian countries. We discuss a matrix of practice examples from a multidisciplinary perspective, including land-use patterns, housing development, transportation, green design, and agricultural and ecological systems to help readers assess the contribution of this book on the achievement of a sustainable urban form while avoiding the vulnerability of the urban system.

Regarding sustainable urban form, many definitions and discussions already exist in the vast body of literature. Jabareen (2006) has presented seven design concepts related to sustainable urban forms: compactness, sustainable transport, density, mixed land uses, diversity, passive solar design (energy saving), and greening.

Moreover, there are four types of sustainable urban forms: neo-traditional development (Furuseth 1997), urban containment (Dawkins and Nelson 2002),

---

Z.-j. Shen (✉) • M. Kawakami
School of Environmental Design, Kanazawa University, Kakuma Machi, Kanazawa City 920-1192, Japan
e-mail: shenzhe@t.kanazawa-u.ac.jp; kawakami@t.kanazawa-u.ac.jp

M. Kawakami et al. (eds.), *Spatial Planning and Sustainable Development: Approaches for Achieving Sustainable Urban Form in Asian Cities*, Strategies for Sustainability, DOI 10.1007/978-94-007-5922-0_1,

compact city (Jenks et al. 2000), and eco-city (Roseland 1997). All of these concepts were originally proposed by researchers in the developed countries of Europe and North America. In this book, instead of discussing the definitions of and differences between those concepts, we argue that this collection of seemingly unrelated case studies about housing, industry, transportation, landscape, ecological environment, and urban vulnerability all contribute to an examination of the "sustainable urban form" in Asian countries and territories.

Usually, spatial planning is considered as a tool for driving the planning policies at regional and national levels (Faludi 2009). In this book, we use "planning policies" to refer broadly to government actions in the process of planning and implementation. At the national level, in order to find a sustainable urban form during the developing period in the urbanization process, many developing Asian cities nowadays are learning from the experiences of European countries. However, most developing Asian cities, such as the cities in China, still pay more attention to economic development and physical planning, following the history of urban sprawl in European, US, or Asian developed cities. Therefore, conflicts are emerging between economic extensions and compact urban areas in most Asian cities. In such situations, we believe that effective planning policies are necessary for reaching a sustainable urban form.

From a viewpoint of urban and regional planning, it is important for local government to set demand allocation patterns, public transportation, and green spaces in the process of urban growth. Meanwhile, setting the balance between social and ecological quality in the economic development process is an important task of local government. For those cities in the process of urban decline, it is important for city government to make effective decisions on public planning policies in order to prevent population decline, improve urban regeneration, and increase centrality. As discussed above, planning control for urban sprawl in both developing and developed countries is important in achieving sustainable urban form.

All urban projects, either development and conservation, start from each building or land parcel; thus, the case studies presented in this book are selected not only from the national and regional level but also from the level of urban districts or buildings for gaining insights on their policy impact on sustainable urban form. For example, new green technologies in building construction should be taken into account, because doing so will assist planners in carrying out public policy on CO2 emissions for global climate change. Moreover, public awareness is important for employing environmental technology to improve ecological functioning in dense urban areas. Public transportation systems may also play an important role in achieving sustainable urban form. For example, public transportation systems for improving access to downtown areas are currently a popular research topic.

We have organized this book into two parts, which reflect the planning issues in urban and rural areas. The first part covers urbanization and sustainable society, and the second part examines landscape and ecological systems for sustainable development. Here, we introduce how each chapter contributes to the book's main conclusions regarding spatial planning and strategies for sustainable development.

1.2 Part 1: Urbanization and Sustainable Society

In Part 1, we will focus on planning issues regarding economic development and urban form in the urbanization process. We present this material in two sections: (1) urbanization and planning approaches and (2) housing and transportation.

1.2.1 Urbanization and Planning Approaches

Urban containment is employed in managing the location, character, and timing of growth to support a variety of planning goals and the efficient use of infrastructure in urban growth management in the United States (Nelson and Dawkins 2004). Urban containment is most simply expressed as maintaining a line that separates urban and rural areas, which could be an urban growth boundary, an urban service limit, or priority growth areas. However, in most Asian countries, the urban growth boundaries set by planners are broken by rapid urban development, such as in cities in Japan. Korea has also experienced this urban phenomenon, and it is now happening in Mainland China.

For urban growth management, establishment of spatial databases and relevant indexes for sustainable development are helpful in analyzing the social and ecological situation of urban growth in the Asian developing countries. In Chap. 2, Ye describes spatial data sharing as the first step in spatial planning toward ultimately measuring and implementing sustainable development, for which public financing is necessary for the establishment and management of the databases. This chapter also reveals the sharing of spatial data and research cooperation as specific goals, taking inspiration from collaborations such as the European Security and Defence Identity (ESDI) and the European Spatial Planning Observation Network (ESPON) of the European Union (EU). Meanwhile, spatial analysis is suggested as a useful technical tool for understanding the planning issues in developing countries.

After decades of urban development, local cities in developed Asian countries, such as Japan and Korea, are now experiencing urban decline from large-scale development projects, which have led to increasing urban sprawl. Strategies for improving centrality and increasing population in city areas have been implemented for preventing this trend. In Chap. 3, Lee et al. have analyzed urban decline in small-and-medium-sized cities in South Korea and have classified the degree of urban decline in different areas. This work will be helpful to the central government of South Korea in deciding the priorities of urban regeneration policies. It is important to prevent economic decline in downtown areas, which is caused by certain social issues, such as an aging population and depopulation. In Japan, less-favored areas, known as "marginal hamlets," have expanded rapidly, and the continuous care of such areas has been an important problem in national land-use planning. In Chap. 4, Yamashita and Ichinose scrutinize how positioning and directionality of the support for marginal hamlets are defined in the National Spatial

Plan and propose three steps to achieve a consensus among stakeholders in discovering good support policy for marginal hamlets.

In this section, we also have focused on planning issues in downtown areas with advanced depopulation and aging societies, which have resulted from urban sprawl in Japan. We have introduced some planning practices to decrease the negative influences of urban decline, such as urban regeneration by implementing appropriate design guidelines and urban facility designs for aging societies. In Chap. 5, Nishino reveals that the management of farmlands has already reached an upper limit and that much cultivated land has been abandoned in Japan. The status of the daily living environment of elderly residents is examined carefully so that local governments can promote communal living centers for the aged population in "marginal hamlets."

Considering the challenges of urban decline and regeneration in downtown areas, the Minato Mirai 21 project in the central quarter of Yokohama City has been effective in converting former shipyards and railroad yards into mixed-use and high-density urban quarters with a working and resident population of 70,000 people today. The case studies of Chaps. 6 show that local governments are endeavoring to take planning measures for preventing urban decline through improving economic regeneration in and accessibility to downtown areas, where local government can extend the service life of existing downtown infrastructures and manage public facilities in depopulation areas. In Chap. 6, Balaban elaborates on the use of indicators to assess the progress achieved in this urban regeneration project toward climate friendly urban design.

Furthermore, environmental planning and urban planning are separated in most planning systems in Asian countries, even in developed countries. We attempt to argue that integration of environmental planning, including landscape master planning, with urban planning and design is very important in planning practice. Hence, a new model is necessary to integrate environmental and urban designs. Chapter 7 describes a 2-year endeavor to implement the Landscape Master Plan and Townscape Renaissance Project in Changhua and Taichung counties, which are located in the central region of Taiwan. Liu argues that at the local government level, it is gradually becoming possible to link landscape, ecological, agricultural, and cultural urbanism with effective urban policies to the development of urban and suburban sustainable form.

1.2.2 Housing and Transportation

Different from urban growth boundary as deliberated in urban containment policy, a compact city was originally defined as a continuous urban area of a size that allowed optimal use of the existing services, employment, and infrastructure (Zonneveld 1991). Such a city is a high-density urban settlement with central area revitalization, high-density development, mixed-use development, sufficient urban services, and facilities.

In the process of urbanization, urban policies on land-use patterns, such as for housing and industry, are challenged under the background of rapid urban development. Many of the traditional dwellings in Asian countries have been abandoned and replaced by modern housing developments. This section tries to explore sustainable housing development in Asian cities. Some case studies have been conducted in urban downtowns and in villages outside the urban areas for this book.

In developing countries, there are also many "marginal hamlets"; for example, the village culture landscape of Northern Shaanxi, China, is challenged by new rural construction and rapid urbanization in Mainland China. Many of the cave dwellings have been abandoned and flat-roof houses adopted instead. Meanwhile, traditional culture is absent. Dang et al. argue in Chap. 8 that village cultural landscape can be protected through developing sustainable eco-tourism. Taking the cave dwelling village, Gaoxigou Village, as a case, they have explored a sustainable-oriented conservation planning method for cave-dwelling villages.

Meanwhile, researchers pursue housing policies through enriching the methodology of housing demand structure analyses and microscale modeling for policy decision making. The objective of Chap. 9 is to reveal the characteristics of sustainable location diffusions of settlements of townhouse projects to relieve housing problems in Bangkok and its greater metropolitan area. In Chap. 10, Gao attempts to find policy solutions via analysis of the housing demand structure, and housing policy issues are addressed by modeling the housing demands of urban families on a microscale. The results provide many useful implications for formulating housing policies in the Beijing Metropolitan Area.

In Chap. 11, Pai describes a case study of the bicycle valley of Dachia in Taiwan. Experiences in and characteristics of the Dachia bicycle valley demonstrate the successful transformation of a traditional industry into a potential development industry. In this case study, the geographic proximity of related industries and academic institutes is a significant factor. Interactions among firms contribute knowledge transmission and spread. Even for small firms that lack R&D capability, such interactions become a channel for promoting production technology.

Meanwhile, bicycles are now considered as a cleaner transportation method than automobiles or fossil-fuel-powered city buses. Many existing research reports focus on public planning policies for transportation and sustainable urban form (Williams 2005). Some of these reports are about strategies for the public transportation system and the city size. However, traffic congestion in many of the growing cities in Asia is serious and continuing to worsen. In this section, we try to organize some unique and new topics on transportation research, which are challenges that have to be dealt with if we are to achieve a low-carbon transport society.

Here we examine representative technological innovation, such as the personal mobility vehicle (PMV). In Chap. 12, Ando et al. argue that a key success factor for introducing PMVs in the future is to understand social acceptability—that is, whether or not the public will accept the new transport innovation. What people think about PMVs and their significant sociopsychological factors can influence the acceptance of PMVs in society. The authors offer conclusions and suggestions for possible directions in pursuing PMV use.

Urban transportation energy as a major part of urban energy consumption has a strong relationship with urban form. Some metropolitan areas in developing countries, such as Beijing, are attempting to develop tools for evaluating plan alternatives in terms of transportation energy consumption and environmental impact. In Chap. 13, Long et al. aim to investigate the impact of urban form—namely, land-use patterns and development density distribution, as well as distribution of job centers, on residential commuting energy consumption. They developed a multi-agent model for urban form, transportation energy consumption, and environmental impact integrated simulation, which is supposed to be important for considering transportation policies in the Beijing Metropolitan Area.

A series of negative consequences brought out by the surge of "car booming," including traffic congestion, traffic accidents, environment pollution, and worsening of personal mobility and quality of life, have become severe challenges for the sustainable development of Chinese cities. In Chap. 14, Wang argued that urban planning and design will encourage walking, thereby encouraging interaction and a greater sense of community and discouraging automobile dependence. Thus, a city-wide evaluation of the walking accessibility and bus availability of urban facilities and public transit has been conducted in Xiamen City. In planning terms, it is also useful to understand where is the most appropriate location for new facilities as well as public transport services according to the walking accessibility and bus availability in urban space.

1.3 Part 2: Landscape and Ecological Systems, Sustainable Development

For considering the interaction between social and ecological environments, we have organized three sections in Part 2, which are (1) green design and landscape, (2) agricultural and ecological systems, and (3) urban vulnerability.

1.3.1 Green Design and Landscape

The concept of the eco-city can be linked to issues ranging from planning and economic development to social justice, to the ecological environment, which is currently widely discussed (Roseland 1997).

In Japan, local governments are also enhancing technology for improving the urban thermal environment and reducing CO_2 emissions by means of green space, such as the use of green curtains and green roof systems, which benefit ecological function and landscape in dense urban areas. Applications of some of those technologies have been integrated with urban design guidelines, and the Japanese government has supported some private construction as national model projects in urban districts. Some case studies have demonstrated that green spaces are promising not only for the ecological functions of urban landscapes but also for hydrological sustainable designs. In Chap. 15, Kato et al. present the results from experiments performed with sample green curtains that exhibit ventilation and solar shading effects. Green curtains made with vines are considered to help improve the thermal environment of indoor spaces during the summer since they shield the walls and windows of buildings. In Chap. 16, Kikuti estimates the effects of thin-layer green roof systems and lightweight storage-drainage boards on storm-water runoffs in urban areas.

In response to interactions with the urban systems, the supply and demand of the natural ecosystems and socioeconomic system should be coordinated, regulations for the urban ecological security system should be designed, and an early warning and emergency response mechanism for ecological security should be formulated when implementing urban planning. In Chap. 17, Xu and Yang presented an assessment and regulation system for integration of ecological security system into urban planning implementation. As mentioned by Roseland (1997), a collection of apparently disconnected public efforts concerning urban planning and design, ecological design, and environmental planning all form a single framework: the "eco-city."

1.3.2 Agriculture and the Ecological System

Rapid economy growth and urbanization have been leading to a series of resource and environmental problems, such as decreasing cropland, environmental deterioration, and traffic jams. The existing agricultural status—including economic profit, labor cost, water resource depletion, and environmental impact—is discussed in this book. In the case study in Chap. 18, Wand et al. examine the outlook for the trend toward agricultural production in the urban shadow of the Beijing metropolitan area. Since 1984, Beijing's rapid economic growth and urbanization have been leading to a series of resource and environmental problems, such as a decrease in cropland and environmental deterioration. Therefore, more space for lawn and trees is needed to improve the urban environment.

In Chap. 19, Zhang and Zhao discuss the dilemma for Beijing: whether to keep cropland or to plant more trees and larger lawns in response to the requirement from the central government for keeping 227,000 ha of cropland for national food security purposes. In actuality, people can use the croplands to serve both purposes—food production and eco-environmental service—because croplands have multiple functions. Besides food production, croplands also have ecological, landscape, cultural, and tourism functions, especially in the mega cities. Chapter 19 proposes an agricultural pattern consisting of four rings that radiate outward from the city center, based on the multifunctionality of croplands. Thus, agriculture not only exerts its production functions, but also employs landscape and ecological functions to achieve a more comfortable and beautiful living environment for Beijing.

As policy measures to limit sprawl by restricting out-of-town development for protecting ecological systems, spatial databases and spatial analysis can be used for managing urban growth boundaries and for zoning restrictions in urban areas. For spatial planning on these issues, Uehara et al. argue in Chap. 20 that with the use of remote sensing data with high spatial resolution and some spatial indicators reflecting the duration of patterns of land use, social patterns—such as land ownership—are helpful for achieving sustainable landscape management. In particular, new planning tools will play an important role in the spatial planning process of developing countries for identifying the conflicts of urban growth and environmental issues. In the study for Chap. 21, based on high spatial resolution products that show even such individual landscape objects as bridges, embankments, and buildings, Kim et al. argue that it is possible to apply the unit CO_2 emission ratio and for constructing CO_2 emission maps.

1.3.3 Urban Vulnerability

Finally, we switch our perspective from agricultural and ecological systems back to urban systems, in particular to the risk of natural disaster in such systems. This section reveals a widespread concern: urban or economic development and the consequent competition for land greatly enhance human vulnerability to natural disasters, such as floods, drought, and pollution. Thus, we examine vulnerability, a concept used in environmental planning, that includes such aspects as environmental capacity, environmental footprint, and environmental vulnerability. Research on vulnerability has grown considerably in the last several years but is still largely limited in definition and scope and characterized by constraints based on interdisciplinary differences. Urban vulnerability has been included in recent climate change and disaster risk research (Lankao and Qin 2011).

In Chap. 22, Pai et al. indicate that Taiwan is among the top-ranking countries with respect to risk analyses of natural disasters and floods that occur from the destruction of the original hydrological environment during urban or economic development and the consequent competition for land. The authors suggest that a pragmatic overall index must be established in order to increase the number of reference values for disaster assessments and disaster preventions based on spatial planning and relevant planning policies.

In Chap. 23, Hung and Lin present a case study in which they researched socioeconomic vulnerability during the reconstruction of a traditional community, Chun-You Street, Taiwan. As a planning practice for pursuing neo-traditional development, the community tried to use its local environmental features to attract tourism, rather than simply throwing government funds into reconstruction in order to build an environmentally friendly, sustainable community and to decrease its vulnerability in the disaster mitigation phase. Historical architecture styles such as "Feng-Huo-Shan-Qiang" (flame-shaped raised gables) can reduce the risk of spreading fire. In strengthening the resilience against disaster, future recovery efforts should also be aimed at improving the physical environment and expressing the historical context by integrating traditional architectural elements. In Chap. 24, Yuhara and Ye present a case study concerning sustainable communities in mountainous areas with heavy snowfall in Japan. Because of the severe population decrease owing to an aging society, many communities in such areas have become quite vulnerable to natural disasters. In addition, public help in disaster reduction has been weakened because of difficulties in public finances. The results verify the importance of residents' associations to strengthen the power of mutual support within local groups.

The case study by Huang and Shen in Chap. 25 contributes to a growing aspect of urban vulnerability analysis by identifying water-related issues in the largest city of developing countries, namely, Chongqing in China. The target of the present study, Chongqing City, belongs to a very different category in terms of water availability, geographical context, topographical features, economical status, and social systems. It is possible that in the near future, urban vulnerability analysis will be accepted as a methodology for measuring sustainable urban form.

Even as cities in Japan are experiencing urban decline, rapid urbanization is happening in many Asian countries. In developing countries, where the approach to urban development with economic growth leads to exploitation and use of natural resources in excess, the spatial planning effort is expected to encourage the development of the region in order to improve the quality of life of the community and social justice in a sustainable environment.

1.4 Conclusion

Overall, this book attempts to provide insight into achieving sustainable urban form through spatial planning and implementation; we focus on planning practices at the level of local cities and certain metropolitan areas in different Asian countries and territories. Most Asian cities have introduced concepts of sustainable urban form from European and North American into their planning practices, such as urban containment, the compact city, the eco-city, and neo-traditional development.

Currently, some cities in developing countries, such as mainland China, are experiencing rapid urban growth, whereas many cities in Japan and South Korea are experiencing urban decline resulting from depopulation and an aging society after the urbanization process. Neither developed nor developing countries control their urban growth boundaries effectively. Thus, sustainable urban form becomes an important topic in Asian countries and territories.

Regarding case studies in mainland China, Taiwan, and Thailand, planning frameworks and demand structures are taken as topics for case studies. However, in planning implementation, even though such concepts of urban containment planning as the compact city and the eco-city are widely accepted in those Asian countries, there are still many uncertainties and difficulties in the urbanization process. It is necessary to investigate whether economic demand can be satisfied in a compact city with regard to such aspects as housing demand structure, industrial parks, and transportation patterns. The experiences from the Asian cities presented in this book show that sustainable planning of traditional housing areas, public transportation systems, and green spaces should be explored carefully in order to allocate reasonably according to economic demand during urbanization. Meanwhile, local governments in mainland China support spatial analysis and geosimulation of urban structure regarding housing and transportation demand.

For the less-favored and downtown areas in the declining cities of Japan, consensus with stakeholders concerning public services and design guidelines is necessary in implementing spatial planning and policies within the limits of public financial aid. On the other hand, margin hamlets in China are attempting to use eco-tourism to help with development and survival.

New public efforts are also being made to integrate environmental planning and urban design in Japan and Taiwan, which will be valuable examples for the world. As a new experiment in the eco-city concept in China, researchers are reconsidering the function of agriculture not only as production but also as an ecological network in the Beijing Metropolitan Area. Meanwhile, natural disasters are an important factor to consider in sustainable urban form, and urban vulnerability is matter for careful discussion and concern in Asian countries. Urban vulnerability analysis may be accepted as one criterion for measuring sustainable urban form in the near future.

From a global warming standpoint, local governments in Japan are now looking for a solution to transportation congestion problems, and PMV is considered a prospective tool for decreasing transportation volume and CO_2 emissions. Green building technologies, such as roof systems and green curtains, have been introduced into urban design in Japan for improving the urban thermal environment. Even the development and testing of these technologies will likely make a great contribution to eco-city construction. In planning, public awareness is recognized as an important factor for application of those technologies in the cities of Japan.

Finally, in discovering solutions in sustainable urban form, local governments in Asian countries and territories are developing spatial databases and indexes in order to analyze the social and ecological situations of urban areas and to simulate possible urban forms. This book presents case studies that focus on public efforts toward sustainable urban form, and we hope that it will be helpful for researchers and students who are concerned with sustainable development in Asian countries and territories.

References

Dawkins CJ, Nelson AC (2002) Urban containment policies and housing prices: an international comparison with implications for future research. Land Use Policy 19(1):1–12

Faludi A (2009) A turning point in the development of European spatial planning? The 'territorial agenda of the European Union' and the 'first action programme'. Prog Plan 71(1):1–42

Furuseth OJ (1997) Neotraditional planning: a new strategy for building neighborhoods? Land Use Policy 14(3):201–213
Jabareen YR (2006) Sustainable urban forms their typologies, models, and concepts. J Plan Educ Res 26(1):38–52
Jenks M, Jenks M, Burgess R (2000) Compact cities: sustainable urban forms for developing countries. Spon Press, London
Lankao PR, Qin H (2011) Conceptualizing urban vulnerability to global climate and environmental change. Review article. Curr Opin Environ Sustain 3(3):142–149
Nelson AC, Dawkins CJ (2004) Urban containment in the United States history, models and techniques for regional and metropolitan growth management. American Planning Association in Chicago, Chicago
Roseland M (1997) Dimensions of the eco-city. Cities 14(4):197–202
Williams K (2005) Spatial planning, urban form and sustainable transport. Ashgate, London
Zonneveld W (1991) Conceptvorming in de ruimtelijke planning: Encyclopedie van planconcepten (Concept formation in spatial planning: Encyclopedia of planning concepts). University of Amsterdam, Amsterdam

Part I
Urbanization and Sustainable Society: Section Urbanization and Planning Approach

Chapter 2
The Possibility of Sharing Spatial Data and Research Cooperation Within East Asia Countries

For Sustainable and Balanced Regional Growth

Kyung-rock Ye

Abstract For the stable growth of East Asia, a common spatial development perspective, which proposes sustainable and balanced development for the whole region, will be necessary in the future. Three East Asian countries in particular—Japan, China, and Korea—have practiced the interchange of people, money, and information in various fields, but they have not realized the common fruit that should be the result of essential cooperation and coordination. This chapter proposes the possibility of sharing spatial data and research cooperation by referring to a case involving ESDI (European Spatial Data Infrastructure) and ESPON (European Spatial Planning Observation Network) of the EU. The first steps in such a cooperative effort would include developing spatial indicators, analyzing trends, and constructing visual representations of the research results, all of which would assist in understanding the East Asian region and each country's situation. Such progress would also serve as (1) the basis for regional cooperation in spatial development planning, (2) a visual indicator showing balanced development in each region, (3) educational material to enhance the Asian identity, and (4) an assessment indicator for concentrated investment in backward regions. As a follow-up to the maps created for this project, several tasks should be carried out for setting up an East Asian Spatial Data Infrastructure: (1) produce a common seamless map from the national to the regional level showing existing conditions, (2) set comparable regional units based on population and territory size, (3) unify statistical units and comparable regional units by collecting data under the same definitions, and (4) increase the range of use by integrating the global map with regional statistical data. To implement these processes, cooperation and collaboration among related nations and research institutions are necessary.

K.-r. Ye (✉)
Graduate School of Frontier Sciences, The University of Tokyo,
Hongo 7-3-1, Bunkyo District, Tokyo, Japan
e-mail: ye.kr92@gmail.com

M. Kawakami et al. (eds.), *Spatial Planning and Sustainable Development: Approaches for Achieving Sustainable Urban Form in Asian Cities*, Strategies for Sustainability,
DOI 10.1007/978-94-007-5922-0_2,

Keywords East Asian community • Spatial development perspective • Spatial data infrastructure

2.1 Introduction

In recent years, some researchers have begun to promote the idea of transnational cooperation in the East Asian region, called the East Asian Community (Takahashi 2010), by means of deepening economic interdependence (Letiche 2000). Under the progress of the international horizontal division-of-labor system, a demand has arisen to strengthen the network to accommodate such innovations as a one-day return business zone in East Asia and next-day freight delivery—in other words, a situation that would enable the seamless flow of people, goods, and information for achieving sustainable economic development (Wang 2004; Kimura et al. 2007; Ahn 2001). Researchers began to discuss an idea they call the Grand Design for Northeast Asia (NIRA Policy Research 2002), which is a vision for comprehensively improving and enhancing the international infrastructure through cooperating in traffic/logistics, energy supply, environment, and information technologies (von Hippel et al. 2011a; Liu et al. 2010; Yun and Zhang 2006). The European Union (EU), which can be recognized as an advanced cooperation model, apprehends the present condition and aspects of spatial development within its territories and is preparing strategic policies for future balanced sustainable development. The EU's decisions are based on the idea that amending regional disparities and preventing excessive competition for infrastructure development among the regions of the EU will enable the territory to enhance international competitiveness; therefore, the EU considers such planning to be an important task in setting out the strategic vision of spatial development for solving the problems mentioned above (European Commission 1999). As a concrete measure to implement vision making, the European Spatial Planning Observation Network (ESPON), which is a collaborative research institute among member states, was established in the EU for investigating and analyzing the direction of spatial development within the territory (ESPON 2005, 2006). At that time, EuroGeographics and Eurostat offered basic map data and statistical data.

On the other hand, in Asia we find an example of a transnational community, called the Association of Southeast Asian Nations (ASEAN), which is initiating various activities in cooperating for the common benefit within the region. However, in northeast Asia—especially in Japan, Korea, and China—such cooperative activities are not yet well established, although the mutual influence and economic dependence caused by geographical conditions have deepened (Hutton 2003). On the contrary, excessive competition involving the development of infrastructures, such as airports or seaports, that have a close relationship with the movement of people and goods, is being intensified as each country wishes to possess the most important hubs in the territory. This complicated transnational competition may weaken the entire East Asian region. To implement mutual prosperity in East Asia while maintaining international competitiveness with the United States of America, European countries, and other powerful states, it is important that neighboring Asian countries cooperate among themselves, have a sound perspective on present conditions, and discern the problems of spatial development within this area (Chung 2010; von Hippel et al. 2011b).

This chapter first reviews data collection and use of spatial information in the EU as a case study, and second, it uses statistical data that are publicly available in Japan, China, and Korea to demonstrate the present condition of spatial development in these countries and to show that it is necessary to development a spatial data infrastructure in East Asia. These demonstrations are achieved through the production of several experimental spatial data maps, based on existing statistical data and ADC world map data.

2.2 Outline of EU and East Asian Region

The EU and East Asia may be contrasted in several ways: the former is economically and socially unified, and the latter is in the discussion phase wherein it is looking toward becoming economically and socially unified. Today, the EU has 27 member states with a population of about 500 million and covering 3,930,000 square km, while Japan, China, and South Korea in East Asia have a population of about 15 hundred million (13 hundred million of which is in China) and a 9,800,000 square km area. Furthermore, as shown in Figs. 2.1 and 2.2, the GDP (per person) indicator for the EU member states is very different from those of Japan, China, and South Korea. Some member states in the EU are certainly below the average of Europe, but most of the member states are nearly equal with the EU average.

As shown in Fig. 2.3, as of 2005, South Korea was positioned nearly at the average of the three countries; all areas of China were below the average, and all areas of Japan were over the average. Therefore, a notable disparity exists among the three countries. Figure 2.3 also shows differences within each country—for example, between the coastal or northeast regions and the inland of China and between the Pacific coastal zone and other areas in Japan. In addition, in comparing 2000 and 2005, we can see a remarkable growth in GDP in the coastal and northeast areas of China. As everybody knows, the recent noteworthy economic growth in China has also accelerated competition among the three countries while at the same time strengthening economic, social, and cultural relations among them.

2.3 An Analysis of the Spatial Structure of the EU as a Case Study

The main purpose of the EU is to achieve a balanced and sustainable economic/social development along with common security and defense. To implement EU policies, the fundamental goals of European spatial policies are (1) economic and

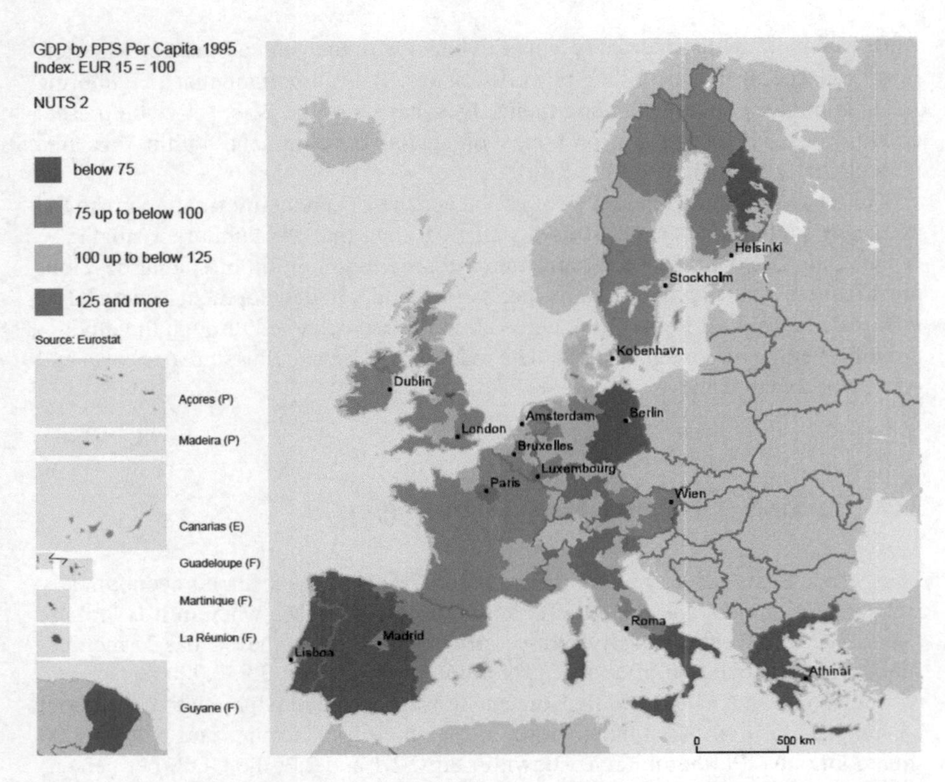

Fig. 2.1 Regional disparities of GDP by PPS per capita (Source: European Commission 1999)

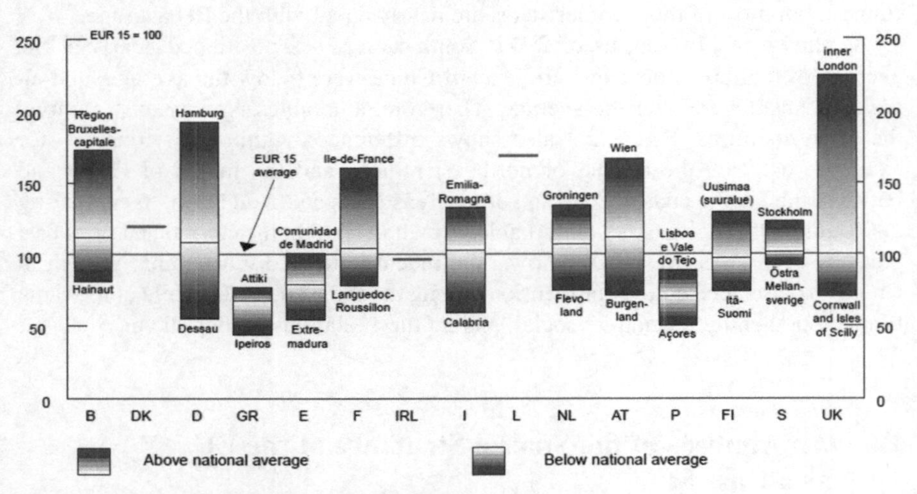

Fig. 2.2 Regional disparities in GDP per capita by member state (Source: European Commission 1999)

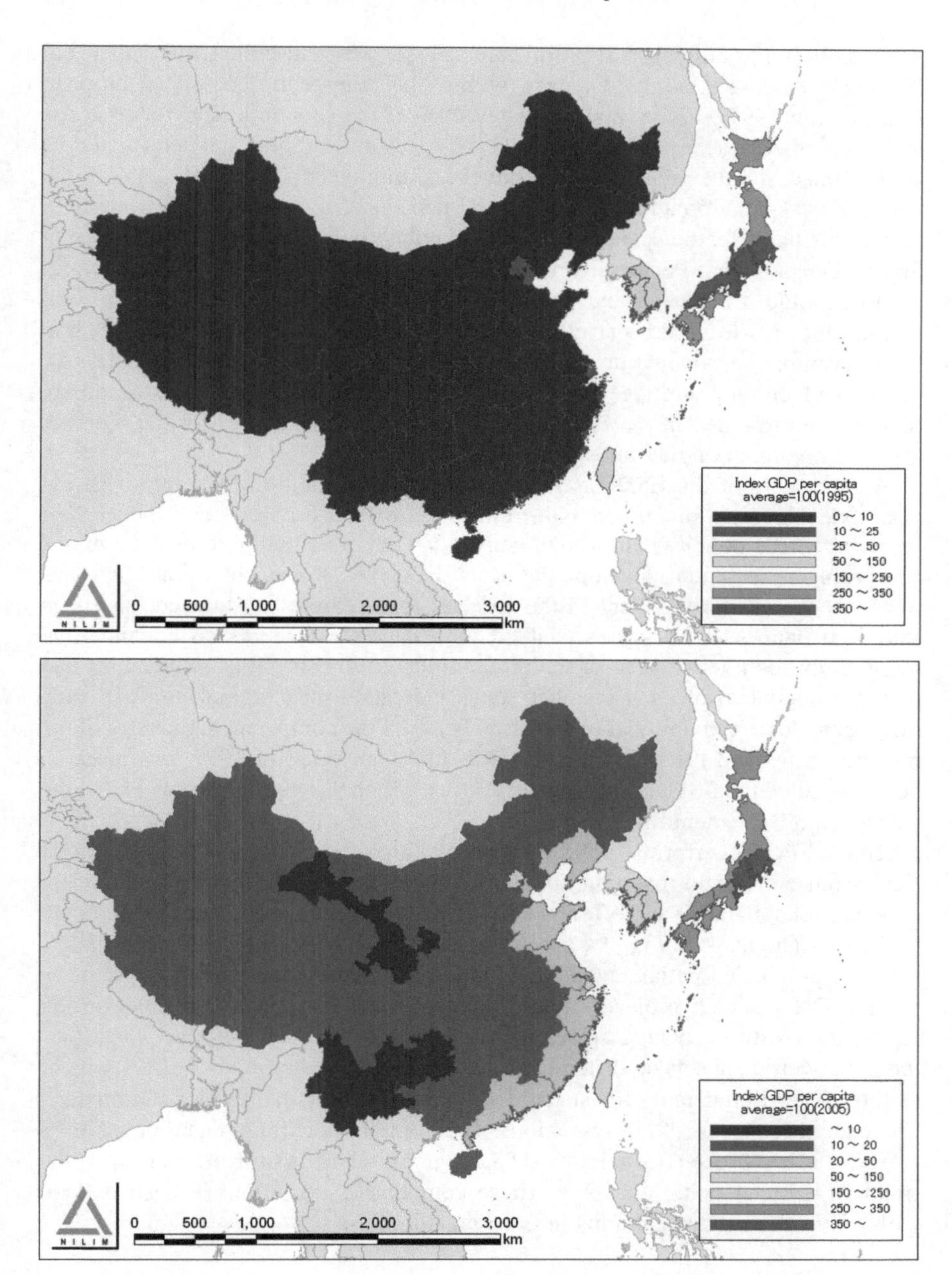

Fig. 2.3 Regional disparities of GDP per capita by East Asian region

social cohesion, (2) conservation of natural resources and cultural heritage, and (3) more balanced competitiveness within the European territory (European Commission 1999). Policy makers believe that if the economic and social needs in the region were achieved, the natural, ecological, and cultural functions would be strengthened and the region's competitiveness improved. These basic goals serve not only the spatial policies but also all EU policies.

For this purpose, the spatial development policy guideline, called the European Spatial Development Perspective (ESDP), was initiated to achieve the following: (1) to develop a balanced and polycentric urban system and a new urban-rural relationship, (2) to secure parity of access to infrastructure and knowledge, and (3) to promote sustainable development, prudent management, and protection of nature and cultural heritage sites (European Commission 1999). Sixty policy options are presented in the ESDP, which function as guidelines for the member states' individual policies and spatial plans.

Another role of the ESDP is to monitor present conditions and trends in the spatial development of Europe and, from those, extract problems and reflect their solutions in EU policies. The ESDP constitutes the first attempt to analyze trends and problems in spatial development at the EU level. It was clear that long-term trends in spatial development in the EU had to be carefully analyzed, based on statistical data. Member states enabled information exchange and collaborative research through networking of the research institutions (such as universities and research institutes) of each member state, and established organizations in each member state to lead political cooperation between the competent authorities of the member states and the EU. This idea was first completed in 1997 as a research network called the ESPON (mentioned above), which was established as one of the measures to implement the ESDP.

The ESPON performs policy proposal through assessing the impact of EU spatial policy and mapping out spatial information for clarifying the concept of territorial unification and offering a database to support harmonized development in the EU. The maps in Fig. 2.4 are examples produced by their research, which is aimed at proposing spatial development policy for improving regional disparity. In the ESPON research projects, many kinds of maps are created through strong cooperation with EuroGeographics and Eurostat. To compare regional situations, the EU government sets up different scales of regional units called NUTS (nomenclature of territorial units for statistics), which consider mainly population size. There are three levels of NUTS regions. The highest scale of aggregation is NUTS 1, which is then broken down into NUTS 2 regions, which in turn are made up of the smallest regional units, NUTS 3. These comparable regions have been defined within European statistical units in Eurostat (Fig. 2.5).

Road network

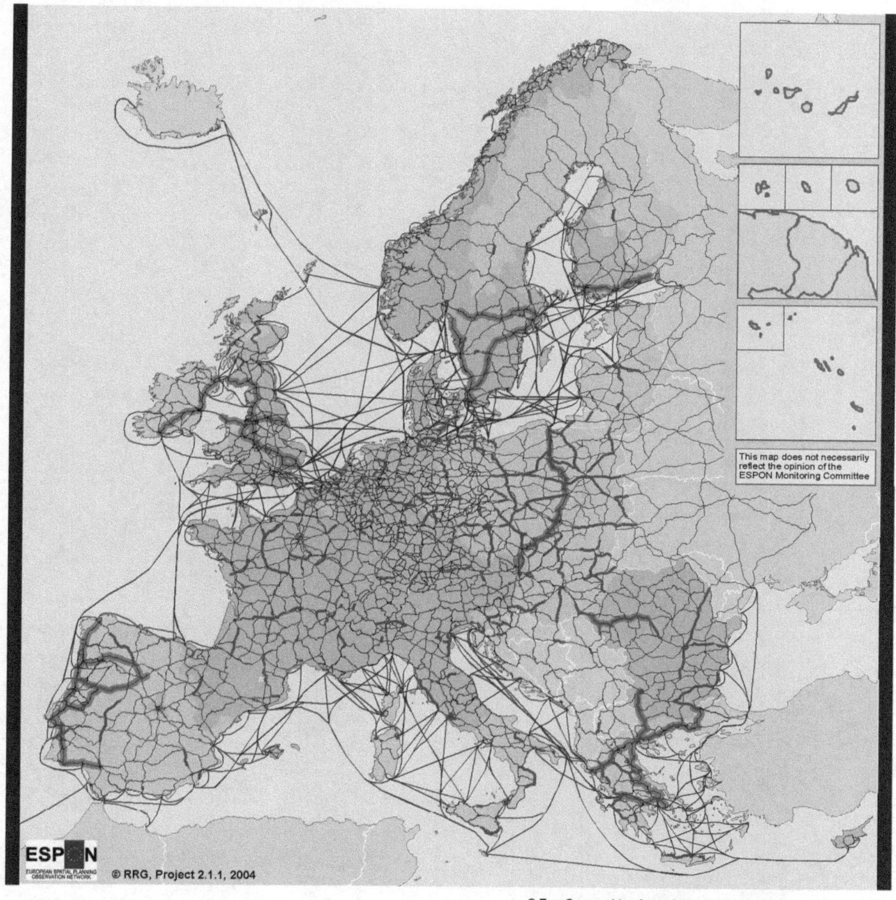

© EuroGeographics Association for the administrative boundaries
© RRG for the network database

- Road links
- Road priority projects (Scenario B1)
- Road projects, non cross-border (Scenarios B2, B3, B5)
- Road projects, cross-border (Scenarios B2, B4, B5)
- Objective 1 areas (Scenario B5)
- Short sea shipping links

Fig. 2.4 Spatial development conditions of the infrastructure in Europe (Source: ESPON_synthesis_report_I, http://www.espon.eu/main/Menu_Publications/Menu_ESPON2006Publications/synthesisreport1.html)

Rail network

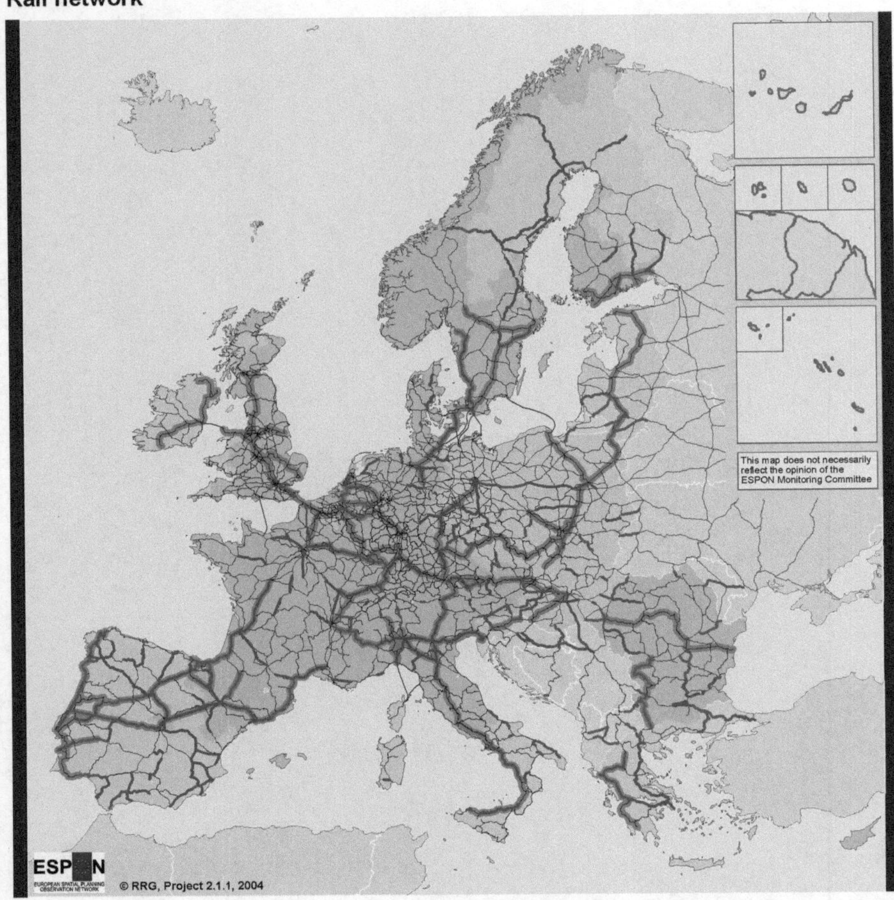

- Rail links
- Rail priority projects (Scenario B1)
- Rail projects, non cross-border (Scenarios B2, B3, B5)
- Rail projects, cross-border (Scenarios B2, B4, B5)
- Objective 1 areas (Scenario B5)
- Rail ferry links

Fig. 2.4 (continued)

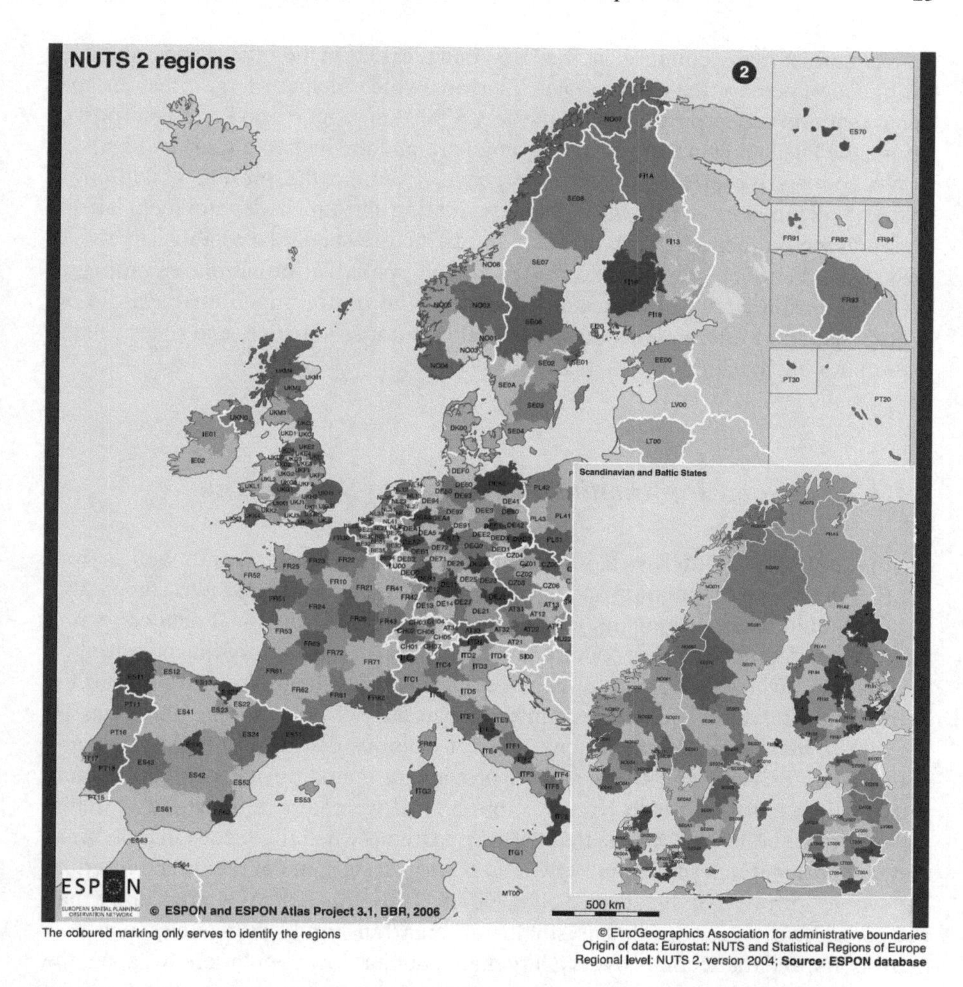

Fig. 2.5 NUTS 2 and NUTS 3 regions in Europe (Source: ESPON Atlas, http://www.espon.eu/main/Menu_Publications/Menu_ESPON2006Publications/esponatlas.html)

2.4 The Present Condition of Spatial Development in Each Region of East Asia

In this chapter, cooperation among East Asian countries according to the EU model can be referred to as "seamless Asia," and it is expected to strengthen the network. Cooperative measures include a 1-day return business zone in East Asia, next-day freight delivery, and Asian broadband; therefore, the viewpoint that recognizes the formation of the East Asian community as a common goal for sustainable economic development has been gradually extended. For example, in the transportation sector, the Asian highway and the Trans-Asian Railway constitute a cooperative project among many countries at the Pan-Asian level. On the other hand, countries strongly compete for the role of Asian gateway, which includes large-scale international seaports and airports, such as the new Shanghai seaport and Pudong airport of China, the Pusan new port, the Kwangyang port, and the Inchon airport and seaport.

No cooperative effort has yet emerged to examine the present condition of spatial development in East Asia, but considering the future development of this region, it would seem that a spatial data infrastructure should be established also in East Asia. For this project, we tried mapping spatial information according to unified criterion among three countries as an initial trial, using Global Map, ADC World Map, and international statistical data on ports, logistics, and other related matters for each country.

2.4.1 Regional Information According to Spatial Unit

Japan, China, and South Korea, which differ greatly from each other as well as from the EU, in terms of population and territory size, need to be divided into comparable regional units. Currently, there is no general standard on statistical data collection that is more detailed than the country level for these three countries, so the setting of the spatial unit may need careful consideration and justification. According to the current statistical data for each country, Japan has 47 prefectures; China has 33 provinces, including the capital city and autonomous districts, with 23 cities and counties in Taiwan, and there are 16 provinces, including the capital city and metropolitan cities, in South Korea (Table 2.1). If we compare the statistical units over the three countries, we see that they are extremely heterogeneous in their areas and population. Each spatial unit of China is very large, some of them far exceeding the whole territory of Japan or South Korea. In South Korea, spatial units may consist of two types: the first is smaller in area but highly populated, such as a metropolis, and the second covers a larger area but has low population. In Japan, the regional spatial units are the smallest among the three countries; therefore, for subnational units, they seem comparably small. Most important, it is not appropriate to map and compare the population scale according to these spatial units among the three countries. However, these spatial units are important classifications, serving as a statistical framework for analyzing the socioeconomic situations of each country. Therefore, the present spatial classification has to be subdivided or integrated into comparable spatial units, based on the character of the three countries, for future analysis. Such implementation would require careful discussion from national governments or related organizations, and the statistical classification should also be subdivided or integrated the same way.

In this chapter, to make comparable mapping units, existing subnational boundaries were integrated into 10 blocks in the case of Japan and into 3 blocks in the case of Korea, as shown in Table 2.1. In Fig. 2.6, I was able to map the comparable regional GDP standard (Index: GDP average = 100). For this region, as of 1995, Taiwan and Korea were positioned nearly at the average of these countries; all areas of China were below the average, and all areas of Japan were over the average; therefore, there is a notable disparity among these countries (see Fig. 2.6). The map from 2005 also shows differences within each country—for example, among coastal, middle, and western inland regions in China and between central zones and other peripheral areas in Japan and Korea. In addition, the comparison of 1995–2005 in Fig. 2.6 shows remarkable growth of the GDP in the coastal and inland areas of China, whereas Japan's GDP decreased during the same period. It is common knowledge that the recent noteworthy economic growth in China has also accelerated competition among the three countries and has strengthened the economic, social, and cultural relations among them. It is possible to some extent to compare or analyze according to the spatial indexes based on unit population, such as population density, economical index, and the ratio of green space per person. Regional classification must be based on a general standard, the settings of the statistical index, and other general data settings for a correct analysis of the difference in development among the regions.

2.4.2 Regional Information According to Infrastructure Development Standard and Quantity of Usage

One effective analysis method is to use the statistics for the main institutions or main cities as different spatial information from the regional classification mentioned above. I constructed a simple map from container traffic data in the main international harbor and from the passenger data at the international airport (see Fig. 2.7). Figure 2.7 shows an increase in air passengers from 1990 to 2000, and the container traffic data shows an increase over 10 years. These statistical data from major sources are fairly easy to acquire, but there are drawbacks: such data can be expensive or the method of data collecting may be different among the countries, making a simple comparison difficult. If each country were to gather its data under the same definition, integrating or updating such data into a common seamless map would be most useful and instructive. East Asian countries could cooperate and coordinate along these lines in the future.

2.4.3 Linear Information Like Road and Railway

Figure 2.8 is based on the Trans-Asian Railway and the Asian Highway. Here a seamless route of the transport in Asia is assumed. These routes include some that are still in the planning stage as well as those not yet constructed or some that are constructed to varying standards, but this idea will strengthen the sense of a unified Asia. Hence, producing a transportation map of this region (including the roads, railways, seaways, and international transportation routes of each country) on a global map or ADC world map is still a monumental task. Because the existing

Table 2.1 Subnational boundary and block areas of four East Asian countries

	Name	Block	Area (km^2)	Population 2005		Name	Block	Area (km^2)	Population 2005
Japan						China			
JA01	Hokka ido	JP1	78,415	5,627,737	CN01	Beijing Shi	CN01	16,808	15,380,000
JA02	Aomori-ken	JP2	9,606	1,436,657	CN02	Tianjin Shi	CN02	11,305	10,430,000
JA03	Iwate-ken	JP2	15,278	1,385,041	CN03	Hebei Sheng	CN03	190,000	68,510,000
JA04	Miyagi-ken	JP2	7,285	2,360,218	CN04	Shanxi Sheng	CN04	156,000	33,550,000
JA05	Akita-ken	JP2	11,612	1,145,501	CN05	Inner Mongolia ZiZhiQu	CN05	1,183,000	23,860,000
JA06	Yamagata-ken	JP2	9,323	1,216,181	CN06	Lianoing Sheng	CN06	145,700	42,210,000
JA07	Fukushima-ken	JP2	13,782	2,091,319	CN07	Jilin Sheng	CN07	187,400	27,160,000
JA15	Niigata-ken	JP2	12,582	2,431,459	CN08	Heilongjiang Sheng	CN08	454,000	38,200,000
JA08	Ibaraki-ken	JP3	6,094	2,975,167	CN09	Shanghai Shi	CN09	6,341	17,780,000
JA09	Tochigi-ken	JP3	6,408	2,016,631	CN10	Jiangsu Sheng	CN10	102,600	74,750,000
JA10	Gumma-ken	JP3	6,363	2,024,135	CN11	Zhejiang Sheng	CN11	101,800	48,980,000
JA11	Saitama-ken	JP3	3,797	7,054,243	CN12	Anhui Sheng	CN12	139,600	61,200,000
JA12	Chiba-ken	JP3	5,156	6,056,462	CN13	Fujian Sheng	CN13	121,400	35,350,000
JA13	Tokyo-to	JP3	2,187	12,576,601	CN14	Jiangxi Sheng	CN14	166,947	43,110,000
JA14	Kanagawa-ken	JP3	2,414	8,791,597	CN15	Shandong Sheng	CN15	157,100	92,480,000
JA19	Yamanashi-ken	JP3	4,465	884,515	CN16	Henan Sheng	CN16	167,000	93,800,000
JA16	Toyama-ken	JP4	4,246	1,111,729	CN17	Hubei Sheng	CN17	185,900	57,100,000
JA17	Ishikawa-ken	JP4	4,185	1,174,026	CN18	Hunan Sheng	CN18	211,875	63,260,000
JA18	Fukui-ken	JP4	4,188	821,592	CN19	Guangdong Sheng	CN19	179,800	91,940,000
JA20	Nagano-ken	JP5	13,585	2,196,114	CN20	Guangxi ZiZhiQu	CN20	236,300	46,600,000
JA21	Gifu-ken	JP5	10,598	2,107,226	CN21	Hainan Sheng	CN21	35,000	8,280,0000
JA22	Shizuoka-ken	JP5	7,779	3,792,377	CN22	Chongqing Shi	CN22	82,400	27,980,000
JA23	Aichi-ken	JP5	5,150	7,254,704	CN23	Sichuan Sheng	CN23	485,000	82,120,000
JA24	Mie-ken	JP5	5,774	1,866,963	CN24	Guizhou Sheng	CN24	176,100	37,300,000
JA25	Shiga-ken	JP6	4,017	1,380,361	CN25	Yunnan Sheng	CN25	394,000	44,500,000
JA26	Kyoto-fu	JP6	4,612	2,647,660	CN26	Tibet ZiZhiQu	CN26	1,228,400	2,770,000
JA27	Osaka-fu	JP6	1,892	8,817,166	CN27	Shaanxi Sheng	CN27	205,600	37,200,000
JA28	Hyogo-ken	JP6	8,387	5,590,601	CN28	Gansu Sheng	CN28	455,000	25,940,000

JA29	Nara-ken	JP6	3,691	1,421,310	CN29	Qinghai Sheng	CN29	722,000	5,430,000
JA30	Wakayama-ken	JP6	4,724	1,035,969	CN30	Ningxia ZiZhiQu	CN30	66,400	5,960,000
JA31	Tottori-ken	JP7	3,507	607,012	CN31	Xinjiang ZiZhiQu	CN31	1,660,000	20,100,000
JA32	Shimane-ken	JP7	6,706	742,223	CN32	Hong Kong	CN32	1,098	6,936,000
JA33	Okayama-ken	JP7	7,111	1,957,264		TeBieXingZhengQu			
JA34	Hiroshima-ken	JP7	8,475	2,876,642	CN33	AoMenTeBeiXingZhengQu	CN33	25	477,000
JA35	Yamaguch-ken	JP7	6,110	1,492,606	Taiwan				
JA36	Tokushima-ken	JP8	4,144	809,950	TW01	Taiwangsheng	TW	35,873	22,690,000
JA37	Kagawa-ken	JP8	1,875	1,012,400	China and Taiwan did not set the block areas.				
JA38	Ehime-ken	JP8	5,675	1,467,815	Korea				
JA39	Kochi-ken	JP8	7,104	796,292	KR01	Seoul-shi	KR1	605	9,820,171
JA40	Fukuoka-ken	JP9	4,968	5,049,908	KR04	Inch ' On-Shi	KR1	994	2,531,280
JA41	Saga-ken	JP9	2,439	866,369	KR08	Kyonggi-do	KR1	10,131	10,415,399
JA42	Nagasaki-ken	JP9	4,091	1,478,632	KR02	Pusan-shi	KR2	764	3,523,582
JA43	Kumamoto-ken	JP9	7,402	1,842,233	KR03	Taegu-shi	KR2	884	2,464,547
JA44	Oita-ken	JP9	6,337	1,209,571	KR07	Ulsan-shi	KR2	1,057	1,044,934
JA45	Miyazaki-ken	JP9	7,734	1,153,042	KR09	Kangwon-do	KR2	16,613	1,464,559
JA46	Kagoshima-ken	JP9	9,186	1,753,179	KR14	Kyongsangbuk-do	KR2	19,026	2,607,641
JA47	Okinawa-ken	JP10	2,266	1,361,594	KR15	Kyongsangnam-do	KR2	10,521	4,105,533
JAB01	Hokkaido Block	JAB01	78,415	5,627,737	KR05	Kwaqngju-shi	KR3	501	1,417,716
JAB02	Tohoku Block	JAB02	79,467	12,066,376	KR06	Taejon	KR3	540	1,442,856
JAB03	Kanto Block	JAB03	36,884	42,379,351	KR10	Ch' Ungch' Ongbuk-do	KR3	7,431	1,460,453
JAB04	Hokuriku Block	JAB04	12,620	3,107,347	KR11	Ch' Ungch' Ongnam-do	KR3	8,601	1,889,495
JAB05	Tsubu Block	JAB05	42,886	17,217,384	KR12	Chollabuk-do	KR3	8,055	1,784,013
JAB06	Kinki Block	JAB06	27,324	20,893,067	KR13	Chollanam-do	KR3	12,073	1,819,819
JAB07	Tsugoku Block	JAB07	31,909	7,675,747	KR16	Cheju-do	KR3	1,848	531,887
JAB08	Sikoku Block	JAB08	18,799	4,086,457	KRB1	S-K Block	KRB1	11,730	22,766,850
JAB09	Kyushu Block	JAB09	42,157	13,352,934	KRB2	K-K Block	KRB2	48,866	15,210,796
JAB10	Okinawa Block	JAB10	2,266	1,361,594	KRB3	C-C Block	KRB3	39,050	10,346,239

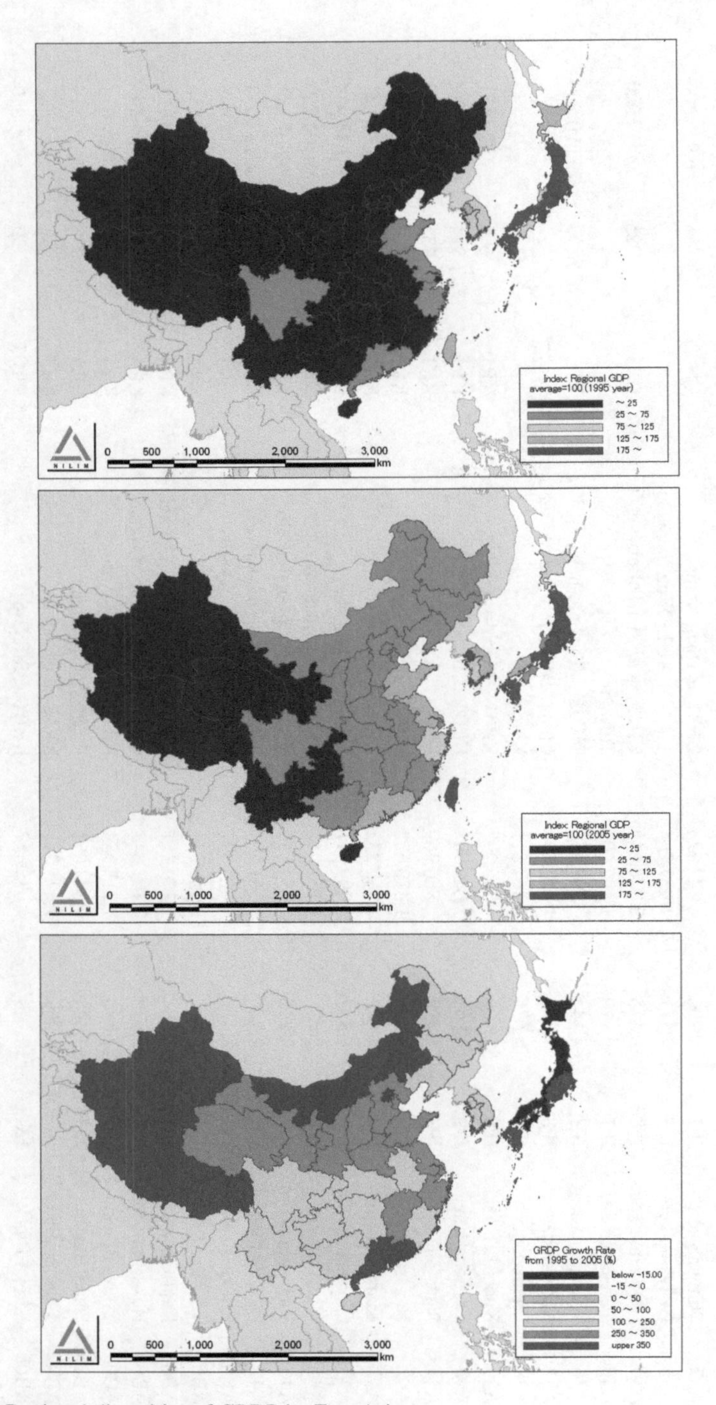

Fig. 2.6 Regional disparities of GRDP by East Asia

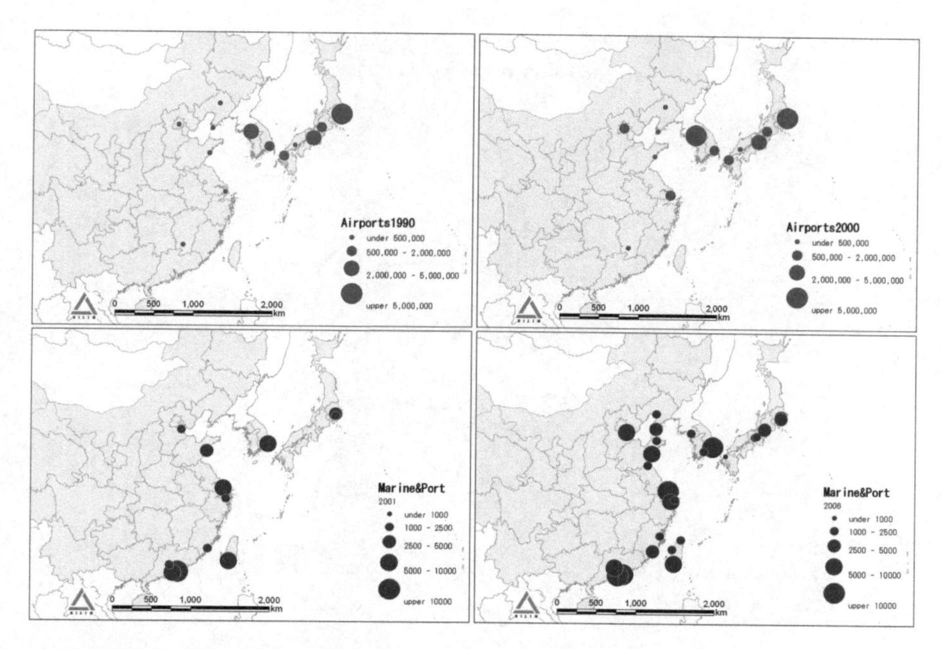

Fig. 2.7 Spatial development conditions and trends of infrastructures

route data on these maps are not produced by general information about the design standard of each route—its hierarchy, accessibility, etc.—difficulties remain in identifying certain routes. In order to evaluate and assess the amount of traffic movement, statistical data still needs to be gathered and integrated.

2.5 Conclusions

The spatial infrastructure development in a state is usually in proportion to its economic situation; thus, according to its GDP gap, the East Asian region must also have a large gap in its infrastructure development. However, the rapid economic growth in East Asian developing countries is speedily advancing their national infrastructure development. As a result, their economic growth is further enhanced. The gap of the East Asian countries will shrink as interdependent relationships gradually deepen and the flow of people, goods, and information expands. For sustainable and balanced development in this region, coordination is necessary between the related countries, which will strengthen international competitiveness and correct regional disparities.

The first steps in such a cooperative effort would include developing spatial indicators, analyzing trends, and constructing visual representations of the research

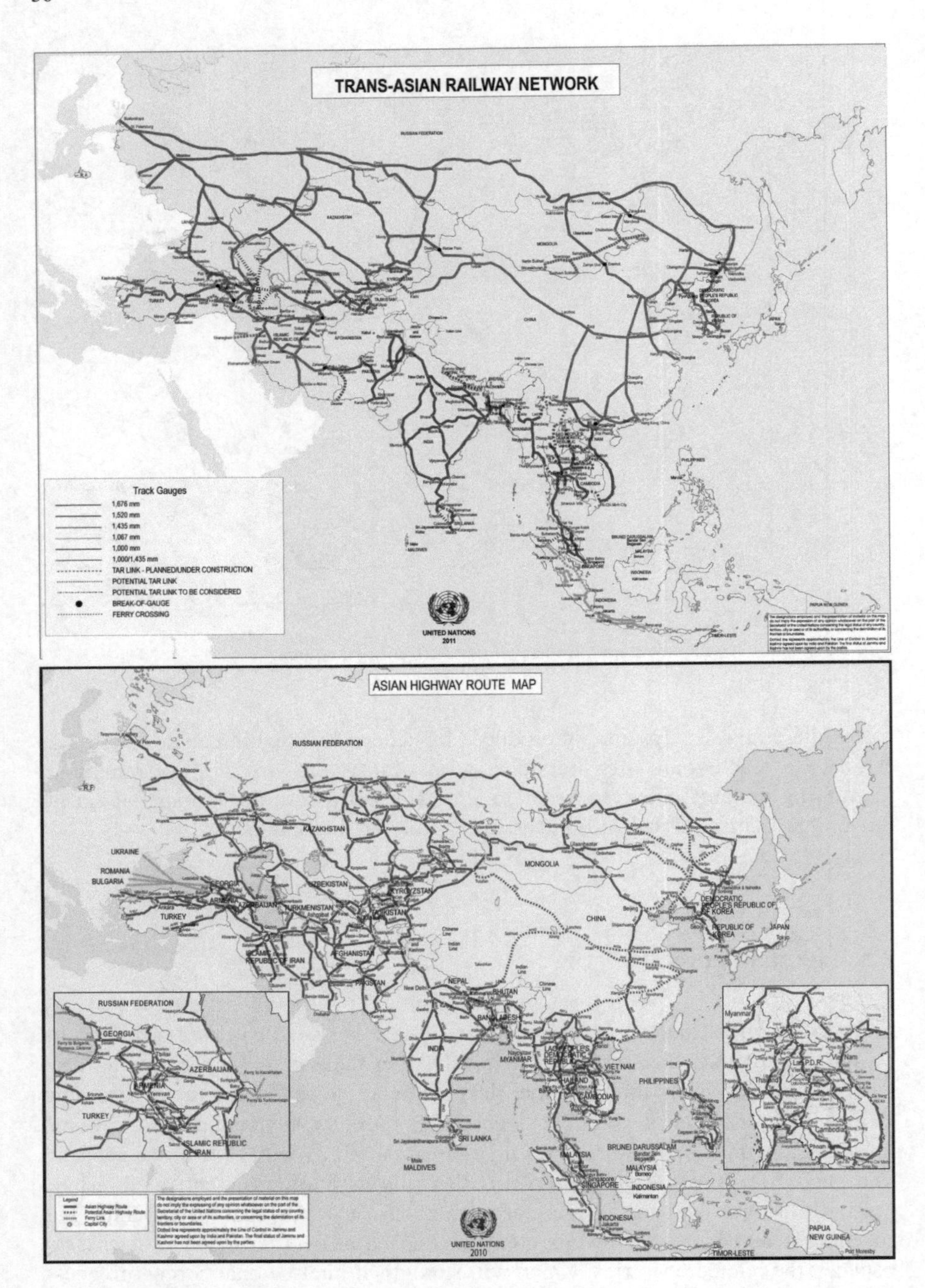

Fig. 2.8 Trans-Asian Railway and Asian Highway provisions (Source: ESCAP http://www.unescap.org/ttdw/index.asp?menuname=asianhighway, http://www.unescap.org/ttdw/common/TIS/TAR/tar_home.asp)

results, all of which would assist in understanding the East Asian region and each country's situation. Such progress would also serve as (1) the basis for regional cooperation in spatial development planning, (2) a visual indicator showing balanced development in each region, (3) educational material to enhance the Asian identity, and (4) an assessment indicator for concentrated investment in backward regions.

As a follow-up to the maps created for this project, several tasks should be carried out for setting up an East Asian Spatial Data Infrastructure: (1) produce a common seamless map from the national to the regional level showing existing conditions, (2) set comparable regional units based on population and territory size, (3) unify statistical units and comparable regional units by collecting data under the same definitions, and (4) increase the range of use by integrating the global map with regional statistical data. To implement these processes, cooperation and collaboration among related nations and research institutions are necessary.

References

Ahn CY (2001) Newly emerging economic integration in Northeast Asia. Int Econ 52:134–135

Chung SY (2010) Strengthening regional governance to protect the marine environment in Northeast Asia: from a fragmented to an integrated approach. Mar Policy 34(3):549–556

ESPON (2005) ESPON 2.1.1, Territorial Impact of EU Transport and TEN Policies. http://www.espon.eu/main/Menu_Projects/Menu_ESPON2006Projects/Menu_PolicyImpactProjects/

ESPON (2006) ESPON ATLAS Mapping the structure of the European territory. The regional setting and the data based representation. http://www.espon.eu/main/Menu_Publications/Menu_ESPON2006Publications/esponatlas.html

European Commission (1999) European Spatial Development Perspective (ESDP). http://ec.europa.eu/regional_policy/sources/docoffic/official/reports/pdf/sum_en.pdf

Hutton TA (2003) Service industries, globalization, and urban restructuring within the Asia-Pacific: new development trajectories and planning responses. Prog Plan 61(1):1–74

Kimura F, Takahashi Y, Hayakawa K (2007) Fragmentation and parts and components trade: comparison between East Asia and Europe. N Am J Econ Financ 18(1):23–40

Letiche JM (2000) Lessons from the euro zone for the East Asian economies. J Asian Econ 11(3):275–300

Liu WJ, Zhang B, Wang ZM, Song KS, Liu DW, Ren CY, Du J (2010) Development of a GIS-based decision support system for eco-environment and natural resources of Northeast Asia. Proc Environ Sci 2:906–913

NIRA Policy Research (2002) A grand design for the Northeast Asia region. Nat Inst Res Adv 15(11):1–83

Takahashi G (2010) The need of East Asian agricultural community and the framework. Agric Agric Sci Proc 1:311–320

Von Hippel D, Suzuki T, Williams JH, Savage T, Hayes P (2011a) Energy security and sustainability in Northeast Asia. Energ Policy 39(11):6719–6730

Von Hippel D, Gulidov R, Kalashnikov V, Hayes P (2011b) Northeast Asia regional energy infrastructure proposals: original research article. Energ Policy 39(11):6855–6866

Wang YJ (2004) Financial cooperation and integration in East Asia. J Asian Econ 15(5):939–955

Yun WC, Zhang ZX (2006) Electric power grid interconnection in Northeast Asia. Energ Policy 34(15):2298–2309

Chapter 3
A Study on Classification of Downtown Areas Based on Small and Medium Cities in Korea

Bum-hyun Lee

Abstract The purpose of this research is to analyze the deterioration of small-to-medium cities and their downtowns, and classify the types of decline. One of this study's features, therefore, is the categorization of the pattern of urban growth, with an accurate demarcation of the downtown area. Another feature is the priority order of the urban regeneration policies to be implemented by the central and local governments, arrived at by analyzing the characteristic indexes. The results of the categorization of the case cities will be used as source materials for urban regeneration projects spearheaded by the local government. The categories—regional-wide, citywide, and downtown-wide approach types—will help decide the priority order of the urban regeneration policies of the central government.

Keywords Urban regeneration • Downtown • Small and medium cities in Korea • Classification of downtowns

3.1 Introduction

3.1.1 Research Background and Purpose

"Downtown"[1] is the urban core area where the commercial, business, and production functions are concentrated. The growth and decline of downtowns have a direct influence on urban development and suburban areas. Recently,

[1] Downtown is the center of rural areas located outside the city and serves as a nexus of the local transportation network. It also plays a role in separating urban areas from rural areas (Christaller 1933).

B.-h. Lee (✉)
Green Territory & Urban Research Division, Korea Research Institute for Human Settlement, 254 Simin-daero, Dongan-gu, Anyang-si, Gyeonggi-do 431-712, South Korea
e-mail: bhlee@krihs.re.kr; hauri1021@naver.com

M. Kawakami et al. (eds.), *Spatial Planning and Sustainable Development: Approaches for Achieving Sustainable Urban Form in Asian Cities*, Strategies for Sustainability, DOI 10.1007/978-94-007-5922-0_3,

however, the population and urban functional decline of downtowns is becoming a serious problem. In particular, downtowns of small- and middle-scale cities in Korea have lost their competitiveness; compared with downtowns of metropolitan cities, their decline is much worse.

Thus, a diagnostic analysis of the downtowns of small- and medium-size cities in Korea should be carried out, considering the changes in industrial bases and urban population. This analytic approach can help urban planners and designers interpret the decline patterns occurring in Korea's small- and medium-scale cities and devise ways of encouraging downtown infill redevelopment and revitalization.

The most important step in gauging the extent of downtown decline is the selection of proper diagnostic indicators and interpretation of analytic results. The results may vary according to the indicators and analytic methods applied. Furthermore, downtown decline can be induced by inherited and external factors. Finally, the spatial scope of downtown decline depends on its spatial hierarchy within urban structure.

In this context, this study suggests types of downtown decline by considering population and employment changes in the specific areas. Moreover, the result of this study will be used as the basis for developing downtown revitalization policies suitable for the characteristics of middle- and small-scale cities in Korea.

3.1.2 *Research Scope and Methodology*

3.1.2.1 Scope

Based on precedent researches (Shin et al. 2004) and relevant urban planning acts,[2] small and medium-size cities[3] can be defined as those with populations ranging from 50,000 to 500,000. However, since cities with less than 100,000 residents do not have sufficiently formed downtowns, such have been excluded. Therefore, the study will examine cities in Korea with populations of 100,000–500,000 to be able to classify the types of downtowns; 39 cities have been considered. Also, changes in population and employment rates between 1995 and 2005 in the cities' downtowns will be used as reference data in classifying such areas. The data were collected at the district (*dong*) level.

In the existing urban structure, downtown is generally defined as the "urban core" area, and the zoning system in Korea designates most of them as "central business" or "general commercial" districts. Based on these general concepts, this

[2] Article 7 of the Local Autonomy Act (Requirements for the Development of the City and the Towns) and Article 11 of Enforcement Decree of the Local Autonomy Act (Exemptions of Official Works for the City with a Population of over 500,000).

[3] According to the academic definition, the small-to-medium city can be determined by its population size. Although recent studies in Korea may differ slightly, the small-to-medium city is defined as one whose population is between 50,000 and 500,000 (Jeongcheol Shin et al. 2004).

Table 3.1 Spatial scope of downtowns in the 39 cities in Korea (examples)

City	Maps
Asan City	
Gwangyang City	

study defines the downtowns of the 39 cities as the existing downtown and urban core areas suggested in each urban comprehensive plan. The downtown boundaries will be determined according to the *dong* system. Table 3.1 shows examples of downtown maps that were selected for this research's analysis.

The temporal range of this study spans 1995–2005, considering the availability of statistical data. This study selected this period because most of the downtowns of small-to-medium cities in Korea experienced a downtown or urban decline during this period. However, some small- and medium-size cities do not have data that describe the characteristics of population or employment changes during this period. In such cases, the study uses available data to plot the trends of population and employment changes.

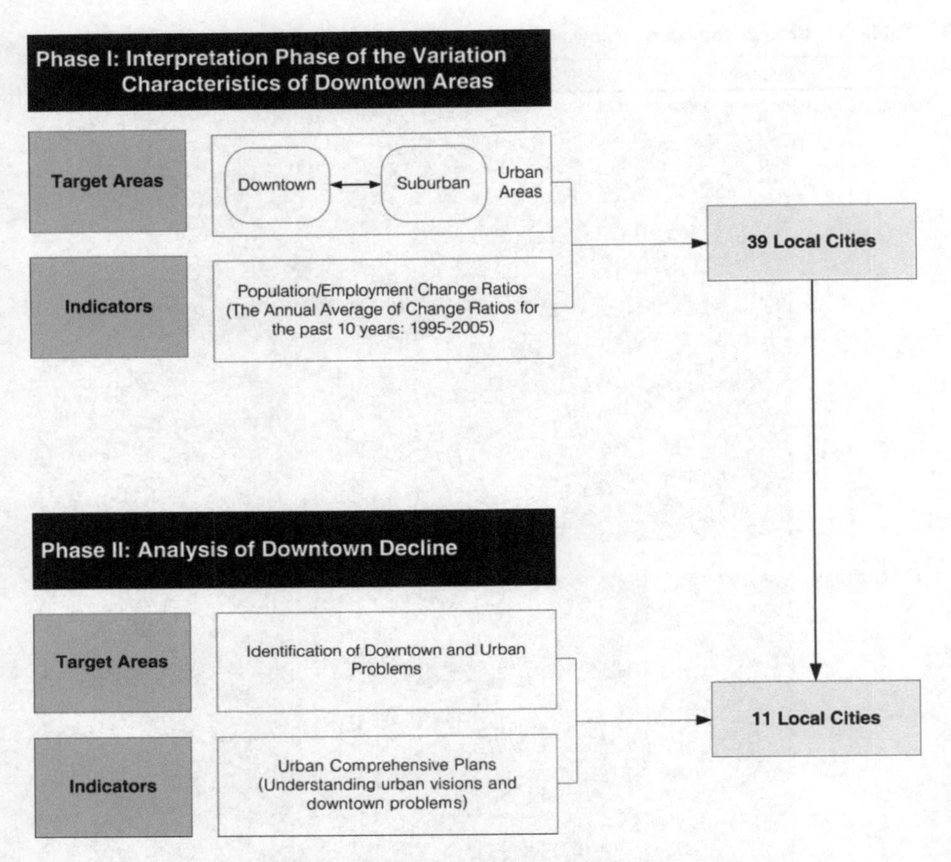

Fig. 3.1 Research flow diagram for setting up types of downtowns in the 39 cities

3.1.2.2 Methodology

This chapter takes both quantitative and qualitative approaches to analyzing the downtowns of the 39 cities and placing them under categories. The classification of these areas is implemented in two analytic phases: the interpretation of their variation characteristics and the analysis of downtown decline. In the interpretation phase, this study draws a comparison between the downtown and suburban areas and the entire urban area in terms of population and employment changes of each city. The cities that present a clear decline in downtown population and employment will be classified in the analysis phase, and their urban comprehensive plans will be taken into consideration (Fig. 3.1).

In the interpretation phase, the population change ratios and number of employees in the downtowns are used as evaluation indicators. Eleven of the 39 cities were selected, having been found to have clear downtown decline problems. In the analysis phase of downtown decline, this chapter focuses on "urban revitalization strategies" in the urban comprehensive plans of the chosen cities, as analyzing the population and employment changes in downtowns merely shows

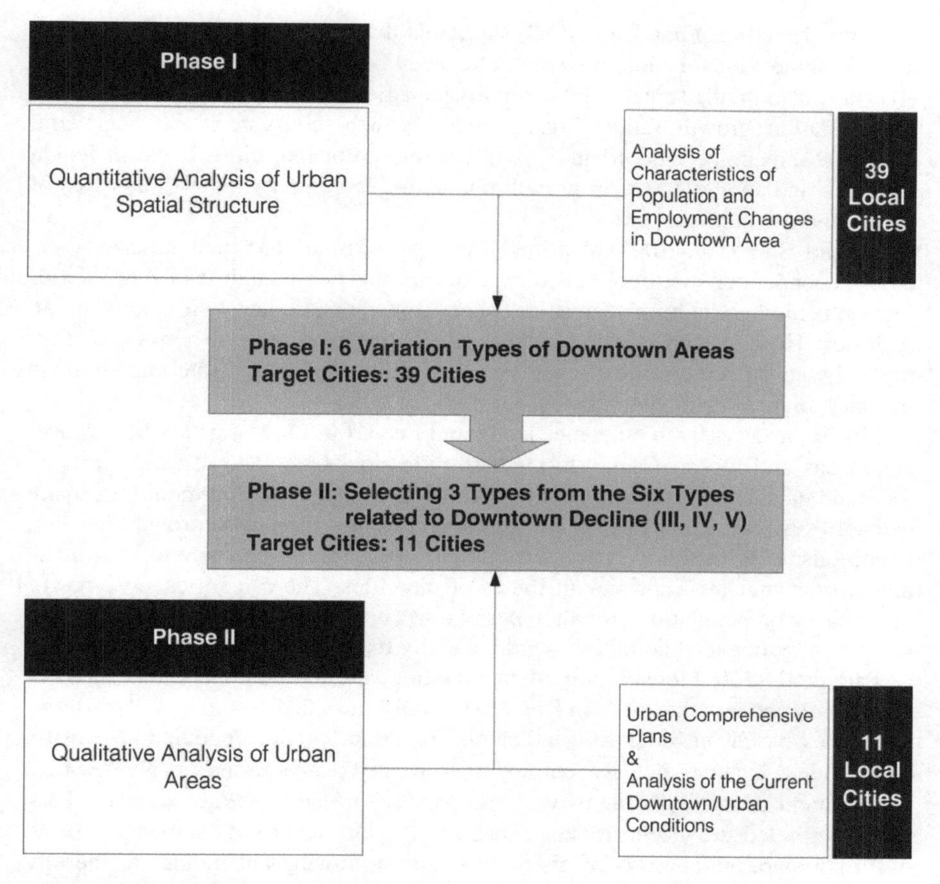

Fig. 3.2 Research framework

the extent of statistical changes in these areas. Therefore, problem recognition and the urban policy responses to downtown decline established by city governments are very important in determining the types of downtowns (Fig. 3.2).

3.2 Research Literature Review on Downtowns

3.2.1 Related Researches

3.2.1.1 Preceding Researches

Precedent researches related to this study can be divided into three main categories: urban functions and typology of small- and medium-size cities, urban revitalization strategies for small- and medium-size cities, and the actual conditions of small- and medium-size cities. First, Lim (1988) suggested that small- and medium-size cities assume some kind of “intermediate” attributes of rural and metropolitan areas. His study also divided small- and medium-size cities into four categories according to population growth ratios: high growth, growth, congested, and backward. Cho (1994) examined the urban density distribution of four cities in North Jeonla Province and suggested urban growth phases for Jeonju City, as well as types of small- and medium-size cities.

Second, Shin et al. (2004) identified four types of small- and medium-size cities: cities dependent on metropolitan cities; cities located behind industrial bases; cities located behind agricultural, forest, or fishery centers; and cities with specific urban facilities. Ha and Kim (1995) grouped small- and medium-size cities into four types, based on self-sufficient characteristics and accessibility: satellite, strategic industry, planned, and general small-to-medium cities.

Finally, as a research outcome, Jeong and Lee (2009) investigated the physical conditions and types of small-to-medium cities in Korea. The research grouped 38 middle-scale cities with populations of less than 300,000 into four main categories by considering several evaluation criteria—population growth, location conditions, administration systems, industrial structure, local economic conditions, and cluster analysis—and the attributes of the cities. The categories are types of cities based on population growth, types of cities based on industrial structure, local city types by financial conditions, and local city types drawn from cluster analysis.

Park et al. (2009) looked into urban structure patterns and population centers in small- and medium-size cities in Korea. The study classified 49 small- and medium-size cities based on diagnostic indicators for urban decline: population density, aging index, business density, employment density per thousand people, medical employment density, and density of the population using welfare services. Park also interpreted urban phenomena related to the city-level indicators and carried out a comparative analysis of distribution and hollowing-out indices of the city populations.

3.2.1.2 Difference Between This Research and Preceding Researches

The purpose of this research differs from the contents of precedent researches in certain points. First, the spatial scope of the small-to-medium cities in this study is defined as cities with populations exceeding 100,000. Furthermore, population and employment change ratios of the downtown, suburban areas, and the entire city are used to determine the final types of downtowns in cities in Korea.

Second, by considering urban revitalization policies and the approaches taken by local municipalities to downtown decline problems, this study tries to deliver a more comprehensive analysis of the setting up of downtown types. After examining the population and employment change ratios of 39 small-to-medium cities, 11 cities were selected and will be categorized according to existing urban comprehensive plans during the analysis phase. This is quite a different approach to understanding downtown decline.

Table 3.2 Characteristics of urban areas according to urban development phases

Urban development phase	Characteristics of population and employment changes in each part of urban area			Re-categorizing downtown development phases
	Downtown	Suburban areas	The entire city	
1st phase	+	−	+	Downtown growth
	+	+	+	
2nd phase	−	+	+	Downtown decline
3rd phase	−	+	−	
	−	−	−	
4th phase	+	−	−	Downtown growth

Refer to van den Berg et al. (1982)

3.2.2 *Brief Review of the Theory of Urban Development Phases*

Van den Berg et al.'s (1982) theory of urban development phases provides a general understanding of a city's characteristics in each phase and how the downtown, suburban areas, and the entire city will change. Berg divided the city into inner (central business district [CBD], core) and ring (outer or suburban) areas and named four phases of urban development: urbanization, suburbanization, anti-urbanization, and re-urbanization. In particular, Berg's theory categorized urban development phases by considering changes in the population and number of people employed within the city, which fits very well into this study's research methodology.

Table 3.2 shows Berg's four phases of urban development. In the 1st and 4th phases (urbanization and re-urbanization), there are overall increases in downtown population and employment. In the 2nd and 3rd phases (suburbanization and anti-urbanization), however, rapid suburban growth and development erode the competitiveness of downtowns. People move to suburbia, which have better service infrastructures and amenities, so that downtown population and employment drop. Therefore, during the 2nd and 3rd phases, downtowns are usually beset by serious social problems, such as slums and deteriorating public safety. Thus, Berg's approach provides useful information for interpreting the development patterns and characteristics of downtowns.

However, in order to construct a simpler analytic model suitable for interpreting downtown decline and growth in each urban development phase, it is necessary to slightly adjust some categories and elements in Berg's theory. As mentioned in the previous section, the purpose of this study is to identify the characteristics of downtowns; the attributes of growth and decline in downtowns are compared with those of suburban areas and the entire city by analyzing the changing ratios of population and employment. Based on Berg's theory of urban development phases, Table 3.2 shows the characteristics of population and employment changes in downtowns, suburban, and entire cities. (Note the changes in the "downtown" column.)

Also, as shown in Table 3.2, the population and number of employees in downtowns increase in the 1st and 4th phases and decrease in the 2nd and 3rd phases. Since this chapter aims to suggest types of downtowns in small- and medium-size cities in Korea, Berg's four urban development phases can be regrouped into two downtown development phases: growth and decline phases. To diagnose downtown decline in the 39 small- and medium-size cities, this study defines Berg's 2nd and 3rd phases as the downtown decline phase and his 1st and 4th phases as the downtown growth phase.

Finally, Table 3.2 divides urban development phases 1 and 3 into two subcategories. This means that the changing characteristics of suburban areas in terms of population and employment can be different even if downtown and the entire city followed the same growth or decline patterns in each urban development phase.

3.3 Typology of Downtowns in Korean Cities

3.3.1 Types of Downtowns According to Population and Employment Changes

Considering the downtown classification model based on Berg's theory and the changing ratios of population and employment in small-to-medium cities, the 39 small- and medium-size cities can be grouped into six variation types. In Table 3.3, "variation types I, II, and VI" show increases in both population and employment within downtowns. In particular, variation types I and II provide different

Table 3.3 Classification of local small-to-medium cities based on population change ratios (1995–2005)

Categories	Characteristics of population changes			Cities
	Downtown	Suburban	Urban areas	
Variation type I	+	–	+	Gangneung, Wonju, Cheongwon, Kongju, Dangjin
Variation type II	+	+	+	Asan, Gwangyang, Kumi, Gyeongsan, Gimhae, Yangsan, Geoje
Variation type III	–	+	+	Jecheon, Seosan, Suncheon, Chilgok, Jinju, Jinhae, Jeju
Variation type IV	–	+	–	Gunsan, Mokpo, Gyeongju, Sacheon
Variation type V	–	–	–	Kimje, Yeosu, Andong, Youngju,Youngcheon, Masan, Miryang, Tongyeong
Variation type VI	+	–	–	Chuncheon, Chungju, Nonsan, Boryeong, Iksan, Jeongeup, Gimcheon, Sangju

changing patterns in suburban areas. Since all three types represent population and employment growth in downtowns areas, they can be regarded as downtown growth types.

Meanwhile, variation types III, IV, and V show decreases in population and the number of people employed in downtowns. Variation type III shows that the population and the number of employees in suburban areas are increasing, in contrast to the decreases registered by downtowns and entire cities. In variation type IV, population and employment decrease throughout the entire city but increase in suburban areas. Variation type V shows a decrease in population and employment in all urban areas. These three types can be classified under downtown decline.

3.3.1.1 Characteristics of Population Changes in the Case Cities

The characteristics of downtown population changes can be analyzed by comparing them with population changes in the suburbs and the entire city. As shown in Table 3.3, the 39 small-to-medium cities can be grouped into six variation types, according to population change ratios from 1995 to 2005. Table 3.3 also shows that 19 of the cities belong to the downtown decline type (variation types III, IV, and V). Seven cities fall under downtown variation type III—a shrinking downtown population. The remaining 12 cities are types IV and V—decreasing population downtown and in the entire city.

3.3.1.2 Characteristics of Employment Changes in the Case Cities

By using the employment change ratio from 1995 to 2005, the 39 cities can be grouped into six variation types. The number of employees decreased in 20 of the cities. There were 14 variation type III cities, including Chuncheon and Jecheon. The remaining six belong to variation types IV and V (Table 3.4).

3.3.1.3 Results of the Variation Type Analysis of the Case Cities

Considering the analytic results of the characteristics of population and employment changes in downtowns, some cities showed a decrease in either population or employment and 11 of them, in both. This study focuses on the latter—the 11 cities—as it may be difficult to determine whether cities with a decrease in only one variable are experiencing downtown decline.

Therefore, the 11 cities will be again categorized into three types of downtown decline. The urban policies of the 11 cities have emerged with the downtown revitalization trend in the urban comprehensive plans (Tables 3.5, 3.6, 3.7).

Table 3.4 Classification of local small-to-medium cities based on employment change ratios (1995–2005)

Categories	Characteristics of employment changes: Downtown	Suburban	Urban areas	Cities
Variation Type I	+	–	+	Chungju
Variation Type II	+	+	+	Kangneung, Wonju, Cheonwon, Asan, Nonsan, Dangjin, Seosan, Kimje, Kumi, Kimcheon, Youngcheon, Chilgok, Kimhae, Yangsan, Geoje, Milyang, Jinhae
Variation Type III	–	+	+	Chuncheon, Jecheon, Kongju, Boryoung, Suncheon, Mokpo, Kwangyang, Kyoungju, Andong, Youngju, Jinju, Sacheon, Tongyoung, Jeju
Variation Type IV	–	+	–	Iksan, Kunsan, Kyoungsan
Variation Type V	–	–	–	Jeongup, Yeosu, Masan
Variation Type VI	+	–	–	–

Table 3.5 The analytic results of the 39 cities based on population change ratios

Categories: Province	City	The rate of increase in population for 10 years (%): Downtown	Suburban	Urban area	Variation type	Downtown decline
Kangwon	Gangneung	0.41	–0.21	0.01	I	
	Wonju	0.61	2.94	1.79	I	
	Chuncheon	0.52	–0.75	–0.01	VI	
North Chungcheong	Chungju	0.59	–0.73	–0.61	VI	
	Jecheon	–4.35	1.67	0.95	III	●
	Cheongwon	5.17	–0.95	0.00	I	
South Chungcheong	Asan	3.19	2.78	2.95	II	
	Kongju	9.06	–0.08	1.51	I	
	Nonsan	1.42	–2.95	–1.60	VI	
	Dangjin	1.33	–0.38	0.05	I	
	Boryeong	8.46	–2.66	–1.02	VI	
	Seosan	–4.08	0.86	0.64	III	●
North Jeonla	Iksan	0.74	–2.02	–0.61	VI	
	Gunsan	–4.57	0.31	–0.67	IV	●
	Kimje	–0.63	–3.72	–2.42	V	●
	Jeongeup	0.19	–3.02	–1.85	VI	
South Jeonla	**Yeosu**	–5.51	–0.28	–1.24	V	●
	Suncheon	–0.43	0.99	0.47	III	●
	Mokpo	–3.51	0.80	–0.12	IV	●
	Gwangyang	1.08	1.05	1.06	II	

(continued)

Table 3.5 (continued)

Categories		The rate of increase in population for 10 years (%)				
Province	City	Downtown	Suburban	Urban area	Variation type	Downtown decline
North Kyung-sang	Kumi	2.48	2.55	2.54	II	
	Gyeongsan	4.33	2.18	3.30	II	
	Gyeongju	−3.01	0.23	−0.29	IV	●
	Gimcheon	0.09	−2.38	−0.87	VI	
	Sangju	1.11	−3.71	−1.60	VI	
	Andong	−1.20	−2.05	−1.85	V	●
	Youngju	−4.73	−0.56	−1.42	V	●
	Youngcheon	−0.30	−1.55	−0.94	V	●
	Chilgok	−0.55	4.01	2.60	III	●
South Kyung-sang	Gimhae	4.27	6.35	5.28	II	
	Masan	−0.75	−0.32	−0.33	V	●
	Jinju	−2.98	0.41	0.19	III	●
	Yangsan	10.63	1.65	2.83	II	
	Geoje	1.23	3.63	2.74	II	
	Miryang	−0.48	−2.09	−1.39	V	●
	Sacheon	−3.55	0.40	−0.63	IV	●
	Jinhae	−0.42	5.04	4.17	III	●
	Tongyeong	−0.96	−0.75	−0.84	V	●
Jeju	**Jeju**	−0.26	4.60	1.86	III	●

Table 3.6 The analytic results of the 39 local cities based on employment change ratios

Categories		The rate of increase in the number of employees (%)				
Province	City	Downtown	Suburban	Urban area	Variation type	Downtown decline
Kangwon	Gangneung	0.57	2.30	1.41	II	
	Wonju	1.78	3.70	2.72	II	
	Chuncheon	−0.92	6.81	1.40	III	●
North Chungcheong	Chungju	0.88	−0.10	0.44	I	
	Jecheon	−12.78	−2.75	−5.98	V	●
	Cheongwon	3.77	5.01	4.88	II	
South Chungcheong	Asan	1.21	5.55	3.98	II	
	Kongju	−2.68	2.10	0.76	III	●
	Nonsan	1.95	3.40	2.78	II	
	Dangjin	4.25	11.49	8.51	II	
	Boryeong	−2.87	2.22	0.72	III	●
	Seosan	0.45	3.98	3.33	II	
North Jeonla	Iksan	−1.79	0.35	−0.68	IV	●
	Gunsan	−5.70	0.74	−1.16	IV	●
	Kimje	1.26	0.82	1.08	II	
	Jeongeup	−0.06	−0.75	−0.40	V	●

(continued)

Table 3.6 (continued)

Categories		The rate of increase in the number of employees (%)				
Province	City	Downtown	Suburban	Urban area	Variation type	Downtown decline
South Jeonla	Yeosu	−4.44	−0.85	−1.68	V	●
	Suncheon	−1.89	2.53	0.18	III	●
	Mokpo	−3.24	2.14	1.05	III	●
	Gwangyang	−0.06	1.07	0.85	III	●
North Kyung-sang	Kumi	−0.01	2.75	2.23	II	
	Gyeongsan	−1.38	0.41	−0.36	IV	●
	Gyeongju	−1.45	3.59	2.19	III	●
	Gimcheon	3.11	4.85	3.55	II	
	Sangju	2.83	−4.10	0.08	I	
	Andong	−0.88	3.29	1.20	III	●
	Youngju	−2.99	1.73	0.15	III	●
	Youngcheon	1.59	0.40	1.12	II	
	Chilgok	1.19	7.47	4.77	II	
South Kyung-sang	Gimhae	2.98	8.92	6.20	II	
	Masan	−4.28	−2.31	−2.57	V	●
	Jinju	−3.53	2.36	1.03	III	●
	Yangsan	8.41	5.19	4.62	II	
	Geoje	3.68	4.68	4.26	II	
	Miryang	0.54	1.35	0.92	II	
	Sacheon	−3.32	2.76	0.92	III	●
	Jinhae	14.96	6.52	7.60	II	
	Tongyeong	−1.48	3.08	0.52	III	●
Jeju	Jeju	−0.08	4.28	1.85	III	●

Table 3.7 The 11 selected cities showing decrease in both population and the number of employees between 1995 and 2005

Types and target cities	Characteristics of population change	Characteristics of employment changes	Remarks
Variation type III	Jecheon, Seosan, Suncheon, Chilgok, Jinju, Jinhae, Jeju	Chuncheon, Kongju, Boryeong, Suncheon, Mokpo, Gwangyang, Gyeongju, Andong, Youngju, Jinju, Sacheon, Tongyeong, Jeju	Decline in downtown areas
Variation type IV	Gunsan, Mokpo, Gyeongju, Sacheon	Iksan, Gunsan, Gyeongsan	Decline in downtown areas
Variation type V	Kimje, Yeosu, Andong, Youngju, Youngcheon, Masan, Miryang, Tongyeong	Jecheon, Jeongeup, Yeosu, Masan	
Cities selected for the analysis of urban decline	Jecheon, Suncheon, Jinju, Jeju, Gunsan, Mokpo, Sacheon, Yeosu, Andong, Youngju, Tongyeong		Decreasing in population and employment

Table 3.8 Downtown decline types

Categories	City	Rates of change in downtown areas	
		Employment (%)	Population change ratio (%)
Downtown decline type I	Suncheon	−1.89	−0.43
	Jinju	−3.53	−2.98
	Jeju	−0.08	−0.26
Downtown decline type II	Gunsan	−5.70	−4.57
	Jecheon	−12.78	−4.35
	Mokpo	−3.24	−3.51
	Sacheon	−3.32	−3.55
Downtown decline type III	Andong	−0.88	−1.20
	Youngju	−2.99	−4.73
	Tongyeong	−1.48	−0.96
	Yeosu	−4.44	−5.51

Type I: Its characteristics are similar to those of downtown variation type III in terms of both population and employment
Type II: The characteristics of population change are similar to those of downtown variation type III, but the characteristics of employment change are similar to downtown variation type IV or V
Type III: Both population and employment exhibit characteristics similar to those of downtown variation types IV and V

3.3.2 Analysis of Downtown Decline

Considering the characteristics of the downtowns of the 11 selected cities, three types of downtown decline can be suggested. Downtown decline type I exhibits decline patterns similar to the downtown variation type III, in which only the downtowns show decreases in both population and the number of employees. Downtown decline type II has characteristics of the variation type III in terms of population, but the characteristics of employment change are similar to variation types IV and V. Lastly, downtown decline type III has characteristics of the downtown variation types IV and V in both population and employment changes (Table 3.8).

In order to make a comprehensive analysis of downtown decline in small-to-medium cities, urban issues related to downtown areas should be investigated along with the interpretation of population and employment change ratios. Downtown decline problems in the urban comprehensive plans of the 11 target cities can be classified as internal (resulting from the downtown itself) and external (city-level concerns).

According to downtown revitalization strategies in the urban comprehensive plans of the 11 cities, downtown decline problems are mostly caused by external factors—in particular, urban structural problems. Also, the 11 local cities are concentrating on developing their suburbs and do not efficiently utilize urban resources.

External factors related to downtown decline can be gleaned by analyzing the industrial and urban structures of the 11 target cities. Table 3.9 summarizes the internal and external problems related to downtown decline in the urban comprehensive plans of 7 cities among the 11 target cities.

Table 3.9 Analysis of downtown problems in the urban comprehensive plans of 7 cities among the 11 target cities

	Downtown decline problems		
Categories	Internal problems	External problems	Cities
Physical problems	Lack of infrastructure Unclear road functions and network Lack of pedestrian paths and space Densely located old buildings Lack of lands for development	Indistinctness of urban structure (weak relationship between urban functions and lack of spatial connectivity)	Gunsan and Jeju
Economic problems	Lack of skilled labor	Cities which do not have any benefits from the National Territory Comprehensive Plans Poor industrial bases within urban areas Lack of financial resources Concentration of development on suburban areas Lack of urban competitiveness compared with other revitalized cities	Jinju, Yeosu, and Mokpo
Social problems	Insufficient and low-quality public service infrastructures	Poor application of tourism resources Poor urban identity and image	Seosan and Jecheon

3.3.3 Analysis Results

Considering the downtown decline problems and urban planning challenges mentioned in the urban comprehensive plans of the 11 target cities, the downtown areas can be divided into three types. This typological approach helps suggest urban planning solutions to downtown decline problems in the urban comprehensive plans as well as identify the current downtown conditions through analysis of downtown decline problems. In particular, urban planning visions and industrial functions of the 11 local cities in response to urban challenges in the urban comprehensive plans served as important evaluation criteria for drawing the final three types of downtown areas. Table 3.10 summarizes three downtown types and their characteristics based on urban visions and downtown decline problems in the urban comprehensive plans of the 11 target cities.

The local cities belonging to the downtown type I have abundant tourism resources within urban areas. However, the cities in this type have problems such as excessive concentration of primary industries and low utilization of tourism resources. For example, Jecheon City in this type focuses on creating the twenty-first century garden city rather than developing other strategic industries apart from primary industries. Also, Jecheon City has a relatively high percentage of primary industries compared with other cities.

Table 3.10 Typology of downtowns in the final 11 local cities

	Characteristics of each downtown type		
Types	Urban visions	Downtown problems	Cities
Downtown type I	Creating a garden city a lot of historic and cultural resources	Monocentric urban structure Densely located old buildings Excessive concentration of primary industries Weak urban industrial bases Low utilization of tourism resources Lack of lands available for future development	Jecheon, Gyeongju, Youngju, Tongyeong
Downtown type II	Creating an urban–rural integration city or an high-tech industrial city	Poor functional distribution between urban and rural areas Low employment rates due to decline of urban strategic industries	Seosan, Jinju Yeosu, Jeju
Downtown type III	Developing logistic and marine industrial cities	Inactive commercial, logistic, and industrial functions in urban areas Lack of urban competitiveness	Gunsan, Mokpo, Sacheon

Downtowns are classified by considering urban visions and downtown problems in the urban comprehensive plans of the 11 local cities

The cities belonging to the downtown type II focus on creating the urban–rural city. However, the type II cities are generally experiencing serious decline of secondary industries like the manufacturing industry. Also, the cities have poor distribution between urban and rural functions and a low employment growth rate because of slow industrial growth. Seosan City is a typical example of this type. Urban and rural functions are poorly integrated in the city, so downtown decline problems caused by the collapse of exiting basic industries are gradually worsening.

The cities in the downtown type III are ones that concentrate on the development of a logistic and marine industrial city. The cities in this type show evident signs of decline in the tertiary industry, so they are falling behind other cities. For example, Gunsan City shows very evident signs of the entire city declining due to a lack of competitiveness with other cities like Jeonju City.

3.4 Conclusion and Implications

By singling out 11 cities of the 39 small-to-medium cities in Korea, whose populations range from 100,000 to 500,000, this study suggests types of downtown decline after analyzing the characteristics of such decline in terms of population and employment changes and urban revitalization policies in the urban comprehensive plans. Urban revitalization policies that reflect the current physical potential of cities and future urban planning values should be developed by both the central and local governments. The typology of downtowns in this study can be useful in developing urban and downtown revitalization policies in this context, based on the clear interpretation of urban and downtown decline problems.

Considering the three downtown decline types, the following urban and downtown revitalization policies can be suggested for each type. The urban policies of cities under downtown type I should focus on conserving the existing historic and cultural assets of downtown areas and maximizing their potential as tourism resources. Moreover, the revitalization of downtowns in this type should be gradual—based on historic preservation, not large-scale clearance redevelopment.

The cities belonging to downtown type II should prioritize urban revitalization policies for improving existing industrial functions. The development of industrial bases in suburban areas should be sublated, and policies that will attract small-scale high-tech industries to downtowns should be drafted.

The cities in downtown type III should come up with strategies for sharing regional functions with the cities around them. Functional sharing with surrounding cities and strategies for reorganizing regional spatial structure should be the most important urban revitalization policies for the cities in this type.

The analysis of the existing urban comprehensive plans of the 11 selected cities, whose downtowns exhibit both population and employment decline, shows that the plans adopt “very similar” urban revitalization policies regardless of the characteristics and potential of the downtowns. Therefore, urban revitalization policies reflecting the physical, social, economic, and cultural conditions of each city’s downtown should be developed by the local government and included in the urban comprehensive plans.

Nevertheless, this study examines only the characteristics of population and employment changes in downtowns for determining three downtown types. In order to develop a typology of downtowns based on a more accurate interpretation of downtown characteristics, future researches should include more factors affecting downtown decline. Also, case studies that are more detailed need to be conducted on cities with evident urban decline. The studies should interpret various statistical data for diagnosing downtown conditions and the local governments’ revitalization policies.

References

Cho JS (1994) Conceptualizing spatial structure in local small-and-medium cities in Korea. J Korea Plan Assoc 29(3):81–99

Christaller W (1933) Die Zentrale Orte in Süddeutschland. Eine ökonomisch-geographische Untersuchung über die Gesetzmäßigkeit der Verbreitung und Entwicklung der Siedlungen mit städtischen Funktionen. WBG, Darmstadt

Ha SK, Kim JI (1995) A study on the spatial functions and role of small- and medium-sized Korean cities for balanced regional development. J Korea Plan Assoc 30(3):35–56
Jeong YW, Lee SS (2009) A study on the current conditions of local small-and-medium cities and their regional development policies. J Korean Reg Dev Assoc 21(1):29–50
Lim CH (1988) A comparative study of small and medium cities. KRIHS Report
Park BH et al (2009) Comparative analysis on the changing patterns population centroid and CBD's decline of small and medium local cities. J Korea Plan Assoc 44(1):61–73
Shin JC et al (2004) A study on revitalization strategy of local small and medium-sized cities. KRIHS Report
Van den Berg L et al (1982) Urban Europe: a study of growth and decline. Pergamon Press, Oxford, p 36

Chapter 4
Significance and Limitations of the Support Policy for Marginal Hamlets in the Strategy of Self-sustaining Regional Sphere Development

Ryohei Yamashita and Tomohiro Ichinose

Abstract In Japan, less-favored areas known as "marginal hamlets" have multiplied rapidly, and the continuous care of such areas has been major problem for national land-use planning. This research aims to scrutinize how the positioning and directionality of the support for marginal hamlets are defined in the National Spatial Plan, the plan's compatibility in solving the problems of a broad area and the requests from individual marginal hamlets, and the plan's limitations. We conducted a field survey at Monzen town in Wajima City, Ishikawa Prefecture, and carried out the study's purpose through a comparative analysis on the future image of bottom-up and top-down processes.

The analysis revealed that many residents had a strong inclination toward settlement; a majority of them stated that they would stay in the town regardless of depopulation. In contrast, farmland management had already reached its limit. When arable land near an area under cultivation was abandoned, it became clear that the resulting external diseconomy would make the idle field impossible to manage.

We propose three steps to achieve a consensus among stakeholders. First, determine how much manpower and funds are needed to support marginal hamlets. Second, find out how these capital resources can be obtained from inside the depressed area. Third, if such capital resources cannot be obtained, determine the necessary revisions in the support policy for marginal hamlets.

Keywords National Spatial Plan • Marginal hamlets • Regional self-sustaining development • Depopulation • Wajima City

---

R. Yamashita (✉)
Tokyo University of Science, 2641, Yamazaki, Noda 278-8510, Chiba, Japan
e-mail: yamashita@rs.tus.ac.jp; r-yama@ishikawa-pu.ac.jp

T. Ichinose
Department of Environment and Information Studies, Keio University,
5322, Endo, Fujisawa, Kanagawa 252-0882, Japan
e-mail: tomohiro@sfc.keio.ac.jp

M. Kawakami et al. (eds.), *Spatial Planning and Sustainable Development: Approaches for Achieving Sustainable Urban Form in Asian Cities*, Strategies for Sustainability,
DOI 10.1007/978-94-007-5922-0_4,

4.1 Introduction

In this chapter, we attempt to clarify the direction and limitations of the current support for marginal hamlets that were first named by Ohno (2005) in designated self-sustaining development spheres in Japan. In accordance with the limited human resources and volition to do agricultural work in marginal hamlets, we try to figure out the endogenous conditions and policy measures in the National Spatial Plan for the sustainable development of marginal hamlets in Hokuriku Region, Japan.

Certain priorities raise important issues related to marginal hamlets. First, as Ohki (2009) mentioned, is the relationship between an urban area and a rural area, which is entrusted completely to the structure of the market economy. From the viewpoint of risk management for a future food crisis in Japan, it is necessary to preserve a portion of the usable farmland. In addition, biodiversity conservation is needed in less-favored areas to ensure the success of spatial conservation and industrial and economic development. Therefore, what kind of supporting policy for marginal hamlets should be characteristic of a wide-area planning policy that aspires to achieve self-sustaining sphere development? More fundamentally, is the breaking point of the continued existence of marginal hamlets taken into consideration at all? Tachikawa (1997) and Hashizume (2004) analyzed this point, which is described as a statistical "rural minimum." Although their researches were elaborately done, the results failed to produce a universal solution; further investigation in an actual sphere is thus necessary.

In 2005, ranged against the above circumstances, the National Spatial Planning Act for Japan was partially revised. Three years later, in July 2008, the National Spatial Strategies for Japan was instituted—a policy engendered by the comprehensive national development plan (1962)—to promote the ground design based on the doctrine of correction of regional divide. It pushed for the construction a multiaxial society for the decentralization of authority. In August 2009, the National Spatial Strategy Regional Plan (NSSRP) was formulated for the local development strategy of each of the eight regional spheres (hereafter, spheres) in Japan in the following decade. Japan's new National Spatial Plan, made by the Ministry of Land, Infrastructure, Transport, and Tourism, serves as a guideline of future spatial strategies. However, notable features of the NSSRP are bolstering liaison with some countries in East Asia and leading the interregional competition. Since the plan covers a vast area, great flexibility is needed. It is also important to carefully consider improving it from the perspective of sustainable rural planning.

In this connection, because the unique development plans for Hokkaido and Okinawa prefectures have been established by each local development bureau, the eight spheres (blocs) are Tohoku, Capital Area, Hokuriku, Chubu, Kinki, Chugoku, Shikoku, and Kyusyu (see Fig. 4.1). This action established the guiding principle of regional development, reflecting a regional characteristic. Moreover, in the NSSRP, the construction of new local new governance that comprised diverse entities (e.g., farmers, nonfarmers, schools, nonprofit organizations [NPOs], voluntary organizations) is strongly promoted.

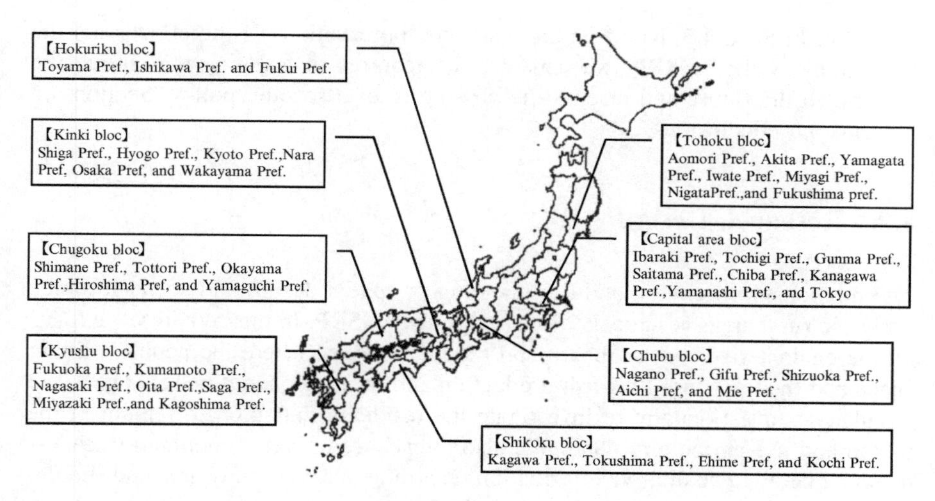

Fig. 4.1 Division of the eight spheres in NSSRP

However, there are many regions that have reached the limit of self-help and self-reliant regional management and can no longer survive without tapping external economic and human resources. In the intermediate stage of the publicly released NSSRP, its priority for urban areas is excessively high and for rural areas much lower (Association of Rural Planning, 2009). In fact, in the majority of less-favored areas, the residents' desire for the rural minimum is still high, as this research illustrates. "Whole nation land care" (in Japanese, *Kokudo no kokuminteki keiei*) refers to a social system in which the public willingly supports the weak regions as their own responsibility; it ought to be a political slogan. Unfortunately, due to the very abstract descriptions of individual project plans, the degree of recognition of the NSSRP is not high among municipal employees who are in close contact with regional residents (Seta 2009).

In this research, we begin by surveying the present state and future prospects of the residents and agricultural land conservation in a case study area. Simultaneously, we extract the description in connection with the regional strategy in the NSSRP and summarize how the NSSRP deals with marginal hamlets. Next, we examine whether the self-sustaining sphere development strategy reinforces the endogenous intention for marginal hamlets and contributes to the rational management of the rural problem. If incompatibilities emerge between intention and reality, a different regional design for spatial planning is required. Thus, we plan to reconsider the methodology for the operation and maintenance of marginal hamlets and submit policy recommendations based on the significance and limitations of the current policy.

This chapter is structured as follows. The study area and survey method are described in Sect. 4.2, and the results of the quantitative questionnaire and the qualitative interview are discussed in Sect. 4.3. The description of the problem and counterplan of the farm and mountain areas in the NSSRP are summarized in

Sect. 4.4. In Sect. 4.5, based on the residents' observations of the effects and the limitations of the NSSRP, we consider the features of a policy that should be adopted in the future and discuss the possibility of a specific policy. Section 4.6 presents our conclusions.

4.2 Research Design

The relationship between land-use planning or spatial design and social resource design in rural areas is scarcely discussed in the NSSRP. In this research, we focus on the characteristics of seaboard and mountain areas, where the people's living space and the tracts used for fish production and farming cannot be separated. We circulated a questionnaire to investigate the residents' intention to remain in the region and use agricultural land. We also interviewed several community leaders to gather detailed qualitative information regarding our case study area and supplement the questionnaire's quantitative findings.

Not many recent Japanese researches on regional policy analyze the inconsistencies and gaps between the national land-use plan and the status of marginal hamlets. In concrete terms, support for agricultural management (Hiraguchi and Morozumi 2010), a quantitative analysis of the human network (Akazawa and Matsuoka 2010), an improvement in the quality of life (Noguchi 2009), the development of community welfare (Koiso 2011), and rural knowledge management (Nakatsuka and Hoshino 2009) are considered crucial in rural planning. However, in rural areas that come under zones covered by urban planning, the political emphasis tends to be on the reconstruction of the local core city, rather than support for a disadvantaged area.

4.2.1 Outline of the Study Area

Monzen village in Wajima City, Ishikawa Prefecture, is our study area (see Fig. 4.2). Wajima is a seaboard tourist city in the northern part of Ishikawa (see Photo 4.1). As of February 1, 2011, after a consolidation of municipalities, Wajima's population was 31,510 and number of households 12,869; Monzen, a seaboard farming and fishing village in western Wajima, had 7,004 people and 3,119 households. Monzen is the most rapidly ageing village in Ishikawa. The Noto railway, once the only train line passing through Wajima City, played the vital role of transporting elders to hospitals and students to school. However, since there were few other users, the line was discontinued in 2005 due to unprofitability.

Although this region was a disaster area in the aftermath of the 2007 Noto Hanto earthquake—Monzen village was the worst hit—reconstruction has proceeded gradually to date. However, because this village and surrounding areas are experiencing rapid depopulation and ageing, it is very difficult to obtain support for only this village.

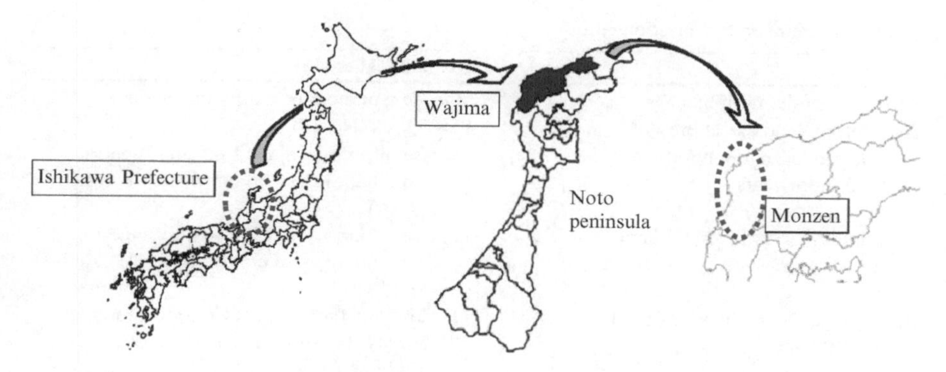

Fig. 4.2 The location of the study area in this research

Photo 4.1 At *left* is the main tourist resource "Asa-ichi" and at *right* the landscape called "Sen-mai da" (a thousand pieces of paddy field) in Wajima

4.2.2 Questionnaire

The important items of the questionnaire are shown in Table 4.1. This questionnaire was delivered as two sheets enclosed in a city report to 3,309 households (excluding empty houses) on January 1, 2009. Meanwhile, the purpose of the interviews was to understand the status of community activity, which is one of the important factors in the durability of rural society. In Monzen village, there are eight districts that are daily activity groups, composed of 10–30 small settlements. Each district selects a leader from among its residents. Because all had been district leaders for over 5 years, we assumed that each of them knew the consensus of the residents and interviewed them in December 2008. The main contents of the interview were the scale and frequency of activities of the community's self-governing organizations, work style, participation in community activities, recent flow of population in the district, internal and intercommunity cooperation (help), regional cooperation with surrounding areas (villages or cities), desire for the administration's support, and the rate of population decrease in the community.

Table 4.1 Items in the questionnaire

Items	Remarks (alternative)
Attributes of the farm manager	Age, sex, number of family members
Agricultural land use at present	(1) Yes, (2) No, (3) N/A
Intention to use agricultural land for 5 more years	(1) Use all farmland, (2) Use only superior farmland, (3) Use no farmland, (4) Other, (5) N/A
Intention to use agricultural land for 10 more years	(1) Use all farmland, (2) Use only superior farmland, (3) Use no farmland, (4) Other, (5) N/A
Reaction when neighboring farmland is abandoned	(1) Continue farming, (2) Consider stopping farming, (3)No farmland around own, (4) Other, (5) N/A
Cognition of the definition of "marginal hamlets"	(1) Know in detail, (2) Know a little, (3) Don't know
Frequency of participation in community activities (e.g., mowing wayside or cleaning the commons)	(1) Each time, (2) Sometimes, (3) Fairly infrequently, (4) Not at all, (5) N/A
Cognition of the minimum numbers of participation for developing and maintaining community activities	(1) Only one household, (2) 20–30% of the households, (3) About half of the households, (4) Most households, (5) Other, (6) N/A
Cognition of the minimum numbers of population as a settlement in this village	(1) Only one household, (2) 20–30% of the households, (3) About half of the households, (4) Most households, (5) Cannot answer because of no idea of depopulation, (6) Other, (7) N/A
Attachment to the village	(1) Strong, (2) Decreasing, (3) Increasing, (4) None, (5) N/A

4.3 Results of the Field Survey

4.3.1 Aggregate Results of Questionnaire

4.3.1.1 Conservation of Agricultural Land

Of the 6,618 questionnaires distributed, 1,724 were collected. About 46.6% of the respondents were males and 53.4% females; the average age was 69.0 years (standard deviation = 11.0). Based on the results of the independent tests of the aggregate value for all eight districts in Monzen village, because Cramer's coefficient of association (*V*) was approximately 0.1 for every question, the differences between districts are not considered. Based on the independent testing of the answers of every age group older than 20, as well as the previous results, Cramer's coefficient of association (*V*) took the value of 0.1–0.2 for all the questions. Therefore, we consider that there are no statistically significant differences within districts or age groups and analyze according to the aggregate values of the entire village.

Table 4.2 Cross tabulation between the intention to use agricultural land for 5 and 10 more years

	5 years later			
10 years later	Use all the agricultural land	Use only superior agricultural land	Do not use all the agricultural land	Not yet determined or other
Use all the agricultural land	75	7	4	0
Use only superior agricultural land	98	404	8	14
Do not use all the agricultural land	17	253	259	46
Not yet determined or other	39	153	18	154

Table 4.3 Answers to the question "If the land surrounding your agricultural land is abandoned, how will you deal with your agricultural land?"

Option for question	Number and rate of answer
Continue farming own land even if the surrounding land is abandoned	401 (32.5%)
Consider stopping the farming of own land if the surrounding land is abandoned	616 (49.9%)
Not applicable or other (alternatives 3 ~ 5 in Table 4.1)	218 (17.6%)

A cross tabulation between the intention to use agricultural land for 5 and 10 more years is shown in Table 4.2. The rough correlation between the durations of the two intentions of agricultural land use and the intention of some farmers to reduce the area of the agricultural land can be observed. Therefore, abandoning the cultivation of agricultural land in adverse conditions will certainly occur in the next 10 years.

In general, respondents also stated that abandoned farmland has an external diseconomy that induces the abandonment of the surrounding land, and Table 4.3 suggests that the farmers' intentions reflect the fact that many of them are influenced by the land use around their own agricultural land. Table 4.3 demonstrates that 50% of the farmers who intend to own and use agricultural land for 10 more years would consider abandoning their own land if any of the surrounding land fell into disuse. Therefore, the abandonment of any farmland may quickly result in the collapse of the village's farmland management.

4.3.1.2 Settlement in the Village and Community Activity

In this questionnaire, we evaluated two important indicators required for community continuation: a sense of the substantivity of community-based resource conservation activity and the intention of settlement indicated by the decreased number of houses. Table 4.4 shows the cross tabulation of the two indicators.

Many residents believe that it would be hard to continue community-based resource conservation activities if the number of households decreases by 50%,

Table 4.4 Cross tabulation between the sense of substantivity of community-based resource conservation activity and the intention of settlement

Intention of settlement (limit of feeling for settlement) \ Resource conservation activity	Continuation is difficult even if only one household moved away from the village	Continuation is difficult if households decreased by 20–30%	Continuation is difficult if households decreased by 50%	Continuation is still possible even if households decreased by more than 50%	Other
Settlement is difficult even if only one household moved away from the village	8	0	3	2	0
Settlement is difficult if households decreased by 20–30%	2	39	11	6	2
Settlement is difficult if households decreased by 50%	8	59	118	26	15
Settlement is still possible even if households decreased by more than 50%	15	103	327	205	80
Other	12	46	102	34	73

but a few think it would still be possible because scale and frequency can be adapted to fewer households (see Table 4.4). In contrast, the majority of the residents indicated their intention to remain in Monzen even if the population shrank to less than 50%.

The results of the questionnaire suggest the following future scenario for this area. First, agricultural land use will decline gradually in adverse conditions, and farm management will continue only on limited superior agricultural land. Second, community-based resource conservation activities will be downscaled and the community ties that are fundamental to accomplishing these activities will weaken. The residents who have great affection for their community will remain settled by adjusting the scale or form of their daily life and activities, although this subset may include those who cannot leave because they no choice about relocation.

4.3.2 *Results of the Interviews*

4.3.2.1 Disposition of Residents

We identify the following present condition from the responses of the community leaders.

In all communities, 90% or more of the younger generation leave the village after graduating from high school, so that their contribution to the villages is declining. In contrast, the older generation of almost all communities has a strong awareness of mutual aid and affection for their community, creating a robust sense of solidarity. Although Monzen village suffered serious damage from the Noto Hanto earthquake in March 2007, its response to the damage is corroborative of the residents' sense of solidarity. Thus, although their successors' attitude was uncertain, many elderly stayed in the village and repaired or rebuilt the houses that had collapsed.

4.3.2.2 Request to the Administration

The residents' primary request to the administration is infrastructure development—road and channel repairs—which will give them comfort and safety. Nearly all the leaders feel that the administration services handling such requests weakened rapidly after the consolidation of municipalities in 2006. However, because they are aware that Wajima City has inadequate financial resources, residents have mixed feelings of depression and resignation.

Each community has several settlements that are having a tough time surviving. Some leaders think that 50% of these settlements will likely lapse into disuse, but others think otherwise. However, all leaders agree that although it is possible for a vigorous settlement to cooperate with a neighboring settlement under the right conditions, cooperation with a noncontiguous geographical community or village is difficult because most residents harbor a psychological resistance to changing the framework of the traditional "bordering community" combinations.

Based on the results of this survey, the profile of Monzen is summarized as follows. First, certain residents, especially the elderly, are closely united in the same community. Second, the network covering the entire community is weak, and the residents want infrastructure development for them to remain in the community.

4.4 Analysis of the NSSRP's Impacts on Marginal Hamlets

4.4.1 Scope of the NSSRP for Marginal Hamlets

The NSSRP is identified as the "main measures that are considered necessary from the large-scale perspective for the region, which consists of two or more prefectures that have some quite close relationships in nature, the economy, culture, etc., and should be governed practically in an integral comprehensive spatial plan" (Article 9 of the National Spatial Planning Act). In the present study, the NSSRP of the Hokuriku bloc contains around 10 years of regional strategies that were determined by the Hokuriku NSSRP Council, comprising the local branch or department of government agencies, Toyama, Ishikawa, Fukui Prefecture, neighboring municipalities, and the

Table 4.5 Description of problem structures in rural areas and coping strategies in the NSSRP

Description of problem structure (including issue + risk)	Coping strategy and breakthrough	Pages in NSSRP
Failure of agricultural land management and drainage, which are the foundation for stable agricultural production and the execution of multiple functions	Promoting improvement in agricultural productivity by preserving and improving agricultural production infrastructure while maintaining harmony with the environment	14
Establishment of a stable supply system of domestic timber	Keeping up the multiple functions associated with forestry, agriculture, and fishery, such as tapping water sources, preserving the natural environment, creating a good landscape, and upholding cultural traditions	14
Failure of the operation and maintenance of wetlands with regard to the role of culture and fishery resources	Developing the fishery industry and strengthening the collection and handling function of marine products to sustain their stable supply while preserving the environment and ecosystem of the body of water	15
The tradition of culture and multiple functions associated with forestry, agriculture, and fishery and the maintenance of the daily means of transportation of the elderly and students	Laying a trunk road network which supports urban–rural exchange and safe local public traffic. In addition, forming a local community that utilizes the power of rural collaboration and establishing new local collaboration through participation in various activities	16
No enough workers for the removal of snow from village roads and roofs	Establishing a disaster prevention organization in case of heavy snow and the resulting isolation of the village due to traffic blockage, and providing such services as snow removal for the elderly	18
Protection of the ecosystem of "Satochi–Satoyama" in the context of an ecological network and promotion of the appreciation of nature. In addition, preservation and utilization of forests or farmland as a place where people share education, nature, and culture	Introducing volunteerism, NPOs, and private enterprises in the maintenance of marginal hamlets' forests, farmland, and fisheries	21

local business community. Additional detailed information on the NSSRP is provided on the website (2009).

The description of the problems and structure and support strategy for farming and mountain areas in the NSSRP are outlined in Tables 4.5 and 4.6. The NSSRP discusses exhaustively several fatal problems confronting a rural area as regards local life, production activity, culture, etc., and proposes a simple solution for each.

Table 4.6 Description of supportive strategies in rural areas in the NSSRP

Principal strategy	Pages in NSSRP
Systematically rehabilitate abandoned farmland, urging the unification and utilization of farmland information for the reservation and effective use of superior agriculture land	29
Preserve agricultural production infrastructure and aim to improve production conditions by promoting collaborative activities—the cooperation of the whole community through the participation of various entities, including the local residents, city residents, and NPOs	29, 50
Establish modern irrigation systems to ensure a stable supply of healthy agricultural produce; in other words, ensure reliable water supply and drainage conditions and efficient stock management in consideration of suitable management and lifespan extension	30
Improve the living environment base in a village through the regular maintenance of farm roads and sewerage infrastructure, which would sustain agricultural production and thereby conserve beautiful rural scenery	30
Promote highly profitable forestry ventures by introducing efficient production systems (e.g., direct delivery to a saw mill of logs from mountain land)	31
Create new natural habitats to address the rising concern of city dwellers about nature, history, and culture, as well as support organizations and structures for the purpose. Furthermore, develop a system for promoting settlement by the relocation (called "U I J turn" in Japanese) of the baby boomers, who seek an alternative lifestyle after retirement	36
Support the synthetic activation of public traffic in the area (e.g., railroad, bus) and maintain and improve the road network, which is indispensable to mobility—commuting, regular visits to the hospital, and emergencies	42

Table 4.5 also describes several ideas for the revitalization of marginal hamlets, as well as a growth strategy. However, even merely maintaining the status quo for farmland and community activities becomes increasingly difficult by the year during a period of depopulation. Financial restrictions on administration are also heavy in a reduced economy. Therefore, it is necessary to carefully deliberate whether the strategies in Table 4.5 will be feasible in the current prolonged depopulation period. Since it is not easy to apply the strategies in Table 4.5 to a concrete project, we believe that a solid strategy needs to be added based on the sufficient understanding of the present phase of marginal hamlets.

4.4.2 Effectiveness of the NSSRP

In sum, judging by the results of our investigation, the NSSRP hardly offers useful indicators to a society, such as Monzen village, which is threatened by depopulation.

Firstly, agricultural and mountainous areas are bases of food production and biodiversity conservation and are thus part of a resource that the National Spatial

Plan should naturally endeavor to conserve. Hence, we can observe a certain significance in the NSSRP, which attempts to address the actual conditions of rural areas more closely than the National Spatial Plan does by dividing planning areas into spheres. However, if the reality is evaluated by the two aspects upon which this research focuses, the uncertainty of the NSSRP's effectiveness is highlighted. We must conclude that the directionality of the NSSRP, which promotes the operation and maintenance of ever-increasing abandoned cultivated land in marginal hamlets by mobilizing successors who are external to the village (e.g., urban residents), does not necessarily function effectively because of the residents' low priority for land conservation and the dwindling population. The framework for the scaling down of farmland use, which should be managed in accordance with depopulation, is also insufficient. Therefore, especially from the standpoint of space management, the effectiveness of the NSSRP's counterplan for farmland conservation in marginal hamlets (4th row in Table 4.1 and 2nd row in Table 4.2) is unclear.

Secondly, with regard to the counterplan's effectiveness for community development and support, the availability of successors and stimulation of community activity—a formative process and an operation of "local governance"—appears to be de-prioritized in plan implementation. The political and economic reality seems to be that, without further legal steps toward achieving this plan's intention, attracting successors by broadening the sphere after regional restructuring is an unlikely prospect in less-favored areas such as marginal hamlets. Therefore, it may also be concluded that the counterplan's effectiveness in broadening the sphere, including the marginal hamlets described by the NSSRP, is insubstantial and requires additional supportive measures.

4.5 Restrictions and Suggestions

Recently, several studies have reported the process of decay and demise of settlements and attempted to clarify the critical issue of the possibility of regional revitalization. For example, Kasamatsu (2005) divided the decay of settlements into three phases—pre-decay, post-decay, and demise—based on the characteristics of community activities and depopulation. Odagiri (2009) analyzed the three phases of deterioration into marginal hamlets qualitatively and defined them as the "hollowing of humans (depopulation), hollowing of villages (abandonment of cultivation and decrease in community-based activity), and hollowing of pride (loss of people's mental power)," as is shown in Fig. 4.3. The present study's area is assumed to be past the pre-decay phase, so that even if it obtains an appropriate counterplan immediately, it is in imminent danger of regional collapse. Therefore, an appropriate support policy is needed, and it should not assume the status quo but rather a shrinking population.

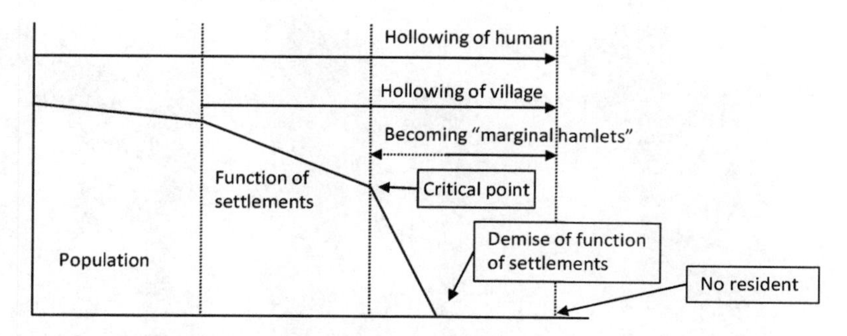

Fig. 4.3 The process of decay of a settlement (Source: Odagiri 2009)

4.5.1 Restrictions on Regional Sustention and Vitalization

As to the depletion of inherently rural human resources, if an alliance spanning a broad area is possible, the sustainability of the rural society will be ensured. It is necessary to reconstruct the broad area as a self-sustaining sphere and improve the durability of the region in the Noto peninsula, where fewer than 100,000 people live. According to the basin-scale population, all the inhabitable area on the Noto peninsula can be classified as two or three divisions at most. However, if the scale of daily life and commuting are taken into consideration, this range is too wide for a realistic sphere. Worse, as the population transition prediction by cohort analysis shows (Fig. 4.4), it appears that the future population will also decrease considerably throughout the peninsula. This issue relates to the premise that realistically, it is difficult to form a sustainable broader-based area. The situation is exacerbated by the sizable psychological barrier to cooperation between communities, rendering it difficult to reconstruct smaller community groups, even though such action may be mandated.

The results suggest that the effectiveness of an alliance between neighboring small areas has limitations, and forging a new intercommunion and cooperation system with other areas is necessary to maintain marginal hamlets. From this viewpoint, the NSSRP strategy proposal was clearly insufficient.

4.5.2 Suggestions for Spatial Planning

Pertaining to the gap between the NSSRP principle and the situation of this study area, which is typical of marginal hamlets, we discuss the possibility of a three-step coping strategy to complement the NSSRP (see Table 4.7).

The first step is to build a consensus for maintaining marginal hamlets and develop a vision by clarifying the total amount of economic and human resources required. We think that the first priority is to undertake a field survey to determine the process and cost for bridging the gap between the present situation and the target

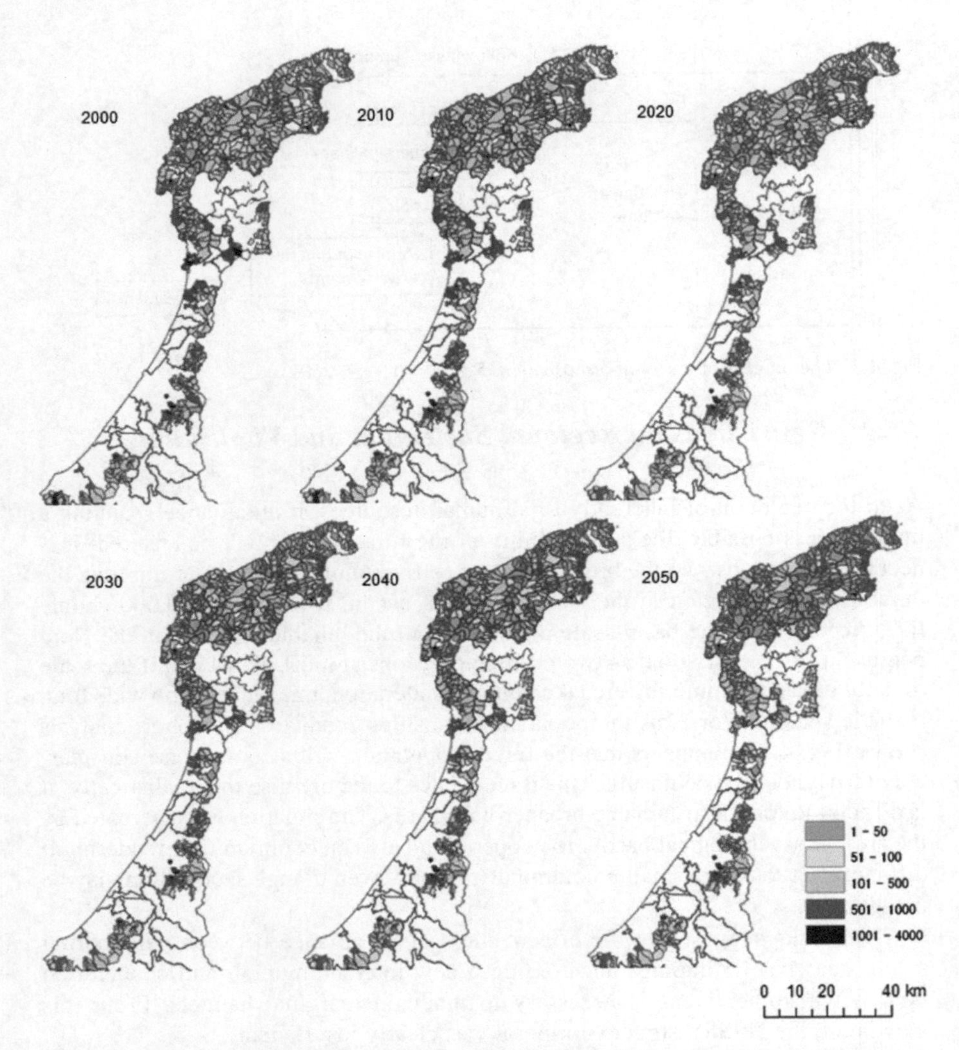

Fig. 4.4 Prediction of the future population of rural areas in Ishikawa Prefecture (*white* portions indicate no-data areas) (Source: Ichinose et al. 2009)

through an official project for the NSSRP's promotion. If the target for support is a small-scale transient event, such as an urban–rural exchange or festival, supplying the necessary human resources would be comparatively easy. However, a radical incentive structure of institutional arrangements is needed to stimulate the human resources' supply for regional management that would affect the continuation of the area.

Meanwhile, according to the population estimation result that the National Institute of Population and Social Security Research released in January 2012, the decrease in population is accelerating faster than the estimated pace at the time of

Table 4.7 Comparison of the NSSRP and proposals in this chapter

Point	Description of NSSRP	Proposal
Step 1: Method and cost for the plan's realization	Despite some objectives that were presented, the path to the plan's realization is not clear	Estimate how many human resources can be aggregated, where to get them, and how to do so
Step 2: Enhancement and diffusion concerns for the plan	There is almost no concrete description	Enhance with migrants from the village and consolidate the social support systems
Step 3: Information and other urgent on-site requirements for the realization the plan	Presupposing the reuse of some abandoned farmlands, there is almost no concrete description	Promptly check and arrange property rights in case of the emergence unowned lands.

the NSSRP's release. Thus, NSSRP projects may not succeed without an adequate understanding of the intentions and concerns of outer entities. The second step, therefore, is to improve institutions in order to attract urban residents, NPOs, and even former inhabitants now living in neighboring cities to regional activities. For example, we propose institutional arrangements such as tax incentives and a special holiday system for connecting marginal hamlets with urban residents.

The third step is to rearrange the proprietary rights to agricultural land as a preventive measure against the demise of the regional community. Because abandoned farmlands with unclear ownership have a lasting negative influence on well-ordered land use, the prime task is to classify abandoned land and its intermediate space as land for either conservation or nonconservation. Research has been conducted on the reconstruction of urban areas which were adapted for population reduction (Altes and Tambach 2008; Chen et al. 2008) and on the strategic shrinking by depopulation in rural areas (Ichinose 2010). However, since there is hardly any case study to integrate them, especially in Japan, the feasibility of separately designed plans for urban and rural areas is uncertain. Thus, we think that the clarification of the gap between the two by analyzing the NSSRP is a good opportunity to reconsider the space planning of a wide area, including the boundaries of urban and rural areas.

4.6 Summary

In this chapter, we demonstrated the effectiveness and limitations of the counter-plan for marginal hamlets in a self-sustaining strategy such as the NSSRP. Regarding the revitalization of marginal hamlets, we cannot expect the plan to be sufficiently effective by itself; as mentioned above, supplementary support and action must be developed independently by leading projects based on the NSSRP. It is noteworthy that Seta (2010), who evaluated the effectiveness of the NSSRP in the Kinki sphere (comprising Osaka, Kyoto, Hyogo, Shiga, Wakayama, and Nara), arrived at a similar conclusion as that of the present study. However, it is also a fact that it will be difficult to provide complete fiscal support for the needs of some marginal hamlets in the long term. It is thus necessary to tackle this problem immediately, based on the evaluation of the NSSRP or other macroeconomic policies, considering the political justification of rural needs.

Lastly, further research should explore why broad regional plans or projects neglect to support the small, less-favored areas. If the restriction on human power or financial capability underlies policy making, it is important to say so openly in order to ensure clear stakeholder communication about the conditions for the development of effective policies.

Acknowledgement We would like to express our gratitude to the residents and officer of Monzen town for their kind cooperation. This work is supported by the Mitsubishi Foundation Research Grant in the Humanities.

References

Akazawa K, Matsuoka N (2010) Quantification and structural analysis of human linkages in marginal rural communities: a case study in a rural community in Shimane by social network analysis. J Rural Life Soc Jpn 54(1):19–23 (in Japanese)

Association of rural planning (2009) Reports of the 2009 spring symposium: proposal to the wide-area regional plan from the Association of Rural Planning: rural planning and the national land sustainability plan 3 (In Japanese). J Rural Plann 2(2):76–102

Chen HY, Jia BS, Lau SSY (2008) Sustainable urban form for Chinese compact cities: challenges of a rapid urbanized economy. Habitat Int 31(1):28–40

Hashizume N (2004) Statistical analysis on the continuance of rural community in hilly and mountainous areas. J Agric Policy Res 7:1–24 (in Japanese)

Hiraguchi Y, Morozumi K (2010) A study about measures against marginalization in hilly and mountainous areas. Spec Iss J Rural Econ 2010:268–275 (in Japanese)

Ichinose T (2010) No-son innovation. Imagine Publishing, Tokyo (in Japanese)

Ichinose T, Azuma A, Harashina K, Hayashi N, Saitoh S, Maekawa H, Yamashita R (2009) Construction of the method for evaluation of community critical point and sustainable basin sphere. Project of research support about national policy 2008, Japan Center for Area Development Research, Japan Center for Area Development Research, Japan, pp 31–50 (in Japanese)

Kasamatsu H (2005) Actual condition of marginal hamlets in intermediate and mountainous area. Q J Chugoku Reg Res Cent 32:21–26 (in Japanese)

Koiso A (2011) Delivery of community-based social services to senior citizens living in aging communities in underpopulated mountainous areas: case study of Yasuoka village, Nagano prefecture. J Welf Soc 8:42–55 (in Japanese)

Korthals Altes WK, Tambach M (2008) Municipal strategies for introducing housing on industrial estates as part of compact-city policies in the Netherlands. Cities 25(4):218–229

Nakatsuka M, Hoshino S (2009) A case study on difference of local knowledge in various residents. Jpn J Farm Manag 46(1):160–164 (in Japanese)

Noguchi S (2009) Logic and policy of the reform in the depopulated areas. Rev Consum Co-op Stud 397:38–45 (in Japanese)

Odagiri T (2009) Problem of revitalization of farming and mountain villages—beyond the problem of 'marginal hamlets'. Iwanami booklet, Iwanami shoten, Tokyo (in Japanese)

Ohki K (2009) What does national spatial policy in the 21st century aim at?: searching for principles which will replace 'well balanced development throughout the country'. Urban Stud 49:86–100 (in Japanese)

Ohno A (2005) Mountain village's environmental sociology marginal villages and cooperative management of basin area. Rural Culture Association, Tokyo (in Japanese)

Seta F (2009) Today's regional disparities and new national and regional policies. J Munic Prob 63(3):89–102 (in Japanese)

Seta F (2010) An empirical study on the geographical specification on national regional plans in Japan—a case study of Kinki area regional plan in national spatial strategies. J City Plan Inst Jpn 45(2):47–53 (in Japanese)

Tachikawa M (1997) Factors of settlement house-number-scale related to sustainability of mesomountainous limitative settlement. In: Farm and rural diversification and a new perspective for the hilly and mountainous areas. Fumin Kyokai, Japan, pp 60–68 (in Japanese)

Chapter 5
Continuity of Relations Between Local Living Environments and the Elderly Moved to a Group Living

Tatsuya Nishino

Abstract "Ageing in place" is becoming a key issue in the ageing society. As such, the location of nursing homes is an important factor in designing suitable facilities for the elderly. It is important for the elderly to maintain their own daily living environments, even if they have to move to group care facilities or nursing homes. However, many facilities for the care-requiring elderly were traditionally built in serene countryside locations. Therefore, to claim that there was significant continuity in the relationships of those facilities' residents to their environments would have been difficult. In this chapter, we examine the continuity of relationships experienced by residents in a group care facility with the physical and human aspects of their daily living environments. As a case study, we selected a group care facility for the elderly with dementia located in a city in which a majority of residents moved in from surrounding communities in Japan. Then, we conducted on-site observations of the behaviors and remarks of residents when they went out. Observations indicated that some form of continuity of relationships with daily living environments was experienced by some residents. We suggest that there is a correlation between one's living-hub history and the status of the continuity of one's relationship with the daily living environment after moving into a group care facility. We then discuss the conditions and significance of this continuity. Our study shows the significance of moving into a facility close to one's former daily living environment in order to maintain a relationship with it.

Keywords Group care facility • Daily living environments • Continuity

---

T. Nishino (✉)
School of Environmental Design, Kanazawa University,
Kakuma Machi, Kanazawa City 920-1192, Japan
e-mail: tan378@se.kanazawa-u.ac.jp

M. Kawakami et al. (eds.), *Spatial Planning and Sustainable Development: Approaches for Achieving Sustainable Urban Form in Asian Cities*, Strategies for Sustainability,
DOI 10.1007/978-94-007-5922-0_5,

5.1 Introduction

Urbanization and population ageing are concurrent global trends. Population ageing transcends the divide between developed and developing countries, because every country hopes to foster independent living and quality of life regardless of life expectancy (Brink 1997).

One of the main issues in the ageing society has been the provision of housing for the elderly (Kose 1997; Katan and Werczberger 1997). For example, in Denmark, known for its highly developed social care system for the aged, formal care was introduced as an alternative to informal care by families (Gottshalk 1999). They have developed new attitudes toward old age and new ways of housing and providing services to the elderly since 1980. Moreover, the policy encouraged the elderly to stay in their own homes as long as possible (Lindstroem 1997). On the other hand, in the Netherlands, a de-linking system of housing and care has been developed, in order to make elderly people more autonomous and force care providers to be more customer oriented (Egdom 1997).

In Asia, most countries will be "aged societies" by 2050, a phenomenon known as "Ageing Asia" (*Nikkei Newspaper* 2011). However, in Asian countries, the elderly have been, and remain, primarily cared for by their own families (Ara 1997; Chi and Chow 1997). Previous studies have pointed out the need to change from traditional family support to public care systems (Kim 1997; Hwang 1997; Harrison 1997). Japan, as the most rapidly ageing country in Asia, has been developing its social welfare system since 1989 (Tsuno and Honma 2009). In 2000, the Japanese social care vision turned toward keeping the care-needing elderly in their own homes with proper formal care services, by introducing "Kaigo Hoken" (The Long-Term Care Insurance Act). It seems to be following the ageing policies of advanced countries such as Denmark.

The notion of "ageing in place" became a key component policy of elderly care and housing in the United Kingdom in the 2000s (Sixsmith and Sixsmith 2008). Older people normally prefer to live in their familiar residence, their home in particular, where memories and special meanings are attached (Tinker 1997; Gitlin 2003; Chui 2008a). The principle of "ageing in place" highlights the need to allow older people to continue to live in the locality with which they are familiar for as long as they wish (Chui 2008b). This principle has the objective of avoiding the risk of older people losing their sense of security when they are faced with removal from a familiar physical and social environment. In Taiwan, for example, with the notion of "ageing in place," the community care model has become a new trend (Chen 2008). The author has also conducted research to show the actual conditions of the community care provided to elderly people continuing to live in their own homes in a historical port town in Japan (Nishino 2010).

Theoretically, the notion of "ageing in place" is based on Lawton's "transplant shock" theory. That is, involuntary relocation causes the elderly negative effects (Lawton 1986). However, there are still some elderly people who would be better served if moved to adequate facilities, such as frail elderly people and those with dementia, even if most can continue to live in their homes.

Therefore, the location of nursing homes for the frail elderly or the elderly with dementia has been an important topic of consideration in facility design. However, many facilities for care-requiring elderly were traditionally built in serene countryside locations, because of the theory that a retreat from the noise, demands, and clutter of the city would be therapeutic (Lawton 1975). This may explain why even now most residents live a self-contained life in a facility far removed from the living environment they once had. In Japan, each local government establishes a sphere of daily existence within which to develop group care facilities and community-based facilities according to the revised "Kaigo Hoken" (Long-Term Care Insurance Act) in 2006. However, the spheres of daily existence set up by each local government are different. In addition, residents of such facilities expect continuity in their relationship with their daily living environment before and after moving into a facility. It is this continuity that has not yet been adequately examined.

Previous Japanese research has examined residents' going out and the structures of their local lives. Kinukawa et al. (2003) focused on the going-out behavior of residents and attempted to clarify the structure of their local living. Furthermore, these researchers target residents who use reverse day service, to examine the emotional impact of going-out behavior (Kinukawa et al. 2005). This study by Kinukawa is based on an awareness of the problem of living a totally self-contained life inside a facility. These researchers conclude that when residents go out in their localities, their quality of life can be expected to improve. On the other hand, based on the principle of "ageing in place," Inoue (2007) shows the actual conditions of residents in group care facilities located in the city. This study shares the same principle. Continuing life in one's locality after moving into a facility, however, and continuity in the same sphere of daily existence before and after moving into a facility are not the same thing. The latter is a main focus of this chapter. This study is thus distinct from the above-mentioned research.

The purpose of this study is to examine the continuity of daily living environments of residents who live in a group care facility and to discuss these conditions and their significance.

In this chapter, we conducted a case study, based mainly on an observational survey, on the going-out behavior of the elderly with dementia living in a group care facility. Firstly, we selected a group care facility for the elderly with dementia located in a city in which a majority of residents moved in from surrounding communities. *The second part of this chapter, "Residents," concerns the elderly with dementia living in the group care facility. Although most residents receive medical treatment, such as regular doses of medicine, the results of observation would nevertheless appear useful, because most elderly with dementia generally receive medical treatment.* Secondly, we conducted research as mentioned in Sects. 5.2 and 5.3. Section 5.4 covers the attributes and living-hub histories of each resident (*permanent-residency, U-turn, I-turn, and I-turn′*). Section 5.5 covers the daily living characteristics of residents, concerning the main living areas in the group care facility and the tendencies of residents to go out and to receive visitors. In Sect. 5.6, through the behavior and remarks of each resident, we examine continuity of daily living environments. Finally in Sects. 5.7 and 5.8, we discuss the conditions and significance of the relationship with one's living environment after moving into a facility and the connection with living-hub histories.

5.2 Methodology

We conducted a case study because it was deemed appropriate to demonstrate the level of continuity of daily living environments of residents in a group care facility. Firstly, we selected a group care facility for the elderly with dementia located in a city in which a majority of residents moved in from surrounding communities. Details of the surveyed region and facility are described in Sect. 5.3. Secondly, to examine the continuity of daily living environments experienced by residents, it was deemed appropriate to apply the Environmental-Behavior Model. We examined it from the point of view of human environments and physical aspects, according to the dimension of Evaluations in Takahashi's model as mentioned below.

5.2.1 Environment-Behavior Model

To obtain a comprehensive understanding of the relationship between humans and the environment, various models are available from the field of Environment-Behavior Studies. Firstly, Altman (1973) tried to find connections between behavioral research and architecture, by considering Environment-Behavior Studies in three dimensions: Behavioral Processes (privacy, personal space, territory, other processes), Places (systemic entities, geographical regions, cities, communities, neighborhoods, hospitals, schools, prisons, homes, rooms), and Design Process (programming design, construction, use, evaluation). Secondly, Moore (1979) constructed three axes for an alternative framework: Settings/Places (nations, regions, city and towns, urban areas, residential areas, complexes of buildings, buildings of various types, parts of buildings, rooms, furniture, equipment, and object), User Groups (different socioeconomic groups, groups with different way of life, infirm, handicapped, elderly, children), and Behavioral Phenomena/Concepts (anthropometrics, proxemics, personal space, territoriality, privacy, perception, cognition, meaning). Later he developed the model to add a time axis. In the new model there are four axes: Places (large scale, intermediate scale, small scale), User Groups (culture, life cycles, lifestyles) and Socio-behavioral Phenomena (internal responses, physiology perception, cognition, social groups, culture, external responses), and Time. Following Altman's model, there have been numerous studies of Behavioral Processes, Behavioral Phenomena, and Socio-behavioral Phenomena.

Takahashi (1997) developed these models for the environmental design theory. He proposed three primary dimensions for a model: Place/Objective of Design,

Method/process of Design, and Evaluation. The elements of the "Place/Objective of Design" dimension are Unknown Objective, New Objective, Renewal of Existing Objective, Simulation, Place of Learning, and Ordinary Place. And the elements of the "Method/process of Design" dimension are General Knowledge, Planning Language, Stimulation of Objective Setting, Design Method, and Decision Method. Finally, the elements of the "Evaluation" dimension are Physical Dimension, Behavioral Settings/Human Environments, Social Environment, Environmental Transition, and Design Process.

5.2.2 Survey Method

To examine levels of the continuity of daily living environments experienced by residents from the point of human environments and physical aspects, three kinds of data were collected: *attributes and living-hub histories of residents*, *daily living characteristics of residents* such as the tendency of residents to go out and receive visits and their main living areas in the group care facility, and *behaviors and remarks when they went out*. Firstly, attributes and living-hub histories of residents are needed to categorize the targeted residents. It is also necessary to examine connections between continuity of living environment and the living-hub history of residents. Secondly, to examine present daily living characteristics of residents, data pertaining to the tendencies of residents' to go out and receive visits are needed to investigate quantitative relationships with residents' daily living environments. Residents' main living areas in the group care facility are also surveyed. Thirdly, knowledge of the actual conditions of residents' behaviors and remarks when they went out are needed to investigate present qualitative relationships with their daily living environments.

We then adopted three survey methods to investigate them and research was conducted in the surveyed region as follows. First, an archival records survey to investigate tendencies to go outside and receive visits was conducted. This survey enabled us to investigate resident attributes and records of going out and receiving visits (37 days from August 26 to October 11, 2006). Second, interviews to investigate living-hub histories of residents were undertaken. Interviews were conducted with the families or relatives of eight residents (excluding resident E). We obtained information mainly about the living history of each resident. We also interviewed clerks from the shops often used by the residents. Third, on-site observations to investigate the continuity status of residents' daily living environments were conducted. On-site observation was deemed appropriate, because it is the most important survey method to grasp the relationship between space and the lives of people (Suzuki et al. 1975). The survey observed the going-out behavior of residents. The survey was conducted from June to December 2006 and covered a total of 20 visits. In the behavior observation survey, we also accompanied the residents and facility workers when they went out. On a local map, we recorded the routes, the remarks and behavior of residents, and the related locations. We obtained a total of 21 survey sheets from recording each separate outing. To make the behavior observation survey participatory, we also asked residents to show us around town. Four separate residents did so. Based on the living history of each resident obtained through the interviews, we asked the residents questions appropriate to each stroll. As we strolled, we recorded the remarks of each resident and the related location. Five residents were excluded from the participatory observation survey. One had difficulty with orientation (resident G), two did not go out often (residents B and I), and there were two others (residents A and D)[1]. We obtained a total of four record sheets from the participatory survey. In the figures, the remarks recorded from the participatory survey are marked with an asterisk.

To examine how residents appear in their group care facility, we conducted a behavior observation survey also on site (September 7–December 6, 2006). For this survey, we used a floor map to record in 10-min intervals the movement and location of residents and workers in the group care facility. In total, we obtained 182 record sheets. Because resident A refused to participate in the survey and resident B did not go out often, observations on them can be found in reference 4.

5.2.3 Time-Space Path Model

To categorize the living-hub histories of residents, the Time-Space Path model in Time Geography was deemed appropriate. Haagerstrand (1970) introduced the Time-Space Path approach, which shows personal activity with geographical space and a time axis spanning a day in a life. The Time-Space Path does not show a free track but a regular track called a "time-space prism." This model is useful to describe the lives of urban citizens. With the Time-Space Path model, the actual conditions of urban lives can be more concretely described and analyzed (Ito 1997). We classified the living-hub histories of residents with the Time-Space Path model, into four types: permanent-residency, U-turn, I-turn, and I-turn′.

5.3 Case Study

5.3.1 Outline of the Surveyed Region

T town in F City faces the Seto Inland Sea. It has been known as a "fisherman's village" since the Kamakura Era. In the latter part of the Kamakura Era, it prospered as a center of marine transportation. During the Edo Period, port facilities were put in place. Even today, it has many Shinto shrines, Buddhist temples, many traditional townhouses, and other old remaining buildings (please see *the* Fig. 5.1). In recent years, the population of seniors has increased. In January 2008, the ratio of

[1] For personal reasons, resident A found the survey difficult, and, because we obtained a lot of data on resident D through nonparticipant observation, he was not a participant himself.

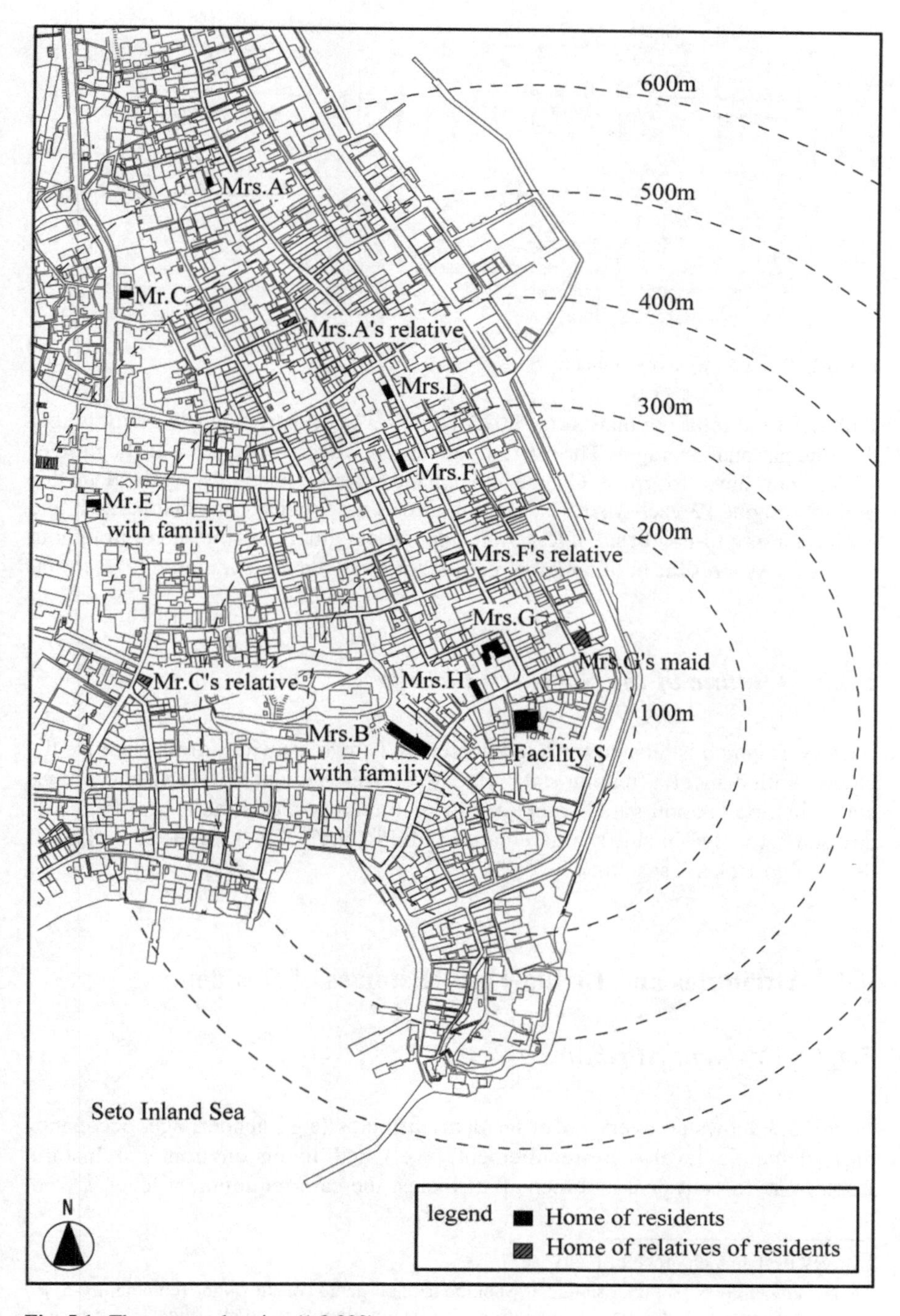

Fig. 5.1 The surveyed region (1:9,000)

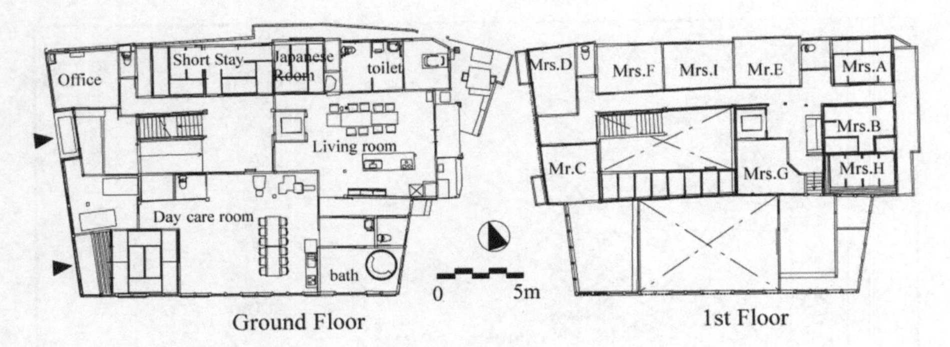

Fig. 5.2 Plan of the surveyed facility S

seniors to the total population of 5,073 was 39.9%, which is considerably higher than the national average.[2] The overall population is decreasing, even as the ratio of seniors continues to grow. On the other hand, various festivals are held in this region – about 12 each year – so neighborhood associations are maintained, and even on a day-to-day basis, interactions among neighbors can be seen throughout the town. As a result, in this region, community ties remain comparatively strong.

5.3.2 *Outline of the Surveyed Facility*

Facility S, which is located in an urban area of T town, provides group care for the elderly with dementia (regular staff: 9), day services (regular staff: 18), short stays, and "Shokibo takinou-gata kyotaku kaigo"(community-based multi-care facilities, day staff: 15; live-in staff: 5; hereinafter "Multi-Care").[3] It operates according to the "Kaigo Hoken" system (Fig. 5.2).

5.4 Attributes and Living-Hub Histories of Residents

5.4.1 *Resident Attributes*

Figure 5.3 shows an overview of resident attributes (e.g., gender, age, occupancy date, dementia level, care-requirement level) and living environment history. Looking at the physical and mental attributes, the care-requirement level is from

[2] Survey by T branch office of F city.

[3] In November 2006, facility S started a "Shokibo takinou-gata kyotaku kaigo" (community-based multi-care) service. At the time of our survey, it was not eligible to be classified as that type of business.

Resident	A	B	C	D	E	F	G	H	I
Gender, age	Female,77	Female,90	Male,77	Female,86	Male,73	Female,84	Female,93	Female,93	Female,97
Occupancy date	1.Apr.04	1.Apr.04	1.Apr.04	12.May.06	1.Apr.04	16.Sep.06	26.Jul.04	1.Apr.04	1.Apr.04
Dementia level and	Ⅱa、Ⅰ	Ⅱb、Ⅱ	Ⅱa、Ⅲ	Ⅱb、Ⅱ	Ⅲa、Ⅲ	Ⅱa、Ⅲ	Ⅲa、Ⅱ	Ⅲa、Ⅰ	Ⅰ、Ⅱ
Type of living in the	My-room type	Living-room type	In-between type	Living-room type	Living-room type	My-room type	Living-room type	Living-room type	My-room type
Type of living hub	Permanent-residency	Permanent-residency	U turn	U turn	U turn	I turn	I turn	I turn	I turn '
Childhood	born in T town went to T primary school engaged in ironworks	born in T town went to T primary school went to women's school	born in T town went to T primary school went to T junior high school	born in T town went to T primary school went to T junior high school	born in T town	born in Osaka (200km away) came to T town	born in F city (12km away)	born in Aichi (320km away)	born in M town (10km away)
Youth		trained for marriage got married worked at electric shop	worked at a cloths store in F city	went to Osaka by marriage (200km away) husband died	went to Osaka (200km away) came to T town	worked as a nurse at F city helped a factory ran by family	came to T town by marriage go to Tokyo	came to T town by marriage taught at primary school	
Their prime			got married and live in O city (28km away) divorced worked at a book shop	back to T town worked at a Japanese hotel	engaged in ironworks at F city engaged in ironworks at T town	closed the factory after father died		worked at primary school	
Old age	attacked entered the hospital moved to another facility	attacked moved to a short term stay	engaged in delivery service attacked back to T town	worked at grocery shop attacked and enter the hospital moved to another facility		attacked and entering the hospital came to day dare in facility S	came to T town attacked employed a housekeeper	retired and enjoyed hobby activities attacked took care at home	
Moving	moved to another facility		back to T town	moved to another facility		came to day dare in facility S	employed a housekeeper	took care at home	came to T town

Fig. 5.3 Overview of resident attributes and living environment history. The *black line* shows periods of living in T town. Dementia level and care-requiring level are those of July 6, 2006, except Mrs. F's data, which is that of on September 16, 2006. Regarding dementia levels, we used the "criterion for showing how independently the elderly with dementia can live." For living environment history, we categorized residents' lives into five periods: childhood, youth, the prime, old age, and after moving into group care facility S

I to III and the dementia level is from I to IIIa. Regarding the place of residence before moving, eight people came from T town and one person (resident I) came from another town. The prior places of residence of the eight residents from T town are all within 600 m of the group care facility S (Fig. 5.1). That is, eight residents maintained continuity with their sphere of daily existence after moving.

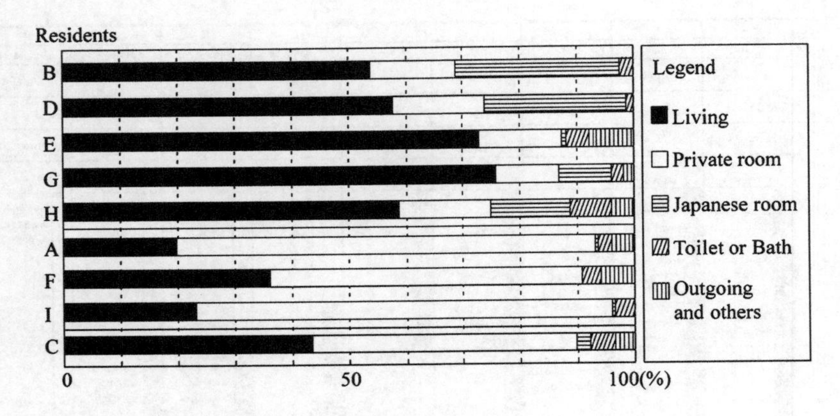

Fig. 5.4 The ratio of main living areas in the group care facility

5.4.2 Tables

We classified changes in the living hub of residents (Fig. 5.3), dividing living hubs into four types. The permanent-residency type indicates a person who has continuously lived in T town since they were born (residents A and B). The U-turn type indicates a person who moved away from T town in their youth or in their prime and later returned (residents C, D, and E). The I-turn type indicates a person who was born and raised somewhere else and moved to T town after she married (residents F, G, H). And the I-turn′ type indicates a person who moved to T town as an opportunity to live in the group care facility S (resident I). *These categories are considered appropriate because the living hubs of residents from elsewhere are at least 10 km away from T town, which is outside of the sphere of daily living by some distance.*

5.5 Daily Living Characteristics of Residents

5.5.1 Types of Main Living Areas in the Group Care Facility

Figure 5.4 shows the main living areas of residents according to the behavior observation survey inside group care facility S. Figure 5.4 shows the average value of two surveys.[4] We classified the main living areas into three types. The living-room type indicates a person who spends a long time in the living room while

[4] Resident D, however, was hospitalized during the second survey. Therefore, data on her is only from the first survey. In addition, during the first survey, resident F was using the group care facility only for a short stay. So her data is only from the second survey.

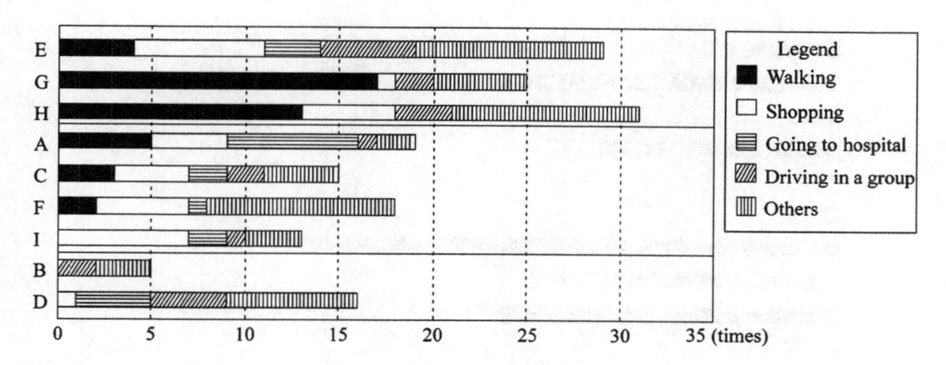

Fig. 5.5 The breakdown of the number of times each resident goes out

at home (residents B, D, E, G, and H). The my-room type indicates a person who spends a long time in his own room while at home (residents A, F, and I). Finally, the in-between type indicates a person who spends an equal amount of time in both rooms. We observed that residents with a lower degree of dementia tended to spend more time in their own rooms.

5.5.2 Tables

Figure 5.5 shows a breakdown of the number of times each resident goes out based on our going-out record survey.[5] We observed individual differences in the number of times residents went out and related details. Residents have the following going-out characteristics. Group care facility workers often called out to people with a high degree of dementia (residents E, G, and H) and encouraged them to go out. Because of that, these residents had a comparatively high rate of going out. People with a comparatively low degree of dementia (residents A, C, F, and I) went out exclusively of their own will. Residents B and D went to the hospital regularly

[5] Figure 5.5 shows the other purposes that each resident had for going out.

A	Festival
B	Home (sometimes staying a night)
C	Friend's shop, drive
D	Favorite shop, beauty treatment, home
E	Personal drive, home
F	Home, attend a grave, temple, bank
G	Attend a grave, home
H	Attend a grave, going out with a daughter, home
I	Drive

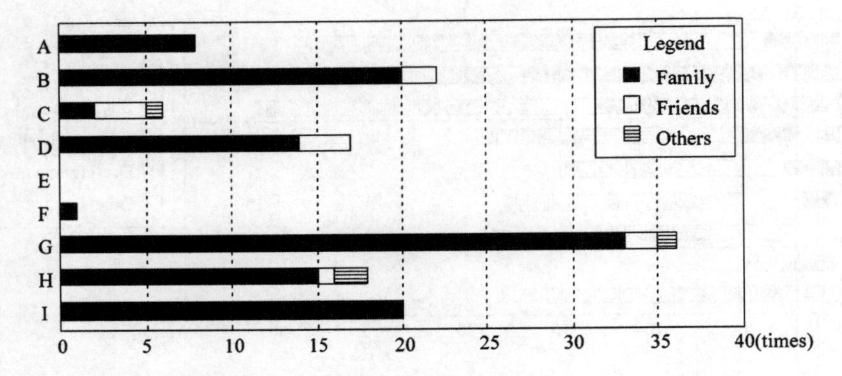

Fig. 5.6 The breakdown of the number of times each resident received a visitor

Table 5.1 Average frequency of receiving visits

Resident		A	B	C	D	E	F	G	H	I
Average in a week	Total	1.51	4.16	1.14	3.22	0.19	0	6.81	3.41	3.748
(times/week)	Family	1.51	3.78	0.38	2.65	0.19	0	6.24	2.84	3.78

and returned to their homes, so they went out when needed. Thus, we observed that the frequency of going out is influenced by the degree of dementia, the activities of daily living, and the intentions and activities of group care facility workers.

5.5.3 Tendency of Residents to Receive Visitors

Figure 5.6 shows a breakdown of the number of times each resident received visitors based on our visitor-reception record survey (*37 days from August 26 to October 11, 2006*). According to this survey, we can see that visits are mostly from family members. Looking at the frequency of visits (Table 5.1), we can see that five residents received two or more visits per week. Resident G received 6.2 visits per week or nearly one visit each day.[6] Table 5.2 shows an overview of the most frequent visitors. Residents B, G, and I received visitors not only on weekends but also on weekdays as well. This is because these visitors lived in T town. These frequent visitors explained their visits in various ways, such as “I was just passing by on the way to somewhere” and “I have something to do.” Thus, we can see that for these visitors visiting had become a habit. At the same time, people who came to visit resident H mainly did so on weekends and holidays. That is because these visitors lived far away from group care facility S. Some residents did not receive many visitors. For example, residents A, C, and F were all alone. Resident E did not have a good relationship with his family. And the visitor to resident F was in poor

[6] The housekeeper who visited resident G the most is counted as a family member.

Table 5.2 Overview of the most frequent visitors

Resident	Relationship with resident	Living place	Average visiting frequency (times/week)	Characteristics of visits
A	Sister	T town	1.32	She visits when she has something to do.
B	Daughter	T town	2.65	She visits when she has a time.
C	Sister in law	T town	0.38	She visits when she has something to do.
D	Sister	F city	1.89	She talks in the facility when she comes to day care services.
E	Daughter	T town	0.19	Family does not visit so often.
F	Sister	T town	0	Sister and daughter do not visit so often.
G	Housekeeper	T town	4.54	She visits in the morning, cooks fish boiled in broth, and spends an hour there, giving massages and exercising.
H	Daughter	Neighbor town	2.83	The oldest daughter spends one in every two months in T town and the second daughter comes on weekends.
I	Daughter	T town	2.27	She visits after three o'clock and talks for an hour.

physical condition. Thus, one condition for being a frequent visitor is living close to group care facility S. Looking at the number of visits by the type of living hub, we can observe that the U-turn type (residents C, D, and E) receives somewhat fewer visitors. We did not find any particular reason, however, so we cannot say that living close and type of living hub have a correlation.

5.6 Examination of Continuity of Daily Living Environments

This section examines the continuity of relationships with the daily living environment.

5.6.1 Continuity of Relationships with Human Daily Living Environments

First, we examine degree of continuity of daily living environments. We often observed people calling out to resident D (U-turn type) when she was going out. Resident D responded to these calls with a nod or other gesture because she had a hard time speaking. The background of resident D's behavior is that she grew up in T town, has childhood friends here, and had a grocery store business after returning to her hometown. Thus, she has many friends. Others, such as residents A, C, E, and F, also had many chances to engage in conversation with friends (Table 5.3). Most of the conversations centered on greetings, exchanging recent news (Table 5.3: circle), reminiscences (Table 5.3: black circle), and even grumbling and vigorous

Table 5.3 Samples of relationships with daily human environments

Resident	Who	Place	Remarks
A	Acquaintance	In the facility S	'We cooked at a festival.' A: 'Yes, it was hard work.' ●
C	Acquaintance	Supermarket	'How are you?' ○
	?	Supermarket	'Oh, she is Mrs. A!' ○
	Classmate	Restaurant	* 'Haven't you got fat?' C: 'Really?' ○
D	Classmate	Hospital	'You look good.' ○
	Acquaintance	Hospital	'It is hot today.' ○
	Acquaintance	Hospital	'Is your daughter well?' ○
	Classmate	Restaurant	'Oh, you worked at the grocery store!' ●
	Classmate	Restaurant	'This is the best season.' ○
	Neighbor	Restaurant	'My grandchild is ...' ○
	Classmate	Hospital	'You ran so fast.' ●
E	Acquaintance	Hospital	'Take care of yourself!' E: 'You too.' ○
	Acquaintance	On the street	'How woeful. Walk and run by yourself!' E: 'I would run if I had a job!' Δ
	Former colleague	On the street	* 'Do you remember that we traveled together?' E: 'Yes, I do.' ●
F	Acquaintance	On the street	'Hello, how are you?' ○

The remarks recorded from the participatory survey are marked with an asterisk

encouragement (Table 5.3: triangle). At first glance, these conversations may seem simple. Upon closer observation, however, it is evident that they would not occur without the participants sharing a common background and knowledge, which are needed to understand the context. From this we can conclude that these residents have experienced continuity in terms of their human environments.

5.6.2 Continuity of Relationships with the Physical Daily Living Environment

Next, we examine the continuity of relationships with the physical daily living environment. Resident F (I-turn type) told us that when going out for a walk, she likes to sit on a certain big rock in a shrine behind her home. We also observed resident D (U-turn type) chatting with the owner of an okonomiyaki (Japanese pancake) restaurant she frequented. These examples show how residents continue to use the same places before and after moving in to the group care facility. That is, these residents have maintained continuity in their relationships with their physical daily living environments.

5.6.3 Continuity in Physical Daily Living Environments from Memory

We also observed residents reminiscing about their physical daily living environment. For example, we often heard resident C (U-turn) remark about how the town

Table 5.4 Samples of remarks about physical daily living environments

Resident	Place	Remarks
A	In front of a cloth shop	Staff: 'Don't you go shopping these days?' A: 'No. Because it costs too much.'
C	Restaurant	*'The okonomiyaki at this restaurant tastes good.'
	In front of a Bar	*'I came here so often.'
	On the street	*'There was a public bath here.'
	On the street	*'This is the last bookshop in this town.'
E	Old downtown	*'I was called for so often to join in with things and enjoy myself.'
	On the street	*'I often enjoyed myself around here but now there's nowhere to enjoy.'
H	Gazing at an island across the shore	'I used to climb to that peak.'

Remarks recorded from the participatory survey are marked with an asterisk

had changed since the distant past. Other residents made similar remarks. While strolling through an old downtown district, resident E commented how he long ago used to come and play and tussle with other young people. While gazing at an island across the shore, resident H commented how she used to climb to a certain peak. These remarks are all based on remembering places habitually visited long ago (Table 5.4). Thus, even in their memory, residents have maintained continuity in their relationships with their physical daily living environments. However, we did not observe any such continuity of relationships with the physical environment through residents' reminiscing in their daily behavior. Rather, it is significant that reminiscing was awakened when residents went out to familiar places, such as when out walking or on a visit to the hospital. In particular, because T town is a traditional city that has not undergone major changes in its streets and buildings, we were able to observe many residents reminiscing on topics similar to the examples above.

We can thus conclude that some form of continuity of relationships with the daily living environment has been maintained with resident A (permanent-residency type) from T town; with residents C, D, and E (U-turn type); and with residents F and H (I-turn type).

5.7 Discussion of the Continuity of Relationships with Daily Living Environments

5.7.1 Correlation Between Living-Hub Histories and the Continuity of Daily Living Environments After Moving

First, we examine the correlation between living-hub types and the continuity status of relationships with the daily living environment.

For the permanent-residency type, in the going-out survey, we observed that residents A and B went out comparatively often. In the behavior observation survey, however, they often stayed in their own rooms. The continuity of their relationships involving other humans was low, with only a few visits from friends. Resident A stayed in her own room out of personal preference. On the other hand, resident B, with a rather high degree of dementia, rarely went out. She mainly stayed in her Japanese-style room, lying down. Even though family often visited her, she did not maintain any continuity in her relationship with her daily living environment.

From the U-turn type, we did not observe any outside human involvement with residents C, D, and E. Resident C, however, was concerned with his childhood and old age. The reason for this is that resident C left T town in his prime and thus has a "blank" period. On the other hand, resident D returned to T town in her prime and helped run the family grocery store. Thus, we observed relationship continuity with the human daily living environment from that point on. In other words, even though two residents may be the same U-turn type, we can observe differences in the continuity of their relationships with the daily living environment based on the period when a person left T town and their local living status after returning to T town. Resident E also reminisced about his childhood and prime. Thus, we can say that U-turn residents, other than for their blank period, do maintain continuity in their relationships with their daily living environments.

From the I-turn type, we observed that resident F maintained both continuity of her daily living environment (human and physical) and continuity of customary practices in returning home on memorial days to recite sutras. The other I-turn types (residents G and H) suffered from progressive dementia. They are discussed in Sect, 5.7.2. We can thus see that one I-turn type (resident F) maintained continuity in her relationship with the daily living environment after she moved in to the group care facility.

From the above, we conclude that there is a correlation between living-hub histories and the continuity of relationships with the daily living environment after moving into the group care facility. In the I-turn type, for example, we observed continuity of daily living environments (both human and physical) between living in T town independently and moving into the group care facility. In the U-turn type, we observed continuity of daily living environments for periods other than when a resident was living outside T town. Residents of the same U-turn type, however, maintained different levels of continuity in their daily living environments depending on whether the blank period was in childhood or in the prime of life. In the I-turn′ type, on the other hand, we observed a breakdown in continuity of daily living environments. That is, the continuity of relationships with the daily living environment after moving into the group care facility is influenced by the period living in T town. For example, when did a resident start living in T town? And during what stage of life did a resident leave T town? Figure 5.7 summarizes the information in this section.

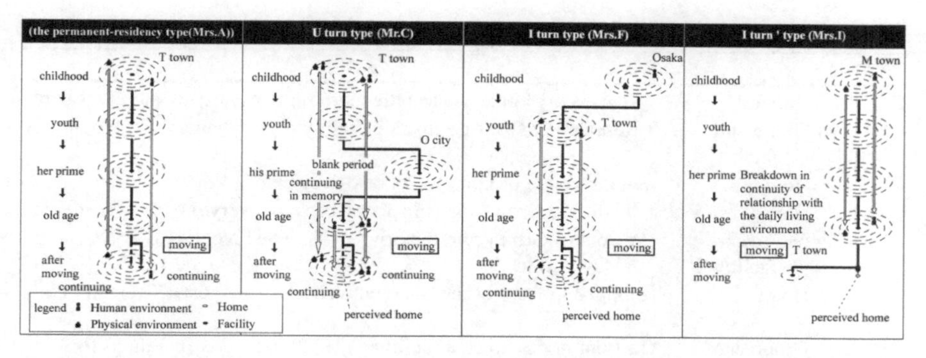

Fig. 5.7 Correlation between living-hub histories and the continuity status of daily living environments after moving

5.7.2 Relation Between Progressive Dementia and Daily Living Environment

Next, we examine residents with progressive dementia and their relationships with their daily living environments. We observed fragmentary reminiscing about places from resident H. But we did not observe remarks and behavior to show that she recognized the relative position of those places. In the remarks of resident F, who had a low degree of dementia, we did observe the ability to recognize the relative position of the hospital in terms of landmarks. We also observed similar remarks from resident C. Thus, we conclude that losing the ability to recognize the relative position of places and failing memory depend on the progression of dementia. In addition, on one occasion when resident G went out for a walk and bumped into a former neighbor, she commented that that face looked familiar. This remark shows that resident G remembered the person as a former neighbor, but no further details such as the name. In both cases, though the memory is not perfect, fragments of names and places remain. These memory fragments may provide a type of mental stability to residents with dementia.

5.7.3 Requirements for Continuity of Relationships with Daily Living Environment (i.e., Continuity in the Sphere of Daily Existence)

In the previous section, we stated how we observed continuity of relationships with the daily living environment in some form among residents of the permanent-residency type, U-turn type, and I-turn type. In the I-turn′ type, on the other hand, we observed a breakdown in such continuity. For all types where we observed relationship continuity with the daily living environment, the distance between the resident's original home in T town and group care facility S was less than 600 m.

Table 5.5 Samples of remarks about home and group care facility

Resident		Remarks
C	Home	* 'This is my home. It has fallen into ruin because nobody lives here.'
	Group care facility S	* Researcher: 'Isn't group care facility S your home?' C: 'No, this is ...'
D	Home	(Seeing inside of home) 'Nobody lives here...'
F	Home	* 'I am ashamed to open my doors, because everything is so scattered.'
G	Group care facility S	(In front of Group care facility S) 'Oh, here? Nobody is inside. May I go in?'
H	Home	(Stopped in front of the house and checked the post) 'There are no letters!'
	Group care facility S	(In front of Group care facility S) H: 'What's this [building] for? 'Staff: 'This is Group care facility S.' H: 'Group care facility S....'
I	Home	'I would like to return to M town.'
		'I rent my house to someone else. So I feel very lonely.'
	Group care facility S	'I cannot relax here.'

* Remarks recorded from the participatory survey.

Thus, though difficult to draw a conclusion from this study alone, we think that a person has a good chance of maintaining continuity of daily living environments if the group care facility he moves into is within 600 m of his sphere of daily existence. Incidentally, the sphere of daily existence in architectural planning is about 500 m from a person's home (Suzuki et al. 1975). So our estimate appears appropriate.

5.8 Significance of Relationship Continuity with Daily Living Environment

In this section, we examine related theories to support the significance of the continuity of relationships with the daily living environment.

First, we examine whether the resident is aware of his own private residence. Table 5.5 shows the main remarks by residents about their homes and group care facility S. With respect to homes, resident I (I-turn′ type) frequently remarked that she wanted to go home. And resident F (I-turn type) returned home whenever she had something to do there. With respect to group care facility S, on the other hand, resident C and others did not know how to respond. They did not clearly recognize the group care facility as their own home. These remarks reflect an awareness of their own private residences and group care facility S. That is, this may suggest that, to residents, the group care facility is only a temporary dwelling and that their perceived home is their own private residence.

We now introduce the concept of "home." According to "Ko-jien," a Japanese dictionary, "home" means one's household, one's private residence, and one's birthplace (Shinmura et al. 2008). According to Tachibana (2005), "home" is a

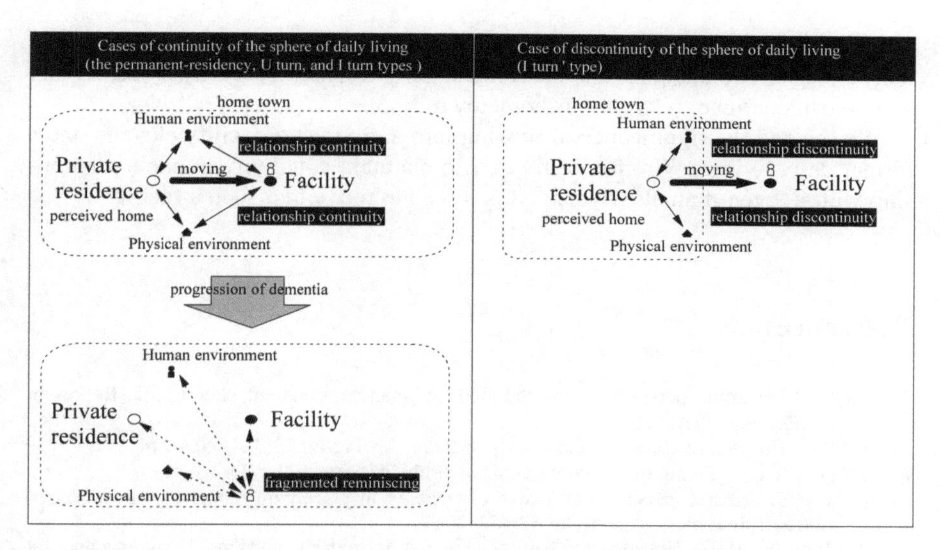

Fig. 5.8 Model of continuity of relationships with daily living environment

multilayered and relative concept. That is, "home" is a perceptual area. If the perceived home is one's private residence, then "home" is a multilayered area that is centered on one's private residence. In particular, if we define one's hometown as one's perceived living area, then even when a resident who comes from T town moves to group care facility S from his own private residence, he maintains the continuity of his perceived hometown through continuity of the sphere of daily existence and with his relationship to the daily living environment. We think this leads to a sense of security.

5.9 Conclusion

In this chapter, we conducted a case study to examine the continuity of relationships experienced by residents in a group care facility with the physical and human aspects of their daily living environments. We then discussed the conditions and significance of this continuity. *Although further study is needed to draw conclusions, we could propose some useful implications of our examination of the continuity of relationships with the daily living environments of residents.* Figure 5.8 summarizes these implications. Firstly, we divided the living-hub histories of residents in the group care facility into four types: permanent-residency, U-turn, I-turn, and I-turn′. Six residents belonging to the permanent-resident, U-turn, and I-turn types showed some form of continuity in their relationships with their daily living environments. We suggest that there is a correlation between the living-hub history and the continuity status of relationships with the daily living environment after moving into a group care facility. We also suggest that it is highly likely that there is a reference condition to maintain continuity of one's relationship with the daily living environment, that is, moving into a group care facility that is within

600 m of a resident's sphere of daily existence. The significance of continuity in daily living environments is that the resident gains a sense of security by perceiving that he can continue to live in his hometown.

We showed the significance of moving into a group care facility close to one's former daily living environment in order to maintain continuity in one's relationship with it, even if an elderly person is forced to move into such a facility.

References

Altman I (1973) Some perspectives on the study of man-environment phenomena. Represent Res Soc Psychol 4(1):111

Ara S (1997) Housing facilities for the elderly in India. Ageing Int 23(3–4):107–114

Brink S (1997) The graying of our communities worldwide. Ageing Int 23(3–4):13–31

Chen Y (2008) Strength perspective: analysis of ageing in place care model in Taiwan based on traditional filial piety. Ageing Int 32:183–204

Chi I, Chow N (1997) Housing and family care for the elderly in Hong Kong. Ageing Int 23(3–4):65–77

Chui E (2008a) Introduction to special issue on 'ageing in place'. Ageing Int 32:165–166

Chui E (2008b) Ageing in place in Hong Kong—challenges and opportunities in a capitalist Chinese city. Ageing Int 32:167–182

Egdom G (1997) Housing for the elderly in the Netherlands: a care problem. Ageing Int 23(3–4):136–182

Gitlin L (2003) Conducting research on home environments: lessons learned and new directions. Gerontologist 43(5):628–637

Gottshalk G (1999) Introduction to: "housing and care for various groups of elderly people". Independent European Housing Forum conference: living condition for the European elderly, Kuopio, 25–26 Sept 1999

Haagerstrand T (1970) What about people in regional science? Reg Sci Assoc Pap Proc 24:7–21

Harrison J (1997) Housing for the ageing population of Singapore. Ageing Int 23(3–4):32–48

Hwang Y (1997) Housing for the elderly in Taiwan. Ageing Int 23(3–4):133–147

Inoue Y (2007) A study on community life in a group home for the people with dementia. J Archit Plan 614:57–63

Ito S (1997) Activity of urban inhabitants. New Urban Geography, Toyo Shorin, Tokyo, pp 143–147

Katan Y, Werczberger E (1997) Housing for elderly people in Israel. Ageing Int 23(3–4):49–64

Kim M (1997) Housing policies for the elderly in Korea. Ageing Int 23(3–4):78–89

Kinukawa M et al (2003) A study on the structure of community life of the group home based elderly with dementia. J Archit Plan 564:157–164

Kinukawa M et al (2005) Effects generated by outgoing behavior on the mental stage of the institution-based elderly with dementia. J Archit Plan 592:17–24

Kose S (1997) Housing elderly people in Japan. Ageing Int 23(3–4):148–164

Lawton MP (1975) Planning and managing housing for the elderly. Wiley Inter-science, New York

Lawton MP (1986) Environment and aging. Center for the Study of Aging, New York

Lindstroem B (1997) Housing and service for the elderly in Denmark. Ageing Int 23(3–4):115–132

Moore GT (1979) Environment-behavior studies (chap. 3). In: Snyder J, Catanese A (eds) Introduction to architecture. McGraw-Hill, New York, pp 46–71

Nishino T (2010) A case study on the place and system of community care for the elderly in a historical port town in Setouchi, Japan. J Int City Plan, The International Symposium on City Planning:357–366

Shinmura I et al (2008) Ko-jien the sixth edition. Iwanami-Shoten, Tokyo

Sixsmith A, Sixsmith J (2008) Ageing in place in United Kingdom. Ageing Int 32:219–235

Suzuki S et al (1975) Architectural planning. Jikkyo-Shuppan, Tokyo

Tachibana H (2005) 'Concept of Home', in Idea of elderly facility from the point of lives. Chuo-Houki, Japan, 124–125

Takahashi T (1997) Aspects of theories in environment-behavior studies. In: Architectural Institute of Japan (ed) Environment-behavior design. Shokoku-sha, Tokyo, p 32

Tinker A (1997) Housing for elderly people. Rev Clin Gerontol 7:171–176

Tsuno N, Honma A (2009) Ageing in Asia—the Japan experience. Ageing Int 24:1–14

Chapter 6
The Use of Indicators to Assess Urban Regeneration Performance for Climate-Friendly Urban Development: The Case of Yokohama Minato Mirai 21

Osman Balaban

Abstract Climate change is one of the greatest challenges in the twenty-first century. Immediate actions are required to slow down climate change and address its impacts on human life and settlements. Cities can play crucial roles in this respect, as they not just contribute to causes of climate change but also are under severe threat from its impacts. Urban regeneration projects can provide opportunities to make cities more climate-friendly and less vulnerable. However, the potential role of urban regeneration in tackling climate change is not sufficiently recognized. In many cities, integration between urban regeneration projects and climate policy is still weak. Besides, limited methods exist to evaluate the performance of urban regeneration projects for climate change mitigation and adaptation. Considering these challenges, this chapter is intended to elaborate on use of indicators to assess the progress achieved in urban regeneration projects toward climate-friendly urban development. The chapter presents the findings of a research on the case of Minato Mirai 21 Project in Yokohama, which is a prominent waterfront redevelopment over brownfield sites. The project has converted former shipyards and railroad yards into mixed-use and high-density urban quarter with a working and resident population of 70,000 people at present. A set of 34 indicators grouped under six performance categories is developed and applied to MM21 project. Research findings not only indicate the extent of achievements in MM21 project toward climate-friendly urban development but also highlight the strengths and weaknesses in using indicators for assessing urban regeneration performance.

Keywords Climate change mitigation and adaptation • Sustainable development • Urban regeneration • Indicators

---

O. Balaban (✉)
Institute of Advanced Studies, United Nations University, 6F, International Organizations Center, Pacifico-Yokohama, 1-1-1 Minato Mirai, Nishi-ku, Yokohama 220-8502, Japan
e-mail: balaban@ias.unu.edu; osman.balaban@gmail.com

M. Kawakami et al. (eds.), *Spatial Planning and Sustainable Development: Approaches for Achieving Sustainable Urban Form in Asian Cities*, Strategies for Sustainability,
DOI 10.1007/978-94-007-5922-0_6,

6.1 Introduction

Climate change is one of the greatest challenges of our time in the twenty-first century. It is likely to have significant impacts on human life and human settlements. According to the Intergovernmental Panel on Climate Change (IPCC), climate change is human induced and unequivocal (IPCC 2007). Even if the most effective mitigation measures are applied, certain degree of global warming and climate change will still take place (Prasad et al. 2009). Therefore, along with mitigation actions to keep climate change at relatively lower levels, adaptation actions are also required to reduce climatic vulnerabilities of human settlements.

Cities are of critical importance in tackling climate change by means of mitigation and adaptation actions. First, cities are among the major causes of global warming and climate change. For example, the 35 largest cities in China are responsible for 40% of the emissions with only 18% of the nation's population (Dhakal 2009). Second, cities accommodate more than half of the world's population and critical economic activities. Therefore, any probable damage by climate change-related events in cites would be immense and severe. Third, despite being part of the problem, cities can, in fact, be the critical part of the solution. Inherent advantages provided by cities, such as offering of economies of scale, are the opportunities to formulate effective and low-cost solutions (Satterthwaite et al. 2007).

Considering the critical links between cities and climate change, it is widely accepted that a significant part of the challenges posed by climate change can be met through relevant policy options at urban level (Prasad et al. 2009). Policy options being argued in current literature can be grouped under three categories: *political*, *sectoral*, and *spatial*. The first group of policy options is related to political processes in cities, particularly to urban governance aspects. *Political policy options*, mostly, refer to the need for reforms in decision-making and enforcement processes, better redistribution mechanisms, strong local governments, and awareness raising among policy-makers and citizens (Satterthwaite et al. 2007; Puppim de Oliveira 2009; Roberts 2008a; Alam and Rabbani 2007; Costello et al. 2009). *Sectoral policy options* aim, in large part, to reduce energy consumption, increase energy efficiency, and improve access to renewable energy by means of policy interventions in energy, transportation, and construction sectors. Besides, construction sector policies can generate adaptation benefits, such as buildings that resist extreme weather events (Roberts 2008b). *Spatial policy options*, which cover a range of issues from regional scale to building scale, principally refer to adjustments in key spatial structures and elements in cities. For instance, promoting sustainable spatial forms like compact cities is argued to provide essential merits for mitigation and adaptation (Ewing et al. 2008). Likewise, more green and blue infrastructure within cities is shown to reduce heat stress, improve air quality, and mitigate flood risks (Gill et al. 2007). Moreover, spatial policy options comprise renewal and rehabilitation of infrastructure systems and promoting public transportation ridership and nonmotorized transportation.

Among the three groups of policy options, the key is *spatial policy options*, as they have long-term effects and can help to achieve mitigation and adaptation goals simultaneously. Yet, introduction of spatial policy options necessitates certain forms of interventions into current built-up areas in cities. Urban regeneration inherently comprises such interventions including renewal, rehabilitation, and redevelopment schemes. Urban regeneration generally refers to a comprehensive process through which economic, social, physical, and environmental conditions of decayed or underutilized urban quarters are improved and associated urban problems are addressed (Roberts 2000). Such a process can also help to introduce many of the *spatial policy options* and hence create climate-friendly urban environments. In other words, spatial deficiencies leading to climatic vulnerabilities in cities can be avoided to a certain extent through urban regeneration projects.

Urban regeneration can contribute to climate change mitigation and adaptation goals in many ways. Regeneration projects on inner-city brownfields and underutilized lands can promote compact urban forms and mixed-use developments (Couch and Dennemann 2000). Low-carbon and climate-friendly neighborhoods, which generate and consume renewable energy, save natural resources via recycling of wastes and storm waters, and restrict use of private vehicles, can be constructed by regeneration projects. Furthermore, urban regeneration projects can also improve public transportation systems and increase walking and cycling opportunities and encourage public transportation ridership in cities. For instance, it has been noted that since the introduction of Urban Renaissance Policy in the UK, a significant increase in investments in public transportation infrastructure, particularly nonmotorized transportation (walking and cycling), has taken place (Urban Task Force 2005).

Nevertheless, there are significant challenges to realize the potential contributions of urban regeneration projects to climate change mitigation and adaptation. First of all, links between urban regeneration and climate change mitigation and adaptation are not yet well established, and the potential role of regeneration projects in achieving mitigation and adaptation goals is not sufficiently evaluated. Linked to this, methodological tools to evaluate the performance of regeneration projects in creating climate-friendly urban development are limited. Currently, indicator frameworks are the most appropriate and common method for communicating the status of certain urban practices, such as plans and projects, toward achieving environmental goals.

Considering these challenges, this chapter sets out to contribute to current literature by exploring the linkage between urban regeneration and climate change. Current research in this area is very limited. So far, the literature on urban regeneration has focused on various aspects including sustainable development, but not much attention has been given to climate change. Therefore, this chapter attempts to point out and fill this gap. In line with the main research purpose, an indicator framework has been developed to assess the regeneration performance in achieving climate change mitigation and adaptation goals. Although the framework has been developed based on the existing research in literature, incorporation of climate change-relevant aspects into the framework by adding new indicators is the contribution of this chapter. The empirical focus of the research is on Minato Mirai 21 Project, located in downtown of Yokohama in Japan. Research findings are used to derive lessons for more climate-friendly urban regeneration projects and more effective use of indicators to communicate the status of regeneration projects.

The chapter is divided into four main sections. The following section includes a literature review on use of indicators to measure the progress toward sustainable development. In Sect. 6.3, methodological aspects of the research and hence steps taken for developing an indicator framework are elaborated. The application of indicators to Minato Mirai 21 Project (hereafter MM21) is discussed in Sect. 6.4. Finally, the Sect. 6.5 presents research findings and lessons learned from the case study.

6.2 The Use of Indicators for Assessing Sustainability Performance

If one aims to assess the performance of a policy or a project toward a set of goals, use of indicators appears to be the most common and appropriate methodology. This is because an indicator is a datum or variable, which has been assigned a role in the evaluation of a phenomenon (Tanguay et al. 2010). The increased attention on environmental concerns has led to the need to use indicators for assessing the impacts on environment (Wong 2000). Currently, it is widely accepted that use of indicators provides internationally an important and a common method to assess the status of a city, policy, or project toward achieving sustainable development (Reed et al. 2006; Moussiopoulos et al. 2010; Shen et al. 2011). In a recent work, authors mention that indicators can enable public administrations to underpin their sustainable development strategies by providing them with tangible assessment and monitoring systems (Tanguay et al. 2010).

A wide range of sustainability indicators and indicator frameworks are in use across different countries, regions, and cities (Shen et al. 2011). However, there is no consensus in literature on which indicators or indicator frameworks to use, and no common set of sustainability indicators that can be used in any city or urban-based practice exists (Tanguay et al. 2010; Hemphill et al. 2004a; Shen et al. 2011; Langstraat 2006). Most of the indicators and indicator frameworks developed to measure sustainability have been city specific (Langstraat 2006). Some authors relate this diversity in indicator frameworks to the ambiguity in definitions of sustainable development, differentiation in needs and goals between localities, and availability and accessibility of data (Tanguay et al. 2010; Shen et al. 2011). The variety of methods and frameworks is also argued to cause conflicts when selecting and applying indicators to certain contexts (Tanguay et al. 2010; Reed et al. 2006; Shen et al. 2011). On that account, in some recent works, there are efforts to formulate a common methodology to develop indicators (Hemphill et al. 2004a; Moussiopoulos et al. 2010), to combine and synthesize the most appropriate indicators into a common framework (Tanguay et al. 2010; Shen et al. 2011) or to summarize the best practice (Reed et al. 2006). However, we are still far from having a widely accepted and a common set of indicators that can be used anywhere.

A part of the current indicator frameworks is designed to be applied at the national level to assess the progress toward achieving sustainable development or improving environmental performance. A well-known example is the Environmental Sustainability Index (ESI), which is a composite index tracking a diverse set of socioeconomic, environmental, and institutional indicators to assess environmental sustainability progress achieved at the national scale.[1] Nevertheless, the emphasis that has been placed on cities and local governments to achieve sustainable development has led to many city-level indicator frameworks. For instance, in a recent work, authors analyzed 17 frameworks of urban sustainable development indicators developed in western countries to propose a common selection strategy and a common set of urban sustainability indicators (Tanguay et al. 2010). The European Commission in cooperation with Eurostat has developed an indicator framework (namely, Urban Audit) to provide 300 statistical indicators for 258 cities across 27 European countries.[2] A similar work on Asian cities has been done by Asian Development Bank to provide a detailed approach to applying policy-based indicator framework in several Asian cities (Villa and Westfall 2002).

Indicator-based approaches have also started to become a common method to measure and evaluate the performance of certain urban projects, especially urban regeneration projects (Hemphill et al. 2004a; Wong 2000). However, project-level indicators and indicator frameworks are very limited, compared to the national- and city-level frameworks. Perhaps the most influential and comprehensive indicator framework at project-level is "Hemphill framework," which tests the performance of urban regeneration projects against a set of sustainability indicators (Hemphill et al. 2004a, b; Langstraat 2006).

6.3 Objectives, Methodologies, and Significance of the Research

The general aim of this chapter is to contribute to the literature on use of indicators to assess urban regeneration performance from a climate change point of view. In line with the general aim, several objectives are intended to be accomplished: (1) to develop an indicator framework by considering the lessons learned from previous research in literature, (2) to evaluate the contribution of MM21 project to climate change mitigation and adaptation by applying the indicator framework, and (3) to derive lessons for improving our understanding of climate-friendly urban regeneration and indicator-based approach to evaluate regeneration performance.

As argued earlier, there is no consensus in literature on which indicators to use to evaluate the environmental performance of urban plans and projects, and no standard method exists for selecting and applying the indicators (Tanguay et al. 2010; Hemphill et al. 2004a; Shen et al. 2011). Besides, majority of current

[1] Please see http://www.yale.edu/esi/. Accessed 25 Feb 2011.

[2] Please see http://www.urbanaudit.org/index.aspx. Accessed 25 Feb 2011.

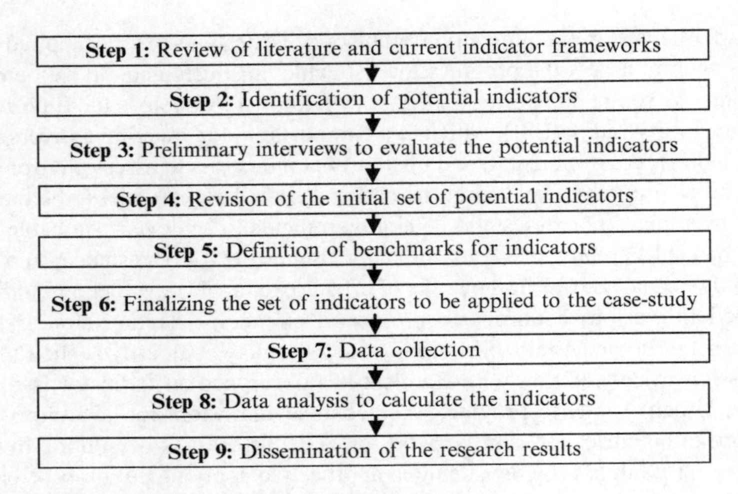

Fig. 6.1 Steps of indicator development and application

indicator frameworks are developed to assess the performance of plans and projects toward sustainable development, and not much work exists particularly on climate change mitigation and adaptation. Therefore, in this research, an own methodology for selection and application of indicators has been developed based on review of similar studies in literature. The indicator framework has been developed and applied through certain significant steps presented in Fig. 6.1.

The 1st step toward developing the indicator framework is the thorough review of literature and readily available indicator frameworks. In this step, the most appropriate indicator frameworks and most commonly used indicators have been identified. This initial step is argued to be effective in achieving the compliance with national and international standards as well as in examining the possible indicators that have already been put into place in similar contexts (Moussiopoulos et al. 2010). Based on the outcomes of the first step, an initial set of indicators that includes all relevant potential indicators has been identified in the 2nd step. At this stage, it is observed that among all of the frameworks examined, Hemphill framework (Hemphill et al. 2004a, b) and Urban Ecosystems Europe (UEE) framework (Berrini and Bono 2007) are the most relevant indicator frameworks for the purposes of this research. Hemphill framework includes a set of 52 indicators to measure sustainability performance of urban regeneration projects. The framework was developed through a comprehensive process including expert panels, Delphi and multi-criteria analyses, and was applied to six regeneration projects in three cities (Hemphill et al. 2004a, b). UEE framework is an integrated assessment of environmental performances of 32 main cities in Europe by employing a set of 25 indicators, most of which are directly related to climate change mitigation and adaptation. As climate change is a core issue in UEE framework, the framework provides essential guidance to develop indicators and benchmarks to measure the progress toward climate change mitigation and adaptation. However, Hemphill framework contains indicators mostly related to sustainable development rather than climate change. On the other hand, as opposed to UEE framework, Hemphill framework has a well-developed system to integrate indicators numerically by combining indicator values to generate a single score for projects evaluated. As both frameworks have some weaknesses and do not fully fit the purposes of the analysis in this research, instead of applying either of them, a new framework has been developed in this research based mostly on Hemphill and UEE frameworks.

As often mentioned in literature, the potential indicators determined in the 2nd step have to be evaluated to select the most appropriate ones in collaboration with relevant experts and local stakeholders (Reed et al. 2006; Moussiopoulos et al. 2010). Therefore, in the 3rd step of indicator development and application process, a number of preliminary interviews have been made with some key actors engaged in MM21 project and with some researchers to get their opinions on indicators. Preliminary interviews have also been useful to obtain suggestions on potential stakeholders and agencies that will be interviewed during data collection process. Such approach is referred to as a special technique, namely, "snowball-sampling techniques" in literature (Reed et al. 2006; Bryman 2001). In the 4th step, initial set of indicators that includes potential indicators has been revised considering the feedbacks from preliminary interviews. The revision has been made either by adding new indicators or by eliminating or redefining some of the initial indicators. Relevance of the indicator and availability of data have been the two major criteria to conduct the revision process.

One of the main steps in indicator development is to define benchmarks or establish baselines from which progress can be monitored (Tanguay et al. 2010; Hemphill et al. 2004a; Reed et al. 2006). So, the 5th step comprises the definition of benchmarks for each indicator by referring to literature and appropriate frameworks, such as Hemphill and UEE frameworks (Hemphill et al. 2004a, b; Berrini and Bono 2007). Benchmarks are determined by taking into consideration the widely accepted targets, ideal situations, and standards in order to make indicators transferable to other cases. For each indicator, a benchmarking scale is determined, and each indicator is assumed to get a benchmarking score from 2 to 10 based on its value. Following the definition of benchmarks, the set of indicators that will be applied to case project has been finalized in the 6th step. This step has ended up with a list of 34 indicators grouped under six performance categories, which are (1) economy and work, (2) buildings and land use structure, (3) transportation and mobility, (4) infrastructure and resource efficiency, (5) energy consumption and efficiency, and (6) community-based issues (Table 6.1). Each performance category is assumed to represent an essential dimension of a regeneration process to promote climate change mitigation and adaptation and also sustainability. The categorization in Hemphill framework has been revised in a way to include aspects directly related to climate change mitigation and adaptation. Such categorization is crucial to come up with a holistic set of indicators (Reed et al. 2006).

Eventually, three kinds of indicators have been included in this framework. The first kind is indicators that are directly taken from existing frameworks. The second kind consists of indicators that are developed by adjusting some of the indicators included in existing frameworks. For instance, Hemphill framework does not

Table 6.1 List of the indicators

Cat	No	Indicator	O.I.	I.V.	B.S.
Economy and work	I_1	*Increase in employment*: total number of jobs created per 1,000 square meters *Benchmarking scale*: 1–25, 2 pts; 26–50, 4 pts; 51–75, 6 pts; 76–100, 8 pts; >100, 10 pts	H.F.	30 Jobs	4
	I_2	*Workplace occupancy*: percentage of workplaces already occupied and active (%) *Benchmarking scale*: <50%, 2 pts; 50–64%, 4 pts; 65–79%, 6 pts; 80–94%, 8 pts; >95%, 10 pts	H.F.	77%	6
	I_3	*Employment for local people*: percentage of employees from local area (%) *Benchmarking scale*: <10%, 2 pts; 10–19%, 4 pts; 20–29%, 6 pts; 30–39%, 8 pts; >40%, 10 pts	H.F.	Above 50%	10
Buildings and land use structure	I_4	*Green space ratio*: percentage of public green spaces within regeneration area (%) *Benchmarking scale*: <5%, 2 pts; 6–10%, 4 pts; 11–15%, 6 pts; 16–20%, 8 pts; >20%, 10 pts	H.F.	19.8%	8
	I_5	*Green space availability*: amount of public green spaces per capita in regeneration area (m^2) *Benchmarking scale*: <5 m^2, 2 pts; 5–6.9 m^2, 4 pts; 7–8.9 m^2, 6 pts; 9–9.9 m^2, 8 pts; >10 m^2, 10 pts	T.R.	5.12 m^2	4
	I_6	*Unsealed surface ratio*: percentage of real vegetation in public green spaces (%) *Benchmarking scale*: <50%, 2 pts; 51–60%, 4 pts; 61–70%, 6 pts; 71–80%, 8 pts; >80%, 10 pts	T.R.	54.5%	4
	I_7	*Reclamation potential*: percentage of contaminated area and underutilized lands reclaimed (%) *Benchmarking scale*: <50%, 2 pts; 50–64%, 4 pts; 65–79%, 6 pts; 80–94%, 8 pts; >95%, 10 pts	H.F.	100%	10

Buildings and land use structure	I_8	*Employment density*: employment density in regeneration area (employees/ha) *Benchmarking scale*: <250, 2 pts; 251–500, 4 pts; 501–750, 6 pts; 751–1,000, 8 pts; >1,000, 10 pts	T.R.	340 emp/ha	4
	I_9	*Net residential density*: number of dwellings per land area devoted to residential purposes (dwelling/ha) *Benchmarking scale*: 0–50, 2 pts; 51–100, 4 pts; 101–150, 6 pts; 151–200, 8 pts; 200–275, 10 pts	H.F.	630 dwl/ha	10
	I_{10}	*Mixed-use structure*: Mixed-use combinations (residential/commercial/recreational) (%) *Benchmarking scale*: 10–20/10–20/>50%, 2 pts; 10–20/40–50/40–50%, 4 pts; 20–30/20–30/40–50%, 6 pts; 40–50/40–50/10–20%, 8 pts; 30–40/30–40/20–30%, 10 pts	H.F.	8–49–43%	4
	I_{11}	*Residential occupancy*: share of houses occupied (%) *Benchmarking scale*: <50%, 2 pts; 50–64%, 4 pts; 65–79%, 6 pts; 80–94%, 8 pts; >94%, 10 pts	H.F.	85%	8
	I_{12}	*Climate-friendly buildings 1*: percentage of buildings with green roofs/green facades (%) *Benchmarking scale*: <10%, 2 pts; 10–19%, 4 pts; 20–29%, 6 pts; 30–39%, 8 pts; >40%, 10 pts	T.R.	3%	2
	I_{13}	*Climate-friendly buildings 2*: percentage of buildings generating/consuming renewable energy (%) *Benchmarking scale*: <10%, 2 pts; 10–19%, 4 pts; 20–29%, 6 pts; 30–39%, 8 pts; >40%, 10 pts	T.R.	10%	4
	I_{14}	*Climate-friendly buildings 3*: percentage of buildings assessed by green building certification programs (%) *Benchmarking scale*: <10%, 2 pts; 10–19%, 4 pts; 20–29%, 6 pts; 30–39%, 8 pts; >40%, 10 pts	T.R.	20%	6

(continued)

Table 6.1 (continued)

Cat	No	Indicator	O.I.	I.V.	B.S.
Transportation and mobility	I_{15}	*Land devoted to roads*: percentage of site area occupied by roads (excl. pedestrian and bicycle) (%) *Benchmarking scale*: >25%, 2 pts; 20–25%, 4 pts; 15–20%, 6 pts, 10–15%, 8 pts; <10%, 10 pts	H.F.	24%	4
	I_{16}	*Land devoted to walking and cycling*: percentage of roads designated to pedestrian and cycling (%) *Benchmarking scale*: <30%, 2 pts; 30–44%, 4 pts; 45–59%, 6 pts; 60–75%, 8 pts; >75%, 10 pts	H.F.	18%	2
	I_{17}	*Work travel habits*: most common mode of transport for work commutes in regeneration area *Benchmarking scale*: private transportation, 2 pts; share a car/taxi, 4 pts; bus/coach, 6 pts; rail transportation, 8 pts; walking/cycling, 10 pts	T.R.*	Rail trans. (76% work trip)	8
	I_{18}	*Leisure travel habits*: most common mode of transport among visitors to the project area *Benchmarking scale*: private transportation, 2 pts; share a car/taxi, 4 pts; bus/coach, 6 pts; rail transportation, 8 pts; walking/cycling, 10 pts	T.R.*	Rail trans. (48% leis. trips)	8
	I_{19}	*Public transportation coverage*: average distance to major facilities from public transport stations *Benchmarking scale*: >1,250 m, 2 pts; 1,000–1,250 m, 4 pts; 750–1,000 m, 6 pts; 500–750 m, 8 pts; <500 m, 10 pts	T.R.*	Below 500 m	10
Transportation and mobility	I_{20}	*Passengers on public transportation*: ratio of annual public transport passengers in project area to that of city	T.R.**	1.3	8

		Benchmarking scale: <0.75, 2 pts; 0.76–1.0, 4 pts; 1.01–1.25, 6 pts; 1.26–1.5, 8 pts; >1.51, 10 pts			
	I_{21}	*Car parking provision 1*: number of car parking spaces per residential dwelling	H.F.	0.62 space	8
		Benchmarking scale: >2 space, 2 pts; 1.5–2 space, 4 pts; 1–1.5 space, 6 pts; 0.5–1 space, 8 pts; car free, 10 pts			
	I_{22}	*Car parking provision 2*: number of car parking spaces per square meter of office and commercial development	T.R.*	1 space: 210 m^2	8
		Benchmarking scale: >1 space, 50 m^2, 2 pts; >1 space, 100 m^2, 4 pts; >1 space, 150 m^2, 6 pts; >1 space, 200 m^2, 8 pts; essential users only, 10 pts			
Infrastructure and resource efficiency	I_{23}	*Improvements in infrastructure*: percentage of infrastructure renewed or rehabilitated in project area (%)	T.R.	Above 90%	10
		Benchmarking scale: <50%, 2 pts; 50–70%, 4 pts; 70–80%, 6 pts; 80–90%, 8 pts; >90%, 10 pts			
	I_{24}	*Water consumption*: per capita daily water consumption in project area (liters/person/day)	T.R.**	175 L/per/day	6
		Benchmarking scale: >250 L, 2 pts; 200–250 L, 4 pts; 150–200 L, 6 pts; 100–150 L, 8 pts; <100 L, 10 pts			
	I_{25}	*Waste generation*: per capita amount of solid waste collected to be disposed (kg/person/year)	T.R.**	260 kg/per/year	10
		Benchmarking scale: >450 kg, 2 pts; 400–449 kg, 4 pts; 365–399 kg, 6 pts; 300–364 kg, 8 pts; <300 kg, 10 pts			
	I_{26}	*Waste recycling*: percentage of buildings participating in paper recycling system (%)	T.R.*	67%	8
		Benchmarking scale: 0–19%, 2 pts; 20–39%, 4 pts; 40–59%, 6 pts; 60–79%, 8 pts; >80%, 10 pts			

(continued)

Table 6.1 (continued)

Cat	No	Indicator	O.I.	I.V.	B.S.
Energy consumption and efficiency	I_{27}	*Renewable energy consumption*: percentage of annual renewable energy consumption in project area (%) *Benchmarking scale*: marginal, 2 pts; <10%, 4 pts; 11–15%, 6 pts; 16–25%, 8 pts; >25%, 10 pts	T.R.**	Marginal	2
	I_{28}	*Renewable energy generation*: share of renewable energy generated on site in total energy supply (%) *Benchmarking scale*: marginal, 2 pts; <10%, 4 pts; 11–15%, 6 pts; 16–25%, 8 pts; >25%, 10 pts	T.R.**	Marginal	2
	I_{29}	*District heating and cooling system*: percentage of buildings connected to district heating and cooling system *Benchmarking scale*: <49%, 2 pts; 50–69%, 4 pts; 70–79%, 6 pts; 80–89%, 8 pts; > 90%, 10 pts	T.R.**	100%	10
	I_{30}	*Energy consumption for heating*: energy consumption for heating in project area (kWh/m^2/year) *Benchmarking scale*: >100, 2 pts; 76–100, 4 pts; 61–75, 6 pts; 50–60, 8 pts; <50, 10 pts	T.R.	52 kWh/m^2/year	8
Community-based issues	I_{31}	*Accessibility 1*: access to key facilities by working and living population in project area *Benchmarking scale*: very low access, 2 pts; less than expected, 4 pts; neutral, 6 pts; positive, 8 pts; very good access, 10 pts	T.R.	Positive	8
	I_{32}	*Accessibility 2*: interest in regeneration area by citizens *Benchmarking scale*: very low interest, 2 pts; less than expected, 4 pts; neutral, 6 pts; positive, 8 pts; high interest, 10 pts	T.R.	High interest (150,000 vst/day)	10

Community-based issues	I_{33}	*Community participation 1*: involvement of community in preparation of regeneration project *Benchmarking scale*: very low involvement, 2 pts; less than expected, 4 pts; neutral, 6 pts; positive, 8 pts; effective involvement, 10 pts	T.R.	Low	4
	I_{34}	*Community participation 2*: involvement of community in management of project area *Benchmarking scale*: very low involvement, 2 pts; less than expected, 4 pts; neutral, 6 pts; positive, 8 pts; effective involvement, 10 pts	T.R.	Very low	2

Abbreviations in the table are as follows: *Cat.* category, *No.* indicator number, *O.I.* origin of the indicator, *I.V.* indicator value, *B.S.* benchmark score, *H.F.* Hemphill framework, *T.R.* this research

*This indicator is developed by the author based on a similar indicator included in Hemphill framework

**This indicator is developed by the author based on a similar indicator included in Urban Ecosystems Europe framework

include an indicator relating to renewable energy, whereas UEE framework has one indicator regarding the presence of solar panels in public buildings. Based on the guidelines and targets concerning renewable energy in UEE framework, two indicators have been developed to assess both generation and consumption of renewable energy in regeneration projects (Table 6.1). Finally, the third kind of indicators refers to new indicators developed in this research by considering the issues ignored by existing frameworks. For example, as neither Hemphill nor UEE framework has an indicator relating to the ratio of real vegetation in open spaces, a new indicator has been developed and included in this framework in order not to disregard the importance of real vegetation or unsealed surfaces in green areas (Table 6.1).

The final three steps of indicator development are generally data collection, data analysis, and dissemination of results (Reed et al. 2006; Moussiopoulos et al. 2010). In line with this sequencing, the data required to calculate the indicator values have been collected in the 7th step. Data used to calculate the indicator values have been gathered from several sources. Interviews with key stakeholders and agencies constitute the first and the major source. 10 interviews have been made with 20 people from 8 organizations engaged in or related to MM21 project. Along with interviews, published and unpublished documents on regeneration project and city-wide data bases have been examined to collect the necessary data. Last but not least, our own observations and calculations constituted another essential source of data to calculate some of the indicators, especially the ones regarding land use structure.

In the 8th step, the data collected have been analyzed, and application of indicator framework to case project has been completed. To communicate the results of indicator-based analysis on MM21 project, the scoring system in Hemphill framework has been adapted and utilized. As presented in Table 6.1, each indicator has a value and a benchmark score from 2 to 10 based on this value. As per the scoring system, the higher the benchmark scores of indicators, the greater the contribution of regeneration project to climate change mitigation and adaptation is. The 9th step is dissemination of research results through different channels, as in the example of this chapter.

Furthermore, a basic issue in indicator development and application is whether or not to apply a weighting structure. Weighting means to attribute a greater value to some of the indicators than the others (Tanguay et al. 2010), and in this sense, weighting is used to reflect the relative contribution of each indicator (Hemphill et al. 2004a). Hemphill framework, for instance, includes a weighting structure for performance categories in order to give some categories (like "transport and mobility") more importance than others (like "resource use"). However, as opposed to Hemphill framework, the indicator framework developed and applied to MM21 project in this research is decided not to have a weighting structure. The reason is that no weighting structure can rationally justify the attribution of a given weight (Tanguay et al. 2010). Therefore, all of the indicators as well as performance categories in our framework are regarded as equally important and designed to contribute to the total score equally.

Another important issue in indicator development is the criteria that are used throughout the selection process of indicators (Tanguay et al. 2010; Reed et al. 2006; Shen et al. 2011). A widely quoted criteria set was put forth by the United Nations Statistical Institute for Asia and Pacific (UNSIAP 2007). According to these criteria, an indicator needs to be SMART: specific, measurable, achievable, relevant, and time related. There seems to be a consensus in literature that an indicator has to be *accurate*, *objective*, *easy to understand and use*, *be limited in number, and scientifically sound* (Tanguay et al. 2010; Reed et al. 2006; Moussiopoulos et al. 2010; Shen et al. 2011). The indicator selection and development process in this research has been performed in line with these criteria.

Last but not least, one important challenge in developing indicator-based approaches is associated with the applicability of indicators to different cases or situations. In order to overcome this challenge and increase the transferability of this framework across cases, indicators are selected and developed with reference to the academic literature and related studies, which are known to be successful. Some of the indicators included in this framework have already been applied to other cases. Besides, benchmarking, which is a critical issue influencing the transferability of indicators, is determined by considering common goals and standards that can apply to many different contexts. Nevertheless, in case some indicators in this framework are found to be difficult to apply to a particular case, they can be replaced with similar indicators that measure the progress toward the same goal. As long as the core structure of the framework is transferred, slight changes can be made to tailor the indicators into particular cases.

6.4 Application of the Framework to MM21 Project

6.4.1 General Information About the Case-Study Area

Minato Mirai 21 (MM21) is a prominent waterfront redevelopment project in the central quarter of Yokohama City (Fig. 6.2). The project has a long history of development, as its initiation dates back to the early 1980s. MM21 was built on former shipyards and railroad yards (60%) and on lands acquired through reclamation (40%) (Kishida and Uzuki 2009; MM21 2010). In this sense, the project has resulted in reuse of brownfields and underutilized lands. The primary objective of the project was to improve and strengthen the central business district (CBD) of Yokohama City and thereby to increase the independence of Yokohama from the rest of the Tokyo metropolitan area (MM21 2010; Kato 1990). A major expected outcome of the project was reduction of work commutes from Yokohama to Tokyo.

MM21 covers an area of 186 hectares (ha) in total. 87 ha of it is designated for private-built investments, and the rest (99 ha) is allocated for provision of public services and facilities (MM21 2010). Today, MM21 is a mixed-use area including

Fig. 6.2 Aerial views of MM21: Chuo district (*left*) and Shinko district (*right*)

office blocks, shopping malls, residential blocks, convention centers, hotels, an art museum, a hospital, green areas and parks, recreation areas, and amusement center (Fig. 6.2). Living, working, and visiting population in the district are able to access a wide range of urban functions. As per the original plan, the project area is designed to accommodate 190,000 working and 10,000 residential populations (Kishida and Uzuki 2009). Currently, there are 63,000 employees in 1,250 companies and 7,000 residents in 4,100 dwellings in MM21 (MM21 2010). It is also estimated that 54 million people visited MM21 in 2009, with an increase of 4 million people after 2007.

MM21 project was developed and implemented through public-private partnership. Public sector's responsibility was to provide the necessary infrastructure, whereas private sector was charged with the construction of superstructure. As part of development and implementation of the project, MM 21 Corporation as a joint organization of public and private sectors was established in 1984 (Kishida and Uzuki 2009; MM21 2010). Since then, Yokohama City and MM21 Corporation work together and share the duties regarding implementation and management of the project.

6.4.2 *Case-Study Results*

6.4.2.1 Economy and Work

There are three indicators that aim to assess the economic performance of regeneration process in this category. Among them, *percentage of employees from local area* showed the best (actually the maximum) performance, as almost all of the interviewees acknowledged the contribution of the project to providing employment to locals in Yokohama. Interviewees noted that significant number of people changed their workplaces from Tokyo to Yokohama, and hence, the number of work commutes from Yokohama to Tokyo has decreased. On the other hand, a poor score is observed on indicator relating to *number of jobs created*, mainly due to the existence of vacant development plots in the district. One of the main features of MM21 project is the delay in its completion, mainly caused by economic recession in the 1990s in Japan. By 2009, out of the 87 ha of development area, development rate is around 80%, even including temporary buildings and uses in the district. This situation also explains the average score on the third indicator relating to *workplace occupancy* (Table 6.1).

6.4.2.2 Buildings and Land Use Structure

This is the largest performance category with 11 indicators developed to assess the performance of regeneration project in terms of providing relevant spatial elements and buildings that bring climate-positive outcomes. The highest scores have been achieved on indicators relating to *reclamation of brownfields, residential density, residential occupancy*, and *green space provision* (Table 6.1). The effective utilization of former shipyards and railroad yards in city center is one of the strengths and contributions of MM21 project. Likewise, several high-rise and almost fully occupied residential blocks in the district secured (although not very high) the existence of a resident population in MM21. However, it needs to be noted that there is an unbalance of uses in MM21 in terms of domination of commercial and recreational uses over residential. This has led to the low score on indicator relating to *mixed-use structure*. MM21 district is rich in provision of green infrastructure, as 20% of the project area is allocated to public green spaces and parks. However, the ratio of real vegetation cover in green areas, which is known to be crucial for storm water infiltration and biodiversity preservation, is around 50%, causing a low score on indicator relating to *unsealed surface ratio* (Table 6.1).

Indicators relating to *climate-friendly buildings* received low scores. This is because there are not many green buildings equipped with sustainable and climate-friendly utilities in MM21. As initiation of the project dates back to the late 1980s, sustainable construction and green buildings have not been a concern in the project from the outset. Nonetheless, the introduction of the CASBEE system (Comprehensive Assessment System for Building Environmental Efficiency) in Yokohama is now an important opportunity to make further progress toward construction of climate-friendly buildings in MM21. Currently in MM21, six buildings corresponding to 20% in total have been evaluated by the CASBEE system, the green building certification system in Japan, and were given high ranks.

6.4.2.3 Transportation and Mobility

Indicators in this category deal with issues associated with improving public transportation systems and walking and cycling networks, encouraging public transportation ridership, and reducing the use of private cars. MM21 project achieved high scores on most of the indicators in this group, reflecting high ridership levels on public transportation and increased access to public transportation systems. 76% of work trips and 48% of leisure trips to and from MM21 are found to be made by rail transportation. Therefore, indicators on *travel habits* received the second highest scores. Design principles regarding allocation of car parking spaces in MM21 also support encouragement of use of public transportation. Car parking spaces are not rich in capacity so as to encourage private car usage. For instance, residential blocks are designed to provide approximately one parking space for two dwellings. This is reflected in the results in form of high scores on indicators relating to *car parking provision*. Conversely, absence of bicycle-only paths and high amount of land area devoted to roads in MM21 appear to be factors leading to low scores on two indicators relating to *land devoted to roads* in this category.

6.4.2.4 Infrastructure and Resource Efficiency

MM21 project indicates significant achievements in infrastructure renewal and resistance. This is because the project was initially designed in a way to provide firm, adequate, and contemporary infrastructure utilities (Kishida and Uzuki 2009). Almost all of the infrastructure utilities in the district are new and resistant against disasters. Indicator relating to *improvements in infrastructure* in this category is calculated to have the maximum benchmark score (Table 6.1). Likewise, remarkable achievements can also be observed in MM21 in waste reduction and waste recycling. For instance, *per capita amount of solid wastes collected to be disposed* in Nishi-ku Ward including MM21 (260 kg in a year) is below the "sustainable waste generation level in a city" (300 kg/person/year) defined in 5th Action Programme of European Union and referred in the Urban Ecosystems Europe framework (Berrini and Bono 2007). MM21 project owes this performance to the G30 Plan of Yokohama City, which includes effective strategies to reduce waste generation and promote recycling in whole city. On the other hand, it has to be noted that there is a particular paper recycling system in MM21 that encourages recycling of used paper and paper products. These findings have led to high scores on indicators about *waste generation and waste recycling* (Table 6.1). The lowest score in this category is calculated for indicator relating to *water consumption*, as very few concerns observed in the project over water consumption and recovery of gray waters.

6.4.2.5 Energy Consumption and Efficiency

There are four indicators in this category selected to assess the performance of MM21 project toward energy efficiency and use of renewable energy. Among the four indicators, the ones relating to *energy efficiency* and *energy consumption for indoor heating* received very high scores. Energy efficiency has been a concern in MM21 project from the outset. There are several measures for energy efficiency including building design techniques and most importantly the central system for heating and cooling. All of the permanent buildings in MM21 are obliged to be connected to the District Heating and Cooling System (DHCS). Interviewees noted that DHCS in MM21 is among the most efficient ones in Japan, as management company provides good maintenance. However, there is very little in the project to promote consumption and generation of energy from renewable sources. Almost all of the buildings in the district were built without renewable energy generation facilities, except some symbolic installments in a couple of buildings, such as the wind turbine on the roof of MM Park Building and PV panels on the roof of the moving walkway between Landmark Tower and Sakuragicho Station. The total amount of electricity generated in 2009 by PV panels on the moving walkway is 54.355 kWh, which is a marginal amount in total electricity consumption in MM21. Therefore, two indicators in this category relating to *renewable energy generation and consumption* have the minimum scores on the scale (Table 6.1).

6.4.2.6 Community-Based Issues

MM21 project shows an average performance regarding aspects covered by community-based issue indicators. The two indicators relating to *accessibility* have very high scores, whereas the other two indicators relating to *community participation* have low scores. MM21 project has been successful in increasing vitality and mobility within the project area and city center. The number of people visiting MM21 for leisure and commercial purposes is in a gradual increase. The number of annual visitors has increased from 50 million in 2007 to 53 million in 2008 and 54 million in 2009 (MM21 2010). On the other hand, MM21 project has been developed and implemented as a closed system, in which landowners, developers, and city government were given all the powers to decide. Citizens and community groups had almost no say in major decisions in the course of project development and implementation. Today, area management system of MM21 shows similar features. The only stakeholders of MM21 Corporation that execute area management are landowners and developers. Even owners of residential units in the district are not represented in area management, just as working population.

6.4.2.7 Overall Evaluation

Table 6.2 presents the overall results of indicator-based analysis of MM21 project. The best two performances among six performance categories are observed for categories titled *infrastructure and resource efficiency* and *transportation and mobility*. Indicators in the first category have scored 85% of the maximum possible total score, reflecting a notable success in terms of improvements in infrastructure

Table 6.2 Overall points scoring summary

Performance categories	Number of indicators	Maximum possible total score	MM21 scoring performance (points)	MM21 scoring performance (percentages)	Rankings of the categories
Economy and work	3	30	20	67	3
Buildings and land use structure	11	110	64	58	5
Transportation and mobility	8	80	56	70	2
Infrastructure and resource efficiency	4	40	34	85	1
Energy consumption and efficiency	4	40	22	55	6
Community-based issues	4	40	24	60	4
Total	34	340	220	65	

systems and services for better environmental performance and resource efficiency. Likewise, transportation and mobility indicators have scored 70% of the maximum possible total score, indicating a good performance for improving the public and nonmotorized transportation systems as well as for encouraging the use of these systems. On the other hand, the lowest two performances are observed for *energy consumption and efficiency* category and *buildings and land use structure* category. The indicators included in these two performance categories have scored 55% and 58% of the maximum possible total score, respectively. The total scoring performance of the MM21 project in terms of percentage of maximum possible total score is calculated to be 65%.

The overall results indicate that further improvements could be made in MM21 project to increase the project's contribution to climate change mitigation and adaptation. Such improvements can be made especially on the aspects and matters encapsulated by the performance categories with low scorings.

6.5 Lessons Learned from the Case Study

6.5.1 Lessons for More Climate-Friendly Urban Regeneration

Large-scale urban projects including regeneration projects are, one way or another, outcomes of the socioeconomic context, in which they have been developed. Therefore, a part of the positive progress in such projects should be related to the *power of the context*. MM21 project owes much to Yokohama city, as the city has functioned as a powerful context in several respects. City government introduced financial incentives to motivate companies in Tokyo metropolitan area to move to

MM21 and invited international organizations in Japan to locate in MM21 via special arrangements. Besides, city officials are well aware of current environmental challenges and their local implications and work to introduce policies and strategies to incorporate contemporary environmental concerns into local agenda. Last but not least, Yokohama has a well-developed urban infrastructure and transportation network to which MM21 has successfully been integrated.

One of the main features of MM21 project is the delay in its completion due to economic recession in Japan in the 1990s. Slow revitalization or delay in completion can be considered as a failure from an economic point of view. However, from an environmental point of view, it is possible to note that *delay in completion* of MM21 project has turned out to be an opportunity to avoid shortcomings in project design with regard to sustainability and climate change mitigation and adaptation. Contemporary concepts like climate-friendly buildings, sustainable construction, and smart grid, which did not exist at the time when the project was initiated, have started to be incorporated into the project recently. An interesting example is the start of the CASBEE system in Yokohama to rate environmental performance of buildings. MM21 is now the main area of implementation of this system in Yokohama, and six buildings in MM21 have already been certified under this system. Therefore, flexibility in design and implementation of urban regeneration projects can help to incorporate the projects with new concepts and approaches over time.

There are many agencies involved in development and management of MM21 district. These agencies include landowners, developers, MM21 Corporation, city government divisions, and service-providing companies. Some of the interviewees mentioned that the number of these agencies is increasing year by year. Each of these agencies has their own agenda and different priorities, and some of them are said to lack the awareness of environmental and climate change-related issues. In order to convert MM21 into a more climate-positive urban environment, joint actions of these agencies are necessary. Nevertheless, an actor or an agency to facilitate the collaboration and cooperation between various agencies in the district is missing. The case of MM21 indicated the need for *a facilitating actor or agency* that establishes the networking of various stakeholders of urban regeneration processes and encourages them to take actions toward a common direction. Moreover, this agency can also mastermind the possible joint actions of stakeholders, as many of the stakeholders may lack the vision and knowledge to steer their efforts.

MM21 project is a prominent example of a *top-down, landowner-led, and closed practice* of urban regeneration. Almost all of the decisions were given by landowners and developers, and these decisions have been imposed to the other actors in the district and hence strictly implemented. Locals in Yokohama as well as living and working population in MM21 were given almost no say over decision-making and area management. Such structure of project management prevented the progressive contributions from other actors concerning issues that have been neglected by managing actors. As a result, mechanisms that enable the participation of community groups and citizens need to be ensured in urban regeneration projects in order to achieve climate-positive outcomes.

6.5.2 Lessons for More Effective Use of Indicators

The application of indicator framework to MM21 showed that indicators are useful and effective to evaluate the performance of urban regeneration projects against a set of climate change mitigation and adaptation goals. However, more efforts and further research are necessary to enhance our knowledge and understanding of development and use of indicator-based approaches. A significant part of future efforts and research should be directed toward increasing the links between indicator-based approaches and climate change mitigation and adaptation. There is an obvious need for objective, scientifically sound, and simple indicators that can be used to assess the progress achieved in cities toward climate change mitigation and adaptation.

Data availability and accessibility are the major challenges in developing indicator-based frameworks. The data problem has several manifestations. First, city governments usually collect data on a sectoral basis that sometimes does not match the theme-based division within indicator frameworks. In such cases, crosscutting data becomes a necessity. Another version of this problem relates to area-based or area-specific data. City governments collect or produce data based on certain administrative divisions that does not fully overlap with project boundaries in many cases. For instance, in MM21 case, it was not possible to obtain data on waste generation within project boundaries. The data on waste management issues is only available at ward level, which was beyond the project area. Third, in some cases, data is not accessible due to concerns over business interests. An example of this in MM21 case is the restriction that was put on data on office vacancy rates. All in all, if necessary data is missing or not accessible, an indicator framework, no matter how comprehensive and well-developed it is, cannot help much. Therefore, city governments should be urged and encouraged to collect and produce data, which is required to employ indicator frameworks, and also make that data publicly available.

Benchmarking appears to be another critical aspect in development and use of indicators. Benchmarking with reference to a standard, norm, or goal is necessary to assess the progress achieved in a particular context. The experience on MM21 case shows that it is not very easy to find relevant and acceptable norms, goals, or standards against which indicators are benchmarked. In this research, benchmarking systems or methods of related previous works have been utilized. Considering this research experience, it has to be noted that city governments should develop standards, identify goals, and determine baselines that can be used as benchmarks for indicators.

Indicators are useful to shed light on only a certain part of reality, that is, how far a project is from the ultimate goals. However, indicators do not tell much about factors that facilitate or hinder the project to achieve the ultimate goals. Therefore, throughout interviews for data collection, researchers should also collect data and information on factors that have been influential on progress achieved in the project. The review of previous research on the same project can help to obtain an understanding of such factors and hence enable researchers to ask relevant questions to interviewees.

Finally, it should be noted that there always remains subjectivity in development and use of indicators. Selection, categorization, and benchmarking of indicators are always open to criticisms. In order to minimize the subjectivity problem, indicator framework and steps, through which the framework has been developed, should be made transparent and strengthened with participatory processes.

6.6 Conclusion

There are two key messages that have been highlighted by this research. First of all, urban regeneration projects can contribute to achieving climate change mitigation and adaptation goals, if they are designed and developed properly. Second, indicators can be useful as an instrument to assess the progress achieved in urban regeneration projects toward climate change mitigation and adaptation. However, there are significant challenges to increase the contribution of regeneration projects to climate-friendly urban development and to use indicators as a method to evaluate urban regeneration performance.

Awareness of how urban regeneration projects can help achieve mitigation and adaptation goals is still limited and hence needs to be raised. The idea of benefitting from urban regeneration projects to create climate-friendly urban environments should be promoted and mainstreamed into urban policy-making. Besides, cities generally lack an overriding urban development vision that integrates environmental concerns and climate change-related actions with urban planning and policy-making. Urban regeneration projects that are not developed as part of such an overriding and integrated vision cannot help much to achieve climate-friendly urban development. Furthermore, indicator frameworks can effectively be developed and used if necessary data is available and accessible in relevant formats and scales. There is also the need for standards and goals that can be used as benchmarks for indicators. All in all, it is primarily city governments who can overcome these challenges and open the way forward to climate-friendly urban regeneration projects.

References

Alam M, Rabbani MDG (2007) Vulnerabilities and responses to climate change for Dhaka. Environ Urban 19(1):81–97

Berrini M, Bono L (2007) Report 2007 urban ecosystem Europe: an integrated assessment on the sustainability of 32 European cities. Ambiente Italia – Research Institute, Milan

Bryman A (2001) Social research methods. Oxford University Press, New York

Costello A et al (2009). Managing the health effects of climate change. In: Lancet and University College London Institute for Global Health Commission Report, 373, May 16, UCL, London, pp 1693–1733

Couch C, Dennemann A (2000) Urban regeneration and sustainable development in Britain: the example of the Liverpool ropewalks partnership. Cities 17(2):137–147

Dhakal S (2009) Urban energy use and carbon emissions from cities in China and policy implications. Energ Policy 37(11):4208–4219

Ewing R, Bartholomew K, Winkelman S, Walters J, Chen D (2008) Growing cooler: the evidence on urban development and climate change. Urban Land Institute, Washington, DC

Gill S, Handley J, Ennos AR, Pauleit S (2007) Adapting cities for climate change: the role of the green infrastructure. Built Environ 33(1):115–133

Hemphill L, McGreal S, Berry J (2004a) An indicator based approach to measuring sustainable urban regeneration performance: part 1, conceptual foundations and methodological framework. Urban Stud 41:725–755

Hemphill L, McGreal S, Berry J (2004b) An indicator based approach to measuring sustainable urban regeneration performance: part 2, empirical evaluation and case study analysis. Urban Stud 41:757–772

IPCC (2007) Climate change 2007: the physical science basis. In: Solomon S, Qin D, Manning M et al (eds) Contribution of working group I to the fourth assessment report of the intergovernmental panel on climate change. Cambridge University Press, Cambridge

Kato Y (1990) Yokohama: past and present. Yokohama City University, Yokohama

Kishida H, Uzuki M (2009) Urban development strategy and project management: the challenge of Yokohama Minato Mirai 21. Gakugei Shuppansha, Kyoto (in Japanese)

Langstraat JW (2006) The urban regeneration industry in Leeds: measuring sustainable urban regeneration performance. Earth Environ 2:167–210

MM21 (2010) Yokohama Minato Mirai 21: plans and projects, vol. 81. Minato Mirai 21 Promotion Division, Urban Development Bureau, City of Yokohama, Japan

Moussiopoulos N, Achilllas C, Vlachokostas C, Spyridi D (2010) Environmental, social and economic information management for the evaluation of sustainability in urban areas: a system of indicators for Thessaloniki, Greece. Cities 27:377–384

Prasad N, Ranghieri F, Shah F, Trohanis Z, Kessler E, Sinha R (2009) Climate resilient cities: a primer on reducing vulnerabilities to disasters. The World Bank, Washington, DC

Puppim De Oliveira JA (2009) The implementation of climate change related policies at the subnational level: an analysis of three countries. Habitat Int 33:253–259

Reed MS, Fraser EDG, Dougill A (2006) An adaptive learning process for developing and applying sustainability indicators with local communities. Ecol Econ 59:406–418

Roberts P (2000) The evolution, definition and purpose of urban regeneration. In: Roberts P, Sykes H (eds) Urban regeneration: a handbook. Sage, London/Thousand Oaks/New Delhi, pp 9–36

Roberts D (2008a) Thinking globally, acting locally-institutionalizing climate change at the local government level in Durban, South Africa. Environ Urban 20(2):521–537

Roberts S (2008b) Effects of climate change on the built environment. Energ Policy 36:4552–4557

Satterthwaite D, Huq S, Pelling M, Reid H, Lankao PR (2007) Adapting to climate change in urban areas: the possibilities and constraints in low- and middle-income nations, vol 1, Human settlements discussion paper series: climate change and cities. International Institute for Environment and Development (IIED), London

Shen LY, Ochoa JJ, Shah MN, Zhang Z (2011) The application of urban sustainability indicators – a comparison between various practices. Habitat Int 35:17–29

Tanguay GA, Rojaoson J, Lefebvre JF, Lanoie P (2010) Measuring the sustainability of cities: an analysis of the use of local indicators. Ecol Indic 10:407–418

UNSIAP (2007) United Nations Statistical Institute for Asia and Pacific, "Building administrative data systems for statistical purposes-addressing training issues and needs of countries". In: Inception/Regional workshop on RETA6356: improving administrative data sources for the monitoring of the MDG indicators, Bangkok, Thailand

Urban Task Force (2005) Towards a strong urban renaissance. An independent report by members of the Urban Task Force chaired by Lord Rogers of Riverside, Urban Task Force, UK. Available at: http://www.urbantaskforce.org/UTF_final_report.pdf

Villa V, Westfall M (2002) Urban indicators for managing cities: cities data book. Asian Development Bank, Manila

Wong C (2000) Indicators in use: challenges to urban and environmental planning in Britain. Town Plan Rev 71(2):213–239

Chapter 7
Imagination and Practice of Collaborative Landscape, Ecological, and Cultural Planning in Taiwan: The Case of Taichung County and Changhua County

Li-wei Liu and Pei-yin Ko

Abstract "Landscape" is an issue normally neglected in the field of urban and regional development in Taiwan. However, an urban development policy, entitled *Urban–rural Landscape Improvement Movement*, was officially launched by the central government of Taiwan in 1999 to improve environmental quality and establish liveable human settlements. Based on the successful *Community Empowerment Movement*, *Urban–rural Landscape Improvement Movement* is intended to demonstrate positive integration among natural ecological preservation, local cultural uniqueness, and rapid urban development. With endeavors to enact and implement *Landscape Law* and *Landscape Master Plan* at the local government level, such an intention to link landscape, ecological, and cultural urbanism with urban development is gradually becoming possible. This chapter demonstrates a 2-year endeavor to implement *Landscape Master Plan* and *Townscape Renaissance Project* in Changhua County and Taichung County, which are located in the central region of Taiwan. It first depicts the course and content of *Urban–rural Landscape Improvement Movement* and reveals policy planning deliberation set by the central government. The interrelationship among landscape consultants, *Landscape Master Plan*, and the basic government level/grassroots organizations constructs a comprehensive mechanism to execute the movement. Second, it describes the process of *Townscape Renaissance Project* at the local government level and places great emphasis on the institutional horizontal integration between counties. Third, it further explores the cross-county collaboration and cooperation. A joint, collaborative project, entitled *Green Station Greenway*, is now subsidized by the central government and endeavors to promote sustainable development in urban–rural planning at the regional level. The *Green Station Greenway* project is intended to turn the historic areas or traditional districts of towns and cities along with Taiwan Railway routes into more sustainable and ecological environments.

L.-w. Liu (✉) • P.-y. Ko
Department of Urban Planning and Spatial Information, Feng Chia University, 100 Wenhwa Road, Seatwen, Taichung 40724, Taiwan
e-mail: lwliu@fcu.edu.tw; pyko@fcu.edu.tw

M. Kawakami et al. (eds.), *Spatial Planning and Sustainable Development: Approaches for Achieving Sustainable Urban Form in Asian Cities*, Strategies for Sustainability, DOI 10.1007/978-94-007-5922-0_7,

Keywords Landscape planning • Ecological planning • Local culture • Collaborative planning • Greenway • Taiwan

7.1 Introduction

"Landscape" is an issue normally neglected in the field of urban and regional development in Taiwan. Vanishing local characteristics usually follow in the wake of rapid industrialization and urbanization. No wonder that *Der Spiegel*, a German weekly, reported the quality of living environments in Taiwan as "living in the pigsty" in its 1995 publication. Apparently, the quality of living environments in Taiwan has already declined a lot during a 40-year period of highly economic growth (Chen 2006). Export-oriented economic policies adopted by the central government in Taiwan have brought about environmental pollution, ecological depredation, and living quality deterioration in this island. Under the circumstances, urban planning became only an instrument for governmental policies and encouraged land speculation (Lee and Fan 2004). Since the national spatial development system has long been dominated by intensive land use to support economic growth, excessive exploitation of natural resources, ignorance of environmental protection, and disregard for the quality of urban–rural landscape have emerged as obstacles to national development and competitiveness (Tseng and Chen 2007).

In order to gradually reform urban–rural landscape and promote quality of living environment, the Construction and Planning Agency, Ministry of the Interior (CPAMI) in Taiwan, drafted a plan under the direction of the Council for Economic Planning and Development (CEPD) regarding a new idea of *New Townscape Creation Action Plan* which was formulated in 1997. A long-term, comprehensive *Urban–rural Landscape Improvement Movement* was then initiated officially in 1999. CPAMI's *Urban–rural Landscape Improvement Movement* encourages public participation and places great emphasis on reforming spatial and living environments of Taiwan. The movement mobilizes many sectors of the central government, local governments, and local community organizations. It also has multifaceted influence on culture, local industries, and ecological considerations. Because the movement has broad influence on the whole society, it is not only a spatial reform process but also a social reform process (Chen 2006; Lee 2003; Lee and Fan 2004). In order to enhance the effect of the movement, CPAMI has carried on a series of mechanisms since 2003. Among them, the landscape consultant, the landscape master plan, and competitive subsidies are of importance. These mechanisms are conducted for the purpose of promoting local culture and characteristics, encouraging ecological sustainability, and improving living environments.

After a general review of related literature, this chapter first depicts the course and content of *Urban–rural Landscape Improvement Movement* and reveals policy planning deliberation set by the central government. The interaction among landscape consultants, the landscape master plan, and local governments or grassroots organizations constructs an executive framework to advance the movement. Second, it describes the process of *Townscape Renaissance Project* at the local government level and places great emphasis on the institutional horizontal integration between counties. Third, it further explores the issue of cross-county collaboration and cooperation. A joint, collaborative project, entitled *Green Station Greenway* and subsidized by the central government, is discussed.

7.2 Literature Review

In its literal meaning, landscape means to adapt visual, cultural, and natural processes to create new territory and thus offers a complex way of seeing, understanding, and shaping environments (Steiner 2011). As major components of urban landscape, parks and open space systems contribute to construct urban amenities. However, American cities in recent decades have seen very little expansion and uneven distribution of parks and open space systems. The shortage and inequity in the distribution of urban green and open spaces have withered the public realm and degraded the liveability of cities (Banerjee 2001). This is because, in traditional urbanism, green spaces were usually relegated to leftover areas, unsuited for building, or used for ornament (Steiner 2011). In fact, urban parks, greenways, and green spaces nowadays play an important role in providing social services and urban sustainability (Chiesura 2004; Fields 2009). Regarding the demand for urban open space systems in the twenty-first century, new landscapes inevitably emerge with changing lifestyles (Antrop 2006). Landscape planning is thus the process of understanding and directing the changing relationship between human kind and nature (von Haaren 2002). Moreover, urban open space is providing space for the expression of diversity, both personal and cultural (Thompson 2002). Landscape changes because they are the expression of the dynamic interaction between natural and cultural forces in the environment. Cultural landscape, reflecting local characteristics, is the result of consecutive reorganization of the land in order to adapt its use and spatial structure better to the changing societal demands (Antrop 2005).

In the face of rapid urbanization and severe climate change in the twenty-first century, landscape planning and management requires a new paradigm and a comprehensive approach to the governance of urban areas. Such a template should be based around multifunctional green networks which can help urban areas to respond more flexibly to environmental challenges (Gill et al. 2008). The Cheonggyecheon project may exemplify the expected template. The Cheonggyecheon elected freeway had been torn down and replaced by an urban stream and linear park in Seoul, Korea, during 2003 and 2005. The CGC freeway-to-greenway conversion project creates a more sustainable urban landscape for Seoul. In addition to recovering and preserving cultural heritage, the spillover benefit of the urban greenway is to mitigate the heat-island effect (Kang and Cervero 2009).

Such a freeway-to-amenity conversion demonstrates an integral consideration combining landscape, ecological, and cultural planning in a single project. In other words, what makes the Cheonggyecheon project interesting lies in the fact it replaced a nuisance mobility asset with an attractive urban amenity, a greenway, and stream. It challenges conventional wisdom that freeways and other high-performance roadway investments have been vital to the economic well-being of metropolitan areas. However, Soul's bold experiment has been proved an unqualified success based on land market outcomes (Kang and Cervero 2009).

The multifunctional template for landscape planning and management reveals three tendencies in professional practice. First, there is a tendency to advance multiple sustainability goals. Development in urban ecology and landscape urbanism as well as a possible synthesis-landscape ecological urbanism open new possibilities to reconstruct urban environments (Fields 2009; Steiner 2011). The term of landscape urbanism was coined by Charles Waldheim who defined it as a branch of landscape ecology, concentrating on the organization of human activities in the natural landscape and highlighting the leftover void spaces of the city as potential commons (Shane 2003; Steiner 2011). The basic premise of landscape urbanism indicates that landscape, but not architecture, should be the fundamental building block for city design (Steiner 2011). While urban ecology addresses the ecology of landscape within urban settings, landscape ecological urbanism suggests an evolution of aesthetic understanding, a deeper understanding of human agency in ecology, and reflective learning through practice (Steiner 2011).

The second tendency for landscape planning and management calls for an intensification of collaborative planning. As linear corridors of green space gradually increase their importance due to resource sharing and landscape integration, landscape planning usually requires policy coordination across multiple political jurisdictional boundaries (Ryan et al. 2006). For example, greenways have the capability to connect communities and protect critical historic, cultural, ecological, and recreational resources within a region (Fábos 1995; Ryan et al. 2006). However, coordination between agencies and organizations is the most troublesome issue in greenway implementation, and governments are generally not known for collaborating with each other or other organizations (Burby 2003; Ryan et al. 2006). In fact, collaborative planning is intended to mediate differences and seeks to democratize planning practice through communication (Abu-Orf 2005; Goldstein and Butler 2010). Collaborative planning should devise creative ways to achieve mutual gains and obtain broader support for implementation (Goldstein and Butler 2010). Moreover, collaboration is more likely to be implemented for proximate cities (Docherty et al. 2004).

The third tendency for landscape planning and management suggests a model for transforming unpleasant urban territories into community assets with multifunctional outcomes (von Haaren 2002; Steiner 2011). Landscape planners have long recognized the link between green space provision in the urban environment and environmental quality. For example, urban green space offers significant potential to help adapt urban areas to climate change through moderating microclimates and reducing surface water runoff (Gill et al. 2008). Moreover, urban green space also provides social and economic benefits, including human health and well-being improvement (Gill et al. 2008). For example, the High Line Project in Manhattan, transforming an abandoned rail line into a linear park as a recreational amenity, becomes a tourist attraction and a generator of economic development (Steiner 2011). In addition to ecological and environmental considerations, culture also plays a strategic role in local sustainable development and forces planners to think about how public spaces should be used (Sacco et al. 2009). Since local competitiveness is one of the important issues in urban studies, culture is a phenomenon that tends to have intensely local characteristics, thereby helping to differentiate places from one another and local culture, and uniqueness definitely build the core competitiveness of a place (Scott 2000). Therefore, in the face of globalization, creating local characteristics is beneficial to ameliorate living environment and strengthen urban competitiveness.

7.3 Overview of Urban–Rural Landscape Improvement Movement

Following a relatively long period of successful economic growth, disordered and stereotyped rural–urban landscape has become the bottleneck in the next stage of national development in Taiwan. Such a situation inevitably reduces national and urban competitiveness in the era of globalization. The disorder of environmental landscape in Taiwan indicates a cumulative consequence of excessive emphasis on economic growth, long-term neglect of ecological environments, slack management of land use, speculative ventures in real estate, and especially a lack of living aesthetics. Local landscape uniqueness is thus gradually disappearing. As a matter of fact, landscape should been seen as a living experience of collective placemaking. Landscape always reflects the interaction between human beings and their environments and expresses itself in a form that integrates natural, social, economic, and cultural elements into physical settings. In order to gradually ameliorate urban–rural landscape and promote quality of living environment, CPAMI drafted a plan under the direction of CEPD regarding a new idea of *New Townscape Creation Action Plan* which was formulated in 1997. A long-term, comprehensive *Urban–rural Landscape Improvement Movement* was then initiated officially in 1999. The movement includes several phases and projects, providing funding for local governments to ameliorate natural, cultural, and artificial landscapes. Figure 7.1 indicates the course and progress of this movement.

CPAMI officially initiated *New Townscape Creation Project* in 1999. *New Townscape Creation Project* split into two projects in 2003, with each having their own funding system. These two projects are entitled *Townscape Renaissance Project* and *Community Renaissance Project*, respectively. In recent years, the momentum of relevant concepts has displayed expansive results, and the projects have also become an indicator of local government policy implementation under

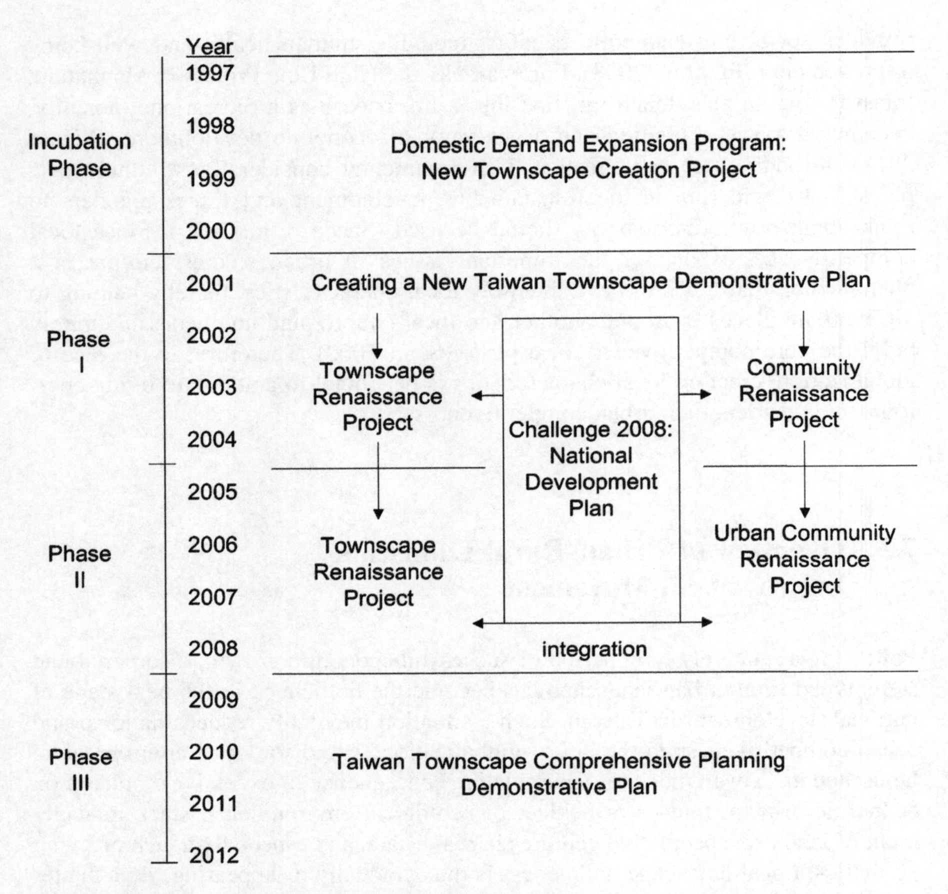

Fig. 7.1 The course and progress of *Urban–rural Landscape Improvement Movement*

public scrutiny. Moreover, in order to keep up with the tide of the times and to make concrete progress toward realizing the vision of Taiwan's development into a "green silicon island," CEPD collaborated with other concerned government agencies to draw up *Challenge 2008: National Development Plan*. Both *Townscape Renaissance Project* and *Community Renaissance Project* have been included in the national development plan since 2002. Among them, *Townscape Renaissance Project* policy focuses on creation of a cultural, green, high-quality new home in Taiwan, with the following six guidelines:

1. Creating new value and paradigm
2. Advancing changes in the townscape development pattern
3. Discovering local characteristics and uniqueness
4. Encouraging public participation and joint venture
5. Discovering local talents and traditional wisdom
6. Executing the policy of administrative reform for governmental sectors.

Townscape Renaissance Project therefore places great emphasis on competition, creation, participation, and learning as execution mottos. It also aims for the synthesis of ecology, history, culture, competitiveness, and sustainable development. In fact, *Townscape Renaissance Project* has become a movement involving community empowerment and plays a major role in serving as the catalyst of local revitalization schemes. Although *Townscape Renaissance Project* is nominally seen as a funding program for tangible environmental improvement construction, in reality it is a social movement with community experience and value at its core. Aside from funding allocation, of more importance is the establishment of relevant systems, accompanying guidelines, community empowerment, and citizen participation. In short, *Townscape Renaissance Project* policy redefines the concept of traditional public work projects and establishes a new model for urban development.

Up to 2008, 21.4 billion New Taiwan dollar (NTD) (roughly, 710 million US dollar) had been distributed to 4,764 projects. Another eight billion NTD (roughly, 260 million US dollar) has been compiled by the central government for the 2009–2012 financial years. Besides, more comprehensive mechanisms have been adopted after 2003. These comprehensive mechanisms are competitive-type funding strategy, *Landscape Master Plan*, and landscape consultants. In accordance to the *Challenge 2008: National Development Plan* policy ratified by the Executive Yuan, CPAMI has adopted a nationwide competitive-type funding strategy since 2003, with a goal serving as landmark projects in local townscape development. The competitive-type townscape renaissance project, usually held every other year, encourages municipal governments to propose relatively large-scale project that is creative, integrative, systemic, and sustainable development in terms of urban–rural landscape amelioration.

In order to establish a more sound and powerful control institution and enhance the benefit of landscape amelioration, management, and preservation, CPAMI started to draw up the Landscape Law (draft) for improving the qualities of environment in 2002. The Landscape Law (draft) requires that the city and county governments should make the *Landscape Master Plan* including urban and nonurban areas to enhance amelioration, conservation, management, and preservation of landscape resources. Due to a time-consuming and complicated legislative process, the draft Landscape Law has not been enacted so far. But because the draft Landscape Law regulates the content and the process of the *Landscape Master Plan*, many city and county governments then begin to draw up their *Landscape Master Plan* according to the draft law. The idea of the draft Landscape Law in Taiwan could trace back to the new institution of urban design in the 1970s. Since the urban design control process was put into practice in the 1980s, some guidelines and control measures regarding landscape reforms were also executed in the urban setting. When *Townscape Renaissance Project* was officially initiated in 1999, a more comprehensive consideration of landscape reforms within city and county limits brought into practice in 2003. Therefore, *Landscape Master Plan* is intended to transform unpleasant living environments into sustainable ecological cities and countries and enhance local characteristics and landscape uniqueness. The procedure of establishing *Landscape Master Plan* can be seen in Fig. 7.2.

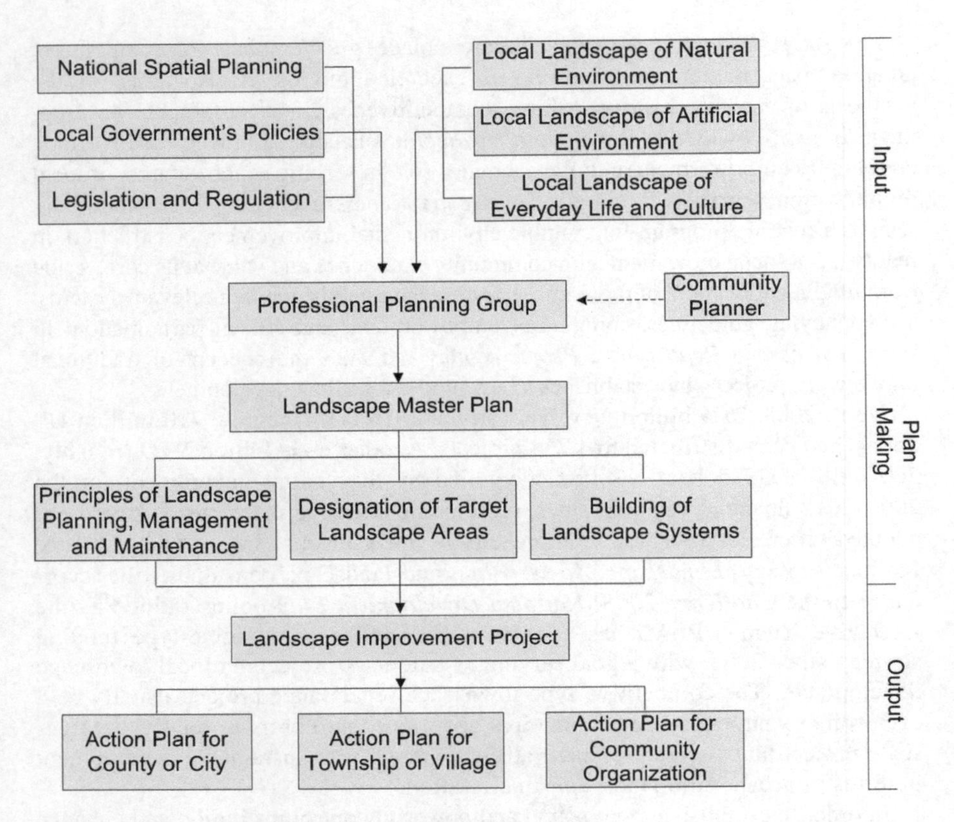

Fig. 7.2 Procedure for establishing *Landscape Master Plan*

The purpose of establishing *Landscape Master Plan* is to guide landscape planning and design for the districts, towns, and villages within city or county limits. Under the circumstances, scattered projects without a systematic consideration are not welcomed. Bureaus and departments at the city or county government are encouraged to propose landscape projects which connect different townships. Besides the establishment of *Landscape Master Plan*, another mechanism entitled *Landscape Consultant* is simultaneously put into practice. *Landscape Consultant* is also an executive program under the *Townscape Renaissance Project*. Usually a group of professionals and academics is, on a yearly basis, commissioned by the city or county government to direct and administer townships and community organizations in executing landscape schemes at the city or county level. The major task of *Landscape Consultant* is, first, to assist local governments in obtaining annual subsidies from the central government to implement the yearly *Townscape Renaissance Project*. Second, the members of *Landscape Consultant* have to control and supervise the implementation of all landscape schemes

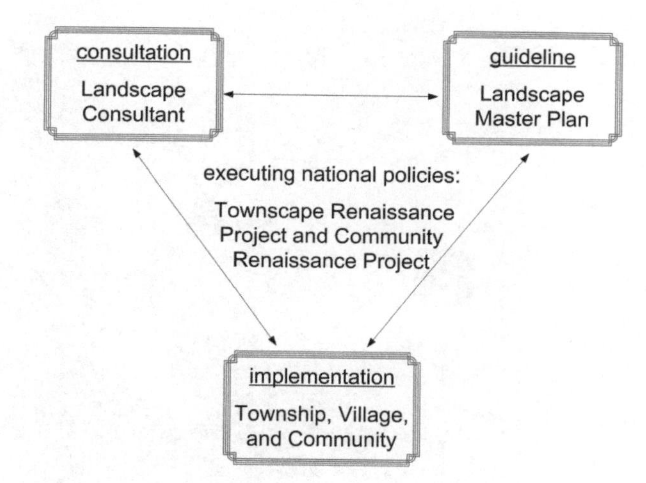

Fig. 7.3 Framework for executing *Urban–rural Landscape Improvement Movement*

subsidized by the central government. Third, the members of *Landscape Consultant* consult for the local government regarding issues of landscape planning and spatial development.

Therefore, *Landscape Master Plan*, *Landscape Consultant*, and the basic executing units (departments in the municipal government, townships, and community organizations) together construct a sound and sturdy framework, as seen in Fig. 7.3, for the practice of *Urban–rural Landscape Improvement Movement*. Such a triangular framework links guidelines, consultation, and implementation together and constructs a complete system for local governments to execute *Townscape Renaissance Project* and *Community Renaissance Project*. Beginning from 2009, *Townscape Renaissance Project* and *Community Renaissance Project* are integrated into *Taiwan Townscape Comprehensive Planning Demonstrative Plan*, and the emphasis is further placed on ecological sustainability, especially on the issues of climate change, energy saving, and carbon reducing. In order to further expand the policy effects of the funding system, the Ministry of the Interior has established "National Landscape Development Awards" since 2006 to recognize outstanding contributions from the implementation of landscape projects and schemes.

7.4 Cooperation and Collaboration Across Counties

More than a decade of *Urban–rural Landscape Improvement Movement* has finished approximately 5,000 projects up to now, but almost all of the projects were conducted within the administrative jurisdiction of a single city or county government. However, Taichung County Government and Changhua County Government had successfully collaborated on a *Townscape Renaissance Project* between 2007

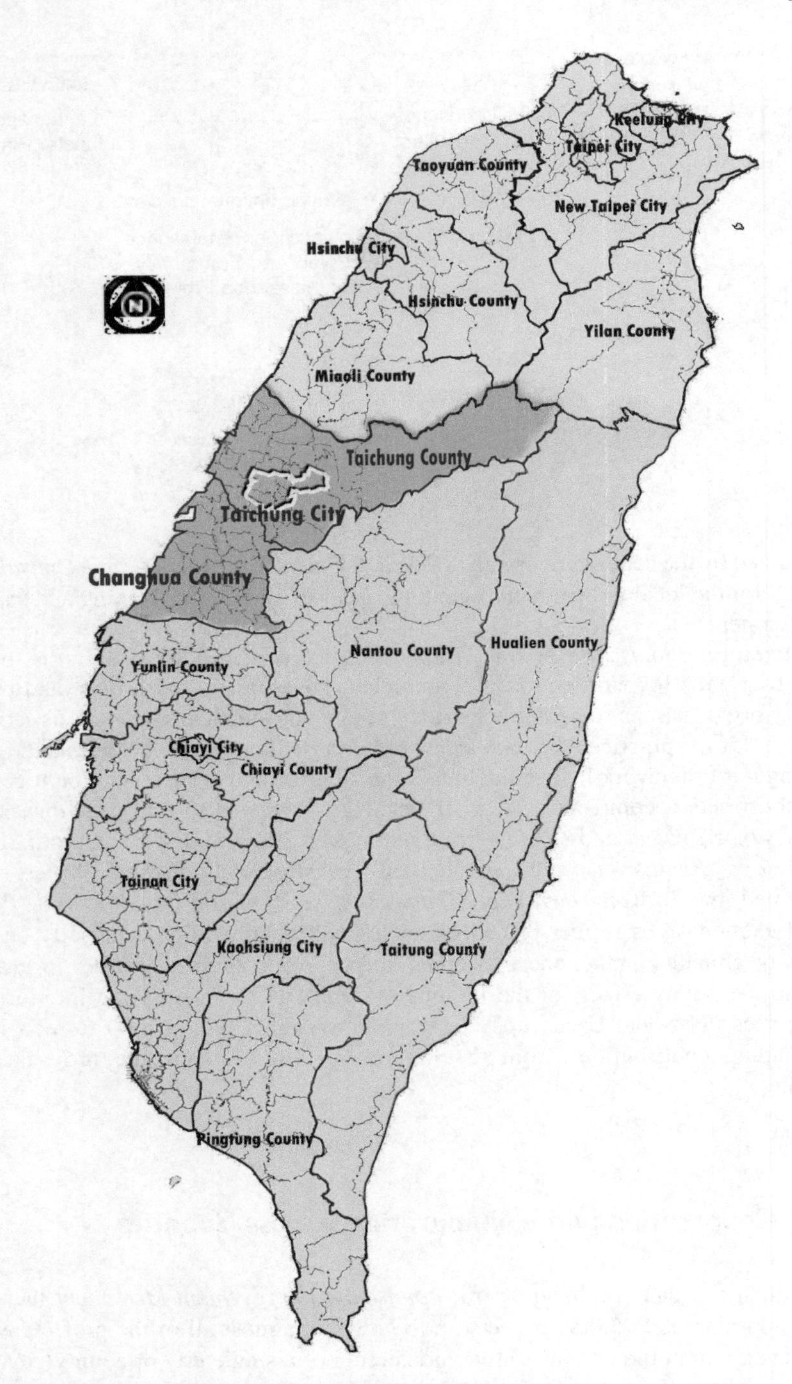

Fig. 7.4 The location of Taichung County and Changhua County, Taiwan

and 2009. Both Taichung County and Changhua County are located in the central district of Taiwan, as seen in Fig. 7.4. Beginning from December 25, 2010, Taichung County and Taichung City have merged into a special municipality, named Taichung City. Therefore, the title of Taichung County does not exist anymore today.

It took about 2 years for the two municipal governments to work together and draw up a joint, collaborative project. This chapter is not intended to discuss the content of the project in details but places the emphasis on the process. First thing for this collaboration is to organize a reciprocal panel for each county. Academic research on regional greenways indicates four major difficulties to the implementing regional landscape planning: (1) lack of coordination between government agencies and organizations, (2) lack of regional governance over local projects, (3) financing of local projects, and (4) public perceptions (Ryan et al. 2006). In fact, government agencies are generally not known for collaborating with each other or other organizations on planning issues (Ryan et al. 2006). City collaboration is usually bound to be a complex and politically difficult process (Docherty et al. 2004). In addition, collaborative planning may take insufficient account of power differences expressed in diverse interests (Abu-Orf 2005). Therefore, collaborative efforts require strong leadership as a key issue to success (Ryan et al. 2006). In order to bridge such a difficulty between Taichung County Government and Changhua County Government, a Landscape Task Force is then established for each side. For each county, the Deputy Magistrate, as a strong or powerful leader, chairs the task force with 8–10 department heads and 10–12 professionals and academics serving as committee members in the panel. A reciprocal institution or organization plays an important part in collaboration and cooperation between local governments.

Furthermore, city collaboration is risky and expensive, particularly since it often requires cities to deal with their mutual suspicions and antipathies, as well as the collaborative benefits are often only seen in the medium to long term (Docherty et al. 2004). Therefore, solutions or shared visions should be adopted by consensus and codified in binding agreements or plans (Goldstein and Butler 2010). A series of forums which invite all members of Landscape Task Force for both sides might be a good start to solve this problem. Then, at least a couple of joint forums were held in this case to reach the agreement and make a plan with shared visions. The first cross-county forum was held in Changhua County on September 28, 2007, and the second one in Taichung on February 22, 2008. A scheme, entitled Prefer X Plan, was proposed at that time as seen in Fig. 7.5. *The Prefer X Plan* consists of two possible collaborative projects: one is the *Dadu River Ecological Improvement*, and the other is the *Green Station Greenway* (GSG). These two projects make the shape of the cross, and they are thus named according to the shape. According to the *Prefer X Plan*, the Taichung County Government leads the first phase scheme and advances a joint, collaborative proposal of the GSG (as seen in Fig. 7.6). In the next phase, the Changhua County will then, in turn, advance another collaborative proposal of *Dadu River Ecological Improvement*. The GSG project focused on ecological redevelopment in the traditional human settlement and asked the central

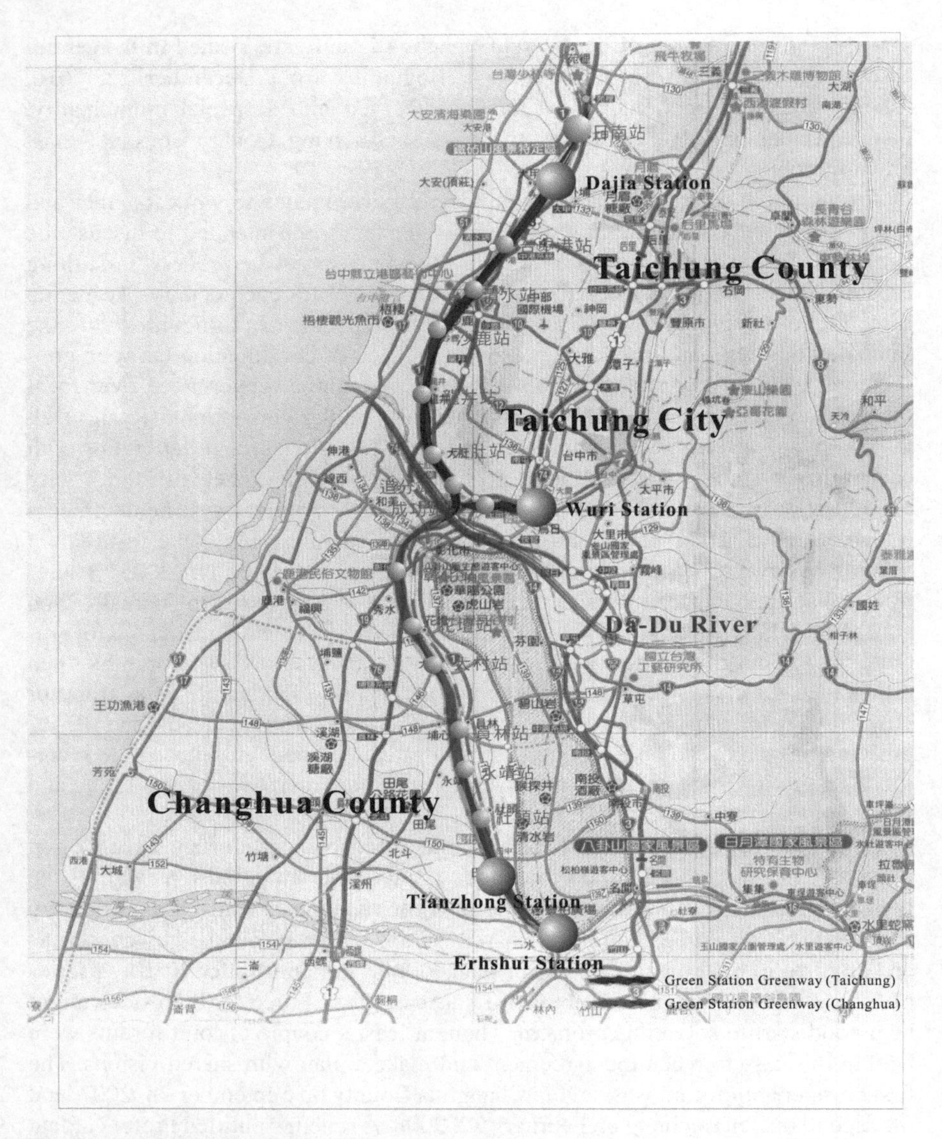

Fig. 7.5 A collaborative *Perfect X Plan*

government for funding in 2008. The competition was very intense, and only 1 out of 18 counties would win the prize. Finally, the GSG project was accepted in 2008. An eight million NTD was offered by CPAMI for the first year to conduct landscape planning and design. Another 130 million NTD may be distributed later for public works construction on conditions.

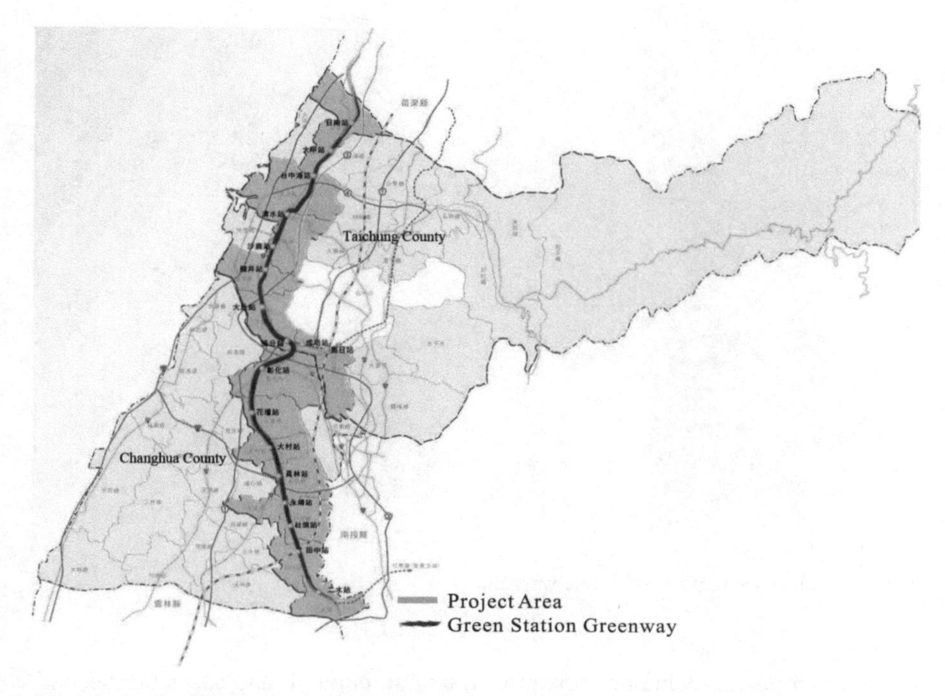

Fig. 7.6 A collaborative *Green Station Greenway* project

The GSG project selects towns and cities located along the western line of the Taiwan Railway routes in Taichung County and Changhua County. It is intended to turn the historic areas or traditional street districts, usually located by the railway station, of these towns and cities into more sustainable and ecological environments. 18 railway stations within these towns and cities are thus selected in this project. However, the GSG project does not directly deal with the landscape amelioration in the extent of the railway station because property rights of the railway station belong to Taiwan Railways Administration. The GSG project then targets local historic areas or traditional street districts which are adjacent to the railway station. These targeted areas are the core of everyday life for local residents, and they are, to a certain extent, places expressing local uniqueness and characteristics. Therefore, the GSG project plans to ameliorate the disordered and unpleasant landscape in the traditional street districts adjacent to the railway station. Besides, another important component of the project is to encourage quality, walkable, and pedestrian-friendly environments linked by public transit. The national railway and local bus system serves as the public transit for residents and tourists.

The concept of networking is heavily applied in this project. First, a network of the historic areas or traditional street districts of these towns and cities are linearly connected through the connection of Taiwan Railway routes that can be planned for

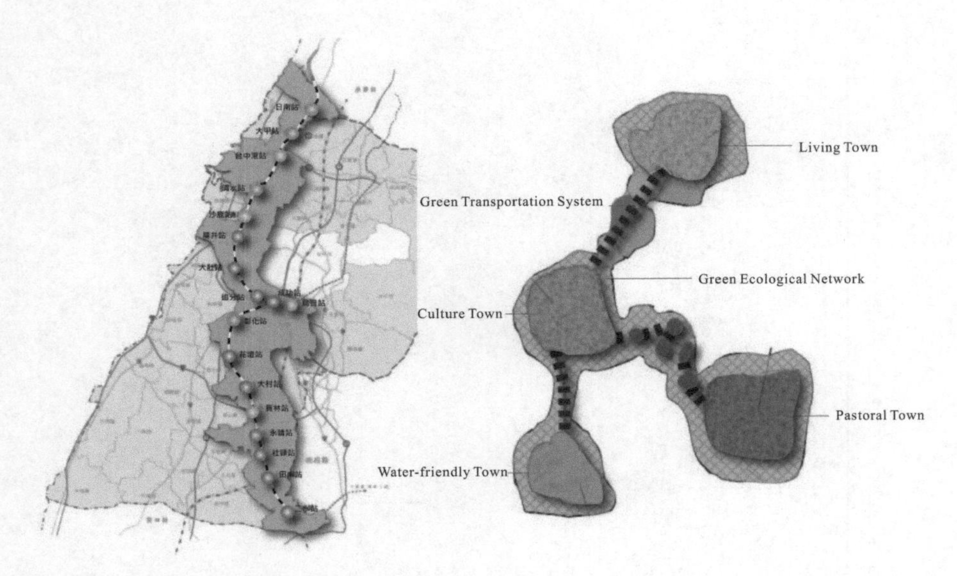

Fig. 7.7 The concept of ecological networking

multiple purposes, including recreational, tourist, cultural, and aesthetic. Second, an ecological network is introduced into the project area, as seen in Fig. 7.7. Green space and public transportation are connecting elements that link ecological landscape together among these traditional street districts.

In the process of collaborative planning, each of the collaborating entities requires a clear, straightforward, and unambiguous answer to the question "what's in it for me?" Collaborating cities would continue to compete with each other, and such continued competition is beneficial not only to a single city but also to the region as a whole (Docherty et al. 2004). In response to the collaborative-competitive model, the GSG project decided to pick four towns and cities as the exemplification areas, distributed equally across Taichung County and Changhua County. At the same time, four prototypes of multifunctional towns and cities are identified: cultural, water-friendly, pastoral, and living. Dajia in Taichung County is designated as the cultural town, Wuri in Taichung County the living town, Tianzhong in Changhua County the pastoral town, and Erhshui in Changhua County the water-friendly town (see Fig. 7.8). Proposed development patterns for these four prototypes are shown in Fig. 7.9. The type of culture-oriented towns and cities will be planned and designed largely from a perspective of tourism, education, and local humanistic history. The type of living towns and cities focuses mainly on tertiary industries, living function, and transportation network. The type of water-friendly towns and cities emphasizes the importance of riverbank landscape, ecological hydrology, and leisure. The type of the pastoral towns and cities highlights the need for rural living milieu, agricultural landscape, and a lifestyle of LOHAS.

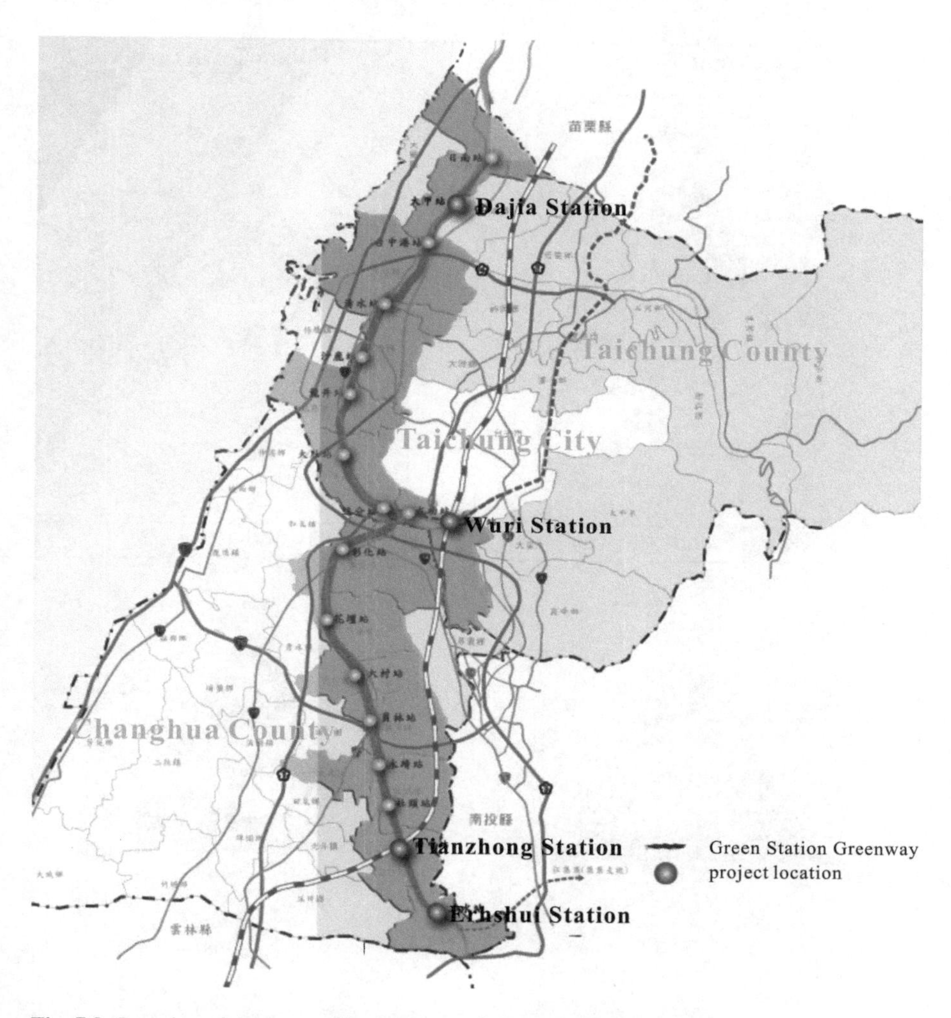

Fig. 7.8 Location of the exemplification areas of *Green Station Greenway*

7.5 Conclusion

In the face of severely global competition, creating a high-quality landscape and promoting quality of life enhance urban competitiveness and become the important task conducted by local governments. More than a decade of *Urban–rural Landscape Improvement Movement* executed by the central government in Taiwan has already established a sound, integral framework for local governments to reform their living environment. However, regional collaboration across jurisdiction boundaries is still rare. This chapter describes a joint, collaborative project conducted by both Taichung County Government and Changhua County

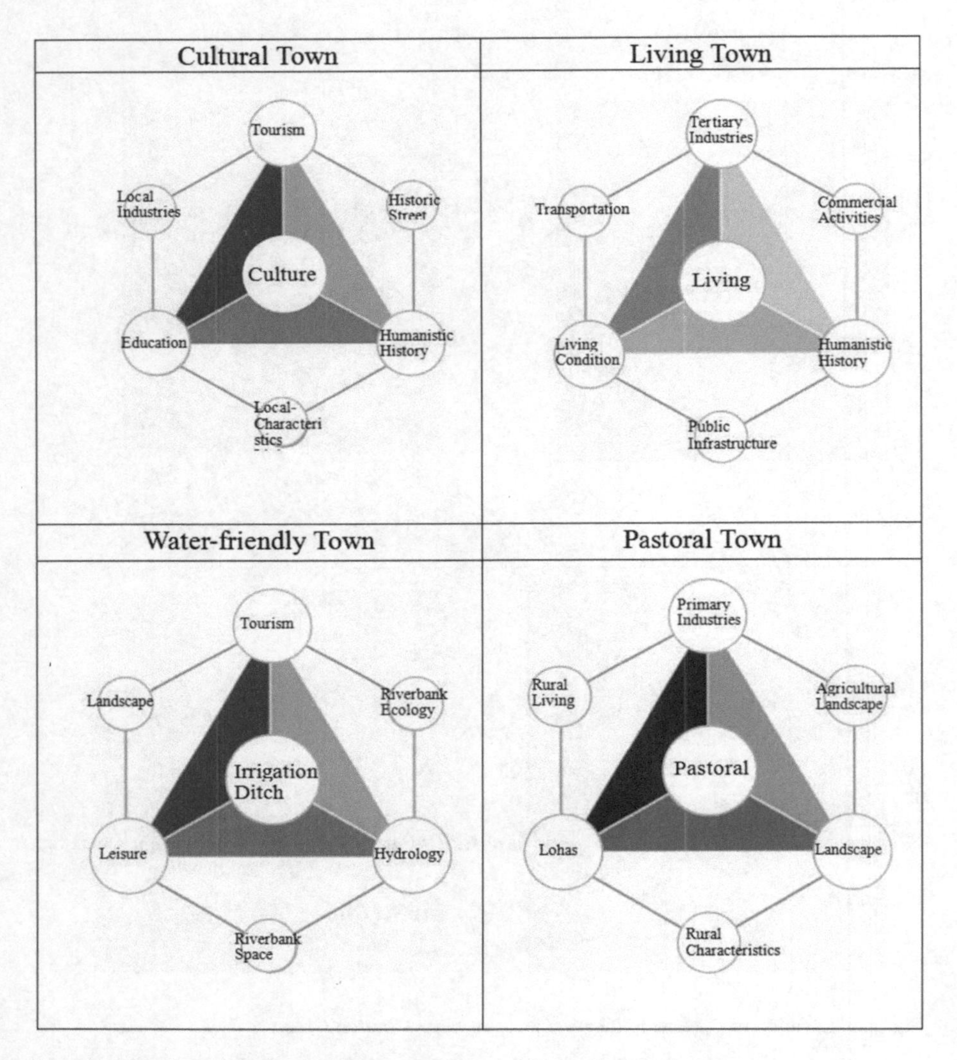

Fig. 7.9 Development pattern for four prototypes of *Green Station Greenway*

Government. Strategic collaboration might offer a means for nearby counties to enhance their competitiveness. Government agencies are generally not known for collaborating with each other or other organizations on planning issues. In order to bridge such a difficulty between Taichung County Government and Changhua County Government, a *Landscape Task Force* is then established for each side. Then, at least a couple of joint forums were held in this case to reach the agreement. Finally, the *Green Station Greenway* project, focusing on ecological redevelopment in the traditional human settlement and asking the central government for funding in 2008, was proposed and successfully received governmental funding for implementing the planning stage.

References

Abu-Orf H (2005) Collaborative planning in practice: the Nicosia master plan. Plan Pract Res 20(1):41–58

Antrop M (2005) Why landscapes of the past are important for the future. Landsc Urban Plan 70(1–2):21–34

Antrop M (2006) Sustainable landscapes: contradiction, fiction or utopia? Landsc Urban Plan 75(3–4):187–197

Banerjee T (2001) The future of public space: beyond invented streets and reinvented places. J Am Plan Assoc 67(1):9–24

Burby R (2003) Making plans that matter. J Am Plan Assoc 69(1):33–49

Chen KS (2006) Creating urban–rural landscape and accelerating to enact landscape law. Constr News Rec 280:47–55 (in Chinese)

Chiesura A (2004) The role of urban parks for the sustainable city. Landsc Urban Plan 68(1):129–138

Docherty I, Gulliver S, Drake P (2004) Exploring the potential benefits of city collaboration. Reg Stud 38(4):445–456

Fábos JG (1995) Introduction and overview: the greenway movement, uses and potentials of greenways. Landsc Urban Plan 33(1–3):1–13

Fields B (2009) From green dots to greenways: planning in the age of climate change in post-Katrina New Orleans. J Urban Des 14(3):325–344

Gill SE et al (2008) Characterising the urban environment of UK cities and towns: a template for landscape planning. Landsc Urban Plan 87(3):210–222

Goldstein BE, Butler WH (2010) Expanding the scope and impact of collaborative planning. J Am Plan Assoc 76(2):238–249

Kang CD, Cervero R (2009) From elected freeway to urban greenway: land value impacts of the CGC Project in Seoul, Korea. Urban Stud 46(13):2771–2794

Lee YJ (2003) Townscape renaissance: review and prospect. Taiwan Archit Mag 29(8):116–119 (in Chinese)

Lee YJ, Fan SM (2004) A study on Taiwan's Townscape Renaissance Movement: from the perspective of regulation theory. J Archit Plan 5(2):132–148 (in Chinese)

Ryan RL, Fábos JG, Allen JJ (2006) Understanding opportunities and challenges for collaborative greenway planning in New English. Landsc Urban Plan 76(1–4):172–191

Sacco PL, Blessi GT, Nuccio M (2009) Cultural policies and local planning strategies: what is the role of culture in local sustainable development? J Arts Manag Law Soc 39(1):45–63

Scott AJ (2000) The cultural economy of cities. Sage, London

Shane G (2003) The emergence of landscape urbanism. Harv Des Mag 19:1–8

Steiner F (2011) Landscape ecological urbanism: origins and trajectories. Landsc Urban Plan 100(4):333–337

Thompson CW (2002) Urban open space in the 21st century. Landsc Urban Plan 60(2):59–72

Tseng TF, Chen TW (2007) New orientation of landscape planning in national spatial planning. J City Plan 34(3):219–240 (in Chinese)

Von Haaren C (2002) Landscape planning facing the challenge of the development of cultural landscapes. Landsc Urban Plan 60(2):73–80

Part II
Urbanization and Sustainable Society: Section Housing and Transportation

Chapter 8
Sustainable-Oriented Study on Conservation Planning of Cave-Dwelling Village Culture Landscape

An-rong Dang, Yan Zhang, and Yang Chen

Abstract The village cultural landscape of China is being challenged by the New Rural Construction and rapid urbanization. One of the dominant theories on the conservation of the village cultural landscape proposes that it can be protected through ecotourism, which is based on the concept of sustainable development. In line with the theory, this chapter discusses the guideline for the sustainable conservation of the village cultural landscape by means of village ecotourism. Gaoxigou, a cave-dwelling village, was used as a case study to analyze the key elements of the village cultural landscape, discuss the resources of village ecotourism, explore the relationship between the elements and resources, and examine and apply the guidelines for conservation planning and design. With the implementation of the guidelines, the author explored the sustainable conservation planning method of the cave-dwelling village's cultural landscape and hopes to provide some references for the future protection of the village cultural landscape.

Keywords Village cultural landscape • Conservation planning and design • Sustainable • Cave dwelling

8.1 Introduction

8.1.1 Significance of Conserving Village Cultural Landscape

The cultural landscape, as the term implies, is fashioned from the natural landscape by a cultural group. Culture is the agent; the natural area, the medium; and the cultural landscape, the result (Sauer 1925). The cultural landscape changes in step

A.-r. Dang (✉) • Y. Zhang • Y. Chen
School of Architecture, Tsinghua University,
Haidian District, Beijing 100084, China
e-mail: danrong@mail.tsinghua.edu.cn; faliya.zhang@gmail.com; c_yang2012@126.com

M. Kawakami et al. (eds.), *Spatial Planning and Sustainable Development: Approaches for Achieving Sustainable Urban Form in Asian Cities*, Strategies for Sustainability,
DOI 10.1007/978-94-007-5922-0_8,

with human activity. Architecture and villages are the most typical examples of local cultural landscape (Sun 2002). Having been an agricultural civilization for more than 7,000 years, China has countless cultural heritages related to the village (Shan 2008, 2009). As the main human settlement of the agricultural population, the village is a unique space that is rich in cultural landscape (Zhang 2008), representing colorful village traditions (Wang 2007). Therefore, the importance of protecting the village cultural landscape is gaining a consensus in China, not only for the villagers' sake but for the whole nation as well.

8.1.2 Necessity of Conservation Planning and Design

With the advent of globalization, industrialization, urbanization, and new countryside construction projects, the village cultural landscape is facing serious threats and challenges (Wu 2010). In recent years, both officials and scholars have been stressing the need to conserve the village cultural landscape because of the serious damage caused by frequent natural disasters (Feng 2009). Therefore, conservation planning and design of the village cultural landscape have been explored and applied in varying degrees, using different methods (Shan 2010; Huang et al. 2011; Zhang 2011). However, most of the studies are focused on minority nationalities and the southwestern region, which abounds in traditional village culture. People rarely pay attention to the cave-dwelling villages located at Northern Shaanxi (Ma et al. 2012). In fact, the village cultural landscape of cave dwellers is unique, considering the Loess Plateau landforms and architecture (Dang et al. 2008).

8.1.3 Sustainable Characteristics of Village Ecotourism

According to the definition and principles of ecotourism established by The International Ecotourism Society (TIES) in 1990, ecotourism is about integrating conservation, communities, and sustainable travel. Therefore, those who implement and participate in ecotourism activities should follow the following principles: (1) minimize impact, (2) build environmental and cultural awareness and respect, (3) generate positive experiences for both visitors and hosts, (4) reward conservation efforts with direct financial benefits, (5) provide financial benefits and empowerment for local people, and (6) raise sensitivity to host countries' political, environmental, and social climate (http://www.ecotourism.org/what-is-ecotourism). Village ecotourism is similarly defined as responsible travel to a rural village to experience its eco-agricultural environment and traditional culture, conserve its cultural landscape, and improve the well-being of local people (Liu et al. 2007; Dong et al. 2009).

8.1.4 Guideline for Sustainable Conservation

The village cultural landscape is a showcase of the long-term interaction between nature and people and contains plenty of physical and intangible culture heritages, which are deemed to have high tourism value and get wide attention. Meanwhile, the village cultural landscape of the Loess Plateau is being challenged: the cave dwellings, seats of traditional culture, have been abandoned and destroyed in favor of flat-roofed houses; their decorative art is vanishing due to the changes in construction materials and architectural style; and the migration and aging of the population make it difficult to hand down culture (Dang et al. 2009). Therefore, this chapter discusses the guidelines for the sustainable conservation of the village cultural landscape by means of village ecotourism. The key factors of the village cultural landscape were analyzed, and the guidelines for conservation planning and design were studied and practiced via case implementation.

8.2 Approach

In order to work out the guidelines for the sustainable conservation of the cave-dwelling village cultural landscape, we developed a three-step approach: first, identify the elements of the village cultural landscape; second, evaluate the resources of village ecotourism; and third, establish the relationship between the said elements and resources.

8.2.1 Identifying the Elements of the Village Cultural Landscape

There are several classification systems of the composition of the village cultural landscape. Generally, it is classified into two categories: physical and intangible elements (Wang 2009; Wu 2010; Ma et al. 2012). The physical elements refer to the natural environment, such as land use and living environment—settlements and architecture—which are fundamental to the village cultural landscape. The intangible elements include the village's evolution history and traditions, which are based on specific physical elements (*see* Fig. 8.1).

Since most of the elements of the village cultural landscape are related to some kind of space, we divided the landscape into living, production, and ecological spaces. The living space includes architecture, the courtyard, the place of worship and equipment, the place of education and activity, public green space, and an administrative area. The production space includes farmland, orchards, fishponds, livestock farms, and related activity. The ecological space is made up of forestland, grassland, bodies of water, geomorphologic sights, and wildland.

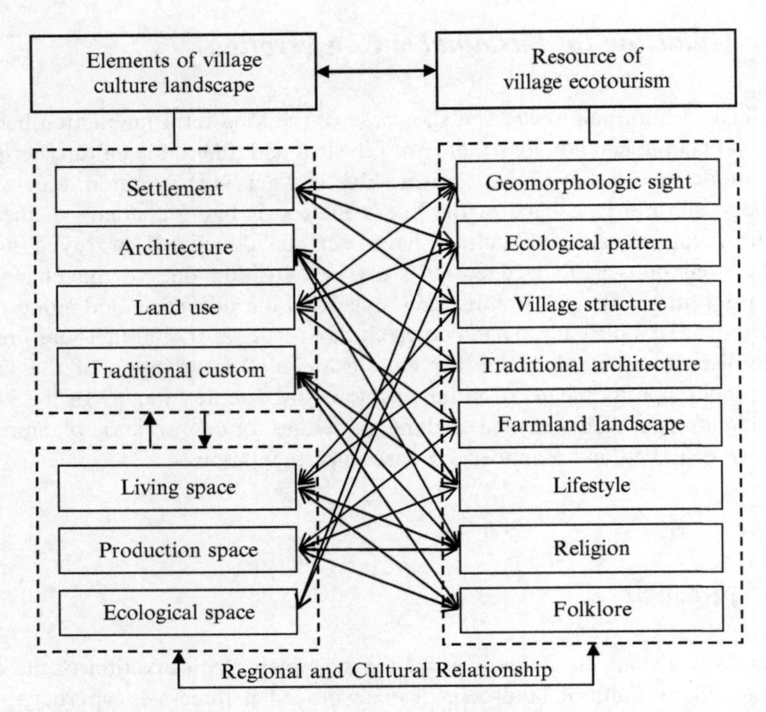

Fig. 8.1 Relationship between the elements of the cultural landscape and ecotourism resources of a village

8.2.2 Evaluating the Resources of Village Ecotourism

The resources of village ecotourism are the attractions to such activity, comprising a combination of natural and cultural factors that have ecological significance (Zhou 2006; Liu et al. 2007; Ren 2011). The natural resources include geomorphologic sights, ecological patterns, etc., while the cultural resources feature farmland landscape, traditional architecture, local food, lifestyle, religion, folklore, and so on (*see* Fig. 8.1). All the resources can be evaluated by means of classification, grading, and zoning.

The evaluation is the basis of village ecotourism development and sustainable conservation planning and design. Owing to the precondition to consider the evaluation, the scientific development of village ecotourism can achieve optimization in (1) combining the natural and cultural resources and recycling materials and energy, (2) the coordinated development of village economy and society, and (3) the protection and bequest of the village cultural landscape. Moreover, sustainable ecotourism profits can be gained by all participants and stakeholders.

8.2.3 Establishing the Relationship Between the Elements of the Village Cultural Landscape and Ecotourism Resources

As described above, the village cultural landscape reflects the regional and cultural characteristics of a traditional village, and tourists are attracted mostly by the unique or scarce natural and cultural resources of the village. The relationship and interaction between these two aspects was established in our analysis of their characteristics (*see* Fig. 8.1).

The relationship between the elements of the village cultural landscape and ecotourism resources is described as follows: (1) the elements of a village's cultural landscape are essentially resources with regional and cultural significance; (2) the resources of ecotourism are extracted from the elements of the village cultural landscape; (3) village ecotourism is a tour activity predicated on ecological culture, which promotes the development and innovation of village culture; and (4) the cultural landscape and ecotourism are interactive and mutually beneficial to a village.

Therefore, classifying cultural landscape elements and ecotourism resources and drawing spatial protection guidelines are vital to the ecotourism-based sustainable conservation of a village's cultural landscape.

8.3 Guidelines

Based on the relationship between a village's cultural landscape elements ecotourism resources, the village area can be divided into living, production, and ecological space, according to the residents' way of living. Meanwhile, the elements and resources can be classified by their spatial characteristics. Therefore, the ecotourism-based conservation guidelines of village cultural landscape take into account the living, production, and ecological spaces, as well as intangible culture, which is more related to people's spiritual space (*see* Fig. 8.2). The detailed description of the four guidelines follows below.

8.3.1 Ecological Space Guidelines

The natural ecological space of the village should be strictly protected by the following guidelines while developing village ecotourism:

1. Village natural elements can be protected with reasonable spatial distribution based on the geomorphologic and climatic conditions.
2. Tourism should be limited to eco-friendly activities, such as hiking, sightseeing, and image recording.
3. Country and ecology museums are applicable to village cultural landscape protection based on the ecological model.

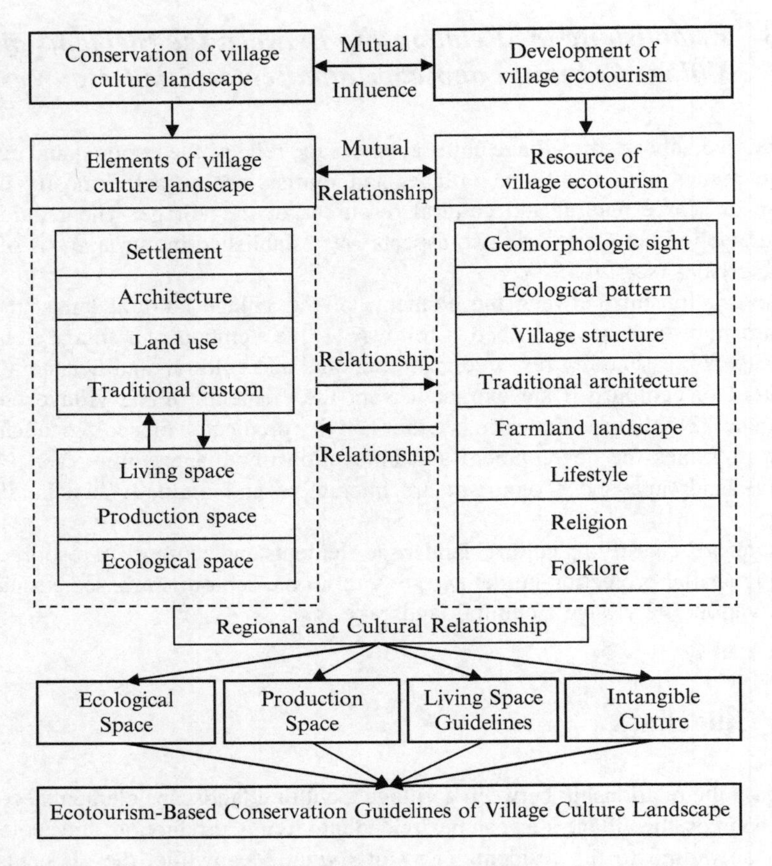

Fig. 8.2 Ecotourism-based conservation guidelines of the village cultural landscape

8.3.2 Production Space Guidelines

The village's production space should be reasonably protected by the following guidelines during ecotourism development:

1. The design of the tourism product should be ecology and production based. It is always better to maintain the agricultural structure and tradition of the village.
2. The scale, mode, and types of village tourism activity should be controlled in order to protect the village's production space. It is better to share and experience the villagers' seasonal farming activities, such as cultivating, weeding, fertilizing, irrigating, and harvesting.

8.3.3 Living Space Guidelines

Living space, such as the village spatial pattern, traditional architecture, courtyard, place of worship, and other public places, should be protected by following guidelines:

1. The village spatial pattern should be protected when developing village ecotourism. It is better to design the tourism product based on improving tourist cognizance of the traditional village landscape and its regional characteristics and enhancing tourists' environmental consciousness pertaining to protection.
2. The scale and style of traditional village architecture should be controlled while maintaining, repairing, and rebuilding. The introduction of modern living facilities should be considered carefully vis-à-vis their impact on the traditional architecture, courtyard, and even the settlement pattern of the village.

8.3.4 Intangible Cultural Guidelines

Intangible culture includes song, music, drama, skills, crafts, and other facets of culture that can be recorded but not touched and interacted with (http://en.wikipedia.org/wiki/Intangible_cultural_heritage). Unlike material culture, which deals with ecological, living, and production spaces, intangible culture is more about the villagers themselves. Therefore, the conservation guidelines should specifically address the intangibility issue, which makes them different from the guidelines related to space.

1. The most important conservation strategy is to protect the inheritors and the transmission of village intangible culture, which pertains to the lifestyle, folklore, religion, and everyday habits of the villagers.
2. Tourism activity related to village intangible culture should be managed based on sustainability and the absence of negative effects. The design, scale, and mode of the tourism activity should consider the interests and participation of the inheritors and villagers. Particular effort should be exerted to cultivate awareness among tourists of the consciousness, self-identity, and self-confidence of the inheritors and villagers.

8.4 Case Study

We used Gaoxigou, the cave-dwelling village in Northern Shaanxi province, for our case study. We proposed a sustainable conservation planning and design for the village cultural landscape based on our study results. According to the conservation approach and guidelines explored above, the case study was completed in the following three aspects. First of all, the elements of the village cultural landscape were sorted under the categories of ecological space, production space, living space, and intangible culture. Then, the resources of village ecotourism were evaluated by means of classification, grading, and zoning. Finally, the planning and design for village ecotourism were worked out by applying the conservation guidelines.

8.4.1 Elements of the Gaoxigou Village Cultural Landscape

8.4.1.1 Elements of the Village Cultural Landscape in Ecological Space

Loess Plateau landscape. Gaoxigou is located on the Loess Plateau in the northern part of Shaanxi province, and it represents the typical landscape of that region. The beautiful images the Loess Plateau combined with hilly terrain give the observer a heady feeling (Fig. 8.3).

Gaoxigou ecological pattern. The effort of over half a century on the Gaoxigou ecological pattern—its primary feature—has transformed the village into a scenic northern frontier (with lush southern-type fields). All of the village's land resources, such as 40 mountains and hills and 21 trenches, have been rationally and effectively exploited. Terraces of excellent quality are nested on the peaks and gentle slopes, while ecological forests with various types of trees and shrubs have been planted on the steep slopes and weirs. The terraces and forests are located in the south dry region. Forestry, as a main part of Gaoxigou ecological construction, achieves a complete ecological pattern (Fig. 8.4).

8.4.1.2 Elements of the Village Cultural Landscape in Production Space

Water-soil conservation achievements. Threatened by soil erosion, Gaoxigou villagers implemented both biological and engineering measures. A first-rate soil-retaining dam (*see* Fig. 8.4), one of the typical engineering measures, irrigates the farmland and thus improves food production. Gaoxigou is proud of owning the first village-level water-soil conservation system. Numerous exhibitions have shown the self-reliant and hardworking spirit of the farmers and their great achievements in conquering nature.

Terraced farmland and orchards. In addition to building a dam to prevent water-soil erosion, Gaoxigou villagers have also been carving terraces on the loess hills and mountains and planting them to crops and orchards. It has taken them about 60 years to shape all 40 hills into terraces. The crops include millet, sorghum, mung beans, soy beans, and potatoes, while the fruit trees include apple, pear, apricot, and jujube (Fig. 8.5).

8.4.1.3 Elements of the Village Cultural Landscape in Living Space

Traditional cave dwellings. Cave dwellings are a unique and traditional form of architecture in Loess Plateau. Gaoxigou villagers have retained their own cave dwellings despite the new countryside construction project. Although Gaoxigou residents have improved their living facilities by installing bathrooms, biogas digesters, and solar heaters, the renovation took place without needing major demolition or construction. The "remodeled" cave dwelling is thus a combination

Fig. 8.3 The Loess Plateau landscape of Gaoxigou village

Fig. 8.4 Ecological pattern of Gaoxigou village

Fig. 8.5 Terraced farmland and orchards of Gaoxigou village

of traditional advantages and modern amenities. Moreover, Gaoxigou has retained its different types of cave dwellings, such as earth cave dwellings and stone cave dwellings (Fig. 8.6).

Tasty local food. Gaoxigou village dishes carry the exciting characteristics of Northern Shaanxi cuisine, such as *qianqianfan*, *helezi*, *yangyuchacha*, *piejie*, *yangzasui*, and *liangfen*. Equally interesting are the unique recipes and special cooking processes (Fig. 8.7).

Fig. 8.6 Earth and stone cave dwellings of Gaoxigou village

Fig. 8.7 Tasty local food of Gaoxigou village

8.4.1.4 Intangible Culture of the Village Landscape

Gaoxigou village owns a rich intangible culture, such as popular religions, folk songs and dances, folk handcraft, and village spirit.

Popular religion. Gaoxigou is located in Yulin Prefecture and is imbued with a strong religious atmosphere. The village is 25 km away from Baiyunguan temple, one of the seven Taoist Holy Places—an important heritage site under state protection. Additionally, there are many temples nearby that combine Buddhism with Taoism, including Qingyun, Daixing, Xiangyun, and Guandi temples. Moreover, the land-god and water-god temples of Gaoxigou are also visited by the villagers to worship.

Folk songs and dances. The folk songs of the village are of the famous Northern Shaanxi variety and are known by their local name, Xintianyou. The folk songs express the villagers' love and aspirations. Folk dances called *yangko* and waist drum are very popular in and around the village; they express happiness and other emotions (Fig. 8.8).

Folk handcraft. Paper-cutting, stone carving, embroidery, and clay sculpture are typical Gaoxigou handicrafts. Villagers are particularly fond of paper-cutting arts, and they glue their products to window panes, edges of *Kang* bed, cabinets, and around wedding rooms to express their hopes for a better life. The villagers also like to weave rugs using homemade equipment.

Fig. 8.8 *Yangko* and waist-drum dancing around Gaoxigou village

Fig. 8.9 Stories of the diligence and spirit of Gaoxigou village

Village spirit of diligence. From the 1950s to the present, Gaoxigou villagers have been trying their best to explore a comprehensive ecological project of soil and water conservation through biological and engineering measures. After 60 years and series of guidelines, their persistent and painstaking efforts have paid off: water and soil erosion were controlled, and land sources have been utilized effectively and rationally. Today, Gaoxigou owes its good living environment to the indomitable spirit of its villagers—their self-reliance, perseverance, and hard work. In over half a century of striving, Gaoxigou has amassed lots of stories about village spirit (Fig. 8.9).

8.4.2 Evaluation of Gaoxigou Ecotourism Resources

8.4.2.1 Classification and Grading of Ecotourism Resources

The classification and grading of Gaoxigou's ecotourism resources took into account their relationship with elements of its cultural landscape. The resources' tourism value, historic value, popularity, integrity, and scarcity were considered.

Table 8.1 Classification and grading of village ecotourism resources

Ecotourism resources		Tourism value	Historic value	Popularity	Integrity	Scarcity	Grade
Ecological space	Loess Plateau landscape	26	26	16	8	6	No. 1
	Ecological pattern	22	22	12	8	6	No. 2
Production space	Water-soil conservation achievements	22	18	12	8	8	No. 2
	Terraced farmland and orchards	22	18	12	8	6	No. 2
Living space	Cave dwellings	26	26	16	8	8	No. 1
	Tasty food	18	22	12	4	4	No. 3
Intangible culture	Popular religion	22	22	16	4	4	No. 2
	Folk song and dance	26	26	16	8	8	No. 1
	Folk handcraft	18	18	10	4	6	No. 3
	Diligent spirit	18	18	10	4	4	No. 3

Experts, tourists, planners, and villagers were asked to participate in the evaluation process. Finally, we came up with grading classification (Table 8.1). There were three first-grade ecotourism resources: the unique Loess Plateau landscape, traditional cave dwellings, and folk songs and dances. Four resources made second grade: ecological pattern, water-soil conservation achievements, terraced farmland and orchards, and popular religion. The three other resources were third grade.

8.4.2.2 Zoning Evaluation of Ecotourism Resources

Zoning evaluation is related to the classification, grading evaluation, and spatial distribution of ecotourism resources. The geomorphologic, ecologic, production space, village customs, and living space resources of Gaoxigou were zoned (Fig. 8.10a, b, c, d, e, f), supported by GIS spatial analysis. Based on the zoning evaluation, the spatial characteristics of Gaoxigou ecotourism resources were summarized as follows: (1) the ecotourism resources—all grades—were distributed over the village territory; (2) the grade of the ecotourism resources depends on the value of their various aspects; and (3) most of the high-grade ecotourism resources were distributed around living and ecological spaces. These characteristics reflect the natural environment and cultural landscape of cave-dwelling villages in the Loess Plateau.

8.4.3 Ecotourism Planning and Design

By integrating the identification of village cultural landscape elements and the evaluation of village ecotourism resources, the sustainable conservation planning and design for Gaoxigou were worked out, along with the ecotourism function

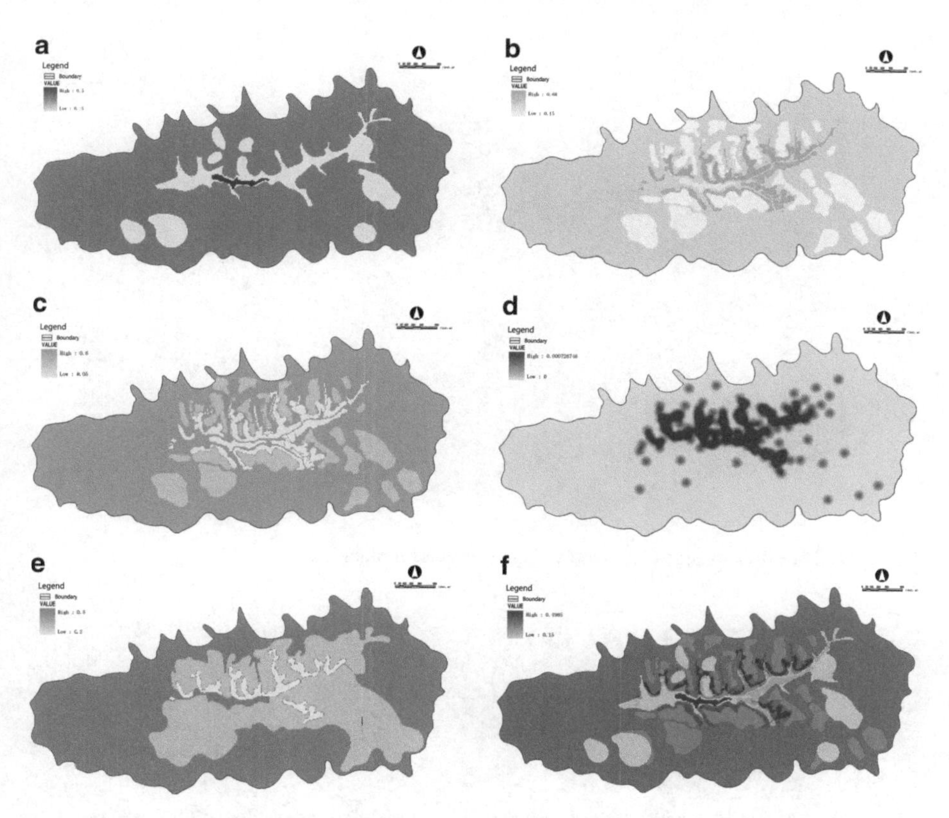

Fig. 8.10 Zoning evaluation of village ecotourism resources. (**a**) Geomorphologic resources zoning, (**b**) Ecologic resources zoning, (**c**) Production space resources zoning, (**d**) Village customs resources zoning, (**e**) Living space resources zoning, (**f**) Ecotourism resources zoning

zoning and product plan, by applying conservation planning guidelines. The main purpose of planning and design is to conserve the village's cultural landscape and develop the village economy through ecotourism in order to arrive at a sustainable development approach.

8.4.3.1 Zoning Plan and Design of Gaoxigou Village Ecotourism

The zoning plan and design of Gaoxigou village ecotourism were constructed in this section based on the analyses and guidelines above. The whole village domain was divided into five divisions: central service, cave-dwelling experience, reservoir side entertainment, agricultural experience, and ecological conservation (Fig. 8.11). The design of each division's functions is described below.

Central service division. Located in the middle of the village, this division plays a role in all the tourism services, such as receiving and managing tourists. Most of the

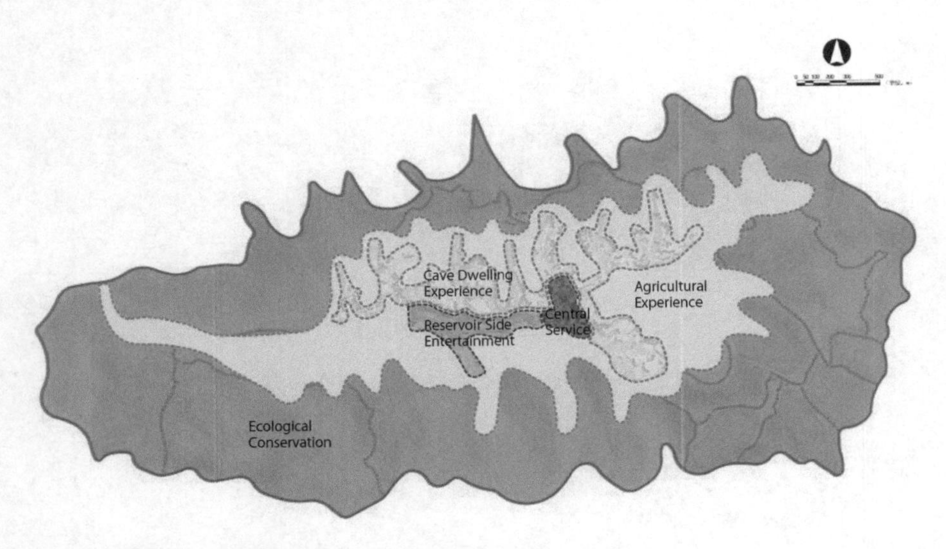

Fig. 8.11 Five divisions of Gaoxigou village ecotourism planning

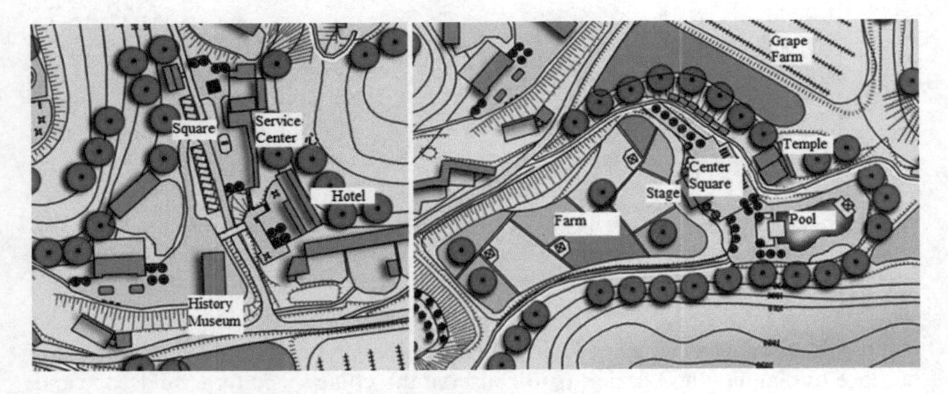

Fig. 8.12 Design of the village ecotourism service center and open space

tourism facilities are found here: restaurants, accommodations, the tourist service center (Fig. 8.12), public open space (Fig. 8.12), shopping street, and so on. The scale, style, color, height, and building materials of all the service facilities will be strictly controlled. The use of cave-dwelling architecture and local stone and bricks is encouraged.

Cave-dwelling experience division. This division includes all the village family yards and traditional cave dwellings. Because of the unique characteristics of cave dwelling, tourists relish the experience of living in them, as well as the chance to engage in artistic activity, such as sketching, drawing, painting, and sculpturing. The traditional cave dwellings have to be maintained, repaired, and rebuilt by this division, although modern architectural styles and building materials are prohibited.

Reservoir side entertainment division. Just beside the cave-dwelling experience division, there is a dam and reservoir that were constructed during 1970s. Tourists can enjoy many kinds of water-related activities on the reservoir side. However, tourism activities that pollute the water and facilities that affect the village cultural landscape are not allowed.

Agricultural experience division. Basically, this division occupies the village production space. Most of the area is farmland, and the majority of traditional agricultural activities occur here. Therefore, visitors can sightsee and experience firsthand the seasonal farm activity of the villagers instead of just watching farm activity shows.

Ecological conservation division. Located in the surrounding loess hills and mountains of the village, the land cover of this division is mostly forests, shrubs, and grass. It is very important to conserve the natural ecologic character of this division for the sake of the other four divisions and the entire region, which includes adjacent villages. The tourism activities of this division are hiking, adventuring, and sightseeing.

8.4.3.2 Products Planning of Gaoxigou Village Ecotourism

Related to the characteristics of village ecotourism resources and the zoning plan (*see* Fig. 8.11), five main ecotourism products were planned for Gaoxigou: ecological science education, cave-dwelling experience, loess landform adventure, seasonal agriculture experience, and hydrophilic leisure activities.

Ecological science education. This product is based on the good ecologic environment of Gaoxigou village and the accompanying water-soil conservation stories, as well as the village spirit behind them. This kind of education is important for all kinds of students and for adults who have no idea of the Loess Plateau environment evolution.

Cave-dwelling experience. There are many types of cave dwellings built from various kinds of building materials, such as earth, stone, and brick. This traditional architecture possesses the consciousness of ecological, green, and lower-carbon architecture, which are very popular modern architectural concepts. It is comfortable for living in, as well as for medical treatment, according to some reports on the experience.

Hydrophilic leisure activities. This product related to the reservoir side entertainment division. The reservoir is a priceless asset for this small loess cave-dwelling village because of the area's dry climate, limited precipitation, and soil erosion. Thus, the hydrophilic leisure activities are most welcome for both local and long-distance tourists.

Seasonal agriculture experience. The agricultural activities of Gaoxigou village are very interesting because of the unique landform and farmland distribution. All the agricultural activities, such as spring cultivation; summer weeding, fertilizing, and irrigating; and autumn harvesting, can accommodate tourist participation. Visitors who join those seasonal activities will learn tremendously from the experience.

Loess landform adventure. In the ecologic conservation division, there are many hills and mountains which have the typical loess landform characteristics. Tourists can join organized adventures to the unique landforms by hiking, sightseeing, and image recording.

8.5 Conclusion

This chapter discusses the guideline for the sustainable conservation of the village cultural landscape by means of village ecotourism. Gaoxigou, a cave-dwelling village, was used as a case study to analyze the key elements of the village cultural landscape, discuss the resources of village ecotourism, explore the relationship between the elements and resources, and examine and apply the guidelines for conservation planning and design.

With the implementation of the guidelines, we explored the sustainable conservation planning method of the cave-dwelling village's cultural landscape and provide some references for the future protection of the village cultural landscape.

With respect to the product planning drawings developed from the suggested design guidelines in this work, three conclusions were worked out:

1. For sustainable conservation planning, the components of the village cultural landscape can be divided into four aspects: ecologic space, production space, living space, and intangible culture.
2. Ecotourism is a sustainable way of conserving the village cultural landscape. To do so, the village's ecotourism resources should be evaluated through classification, grading, and zoning.
3. Four categories of conservation planning guidelines were proposed and practiced by the case study: ecologic space guidelines, production space guidelines, living space guidelines, and intangible culture guidelines.

References

Dang A, Lang H, Feng J (2008) Variation and conservation of cave dwelling architecture in north of loess plateau. World Archit 9:90–93

Dang A, Lv J, Zhao J (2009) The development and conservation of cave dwelling. China Homes 2:89–95

Dong N, Xu S, Qiu H et al (2009) Study on the deep exploitation model of rural ecotourism in Guangzhou. J South China Norm Univ (Nat Sci Ed) 1:116–120

Feng J (2009) Conservation of the ancient village is the most important work for culture protection. Consult Forum 1:18–19

Huang S, Lu W, Huang Z (2011) Conservation and transmission of Gaoding Dongzhai village culture landscape in Guibei. Shanxi Archit 04:1–2

Liu J, Li B, Lin X (2007) Direction of country eco-tourism of Yunnan province amalgamation of nationality culture and tourism geological resource. Yunnan Geogr Environ Res 01:112–117

Ma Q, Dang A, Zhao J (2012) Research on conservation planning and designing of village cultural landscape. Urban Flux 1:37–42

Ren J (2011) A survey of cultural landscape tourism. J Shijiazhuang Univ 06:82–86

Sauer C (1925) The morphology of landscape. Univ Calif Publ Geogr 22:19–53

Shan J (2008) Idea and method for vernacular architectural heritage conservation (Part I). City Plann Rev 12:33–39

Shan J (2009) Idea and method for vernacular architectural heritage conservation (Part II). City Plann Rev 1:57–66

Shan J (2010) Exploration and practice of country cultural scenery heritage protection. China Anc City 4:4–11

Sun X (2002) Landscape architecture from garden craft, garden art, landscape gardening – to landscape architecture, earthscape planning. J Chin Landsc Archit 4:8–13

Wang J (2007) Conservation of traditional village and village culture along with the new-rural development. China Village Dis 5:104–107

Wang Y (2009) Local identities of landscaped garden: interpretation of traditional territorial-culture landscape. Archit J 12:94–96

Wu X (2010) Study on the village cultural landscape: conception, constituents and influence. Anc Mod Agric 4:84–93

Zhang J (2008) Ecotourism. China Travel & Tourism Press, Beijing

Zhang W (2011) Conservation of ethnic group village cultural landscape. China's Ethn Groups 4:58–59

Zhou N (2006) New challenges of world heritage conservation: cultural landscape. Hum Geogr 5:61–65

Chapter 9
Characteristic of Sustainable Location for Townhouse Development in Bangkok and Greater Metropolitan Area, Thailand

Siwaporn Klimalai and Kiyoko Kanki

Abstract This study examines the relationship between the location and development of townhouses in the Bangkok Metropolitan Region, which has expanded considerably to cover a greater area. This rapid growth has caused many problems, particularly in the environment and quality of living for residents. The townhouse is an alternative residential type that is a response to the needs of the middle class. The objective of the study is to examine location characteristics for optimal townhouse development. The analysis explored the proper location for townhouses to provide understanding of the relationship between location and the time periods of housing development since 1967. The research revealed that each area has different characteristics that influenced housing settlements, and the numbers of projects differ significantly in each part of the city, depending on the urban context and the social and economic situations. Moreover, townhouse settlements have been diffused from inner-city Bangkok to the suburbs, while the suburban fringe has been in need of proper residential areas from 1977 to 2006 (the end date of the study). This movement demonstrates trends in the location of townhouses or housing developments that may help private developers in choosing potential sustainable locations for their new projects. This research may also aid city planners in formulating urban regulation guidelines to control or support townhouse development in the future.

Keywords Sustainable location • Location diffusion • Townhouse projects

S. Klimalai (✉) • K. Kanki
Graduate School of Engineering, Kyoto University, C-Cluster C2-414,
Katsura Campus, Nishikyo-ku, Kyoto 615-8540, Japan
e-mail: sklinmalai@gmail.com; kanki@archi.kyoto-u.ac.jp

M. Kawakami et al. (eds.), *Spatial Planning and Sustainable Development: Approaches for Achieving Sustainable Urban Form in Asian Cities*, Strategies for Sustainability,
DOI 10.1007/978-94-007-5922-0_9,

9.1 Introduction

The objective of the study is to examine location characteristics in the development process of townhouses in the BMR through looking at the relationship between socioeconomics and physical housing development according to timeline, area classification, and townhouse diffusion. We aim to clarify the characteristics for townhouse location that are necessary in helping private developers choose potential and sustainable locations for their new projects. The results may also aid city planners in making urban regulation guidelines to control or support this development. Thus, the quality of living and future urban environment may be improved[1] (The Government Gazette for Residential Design 2007).

Appropriate residential location is the key to housing development in Bangkok and the greater area. It is crucial for stakeholder satisfaction—that is, residents, private developers, and city planners—and it also promotes a good urban environment and improves living quality. Therefore, locational studies in residential development constitute an approach to understanding the tendencies and characteristics of housing locations in such a large capital city as Bangkok.

In this chapter, we describe a unique framework for discussion on residential location issue as shown in Fig. 9.1. Land acquisition is the initial stage in the housing development process before developers can do a feasibility study and proceed with design, construction, and sales. The *location* is as an important factor in the accomplishment of a *townhouse* project: if the location is not appropriate, the projects will be unsuccessful and will result in many further problems. For instance, *developers* can lose their profit margins (an economic problem) and neglect projects; *consumers*, namely, buyers or residents, may lose their money and be unable to move into the houses (a social problem); and finally, the *urban* area will become unsustainable, and waste urban developed areas (an urban problem) incur crime risks on the land (a social problem).

A *townhouse project* consists of many townhouse blocks that contain many townhouse units. Not only does the residential building code apply to townhouse design, but also residential environment was established for providing basic quality of life for residents' community living with sound environmental design for housing and townhouse projects. *Sustainable location* in this study means a location supporting sustainability and involving stakeholders. For instance, considerations would include convenient access for residents, development projects that would be successful for developers, and a reduction in social and economic problems resulting from abandoned projects. However, this research does not cover sustainability evaluation, and the results can assist in proposing guidelines for classifying housing locations and achieving sustainable location through further study.

---

[1] The Government Gazette for Residential Design (2007); Designated areas for garden, playground, and sports field are a minimum of 5% of the salable area; they should be of suitable size, location, and shape and easy to access. Land width is minimum 10 m and not allowed to be split into many small areas unless the area is larger than 1 rai (1,600 m^2). Perennial plants should be provided for more than 25% of the garden area according to landscape theory design.

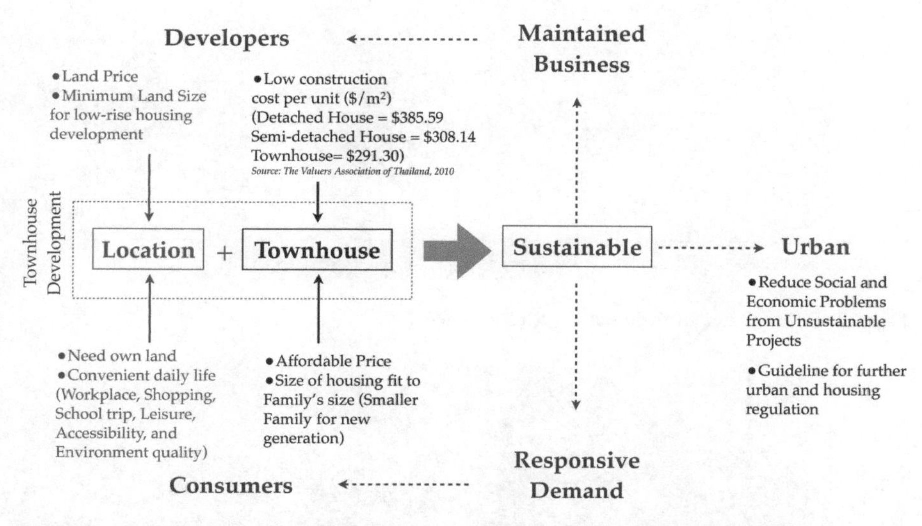

Fig. 9.1 Relationship between townhouse development and location to sustainability

This chapter is structured as follows. The background of the study area and research design are described in Sect. 9.2, and the history of the townscape development projects is analyzed in Sect. 9.3. The analysis of location characteristics is discussed in Sect .9.4. Section 9.5 presents our conclusions.

9.2 Background and Research Design

9.2.1 Bangkok Metropolitan Area

As a metropolis of Thailand since 1782, Bangkok is located in the central region of the country. This city has expanded considerably to cover a large area, beginning from the center of the city beside the Chao Phraya River (Fig. 9.2). The high rate of expansion has caused many problems in residents' living and housing developments: heavy traffic, high cost of living and traveling, high density of residential buildings, and land-size limitations. Because the first official urban city-planning act of the Bangkok Metropolitan Area was established in 1975 (Department of City Planning Bangkok Metropolitan Administrator 2004), the infrastructure is inadequate in many areas, and the plan has improved more slowly than urban development has occurred (Fig. 9.3).

Given these problems in the urbanization of Bangkok, middle- to low-income people who still work in the city have difficulty finding housing that they can afford. Townhouses offer an alternative that may be responsive to the enumerated

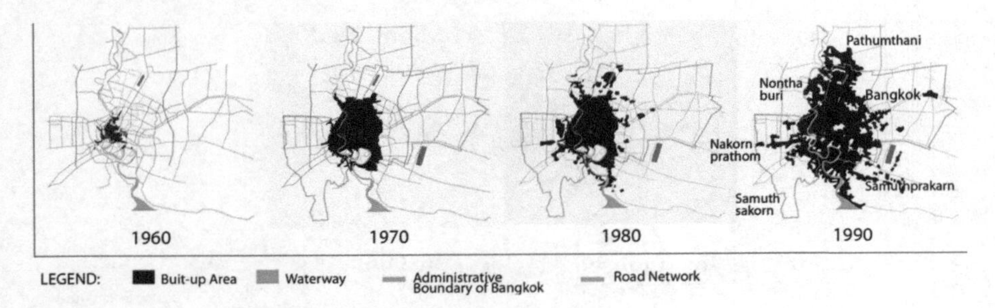

Fig. 9.2 Built-up area diffusion in BMR (1960–1990)

Fig. 9.3 Sample abandoned townhouse projects

limitations and to the demand from urban residents. Townhouses are also a part of the mainstream housing market in the Bangkok Metropolitan Region (BMR). Typical housing in the BMR consists of two main types: low-rise housing (detached houses and townhouses) and high-rise housing (apartments and condominiums). Low-rise housing accounts for 39.62% in the Bangkok vicinity and 36.84% in Bangkok City, while high-rise housing constitutes 4.7% in the vicinity and 18.84% in the city proper (Real Estate Information Center 2010). The typical townhouse in Thailand has the following qualifications per unit: a land size of 64–160 m^2; 2–4 stories; a townhouse block width of not over 40 m; minimum width and length per unit at 4 and 24 m, respectively; minimum open space at 30% of land size; and offset in the front and rear at 3 and 2 m, respectively (Prasarnthai and Sani 2003).

In the next subsection, we will discuss the research design.

9.2.2 Research Approach

This chapter is based on physical evidence of the influences on the location diffusion of townhouse projects in Bangkok from 1977 to 2006. In addition, this study was conducted by statistical methodology, using stratified random sampling, with three indicators: (1) the period of townhouse development, (2) townhouse

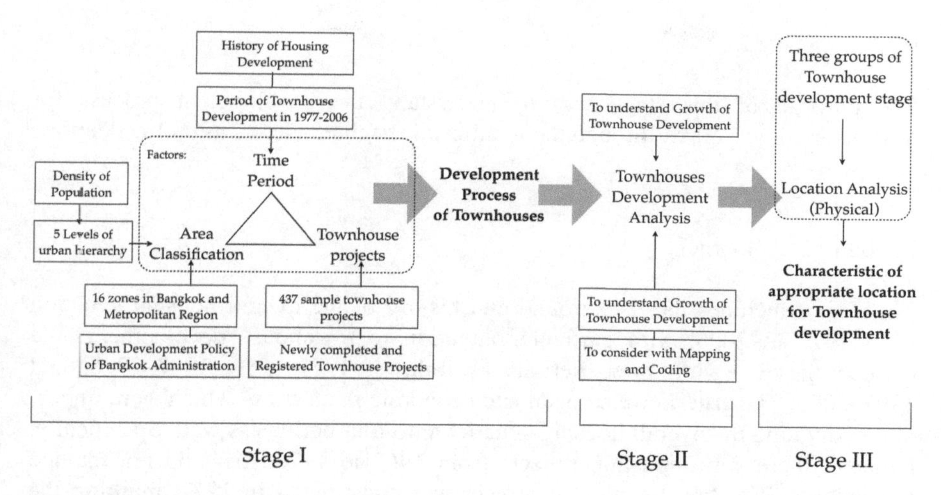

Fig. 9.4 Research framework

projects' diffusion, and (3) the zoning classifications of the Department of City Planning in Bangkok, by which the administrator divided these areas into 17 zones for investigation from 1999 to 2004 according to different purposes of development. The samples were 437 projects, part of the total number of townhouse projects in Bangkok, selected from newly completed and registered townhouse projects. Our data sources were the Real Estate Information Center (REIC),[2] the Bangkok Metropolitan Administrator (BMA) database in each district, and the National Housing Authority. These organizations collect statistical information on every housing type in the BMR, and it is published online yearly and quarterly.

The research framework is shown in Fig. 9.4. We proceeded with this study along three main lines, as detailed in the following sections: first, the development of low-rise housing in the BMR was examined by factors influencing the development, and the areas in the BMR were also classified with factor area classification. Second, townhouse project diffusion and grouping of locations was assembled by means of the results from area classification and the factor statistics of the townhouse projects by mapping and dotting in stage 2. Finally, location characteristics for townhouse development in the BMR were discovered by sample selection from each group of townhouses in stage 3, as derived from the results in stage 2.

[2] The registration system of housing projects in Bangkok: normally, housing projects were completely constructed, and developers or contractors had to inform the Bangkok Metropolitan Administrator in their districts. The Real Estate Information Center (REIC) was established by the Ministry of Finance (MOF) and initially operated by the Government Housing Bank (GHB) to collect and gather real estate data on the city, as well as national information. Moreover, they publish real estate statistical information services; for instance, they provide online information such as a real estate journal, news, research reports, and yearly and quarterly real estate indexes.

9.2.2.1 Stage I

The purpose of this stage was to understand the development process for townhouses in the BMR through the relationship of three factors as outlined below.

Period of Development

This factor includes the background and history of the overall housing situation from 1977 to 2006, with particular attention to townhouse development. We divided the time into three periods as follows: 1977–1986, 1987–1996, and 1997–2006. The criteria were social and economic fluctuation, which were important in dividing the overall housing situation into four periods as well. Specifically, although there were housing projects from 1967 to 1976, they did not include townhouses. The first townhouse project was constructed in 1977, marking the beginning of townhouse development.

Area Classification

This factor was considered in both sub-factors. First, Bangkok City was classified into 12 zones according to the urban development policy of the Department of City Planning Bangkok Metropolitan Administrator in 2004 for improving each part of the city. However, the vicinity around Bangkok, that is, the greater metropolitan area, is also in the scope of this study because townhouses have spread out from the city into the surrounding areas. They were divided into five areas according to administrative districts for a total of 17 zones. Second, those 17 zones were classified according to an overlay of five levels of urban hierarchy based on population density, namely, inner city, urban fringe, suburban fringe, suburban, and further suburban. This criterion relies on the urban planning guideline of the Bangkok Metropolitan Department of City Planning.

Statistics for Townhouse Projects

The number of townhouse projects in this study was limited by the scarcity of records concerning projects before 1982 because there were not enough systematic real estate reports to provide information. Thus, we derived 437 projects for the sample from 1977 to 2006 according to this qualification: newly completed and registered townhouse projects by a public or private organization that took the responsibility to collect statistics for housing estates. For instance, the projects were certified with REIC and BMA. In addition, they were collected from the National Housing Authority for 1985 because of the limitation of data sources.

In order to clarify the relationships among the three factors above, a mapping and coding method was used for interpreting the overlapping of the three layers of factors so that the diffusion of townhouse projects would be revealed.

9.2.2.2 Stage II

This stage included an analysis of the figures and tables derived from data collection for understanding the development process of townhouses. This part of the research fulfilled three purposes as follows:

To Display the Results of Mapping and Coding

We present the townhouses' location diffusion according to the statistics from the townhouse project area classification and the periods of townhouse development. Moreover, it showed the location representing the maximum diffusion of townhouses and revealed the areas of the highest aggregation of townhouses.

To Understand the Growth in Townhouse Development

For studying the continuity of townhouse development growth in every zone of the BMR, the results were gained from establishing a relationship between the five urban hierarchies combined with the three periods of development, and the statistics of the townhouse units as well as adding characteristics from each zone reveal the emphasis in each area.

To Analyze the Tendencies of Townhouse Development

The criteria here included the number of projects and continuity of the projects' construction across three periods of time. We discovered the prospective area that was most suitable for developing townhouses in Bangkok.

9.2.2.3 Stage III

The purpose of this stage was to develop a location analysis of the prospective area selection to derive the characteristics of the most appropriate locations for townhouse development.

The sample area was selected by synthesizing the outcomes from the first and second stages, combined with a literature review of residential location theory and related research in order to account for such factors as transportation, density

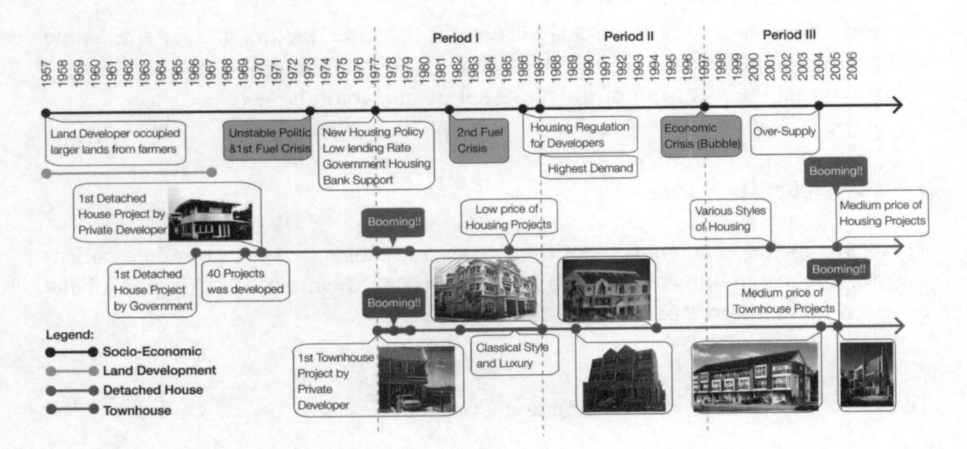

Fig. 9.5 Low-rise housing development in Bangkok and metropolitan region

of population, and urban facilities. A summary of the results might reveal characteristics among townhouse development, the location of projects, and time periods. Moreover, the results could be used as partial considerations for future housing projects in both the public and private sector.

9.3 The Development Process of Townhouses in BMR

9.3.1 Period of Development

In a primitive capital city like Bangkok, every part of the town has many types of accommodations. Townhouses have been in both urban and suburban areas since 1977, after the end of the land development era in 1976 and the beginning of detached housing projects by private developers in 1970 (Fig. 9.5). The timeline of the low-rise housing development reveals significant influences, from socioeconomic conditions to housing developments. Since a new housing policy was launched with financial support from the government housing bank in 1977, housing experienced a boom until 1985 when the price of housing declined because of the effects from the fuel crisis. While the townhouse styles became more varied and modern in the third period because the economic crisis in 1997 led to an oversupply in the housing market, new projects must satisfy customer needs and still be highly competitive. Therefore, socioeconomic factors directly influence the physical style and price of housing developments, as shown in Fig. 9.5.

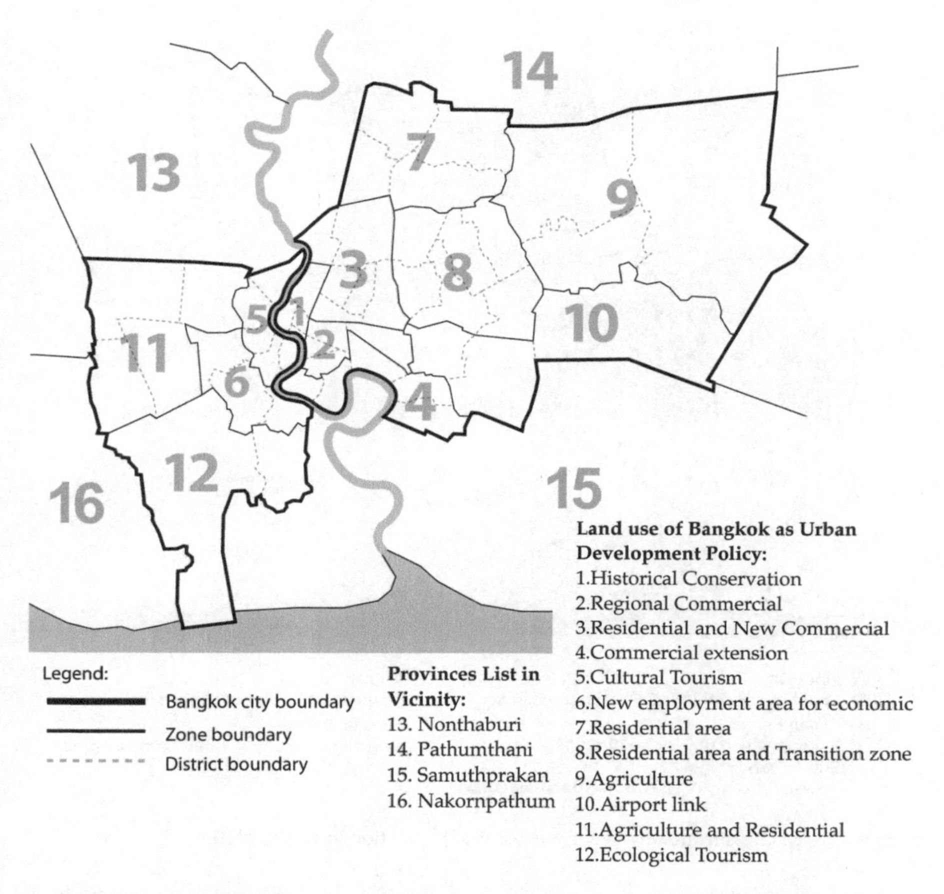

Fig. 9.6 Area classification by urban management policy (Department of City Planning Bangkok Metropolitan Administrator 2004)

9.3.2 Area Classification

In the past, townhouse projects have been built repeatedly in the same zones. Nowadays, however, housing in Bangkok City has changed in both style and location from the past. Every part of the city has been developed under the administration of the Department of City Planning for Metropolitan Bangkok since 2004. The 17 zones previously mentioned have different goals: the north of the city, for example, Zones 7 and 8, is intended mainly for residential areas (Fig. 9.6). This plan promotes urban management and is supposed to be a compact city that conforms to the area classification of the new land use policy. Moreover, the Department of Provincial Administration also collected data on population density in 2010 that may be used to divide the city according to land classification: inner city, urban fringe, suburban fringe, suburb, and further suburb. This approach can relate population density to land classification for identifying the classes of land in each zone, as shown in Fig. 9.6.

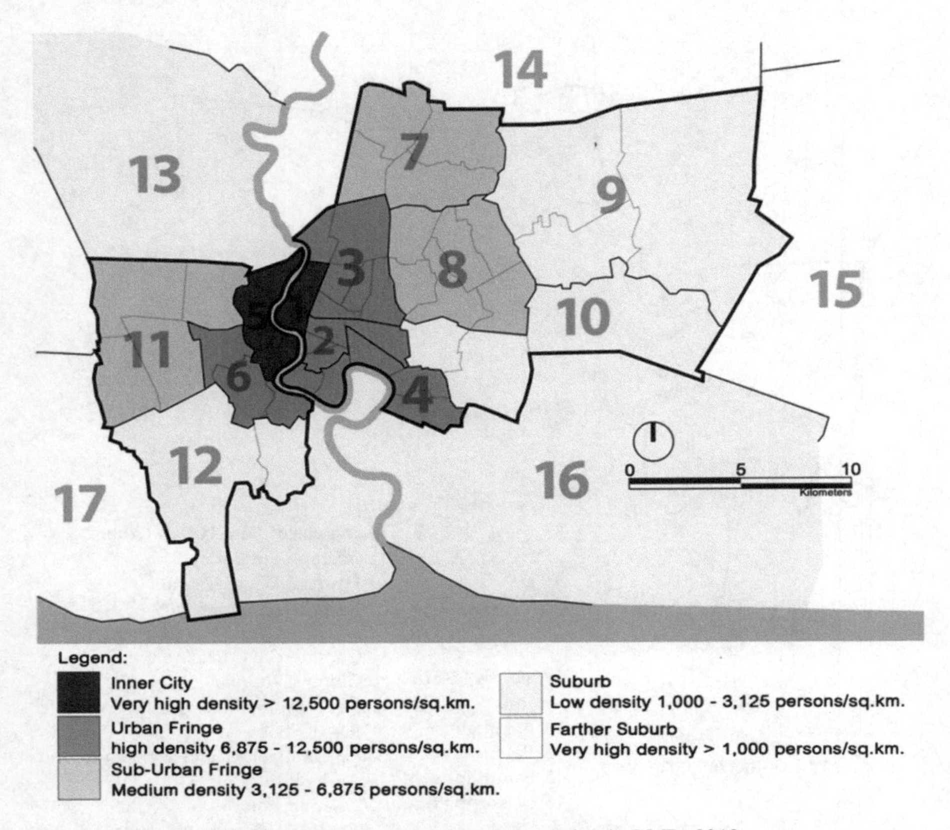

Fig. 9.7 Area classification by average density of population in BMR, 2010

Population density can reflect the urban hierarchy levels in the BMR. The city has already identified the population average density at each level of the hierarchy, as shown in Fig. 9.7. When the map of area classification by urban policy is overlaid on the map of population density, the resulting configuration reveals that the appropriate areas for zoning as residential are on the urban and suburban fringes of the BMR, which has population densities of 3,125–12,500 persons per square km.

9.3.3 Diffusion of Townhouse Projects

Some uncontrollable developments of the city have caused such problems as pollution, congestion, and overcrowding. Nevertheless, the demand for residences in town is still increasing. Although the Real Estate Information Center (REIC) found that the rate of overall housing construction in the BMR in January 2010 had

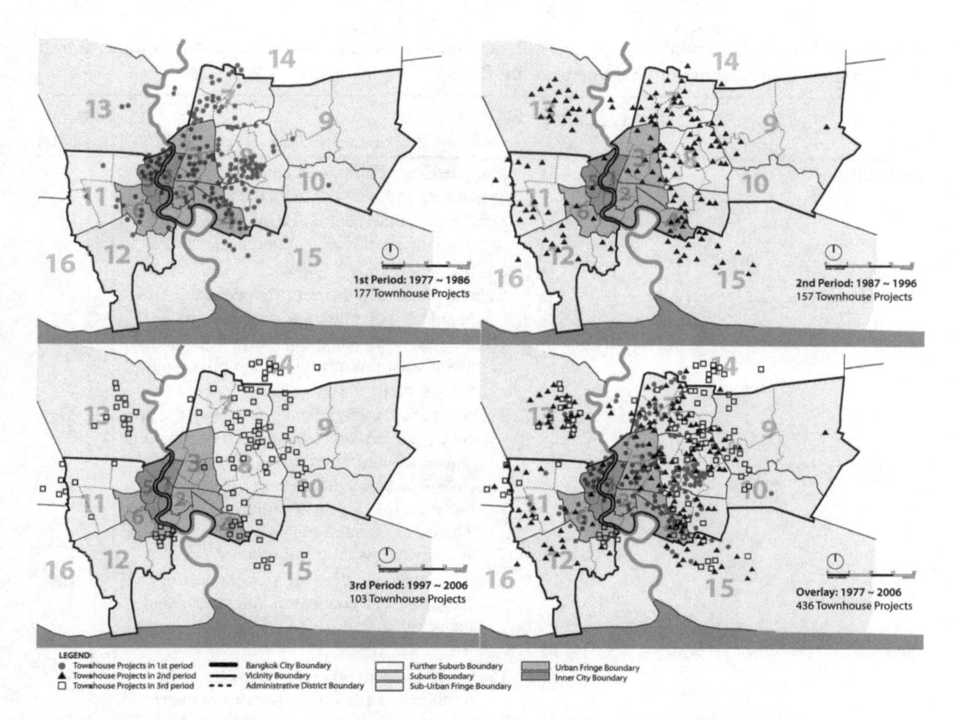

Fig. 9.8 Number of samples of townhouse projects in 17 zones during 1977–2006

decreased 41% from December 2009, townhouse construction peaked at the same period, especially in the southern, eastern, and northern parts of the city, in that order. Although the housing estate has been developed since 1967, the first townhouse projects were built in 1978 in the Bangkapi District, Zone 8 (Horayangura et al. 1993). Townhouse projects have been expanding from the original location in the first period, with new projects gradually located farther away from the center. As reported by the statistics from townhouse projects, area classification, and time periods, this phenomenon was emphasized in an empirical study about the general physical characteristics of zones according to a field survey (Klinmalai 2006). Townhouse diffusion occurred most significantly from the inner city (Zones 1 and 5) and a part of the urban fringe (Zones 2 and 6) because of a very high density of population and historical buildings (Fig. 9.8 and Table 9.1) that prevented new townhouse projects.

Similar conditions prevailed in the further suburb area (Zones 15 and 17), in a part of the suburban fringe (Zone 11), and in part of the suburb area (Zone 12). In these areas, the main land use was still agricultural, and housing development could be uncertain because of economic fluctuation. The coherence of townhouse development and the settlement of townhouse projects can be seen in Table 9.1.

Table 9.1 Number of townhouse projects in each location and according to each phenomenon of urban context and housing development in the BMR

Location	Zone	Period I	II	III	Characteristic of zone (field survey)	Total (units)
Inner city	1	2	0	0	Many heritage buildings, center of valuable culture and old commercial shop houses	2
	5	15	2	0	Similar to Zone 1, cultural conservation area, many original local communities	17
Summary		17	2	0		19
Urban fringe	2	12	0	0	Center of business district (high density of population and buildings); many office buildings, big shopping malls, apartments, and condominiums; high land prices and narrow main streets	12
	3	23	13	1	Support for commercial extension from the city center, mixed-land use between commercial and residential buildings	37
	4	13	9	6	Nearby Chao Phraya River and industrial ports, logistics area for distributing goods eastern Thailand; "slums" and wide streets to support industrial transportation	28
	6	8	3	7	Center of transportation on western side of Bangkok; can connect to industrial area in the south	18
Summary		56	25	14		95
Suburban fringe	7	27	25	15	Connects and supports city extension to the north and northeast; sub-streets network connects main roads; many kinds of accommodation	67
	8	41	26	19	Similar to Zone 7; many facilities for communities, but a weakness is congestion from high use of road network	86
	11	4	11	3	Mixed-use area between agricultural and residential area; most accommodations are detached houses >400 m^2.	18
Summary		72	62	37		171
Suburban	9	3	8	12	Half of this area still natural environment and used for agriculture; low density of population	23
	10	15	11	8	Mixed use between commercial, residential, and industrial	34
	12	2	10	1	Ecotourism area, many mega-housing projects	3
	13	2	20	14	Most areas used for agriculture; urbanization is next to Bangkok	36
	16	6	13	7	General buildings are industrial factories; some fields	26
Summary		28	62	42		132
Further suburban	14	4	5	10	Government sector investment area; many kinds of accommodation, infrastructure, and facilities	19
	15	0	0	0	Lack of infrastructure and less urbanization; main land use is agriculture	0
	17	0	0	0	Lack of infrastructure and less urbanization; main land use is agriculture	1
Summary		4	6	10		20
Total		177	157	103		437

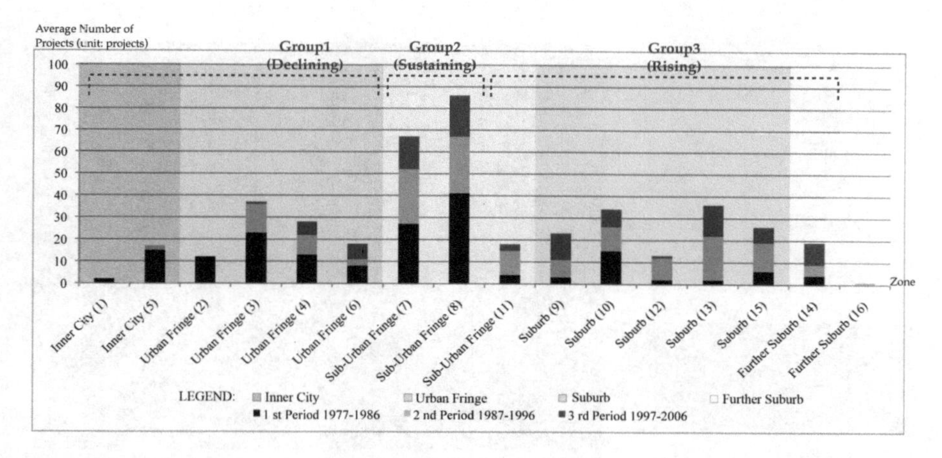

Fig. 9.9 Average number of projects in five location and three periods

9.4 Location Characteristics

An analysis of the number of housing projects in each zone and the urban hierarchy (Fig. 9.9) showed that there are three significant differences between townhouse development rates during different periods. First, a high growth in townhouse development occurred in the first period (1977—1986). Development growth dramatically decreased in the second and third periods in Zones 1–6 (inner city to urban fringe); this group comprises the declining area of townhouse development. Second, the growth of development is higher in Zones 7 and 8 than in the other zones in every period, and the number of projects gradually changes between periods. Zones 7 and 8, the suburban fringe, constitute the group of sustained townhouse development. Finally, the growth rate of townhouse development peaked in Zones 9–16 during the second period, and the number of projects increased. Zones 9–16 are the suburban fringe, suburbs, and further suburbs. Therefore, these groups are used to analyze location characteristics.

For the location character, Stegman (1969) proposed that accessibility could no longer be considered in one-dimensional terms as minimizing distance from the center business district. Senior and Wilson (1974) similarly demonstrated that the connection between residence and workplace need not depend on the relationship between the central business district and radial locations but could assume a crisscross pattern between any pair of zones. A more broadly defined theory of residential location (Harry 1977) considers the following factors: differences among household types, location of utility function (workplaces, shopping, school trips, or leisure), time for accessibility to work or utility functions, price of land and construction, and income earning. Hence, we recognize the above factors that can distinguish the character of location in different groups.

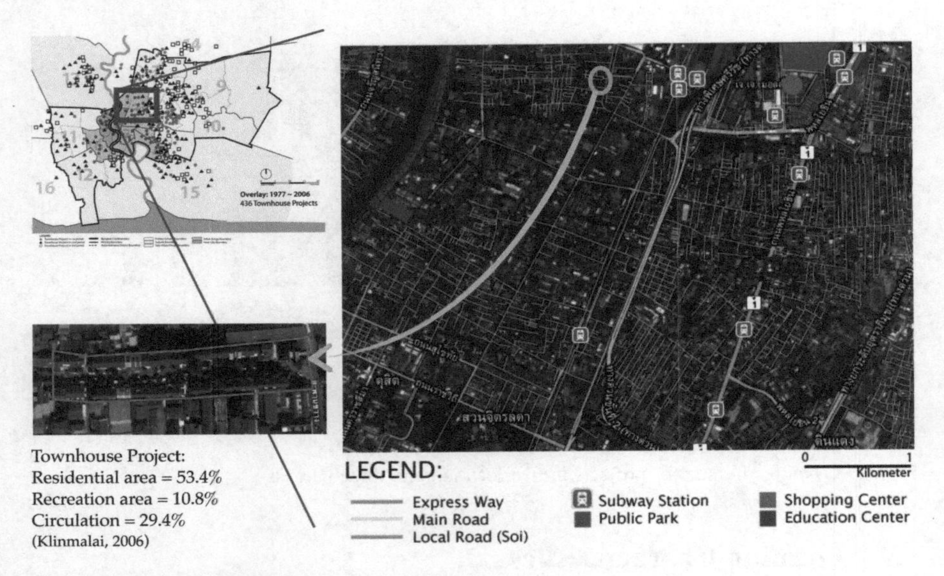

Fig. 9.10 Characteristics of location in Group 1

9.4.1 Group 1: Declining Area of Townhouse Development

In this group, the average density of population ranges from 6,875 persons/km^2 to over 12,500 persons/km^2. It covers six zones in the inner city and urban fringe that are devoted mostly to economic and historical areas and to new employment. When we consider the sample area in group 1 on Fig. 9.10, this area contains many small patches connected with local roads, with few main streets or expressways. There are many subway stations and enough public amenities, such as public parks, shopping malls, and schools. The road network has a grid pattern. Moreover, the shape of sample townhouse projects is rectangular and small scale. The land use area of the project is about 53.4% for residences, 10.8% for recreation, and 29.4% for circulation.

9.4.2 Group 2: Area of Sustained Townhouse Development

According to the previous section, Zones 3, 7, and 8 have a population density ranging from 3,125 to 8,000 people per square km, the same as Zone 11 on the western side of Bangkok. Although both areas have the same population density range, the growth of townhouse projects is dramatically different, being four times higher in Zones 3, 7, and 8 than in Zone 11. Therefore, we can see that other factors must be considered for townhouse location, such as infrastructure like main roads, expressways, and rail systems, which cut residents' commute time to work. Even

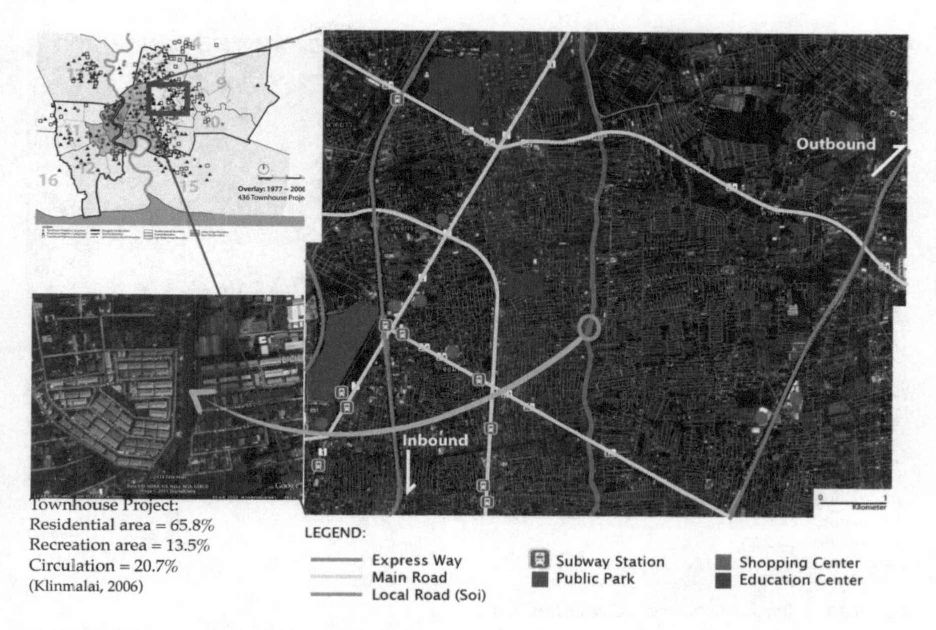

Fig. 9.11 Characteristics of location in Group 2

though the prospective zones do have not much access to skytrain or subway, the Bangkok Mass Rapid Transit (BTS) is planning for a great deal of expansion in the future. These projects were built before the skytrain plan. Added to the main transportation of these zones are roads and small streets inside the blocks (Fig. 9.11). Some of the geography inside the blocks is rough, with poor streets. Such small streets are a waste of time compared to the highway or BTS, but most of the projects are still located in this zone. However, utility functions, such as educational and commercial centers, are still the most important factors for residential location. According to Fig. 9.11, shopping centers are distributed in every big block of the zone, as well as several education facilities and places to work. Moreover, the plot shape of the project is polygonal, with 65.8% of it being residential and 13.5% recreational, with 20.7% for circulation.

9.4.3 Group 3: Area of Townhouse Development Growth

In this group, the average density of the population is under 3,125 persons/km^2, and it is located on the western suburban fringe and in the suburban and further suburb area. The plot composition has an organic form according to a natural waterway, which conforms to the mainland use policy as agriculture and ecological tourism. There is no mass transportation, such as subway or skytrain, but there are many main streets. Therefore, people who live in this area need to use private cars or

Fig. 9.12 Characteristics of location in Group 3

buses. However, it lacks public amenities, such as public parks and education centers. It consists of eight zones and covers a very large area. The tendency toward townhouse development is increasing, as shown in Fig. 9.12. The plot shape also follows the geographical pattern in an organic or free form. The residential area is 53.6% of the plot, the recreation area 3.88%, and the circulation 27.2%.

9.5 Conclusion and Discussion

Townhouse construction in Bangkok has rapidly changed in location from the inner city to the further suburban areas in every direction, especially north and east of the town. In each time period, social and economic situations directly influenced the housing estate development. Therefore, the result could reflect the townhouse development via townhouse location analysis and social or economic impact because location is at the top of residents' and developers' decision-making criteria in buying, living in, and producing townhouses. As seen in Table 9.1 and from the townhouse development and location diffusion in the study area mentioned above, Figs. 9.8 and 9.9 portray of the conclusions of our analysis. The results may be summarized according to the location of townhouse development in Table 9.2.

1. The suburban fringe as a prospected area continues to show suitable characteristics for developing townhouse projects, but its suitability may decrease in the future, as the line graph shows. According to Fig. 9.8, the number

Table 9.2 Comparison of the characteristics in the three groups of townhouse development by location

Group	Number of townhouse projects	Location	Townhouse project Res.	Rec.	Cir.	Characteristic
Declining	Considerably decreased after period I	Inner city, urban fringe	53.4%	10.8%	29.4%	Grid pattern of road network, many subway stations, small projects
Sustaining	Gradual change	Suburban fringe	65.8%	13.5%	20.7%	Various alternative accessibilities; enough public utilities facilities
Rising	Considerably increased after period I	Suburban, further suburb	53.6%	3.88%	27.2%	Many undeveloped areas; insufficient infrastructure and utilities; townhouse projects located near the main roads

Res. residential area, *Rec.* recreation area, *Cir.* circulation

of townhouse projects, which were developed through the three periods, depends on location—that is, how far away they are from the inner city. The greater the distance from the center of Bangkok City (the inner city), the greater the number of townhouse projects until the suburban fringe and, afterward, a significant decrease in the suburbs and further suburbs.

2. Most townhouse development was at first located near the inner city of Bangkok, whereas the results show the largest amount of more recent development in the BMR to be toward the suburban fringe. In addition, the city center had the longest background of townhouse development. This phenomenon could be referred to as an appropriate characteristic and environment for living through any crisis condition. Therefore, a potential location that is similar to the suburban fringe area of BMR may be a good prospect as a location for sustainable living.
3. The overall tendency for the number of townhouse projects in every zone was to decline in every period except for the further suburbs, in which the number increased slightly. However, during the first and second periods, townhouse development in the outbound suburban fringe area was the inverse of that in the inbound (Fig. 9.13). Economic fluctuation, the high price of townhouses in 2005 (Table 9.1), and the density of the population in different zones (Fig. 9.7) greatly reduced the numbers of townhouse projects.
4. The characteristics of sustainable locations for townhouse development in Bangkok are as follows: (a) not too close to mass rapid transit, such as BTS or subway, because of the limitation of high land prices, but multi-alternative in terms of accessibility by means of shortcut roads; (b) population density in the middle to low range; (c) enough large blocks for residential development with appropriate distance between two main roads (Fig. 9.12); and (d) enough utility functions and reliable distribution of goods in the area. This qualification is necessary for sustaining townhouse development and residential living.

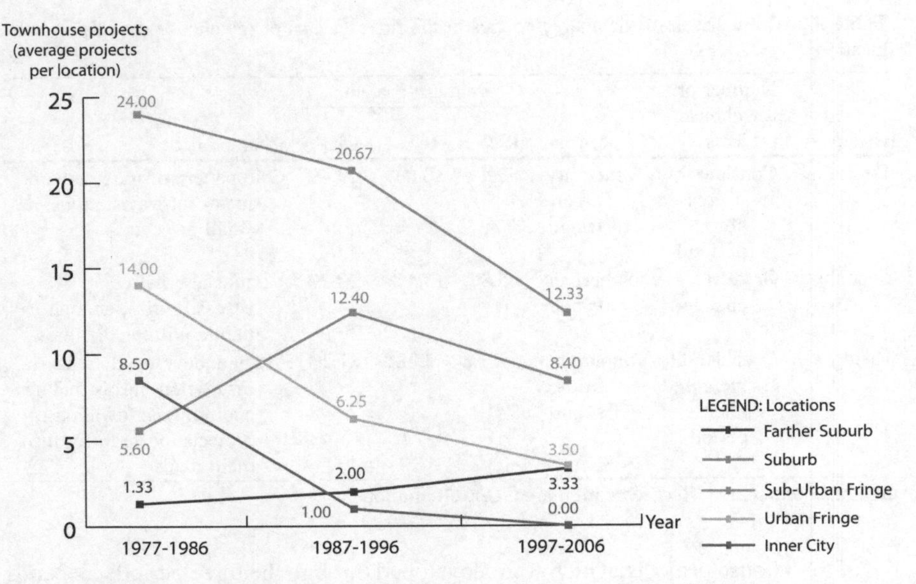

Fig. 9.13 Line graph of total number of projects in each location during 1977–2006

Finally, developers and the public are able to recognize the appropriate situations (in terms of the economy) and potential locations for generating a new townhouse project. This summary could be a guideline or a part of the criteria for consideration concerning housing development in the public or private sector in order to understand the context for residences in Bangkok.

References

Bangkok Land Development Commission (2007) The government gazette for residential design. Department of Land, Bangkok, pp 62–63

Department of City Planning Bangkok Metropolitan administrator (2004) Bangkok: delightful city development policy. Ammarin Printing, Bangkok

Harry WR (1977) A general of residential location theory. J Reg Sci Urban Econ 7:251–266

Horayangura V, Intaravichit K, Santi C, Inpantang W (1993) Development of architectural concept past present and future. Ammarin Printing, Bangkok

Klinmalai S (2006) Development and trends of townhouses for urban residents in Bangkok and Greater Metropolitan Area. Thammasat University, Bangkok

Prasarnthai S, Sani W (2003) Townhouse. Ammarin Printing, Bangkok

Real Estate Information Center (REIC) (2010) Housing report. http://www.reic.or.th. Accessed 18 June 2010

Senior ML, Wilson AG (1974) Disaggregated residential location models: some tests and further theoretical developments. In: Crippes EL (ed) Space-time concepts in urban and regional models. Pion, London, pp 141–172

Stegman MA (1969) Accessibility models and residential location. J Am Inst Plan 35:22–29

Chapter 10
Modeling Housing Demand Structure: An Example of Beijing

Xiao-lu Gao

Abstract In this chapter, a housing demand model is developed for classifying households with different needs and estimating the appropriate dwelling size for different households. The model is then applied to explore the housing demand of Beijing residents with questionnaire survey data. With a "surplus demand" indicator being incorporated to the model, the "true demands" of various families for dwelling size are predicted. The results contribute to the decision support of housing policies.

Keywords Housing demand • Housing consumption • Dwelling size • Interaction model • Partition analysis

10.1 Introduction

Housing policies serve the purpose of providing housing corresponding to the actual demand. Through appropriate means of planning, regulation, taxation, and providing public housing, it aims to ensure the balance of supply and demand so that the reasonable demand of families can be met. A clear understanding of the actual housing demand is the basis for the management of housing demand. In particular, how to define demand groups, and how to determine the suitable levels of housing provision for different people, are issues of great importance. However, existing studies on these issues have been inadequate.

So far, the impact of macroeconomic and geographical factors on housing demand and housing consumption has been emphasized, for example, per capita GDP, levels of urbanization, natural environments and dwelling customs, average

---

X.-l. Gao (✉)
Key Laboratory of Regional Sustainable Development Modeling, Institute of Geographic Sciences and Natural Resources Research, Chinese Academy of Sciences,
11A Datun, Chaoyang, Beijing 100101, China
e-mail: gaoxl@igsnrr.ac.cn

M. Kawakami et al. (eds.), *Spatial Planning and Sustainable Development: Approaches for Achieving Sustainable Urban Form in Asian Cities*, Strategies for Sustainability,
DOI 10.1007/978-94-007-5922-0_10, © Springer Science+Business Media Dordrecht 2013

income levels, financial circumstances, economic cycles, and so on (Jin and Zhang 2003; Wang and Ma 2005; Han and Xiao 2006; Xiong 2007; Shen 2007; Zhang 2007; Liu 2002). Many studies have proposed models for predicting the average level of future housing demand, and the total level of the housing demand was forecasted by multiplying the average level with the expected population (Lavender 1990; Tse et al. 1999; Akintoye and Skitmore 1994; Goh 1996, 1999; Thomas et al. 2008). The differences in demands across people groups were commonly perceived; however, the demands were naively separated in most cases, for example, into demands of high-income, middle-income, and low-income families; demands for survival, development, and enjoyment; demands for consumption and for investment; and demands for shelter, basic function, comfort, and luxury. Then the needs of each kind for housing size were decided based on expert knowledge or experience. Obviously, these methods are based on many subjective assumptions.

In recent years, many empirical studies of housing preference have been conducted with microeconomic approaches. Many studies examined the impacts of income, price, and living environments on housing choice behaviors, with particular interests on tenure and location choice, willingness to pay, etc. (Geoffrey 1973; Haurin and Lee 1989; Börsch-Supan and Pollakowski 1990; Zhang et al. 2005; Ioannides and Zable 2008). Most of the developed models have neglected the difference of consumption behaviors of different people; thus, the actual structure of housing demand was greatly simplified.

In practice, three kinds of approaches are common for determining the housing needs of people: to investigate the desired accommodations by questionnaire, to study the necessary dwelling space with an architectural approach, or to examine the levels or standards adopted by comparable countries and regions. Because it is quite difficult to justify the reasonableness of the results, the selection criterion of policy target strata and relevant policies are often criticized (Wang 2007).

This chapter attempts to solve these problems by modeling the housing consumption behaviors of families, whereby the demands can be classified and the respective size demands can be identified.

10.2 Theoretical Model

Housing demand is influenced by various factors including socioeconomic circumstances, geographic environments and living customs, market conditions, family size, income level, and previous dwelling status. From the perspective of housing consumption, the effects of geographic environments, living customs, and market conditions can be omitted if they are homogeneous over space and time. On the other hand, the impacts of some uncertain factors on the behaviors of families ought to be considered. For example, high-income families do not necessarily find large houses desirable. In fact, many people want to buy additional small houses for their children or parents or for investment purposes. This reflects an implicit need due to investment activities or expectations of changes in family structures.

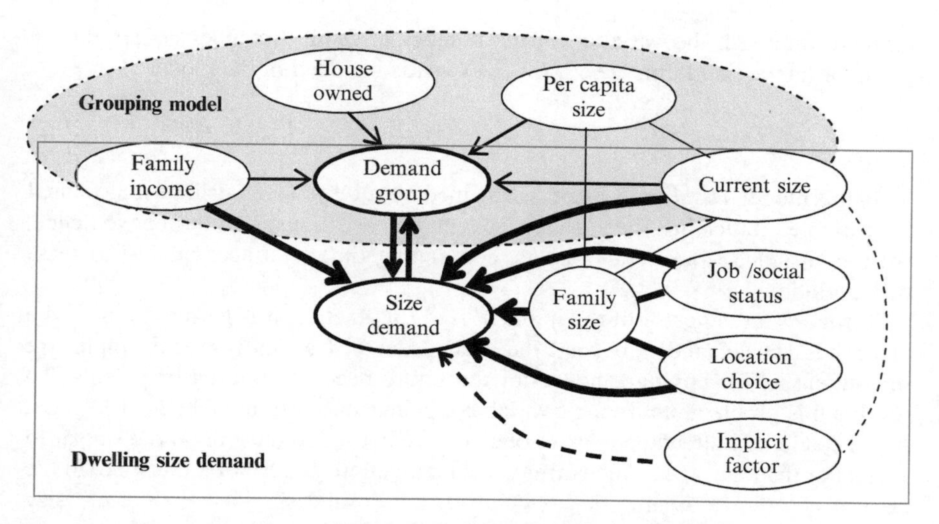

Fig. 10.1 Influencing factors of housing demand and their relationships

The housing consumption behaviors of urban households are determined by both the explicit factors, including family size, income level, and previous dwelling status, and the implicit indicator of investment and expectation of changes in family structure.

The consumption behaviors of urban families can be reflected by such indices as the willingness to buy houses and expected housing sizes. Assuming that the behaviors are based on rational decisions, it is possible to model housing demand with the explicit and implicit factors together. To represent the relationships, a two-level housing demand model is proposed (Fig. 10.1).

The upper level is designated as the housing demand grouping model. We describe the housing demand group as a function of the current dwelling status, family income, and family size as follows:

$$D_c = \mathrm{f}(n,\ s,\ I,\ s/m) \tag{10.1}$$

where D_c is the index of the housing demand group and n, s, and I denote whether a family owns a house or not, the current size of accommodation, and family income, respectively. Presumably, family size m is correlated to the size of the family's current accommodation, so the effect of family size on D_c is replaced by that of s/m (per capita dwelling space).

The lower level is termed the housing size demand model. In addition to demand groups (D_c), the observed size demand (D_s) is assumed to be affected by family size (m), the size of the current accommodation (s), family income (i), job rank and social status (j), location choice (l), and a term indicating the likelihood of investment or expectation of changes in family structure, which are indicted by c. It is assumed that family size m interacts with the other terms; that is, the impact of other terms differs with the variation in the number of family members. Therefore, a group of interaction terms (D_c, s, I, j, l, c) $\times m$ is included in the model:

$$D_s = \mathrm{g}(D_c,\ s,\ i,\ j,\ l,\ c,\ m,\ (D_c,\ s,\ i,\ j,\ l,\ c) \times m). \tag{10.2}$$

In this model, n, s, i, m, j, h can be obtained by survey. Term l can be represented by variables indicating the location advantages and transportation convenience, such as distance from urban centers, commuting time, and accessibility to mass transportation.

Term c is an indicator of the implicit need. In practice, it is hard to judge if the observed size demand is beyond the current need of a family, for example, for investment or for buying a house for the future need of children or parents. To resolve this problem, an implicit variable h is introduced in model (Eq. 10.2). Let $h = 1$ if a family already possesses one or more houses, and the observed size need is smaller than the size of the current accommodation, and $h = 0$ otherwise. In the case of $h = 1$, we assume that the sum of the desired size and the current size corresponds to the family size. Then, model (Eq. 10.2) becomes

$$D_s = \mathrm{g}(D_c, s, I, j, l, h, m, (D_c, s, I, j, l, h) \times m, h \times (D_s + s)). \tag{10.3}$$

Assuming that $D_s = \mathrm{g}(.)$ is linear, Eq. (10.3) simply turns to

$$D_s = \mathrm{g}(D_c, s, I, j, l, m, (D_c, s, I, j, l) \times m, h \times s). \tag{10.4}$$

With Eqs. (10.1) and (10.4), it is possible to classify housing needs and estimate the housing size needs of people belonging to respective groups. In Fig. 10.1, the interactions between the factors are indicated by line segments, and the uncertain relationships of investment and family structure with "current size" and "size demand" are shown by dotted lines.

10.3 Housing Size Demand in Beijing

10.3.1 Housing Demand Survey

Based on the theoretical model, we empirically studied the housing demand behaviors of families in Beijing. The data come from a large-scale questionnaire completed by a group of urban residents in 2005 (Zhang et al. 2006). The sample area covered the developed areas of eight central districts, new suburban towns, and residential areas (Fig. 10.2).

There were 117 subdistricts in total. The sampling was based on 1/1,000 of the population of subdistricts, and residents were randomly selected and interviewed in

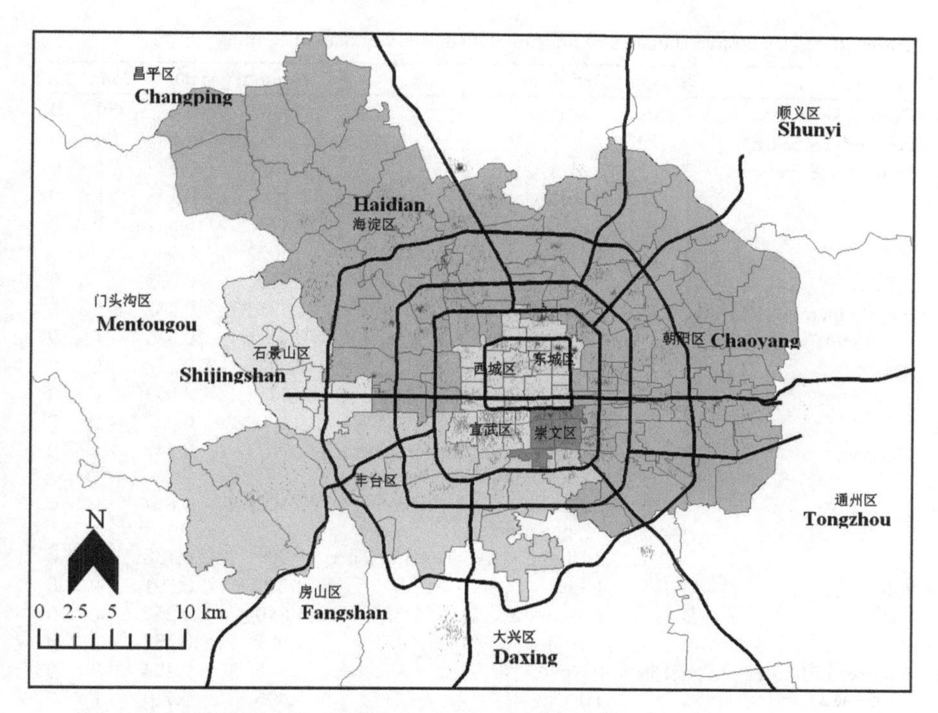

Fig. 10.2 Sample distribution of housing demand survey in Beijing

residential blocks. Approximately 10,000 residents were interviewed, and the valid sample size was about 7,000.

The survey items included current size of accommodation, family size, family income, job type and job rank, number of houses owned, plan to buy house(s), expected housing size, and maximal tolerable commuting time. Table 10.1 gives the statistics of them.

The average size of the current accommodation was 75.87 m^2, and the average per capita size was 34.63 m^2. Among the sample families, 12.4% did not own a house, 71.9% owned one, 14.4% owned two, and 1.4% owned three or more. These data are a good representation of the real situation of Beijing.

10.3.2 Analysis of Housing Demand Groups

A partition analysis was conducted based on Eq. (10.1), with "plan to buy house" being the judging criterion of grouping and with "current size of accommodation," "per capita size," "family income," and "number of houses owned" being the independent variables. As a result, eight groups were obtained, namely, G1 to G8 (Fig. 10.3).

Table 10.1 Descriptive statistics of housing demand survey data in Beijing

Item	Sample	Definition	Amount	Mean	Max	Min
Current size (m^2)	7,156			75.87	450	0
Per capita size (m^2)				34.63	300	0
Family size (persons)	7,647	1: = 1	1,265	0.177	1	0
		2: = 2	1,164	0.163	1	0
		3: = 3	3,877	0.542	1	0
		4: = 4	602	0.084	1	0
		5:≥5	251	0.035	1	0
Family income (RMB/month)[a]	7,159	1: < 3,000	1,875	0.262	1	0
		2:3,000–4,999	2,754	0.385	1	0
		3:5,000–9,999	1,992	0.278	1	0
		4:10,000–14,999	378	0.053	1	0
		5:≥15,000	160	0.022	1	0
Job type	7,148	1:Commercial, manufacturing, retired, unemployed	1,767	0.247	1	0
		2:Finance, IT, professional, government	4,392	0.614	1	0
		3:High-end services, freelance	989	0.138	1	0
Job rank	7,159	1:Low	4,232	0.591	1	0
		2:Medium	1,840	0.257	1	0
		3:High	1,087	0.152	1	0
Number of houses owned	7,159	0:None	945	0.124	1	0
		1:1	5,495	0.719	1	0
		2:2	1,097	0.144	1	0
		3:3 or more	107	0.014	1	0
Plan to buy house	7,647	0:No	4,286	0.561	1	0
		1:Yes	1,688	0.221	1	0
		2:Unsure	1,673	0.219	1	0
Expected size (m^2)	7,244			109.53	300	10
Expected per capita size (m^2)	7,244			51.11	300	3.33
Tolerable commuting time (min)	7,159			58.75	150	15

[a]The five categories correspond to family incomes of the lowest 25, 25–65, 65–90, 90–95, and 95–100%, respectively

Concerning the plan to buy a house, let the probability for urban families to choose "no," "yes," and "unsure" be p_0, p_1, and p_2. We define the probability of buying housing *comp* as

$$comP = p_1 + p_2/2, \ (comP \in [0, 1]).$$

Then, the intensity of the demand D_r is defined for each demand group:

$$D_r = comP/comP_0$$

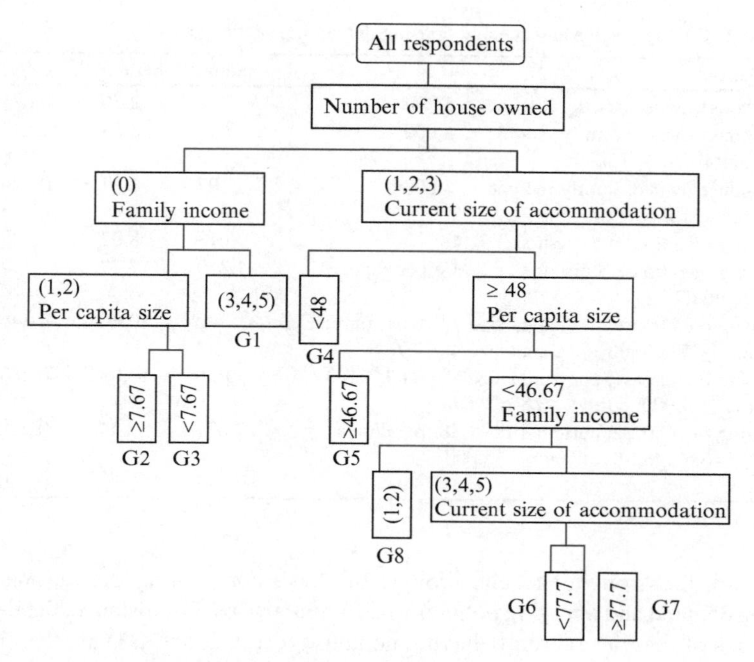

Fig. 10.3 Grouping of families w. r. t. plan to buy house

where *comp* is the probability of buying housing for each demand group and $comP_0$ is the average probability. The characteristics of each demand group, their probabilities of buying a house, and the intensities of demand are summarized in Table 10.2.

First, the grouping results give the threshold values for splitting housing demand groups. It was identified that people with houses significantly differ from those without their own houses. The monthly income of 5,000 RMB is the separating line for family incomes. 48 and 77.7 m^2 are thresholds for current size of accommodation, and so are 7.67 m^2 and 46.67 m^2 for per capita size. The architectural and policy meaning of these values is clear. In Beijing, 50 m^2 and 80 m^2 are approximately the average sizes of one-bedroom and two-bedroom apartments with full functions. The per capita size threshold of 7.67 m^2 is applicable to families who own their own housing, and it is very close to 7.5 m^2, the per capita size criterion used in the public housing program for lowest income families in Beijing. Moreover, for a three-person family, the per capita size of 46.67 m^2 is equivalent to a total of 140 m^2, and below 140 m^2 is the criterion of normal-class commercial housing whereby preferential tax rates are applied.

The characteristics of the eight groups are distinctive. For instance, G1 is composed of families who do not own a house but are economically capable of buying one. G2 and G3 are the housing-poor and housing-poorest families. G5 are wealthy families, and G8 are normal working families. As the demands significantly differ, the eight group classification results provide a useful standard of the eligible policy stratum in practice.

Table 10.2 Characteristics and specific housing demand of eight groups

Demand group	Sample	Sample %	*ComP*	D_r
G1: No house owned; family income ≥ 5,000	211	2.76	0.671	2.03
G2: No house owned; family income < 5,000; per capita size ≥ 7.67	637	8.33	0.499	1.51
G3: No house owned; family income < 5,000; per capita size < 7.67	97	1.27	0.526	1.59
G4: Having own house; current size < 48	613	8.02	0.388	1.17
G5: Having own house; current size ≥ 48; per capita size ≤ 46.67	1,359	17.77	0.337	1.02
G6: Having own house; per capita size < 46.67; family income ≥ 5,000; 48 ≤ current size < 77.7	800	10.46	0.346	1.05
G7: Having own house; per capita size < 46.67; family income ≥ 5,000; current size ≥ 77.7	1,059	13.85	0.283	0.86
G8: Having own house; current size ≥ 48; per capita size < 46.67; family income < 5,000	2,871	37.54	0.259	0.78
Total	7,647	100	0.330	1

Second, the grouping results provide the basis for judging the rationality of housing demand and applying policies to different groups. For instance, the demand intensities of families presently having no house (G1, G2, and G3) are the highest. The demand of G4 is also significant because the houses of these families are very small (less than 48 m^2), so the functions of the houses are likely to be inadequate. Group G6, which belongs to the middle- and high-income strata, also has strong demand, as the size of current accommodation is limited (48–77.7 m^2) in relationship to their incomes (≥5,000 RMB per month). We think that the demands of the above groups are reasonable.

On the other hand, the intensity of demand of families belonging to G5 (=1.02) is higher than the average level. As these families already own one or more houses and the per capita size is fairly large (≥46.67 m^2), the demand embodies the desire for a luxurious life or for investment. At present, this group composes a significant percentage of urban families (nearly 18%). They should be the policy target of regulations on irrational housing consumption and speculation.

Interestingly, the relationship of income and expected housing size was suggested by the analysis results. As Table 10.2 shows, the demand intensities of G7 and G8 are relatively low, indicating that the current accommodations are satisfactory. Scrutinizing the characteristics of the two groups, we may find that for families whose monthly income is above 5,000 RMB, houses of no smaller than 77.7 m^2 (approximately 80 m^2) are satisfactory; for families whose monthly income is less than 5,000 RMB, houses of no smaller than 48 m^2 (approximately 50 m^2) are satisfactory. These results have added to our understanding of the appropriate size demand of urban families.

The partitioning of the eight demand groups is pragmatically important because it enables the stipulation of policies corresponding to the demand of different family groups; at the same time, it provides convincing criteria for selecting which families are eligible for public aid. Specifically speaking, subsidized housing and public housing programs should be applied for groups G2, G3, and G4; the needs of groups G1, G6, G7, and G8 should be solved through the housing markets; and the demands of group G5, especially which are a result of speculation, should be subject to control. The figures in Table 10.2 suggest that the present shares of the three groups of families in Beijing are 20, 60, and 20%, respectively.

10.3.3 The Needs for Housing Size

A multivariable linear regression model was established on the basis of model (10.4). Regression was run on the expected per capita size (D_s/m), so the variables at the right side of model (10.4) were multiplied by $1/m$. The model takes the following form:

$$D_s/m = \text{intercept} + \sum_i a_i x_i + \sum_i b_i(x_i/m) + c(1/m) + d \cdot hS/m \qquad (10.5)$$

where x_i are variables including demand group, current size, family income, job type, job rank, and tolerable commuting time. m is family size, S/m is per capita size, and h is a dummy variable indicating the implicit needs of investment and expectation of family structural change. a_i, b_i, c, and d are coefficients to be estimated.

The results of stepwise regression are given in Table 10.3. The R-square is 0.797, showing that the model provides a satisfactory interpretation of the situation. Eleven significant variables are identified by the regression, including demand group (1), family size (2), current size and its interaction term with family size (3, 4), job type (5), job rank and its interaction term with family size (6, 7), tolerable commuting time and its interaction term with family size (8, 9), and the implicit term and its interaction term with family size (10, 11).

In the regression model, family income has been omitted due to strong correlation with the demand group variable. The correlations of some other variables are strong (>0.5) in the model, but the regression coefficients are very stable when the variables are repetitively replaced; thus, multicollinearity was not deemed a big problem. As such, the results indicate that the assumptions made about the critical factors of housing demand and their interactions proposed in Fig. 10.1 are reasonable.

10.4 Implications for Housing Demand Management

The analysis results of housing demand structure have provided essential information for housing demand management and further analysis of housing policies.

Table 10.3 Regression results on anticipated per capita housing size (m^2)

No	Item	Definition	Coeff.	Std. error	t	P
	Intercept		11.297	1.272	8.88	0.000
1	*Demand* {3&8&6&7&4&1&2-5}	Demand group[a]	−4.518	0.352	−12.85	0.000
	Demand{3&8&6&7&4-1&2}		5.509	0.393	14	0.000
	Demand{3&8&6-7&4}		−2.568	0.432	−5.94	0.000
	Demand{3-8&6}		−6.596	0.851	−7.75	0.000
	Demand{8–6}		−0.690	0.270	−2.56	0.011
	Demand{7–4}		3.186	0.409	7.78	0.000
	Demand{1–2}		1.391	0.536	2.6	0.010
2	$1/m$	1/family size	65.973	1.652	39.94	0.000
3	S	Current size (m^2)	0.095	0.011	8.76	0.000
4	S/m	Per capita size (m^2)	0.242	0.020	12.27	0.000
5	*Job type*{1-2&3}	Job type	−1.219	0.196	−6.22	0.000
	Job type{2–3}		−1.034	0.229	−4.52	0.000
6	*Job rank*{1&2-3}	Job rank	−1.081	0.216	−4.99	0.000
	Job rank{1–2}		−0.965	0.186	−5.19	0.000
7	(*Job rank*{1–2}−0.334)[a]($1/m$−0.467)	Interaction term	−3.282	0.729	−4.5	0.000
8	*T-commute*	Tolerable commuting time (min)	0.037	0.005	7.45	0.000
9	(*T-commute*−58.73)[a] ($1/m$−0.467)	Interaction term	0.043	0.019	2.28	0.023
10	h	1: implicit demand; 0: otherwise	−14.758	0.941	−15.68	0.000
11	$h^a S/m$	Interaction term	−0.065	0.015	−4.47	0.000

[a]The coefficient of $x\{a - b\}$ is β: when $x = a$, level $= \beta$; when $x = b$, level $= -\beta$; and when $x =$ other values, level $= 0$

10.4.1 Differential Needs of Families for Dwelling Size

Keeping all other variables the same, we may calculate the expected per capita size with regard to demand groups, the current size, and per capita size levels of the eight groups. The differences among them are fairly large. For instance, a three-person normal working family (G8) presently living in an 80-m^2 house would like a 119-m^2 house; while a similar wealthy family (G5) wishes to have a 133-m^2 house.

The desired sizes of housing are also influenced by job and tolerable commuting time. The interaction terms in Table 10.3 reveal that the expected per capita housing size of families in high-end services, IT, professional, and government jobs are 2–4 m^2 larger than that of people working in commercial and manufacturing jobs. The difference expands with the increase of family size. A similar trend can be identified with job rank.

By contrast, differences with the effect of tolerable commuting time decrease by family size. That is, the effects on a larger family tend to be smaller. For instance, let us compare the desired per capita housing sizes in some places where commuting times are 15 and 120 min. For a single-person family, the elongated commuting time can be compensated by an increase in size of 6.2 m^2, but for a 4- or 5-member family, an extra 3 m^2 per person would be enough. This result can be explained by the comprehensive transportation cost of the whole family. For large families, a longer commute increases the total transportation expense more; thus, the compensable per capita willingness to pay for housing size is decreased.

From the marginal effect of the demand group in Table 10.3, the expected per capita sizes in relationship to the average level were computed. It was found that the expected per capita sizes of the groups comprising people who do not own houses (G1, G2, and G3) were 8–12 m^2 lower than the average level. Among the families who own houses, the level of G4 (currently living in houses less than 48 m^2) is close to the average level, and those of the other groups are generally 4–7 m^2 higher than the average. The differences reflect the rule of stepwise consumption in reality.

10.4.2 The Ideal Dwelling Size of Families

Table 10.4 shows the average desired housing size with respect to the current housing size and family size. It was found that the desired size and current size converge at a certain point. Beyond this size, the needs for a larger size are no longer evident. For families with two, three, four, and five or more members, the convergence points are approximately at 100, 110, 140, 170, and 230 m^2, respectively. These sizes imply the "ideal size" of accommodation for different-sized families.

10.4.3 Implicit Term and the "True Demand" of Families

Based on the coefficients of terms (10) and (11) in Table 10.3, the effect of h on desired per capita housing size was estimated and shown in Table 10.5. The desired per capita housing size of families buying houses for investment or other implicit reasons is 15–20 m^2 less than would otherwise be the case for members of the same group.

No doubt, for those making housing development plan decisions, the true size demands are much more valuable than just the ideal size. However, it is difficult to identify the true housing size needs of families simply through a questionnaire because the respondents usually indicate a larger size than they truly need.

To overcome this problem, we developed an adjustment method with the information given by the effect of implicit variable h. Consider the situation of $h = 1$, where a family plans to buy a surplus small house for their children or parents. We assume that the expected per capita size (= (current size + expected

Table 10.4 Desired housing sizes w. r. t. current sizes (m^2)

Current size (m^2)	Family size (persons) 1	2	3	4	5
30	76.08	78.93	81.78	84.63	87.48
50	82.82	87.57	92.32	97.07	101.82
70	89.56	96.21	102.86	109.51	116.16
90	96.30	104.85	113.40	121.95	130.50
100	99.67	109.17	118.67	128.17	137.67
110	103.04	113.49	123.94	134.39	144.84
120	106.41	117.81	129.21	140.61	152.01
130	109.78	122.13	134.48	146.83	159.18
140	113.15	126.45	139.75	153.05	166.35
150	116.52	130.77	145.02	159.27	173.52
160	119.89	135.09	150.29	165.49	180.69
170	123.26	139.41	155.56	171.71	187.86
180	126.63	143.73	160.83	177.93	195.03
190	130.00	148.05	166.10	184.15	202.20
200	133.37	152.37	171.37	190.37	209.37
210	136.74	156.69	176.64	196.59	216.54
220	140.11	161.01	181.91	202.81	223.71
230	143.48	165.33	187.18	209.03	230.88
240	146.85	169.65	192.45	215.25	238.05

Table 10.5 Impact of variable h on anticipated per capita housing size

Current size (m^2)	Effect of implicit term (m^2)
20	−16.06
30	−16.71
40	−17.36
50	−18.01
60	−18.66
80	−19.96
100	−21.26

size of the extra house)/number of family members) is close to the true demand. This assumption is reasonable, as people planning to purchase an extra house have usually thought about more than just a larger size. This means that the "true size demand" can be derived by deducing 15–20 m^2 from the "ideal size."

Based on this assumption, the true housing demands of urban families with different sizes were calculated (Table 10.6). The calculation results imply that the appropriate housing sizes for 2-person, 3-person, 4-person, and 5- or more person families are 70–80, 80–95, 90–110, and 90–155 m^2, respectively.

In practice, there are always some differences between the true demand and the ideal size. Preferential policies should be made to encourage the provision of houses corresponding to the true demand. On the other hand, a size demand significantly beyond the true demand is unlikely, though the demand might be "ideal," because it lakes the sustainability of true demand. Therefore, meeting the true demand of people should be the goal of housing supply in terms of size distribution and unit types. Those overlarge houses exceeding the true demand should be limited with taxation policies.

Table 10.6 Adjustment of housing size demands by demand groups

Family size[a]	"Ideal size" (m^2)	Adjustor	"True demand"
2	110	$-(15–20\ m^2) \times 2$	70–80 m^2
3	140	$-(15–20\ m^2) \times 3$	80–95 m^2
4	170	$-(15–20\ m^2) \times 4$	90–110 m^2
5–7[b]	230	$-(15–20\ m^2) \times (5–7)$	90–155 m^2

[a]Single-member families are omitted as the sample families rarely desire surplus houses
[b]Few families have more than 7 persons, so the size demand is estimated for 5- to 7-person families

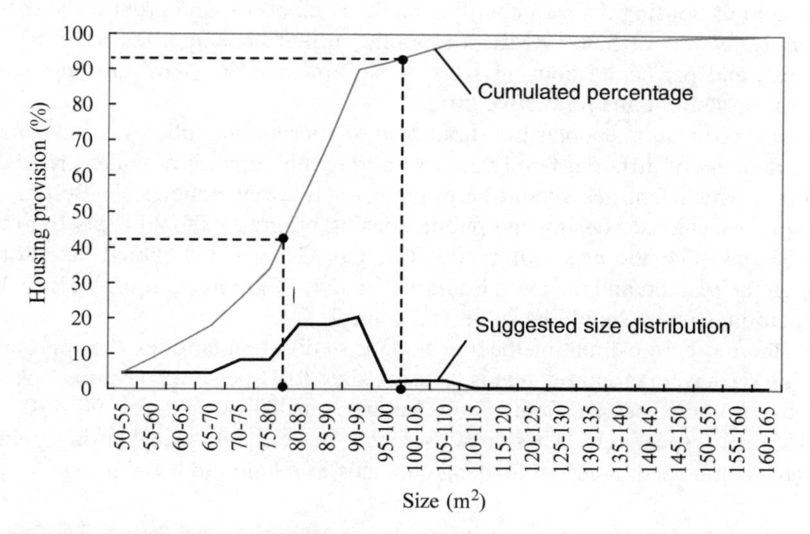

Fig. 10.4 Distribution of predicted housing size demand

10.4.4 Prediction of the Total Housing Demand

Based on the analysis results regarding the true demands of different family groups, we can predict the housing demand accurately with the proportions of each group. The results provide an objective basis for the assessment of relevant policies.

For example, we estimated the necessary amount of housing with respect to housing size in Beijing (Fig. 10.4). The true size demands of each family group in Table 10.6 and the ratios of each group in the sample in Table 10.2 were used. The results show that at present, approximately 45% of total families need houses smaller than 80 m^2, 70% need houses below 90 m^2, and 8% need houses beyond 100 m^2. These results are useful for the evaluation of housing development plans. In particular, they suggested that the housing needs of about half of the families are within the scope of 80–100 m^2; therefore, such housing should be the main body of housing stocks to be enhanced.

10.5 Conclusion

The consumption behaviors of urban households provide a good perspective for exploring the classification of housing demand. The analyses in this chapter demonstrated that the proposed model is a good representation of the actual demand. It facilitates the partition of the housing demand empirically, and the differences of necessary housing size demand between the groups are clarified. This information is very significant for the management of housing demand in practice.

The analysis results with the Beijing questionnaire data suggested that the prediction of housing demand could be made by eight demand groups; the indices of family income of 5,000 RMB per month, current housing sizes of 48 m^2 and 77.7 m^2, and per capita housing sizes of 7.67 m^2 and 46.67 m^2 are appropriate criteria for defining the respective groups.

This classification enables the stipulation of appropriate policies corresponding to the demand of different family groups; at the same time, it provides criteria for selecting which families should be eligible for relevant policies. In Beijing, for example, subsidized housing and public housing programs should apply to groups G2, G3, and G4; the needs of groups G1, G6, G7, and G8 should be resolved through the market; and the extraordinary demands of group G5 result mostly from speculation, so they should be subject to control.

Furthermore, by estimating the true demand of different families, the appropriate housing sizes were proposed, that is, housing sizes for 2-person, 3-person, 4-person, and 5- or more person families in Beijing are 70–80, 80–95, 90–110, and 90–155 m^2, respectively. These results enable the evaluation of housing policies and an accurate estimation of housing demands in relation to housing size.

Acknowledgements This study was supported by the National Natural Science Foundation of China (No. 41171138) and the Knowledge Innovation Project of the Institute of Geographic Sciences and Natural Resources Research. The author is grateful to Professors Yanmin Zhou, Yasushi Asami, Toru Ishikawa, Jiabin Lin, and Tian Chen for their valuable comments.

References

Akintoye AS, Skitmore RM (1994) Models of UK private sector quarterly construction demand. Constr Manag Econ 12(1):3–13

Börsch-Supan A, Pollakowski HO (1990) Estimating housing consumption adjustments from panel data. J Urban Econ 27(2):131–150

Geoffrey C (1973) Income elasticity of housing demand. Rev Econ Stat 55:528–532

Goh BH (1996) Residential construction demand forecasting using economic indicators: a comparative study of artificial neural networks and multiple regression. Constr Manag Econ 14:25–34

Goh BH (1999) An evaluation of the accuracy of the multiple regression approach in forecasting sectoral construction demand in Singapore. Constr Manag Econ 17(2):231–241

Han KF, Xiao ML (2006) Demand elasticity of housing market. Bus Econ 3:35–36 (in Chinese)

Haurin DR, Lee K (1989) A structural model of the demand for owner-occupied housing. J Urban Econ 26(3):348–360

Ioannides YM, Zable JE (2008) Interactions, neighbourhood selection and housing demand. J Urban Econ 63(1):229–252

Jin DH, Zhang SL (2003) Research on housing demand income flexibility in Shenzhen. J Chongqing Archit Univ 3:82–85 (in Chinese)

Lavender SD (1990) Economics for builders and surveyors: principles and applications of economic decision-making in developing the built environment. Longman, New York

Liu MX (2002) Housing consumption of urban residents. China Land Press, Beijing (in Chinese)

Shen YZ (2007) The demand flexibility analysis about the housing consumption of Chinese town residents. Econ Manag 3:38–41 (in Chinese)

Thomas NS, Skitmore M, Wong KF (2008) Using genetic algorithms and linear regression analysis for private housing demand forecast. Build Environ 43(6):1171–1184

Tse RYC, Ho CW, Ganesan S (1999) Matching housing supply and demand: an empirical study of Hong Kong's market. Constr Manag Econ 17(5):625–634

Wang SF (2007) Application of multivariate statistical analysis in urban residential housing consumption in Tianjin. J Tianjin Polytech Univ 1:81–84 (in Chinese)

Wang YC, Ma ZL (2005) Evaluation of the reasonableness of urban residents' housing consumption in China. Econ Manag 12:43–45 (in Chinese)

Xiong SF (2007) Economic analysis of inter-period housing consumption. J Postgrad Zhongnan Univ Econ Law 4:25–29 (in Chinese)

Zhang H (2007) Analysis of urban households' housing demand and motives model. J Cap Univ Econ Bus 1:116–120 (in Chinese)

Zhang WZ, Liu W, Meng B (2005) On location advantage value of residential environment (LAVRE) in the urban and suburban areas of Beijing. Acta Geogr Sin 1:115–121 (in Chinese)

Zhang WZ et al (2006) A study of livable city in China: Beijing. China Social Sciences Press, Beijing (in Chinese)

Chapter 11
The Role of the Knowledge Community and Transmission of Knowledge: A Case of Bicycle SMEs in Taiwan

Jen-te Pai and Tai-shan Hu

Abstract In the era of the knowledge economy, knowledge comprises the core of economic growth, making the exploration of the spatial organization of knowledge generation an important issue. This study clarifies the knowledge generation process. In terms of the development mode, the Taiwanese bicycle industry can be geographically divided into northern, central, and southern regions. Each area possesses its own independent system with related and cooperative factories. Dajia, located in central Taiwan, is the most mature of the three. The geographical closeness of and industrial cooperation between community structures allow them to share resources, reduce deal prices, and promote professional investment. Even social networks can be exploited to produce and promote creativity. Therefore, this investigation proposed utilizing knowledge groups to understand spatial organizational style, with particular reference to villages. The study results demonstrate that the knowledge activities conducted in Dajia bicycle village are determined by the composition of the village, cooperation within the industry, and effective support from industrial and social resources.

Keywords Industrial cluster • Knowledge community • Bicycle industry

---

J.-t. Pai (✉)
Department of Land Economics, National Chengchi University, No.64, Sec.2, ZhiNan Rd., Wenshan District, Taipei City 11605, Taiwan
e-mail: brianpai@nccu.edu.tw

T.-s. Hu
Department of Architecture and Urban Planning, Chung Hua University, 707, Sec.2, WuFu Rd., Hsinchu 30012, Taiwan
e-mail: hts@chu.edu.tw

M. Kawakami et al. (eds.), *Spatial Planning and Sustainable Development: Approaches for Achieving Sustainable Urban Form in Asian Cities*, Strategies for Sustainability, DOI 10.1007/978-94-007-5922-0_11,

11.1 Introduction

Since the 1980s—the rise and formation of the knowledge-based economy—globalization, increased international competition, and innovation have become increasingly critical for local economies (Camagni 1995; Feldman 1994; Malmberg and Maskell 2002; Porter 1990; Ritsila 1999; Storper 1995). Consequently, the significant advantages of specific industrial clustering have been broadly recognized. There are two main sources of knowledge workers for recently established firms: high-level educational institutions and well-established enterprises in relevant industries. Therefore, new corporations tend to locate around universities, research organizations, and existing companies. Besides the central elements, other considerations include technology transfer, spin-offs, and technological infrastructure. Empirically, clustering is a major catalyst for innovation (Baptista and Swann 1998; Breschi 2000; Wu and Chen 2001). In contrast, despite building their plants close to one another, firms in traditional industries have failed to effectively inspire innovation.

Knowledge is the key driver of economic growth in the knowledge economy age—an "internal growth mechanism." The capacity to generate and employ knowledge and the effectiveness of knowledge creation and utilization have become essential for supporting industrial clustering and continuous economic development and growth. The Taiwanese bicycle industry has a development history of over 50 years; it used to be a traditional industry reliant on government support and protection. Following changes in the global economic situation, the open economic policies of Taiwan, and heightened competition from China, the industry changed its strategies and carved a new impression in the global bicycle industry.

In addition, environment protection concerns recently have indirectly promoted the use of bicycles for daily transportation. Therefore, the demand for bicycles is increasing, particularly at the high-end segment of the market. During the early 2000s, sports and leisure became a major trend, and developed countries became concerned about the ecology and green business. This gave bicycle producers the opportunity to maintain healthy production levels. Thus, government influence remains the key emerging force propelling creativity and growth in the Taiwanese bicycle industry. Although commuting by bicycle has yet to make a breakthrough in Taiwan, integrating bicycles with the development of recreational areas has already become quite popular. This trend should be encouraged since it provides substantial health benefits and enjoyment for the riders, awakens environmental consciousness, and generates economic benefits with a low negative environmental impact (Chang and Chang 2009). With current global concerns about sustainable development, recreational cycling must be studied further and encouraged in science cities.

As a traditional industry cluster and the most mature bicycle cluster in Taiwan, the Dajia bicycle cluster led the transformation, which was characterized by knowledge activities—the presentments of a knowledge city. Having been born as a traditional industry cluster and now transforming itself to a knowledge community, Dajia comes equipped with a special knowledge-community background. By understanding the industrial development process and conducting the questionnaire survey and interviews to bicycle firms, this study uses Dajia bicycle valley as its research object. The following questions are considered: Do generation, transmission and transformation occur in the cluster? Do the characteristics of high-tech industry knowledge communities differ from those that are not? How does the bicycle industry break through the limitations of traditional industry?

The final product of this study is the main features of the knowledge community at Dajia bicycle valley. The experience of this cluster serves an example of the transformation of traditional industries to tackle today's highly competitive global economy.

11.2 Literature Review

Innovation is crucial in the current economic environment, and obtaining competitive advantage is a growing struggle. In particular, the localization of industrial learning is causing the convergence or clustering of high-tech industries and the agglomeration of new knowledge-based industrial activity. Most significantly, high-tech companies are actively approaching or seeking links with sources of new knowledge production, with the aim of facilitating knowledge transfer to achieve competitive advantage. This situation indicates that the tendency of traditional industry firms to cluster has failed to stimulate innovation.

Nevertheless, clustering is a catalyst for innovation, enabling firms or individuals to successfully commercialize new knowledge, for example, by applying it to production. In cluster, knowledge can generate, transform, and spread, and form a knowledge community (Yigitcanlar et al. 2008).

This work assumes that competitive advantage in innovation-dependent industries hinges not only on participation in the local technological infrastructure and the establishment of an environment conducive to innovation, but also on interactions within the technological community that encourage local innovation and knowledge accumulation. This study reviews the literature on the formation of the local innovation environment to clarify the basis of the interaction between local industrial clustering and the technological community.

11.2.1 Proximity, Interaction, and Innovation

In local industrial clustering, scientific and technological interactions between academic and research organizations, and firms involved in the high-tech industry represent part of the technological infrastructure. Such infrastructure facilitates the production, transfer, and application of knowledge, information, and technology.

Regarding useful and feasible mechanisms, academic research has focused on the industrial clustering that follows spin-offs in science parks. Science parks have more mature and proactive mechanisms for building the basic infrastructure, which helps establish and strengthen the interaction between academic and research institutes and firms, and within the high-tech talent communities (Bell 1993). It also enhances the dimensions of nonphysical proximity (Boschma 2005).

Regional technological infrastructure is central to technological system innovation and analysis. Specifically, universities, research institutes, and other regional technological infrastructures, as well as interfirm interaction, create a localization effect. Therefore, although advanced information and telecommunications technologies can reduce transaction costs, innovation still requires geographical proximity to facilitate face-to-face communication. Given such spatial clustering, local competition within the wider environment increases after agglomeration, information acquisition, and the development of mature technological infrastructure. Competition is also facilitated by the firms' need to meet the challenge posed by the steadily shrinking periods during which they can maintain their competitiveness.

Considering the success of science parks within industrial clusters, most of the existing research and literature stresses the importance of academic and research institutes (Monck et al. 1988; Massey et al. 1992; Westhead and Batstone 1998). Differently put, most of the current and past research in Taiwan and elsewhere has examined the links between firms within science parks and neighboring academic and research institutes from the perspective of physical proximity. It has done so by comparing firms inside and outside science parks and clarifying the interactions between firms and academia within specific areas. Furthermore, these studies have clearly demonstrated the establishment of links between industry and academia, particularly links formed in industrial clusters in and around science parks.

Accordingly, science parks are crucial for improving expected relationships and mechanisms. However, Massey and others have noted that the strength of these links is unclear (Massey et al. 1992). The geographical proximity of firms and academic institutions is important in maintaining and promoting industrial innovation, both within science parks and in the surrounding areas. The other dimensions of proximity—cognitive, organizational, and institutional (Torre and Gilly 2000; Torre and Rallet 2005)—can both substitute for and complement its physical aspect (Boschma 2005; Torre and Rallet 2005).

To summarize, the above literature review demonstrates that from the perspective of regional economic innovation, science-based parks and the industrial clusters that evolve around them can stimulate local economic development. Furthermore, the key determinants of the influence of clustering are interactions within the talent community and network links. In turn, these depend on the establishment of initial basic infrastructure and input from sustainable technological infrastructure. Moreover, knowledge generates, transmits, and transfers within the clusters or certain areas, and molds the clusters into the knowledge community (Yigitcanlar et al. 2008).

11.2.2 Spin-Offs and Technology Spread

Firms in local systems are frequently said to share, imitate, or obtain advantage from structures that have similar cultures and mechanisms. Furthermore, economic globalization has been driven partly by a reduction in the importance of the elements of traditional localized production and the progressive increase in the returns obtained by parts of knowledge-based and learning-type economies (Malmberg 1997). Technological interdependence denotes a form of mutually beneficial structural interaction between technologies or obstacles to technological development. The path-dependence model considers situations where competing technologies already exist and where decisive technological developments are focused within industries (Cowan and Hultén 1996). Consequently, economic and technological connections exist between various industries.

In the present era of the knowledge-based economy, the establishment of local networks, research and technological development, and collective learning are all essential to future local and regional development and attractiveness. Regarding the spin-off and acquisition of related technologies, small new technology-based firms possess two key sources of scientific and technological human resources: institutions of higher learning and well-established firms (Oakey 1985). Consequently, new companies will likely sprout around universities, research institutions, and existing firms. This situation usually generates stable, but often uneven, growth across regions. Regions with internally undertaken technological development enjoy relatively rapid but stable growth. In contrast, other regions grow more slowly.

Changes during technological expansion significantly influence economic growth. Regions that lag in the development and application of new technologies will suffer industrial decline. However, during technological expansion, the application or use of technology is not a mere function of knowledge; it also involves the firm's capacity to replace its current technologies and depends on assessment and testing. Second, during the application of innovation technology, most of the required information is transferred via individual interpersonal contacts that support the flow of innovation. Consequently, spin-offs and the mobility of skilled manpower accelerate technological expansion. Notably, spin-offs are more frequent and significant than instances of conscious, formal cooperation.

11.3 The Emergence of the Bicycle Industry Valley

From 1958 to 1968, the market contained numerous small firms, and the vicious price competition between them caused an industry recession. Meanwhile, the motorbike industry surged, as motorcycles became an alternative form of transportation that challenged the dominance of the bicycle. To survive, merchants started to export bicycles to the United States of America (USA) in the 1960s, and sales posted consistent annual increases. In 1970, export sales exceeded 1 million bicycles for the first time. In terms of global labor distribution, 1960–1970 was the era of the original equipment manufacturer (OEM) for the Taiwanese bicycle industry. Firms primarily produced products for known brands overseas. By the end of the 1980s, the industry had matured, and during the 1990s, the number of products and their overall quality steadily grew and developed. According to Taiwan's economic ministry, 45.06% of the bicycle manufacturing industry and 91.51% of bicycle parts factories are made up of small and medium enterprises (SMEs).

In 2001, exports accounted for 84% of the increase in sales. International trade in the bicycle industry is conducted primarily via OEM, foreign trading businesses, agencies, or overseas wholesalers. Taiwan ranked among the top five bicycle exporters worldwide in terms of export dollars (US$) earned and was followed by the USA (28.84%), Japan (15.77%), England (8.19%), Belgium (7.15%), and Holland (7.04%).

The bicycle manufacturing industry is aggregated in northern, middle, and southern Taiwan. The northern area lies around Taipei, with some factories in Taoyuan (Fig. 11.1). The central area has the biggest concentration of bike makers, mostly located in Taichung and Changhua. Chiayi and Tainan are the main sites in the southern area. Bicycle manufacturers in the middle area comprise roughly 69% of bicycle factories in Taiwan and those in the south, 10%. Most of the bike companies in Taichung County are clustered in Dajia district, and this cluster accounts for 65.94% of the total number of bicycle firms in Taiwan (Table 11.1). More than 92% of the firms in Dajia are SMEs. From 1996 to 2001, the number of manufacturers in Dajia doubled, intensifying industrial development in the area. Furthermore, central Taiwan has become a favorite destination for producers of precision instruments. In fact, 70% of the metal-cutting machinery produced in Taiwan comes from this area (Chuang 2005), proof that it is indeed the center of the local machinery industry. Bicycle manufacturers and assemblers must also utilize parts and machinery developed by mechanical firms. These conditions have helped the bicycle industry establish a natural home in central Taiwan.

Manufacturers in the Taichung area have set up their operations surrounding Giant Manufacturing, a complete bicycle production company (Fig. 11.2). Giant is surrounded by parts factories, which include the following: (1) bicycle frame firms—Alu Mega Industrial Co., Ltd., Maxway Cycles Co., Ltd., and Hsu Sheng Bicycle Ind. Co., Ltd.; (2) front fork firms—Kinesis Industrial Co., Ltd. and Aprebic Ind. Co., Ltd.; (3) pedal suppliers—Wellgo Ind. Co., Ltd. and Jendel Co., Ltd.; and (4) makers of other parts. However, although Giant is the biggest company in the area, neighboring Youth Industrial Park sits on a bigger expanse of industrial land. It is also a very important destination for certain parts factories, including Apro, Wellgo, and Kinesis.

The manufacturers in Dajia were established between 1981 and 1990 and comprise 61.3% of all bicycle factories in Taiwan. The data also indicate that the Taiwanese bicycle industry developed most rapidly during the 1980s. Simultaneously, the bicycle manufactures in Dajia also established numerous bicycle parts factories. After the 1990s, the increase in the number of factories slowed down. Moreover, due to globalization, most large capital investors no longer wish

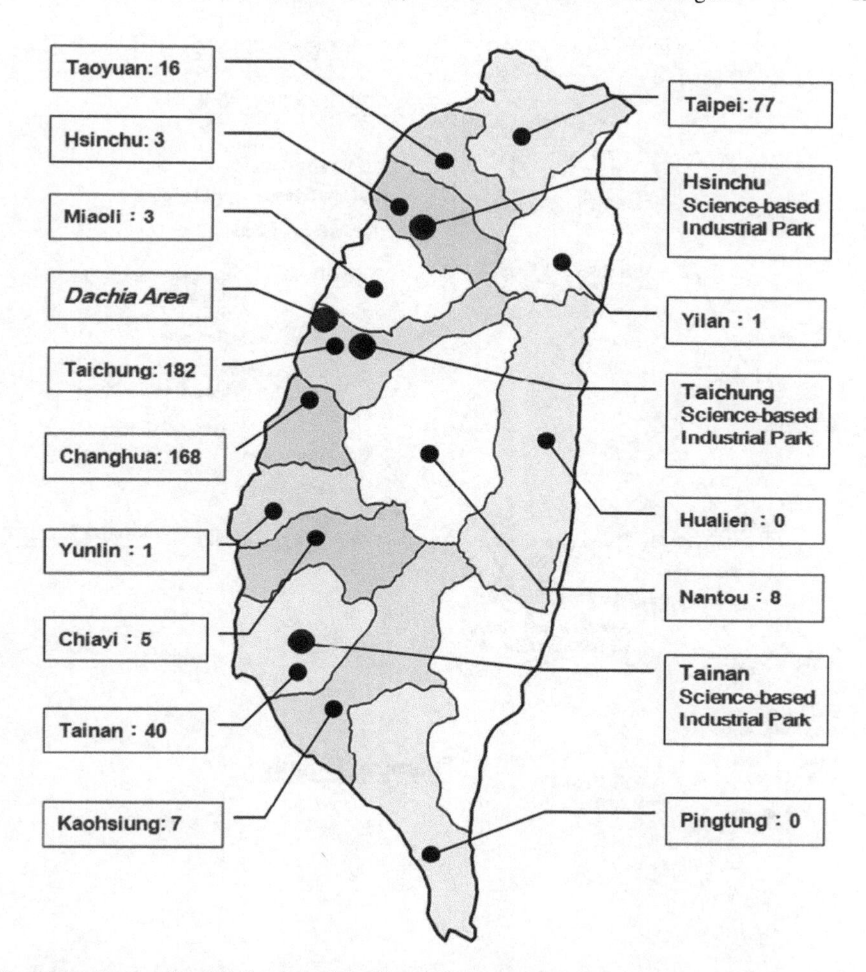

Fig. 11.1 Geographic distribution of the bicycle industry in Taiwan

Table 11.1 Total number of bicycle companies around Dachia from 1996 to 2006

	1996		2001		2006		The growth from 1996 to 2001		The growth from 2001 to 2006	
Township/county	No.	%	No.	%	No.	%	No.	%	No.	%
Yuanli, Miaoli	14	8.00	10	5.75	10	5.62	−4	−2.25	0	−0.13
Dachia, Taichung	101	57.72	113	64.94	118	66.29	12	7.22	5	1.35
Waipu, Taichung	36	20.57	27	15.52	28	15.73	−9	−5.05	1	0.21
Da-an, Taichung	24	13.71	24	13.79	22	12.36	0	0.08	−2	−1.43
Subtotal	175	100	174	100	178	100	−1	0.00	4	0.00

Source: The Report on 2006, 2001 and 1996 Industry, Commerce and Service Census, Directorate-General of Budget Accounting and Statistics, Executive Yuan, Taiwan

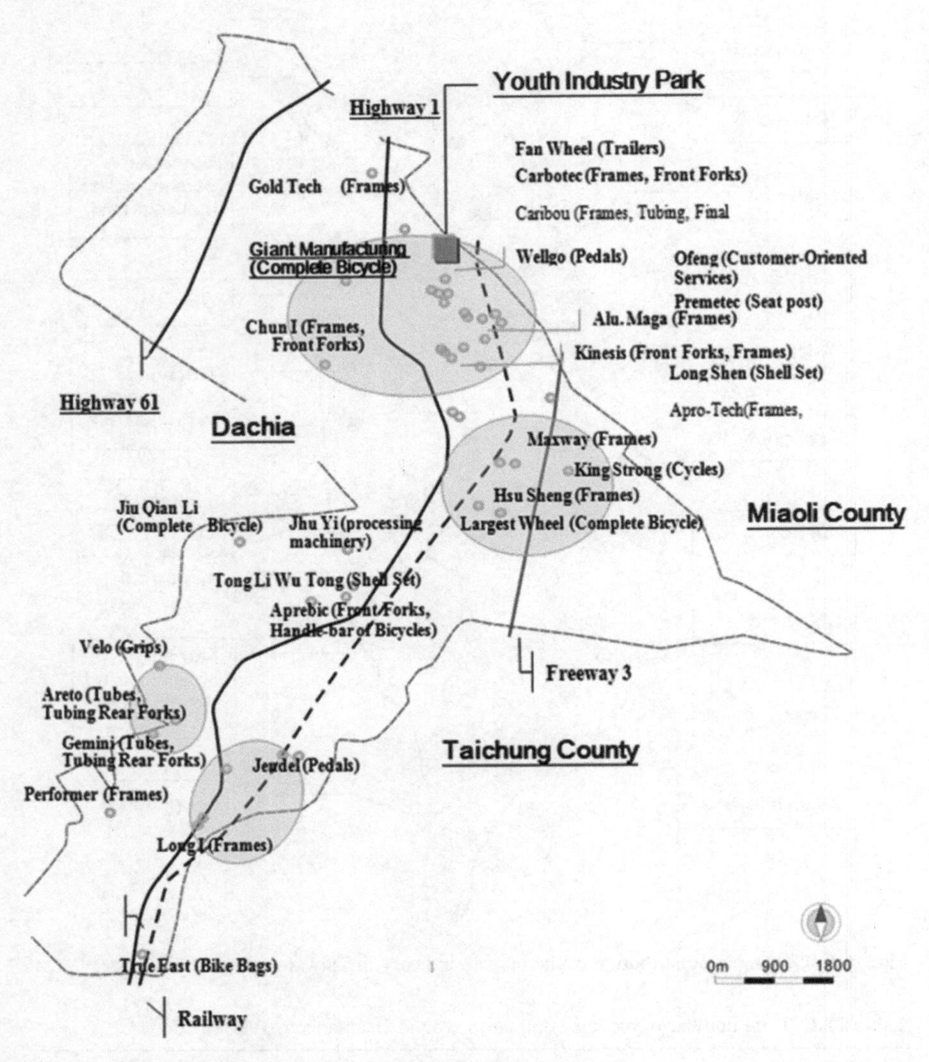

Fig. 11.2 The geographic distribution of the bicycle industry in Dachia district, Taichung (as of the end of 2005)

to establish factories in Taiwan and are instead relocating to countries with lower costs and cheaper labor.

The global economic situation and lure of other countries' lower production costs have left the Taiwanese bicycle industry no choice but to innovate in order to survive in the intensely competitive market. During the early 2000s, bicycle production firms achieved enhanced benefits and improved production efficiency, and Taiwanese-made bicycles cornered a growing share of global production. The Dajia bicycle cluster led this change. As mentioned earlier, the Dajia bicycle cluster surveys the bicycle and manufacturing industries, and now it leads the development strategies of Taiwan. This study uses the Dajia cluster to examine the bike industry's development, composition, and knowledge activities, and understand the characteristics of its transformation.

11.4 Creating a Knowledge Community

The developmental mode of the bicycle industry cluster in Taiwan geographically divides into the northern, central, and southern areas. Each area has its own independent system of various interrelated and cooperating factories. In a particular area, most factories implement multiple central corporate synergy systems. Even social networks produce and promote creativity. Within the communities, factories experience level production competition; examples are Giant and Merida. Although both are in central Taiwan, the factories function as separate unit centers and competition between them persists. This kind of level competition and vertical cooperation with parts factories makes them stand out among the three focuses of the bicycle industry and allows them to maintain an advantageous position in the highly competitive international market. Therefore, this research proposed utilizing the knowledge group concept to understand spatial organizational style, with specific reference to villages.

In the following sections, the authors discuss the results of a survey conducted on bicycle companies in Dajia. Meanwhile, for better understanding, they also present the results of interviews with managers of bicycle companies, the chairman of the association, and other related personnel. The results illustrate the form of the industry cluster and related interactions, and examine knowledge activities.

11.4.1 Analysis of Survey Results

This study administered a questionnaire to managers within the bicycle industry village. The fields that were covered included reasons for factory clustering, interaction between factories, involvement of manufacturers in research and development, and sources of impact. Around 83.33% of the questionnaires were obtained from within Taichung County, Dajia Township, and 16.67% from the surrounding area. There are 31 valid questionnaires. The distribution of the questionnaires corresponds to the main distribution of bicycle industry factories.

11.4.1.1 Reasons for Manufacturer Clusters

Local Background of Involved Businessmen

Fourteen of the businessmen interviewed stated that they established factories in Dajia because it is their hometown and they are familiar with the local environment. Furthermore, their businesses are closely related to local industries.

Convenient Transportation

The businessmen interviewed saw convenient transportation as another reason to place their plants in the area:

> The nearby Taichung Harbor and highway provide rapid transportation for shipping. We also look forward to the convenience of direct flights to China.
>
> —Mr. Chien, manager

Taichung Harbor, railway access to the airport, and Highway No.1 save time and reduce transportation costs. Ten of the survey respondents saw this as their primary reason for establishing their factories in Dajia. This consideration also reflects the geographical distribution of manufacturer clusters. Most factories are clustered around Youth Industry Park and along Taiwan's Highway No 1.

Chain Parts Factories in the Area

Eight businessmen indicated that the proximity of parts suppliers is a key reason for their choice of location. Parts factories flock around the core factory (e.g., Giant) and are dependent on it:

> The cost of bicycle parts of is high, and most firms prefer to buy the parts locally, not abroad. When we think about where to set up a bicycle firm, we have to consider whether there are chain parts factories nearby. Few bicycle firms would choose a place without suppliers in the vicinity.
>
> —Mr. Liu, chairman of Taiwan Bicycle Export Association

Easy Access to Materials and Abundant Energy Resources

Easy access to material suppliers (seven companies) and abundant energy resources (five companies) are also main reasons for the existence of the factory cluster. The well-developed industry at Dajia uses electricity supplied by the Taichung Thermal Power Plant, and the trains and highways of Taichung Harbor provide easy access to materials. Consequently, firms located in this area have abundant energy resources that they can tap (Table 11.2).

11.4.1.2 Marketing Methods

According to this study, the main marketing methods used by bicycle manufacturers in Taichung are as follows: (1) Follow customer blueprints, and design and process the products (12 OEMs) and (2) follow customers' requests, and design and produce the products (9 original design manufacturers [ODMs]). Factories in this area rely primarily on commission sales to attract business; self-marketing is not a popular business model. This phenomenon likely exists because parts factories are heavily reliant on the major manufacturers of complete bikes for their sales.

Table 11.2 Location determinants of bicycle companies (%)

	Small-sized firms	Medium-sized firms	Large-sized firms	Total
Energy supply	2.1	4.1	0	6.2
Learning with the related industries	0	4.1	6.0	10.1
Material supply	0	10.2	4.1	14.3
Near to upstream bicycle parts companies	2.0	6.3	8.2	16.5
Preference of entrepreneur	6.0	10.2	12.2	28.4
Related business service	0	2.0	2.2	4.2
The proximity to university, R&D institutes, or technology-leading large firms	0	0	0	0
Transportation accessibility	4.1	6.0	10.2	20.3

Denotes small firms have capitalization of less than US$30,000; medium firms, US$30,000 and US $300,000; and large firms, US$300,000

11.4.1.3 Technical Cooperation and Operation Interactions

The results of the survey on interfirm interactions in relation to operations demonstrate that firms primarily have dealings with material and parts suppliers, as well as with legal and financial services firms. Cluster firms normally interact with other firms located within 30-min travel time, although they also do so with firms that are up to 2 h away. This phenomenon demonstrates the proximity of clustered bicycle firms to each other (Tables 11.3 and 11.4).

This study found that 30% of business owners have participated in technical cooperation and interaction, while the remaining 70% never have (Table 11.5). This result indicates that technical cooperation and interaction between the factories in the bicycle industry remain limited, and smaller enterprises have relatively low motivation and resources. However, the human resources who were employed in previous mechanical industry development—including technical personnel from Giant and Merida and some parts makers who joined later—formed a production chain that was based on close cooperation. Additionally, past research noted that high-tech industry personnel move around the cluster, so knowledge and skills are rapidly expanding in this area (Hu et al. 2005; Saxenian 1994).

11.4.1.4 Research and Development (R&D) Activity

This study found that in Dajia, few firms with minimal investment capital have invested in research (Table 11.5). Meanwhile, R&D increased in proportion to investment in the firm. Investments exceeding US$3 million account for a higher percentage of total R&D activity. The next issue is that most firms claim they spend less than 3% of total operational funds on R&D, and research staff comprises a maximum of three people (Table 11.6). This denotes that companies do not invest sufficiently in R&D.

Numerous factors can affect R&D activity in these firms. The data presented in this study indicate that the research materials and information they use are for

Table 11.3 Interactions between firms and firm size (%)

Object of interactions		Small-sized firms	Medium-sized firms	Large-sized firms	Total
Materials	Have interactions	66.7	80	85.7	80
	No interactions	33.3	20	14.3	20
Parts	Have interactions	66.7	60	71.4	65
	No interactions	33.3	40	28.6	35
OEM	Have interactions	100	10	42.86	35
	No interactions	0	90	57.14	65
Outsourcing	Have interactions	0	30	42.9	30
	No interactions	100	70	57.1	70
Logistics	Have interactions	0	40	28.6	30
	No interactions	100	60	71.4	70
Marketing	Have interactions	0	10	28.6	15
	No interactions	100	90	71.4	85
Legal service	Have interactions	0	20	42.9	25
	No interactions	100	80	57.1	75
Financial service	Have interactions	33.3	90	57.1	70
	No interactions	66.7	10	42.9	30
Accounting	Have interactions	33.3	80	57.1	65
	No interactions	66.7	20	42.9	35
Partnership	Have interactions	0	0	28.6	10
	No interactions	100	100	71.4	90

self-R&D, and the trade guidelines based on these materials and information are outdated. Firm R&D is focused on new products and promoting product quality (Table 11.7). The data does not clearly reflect the role of business managers. The present data does not represent R&D because the sample comprises small businesses with investment capital of US$1,000–10,000. Research organizations tend to find larger enterprises with investment capital of US$300,000, such as Apro Tech or Giant. These companies have all worked with the Mechanical Industry Research Laboratories and Industrial Technology Research Institute (ITRI), which provide most of the technologies that the firms adopt. In fact, these firms are sources of new technology. Moreover, the bicycle industry in Dajia faces a shortage of technical manpower, which dampens its competitiveness and slows down its transformation.[1]

[1] Bicycle production is a traditional industry which requires creative designs and research. For example, a large factory cannot survive by making bicycles alone; it needs to undertake research and develop related products, such as electrical bikes and leisure bikes. When they face a shortage of technical know-how, their multiproduct operations encounter difficulties. Furthermore, without new batches of professionals to help raise the quality of overall performance, the quality of the workforce as a whole comes into question. This is also an important task for bicycle industry.

Table 11.4 Travel time of interacting objects (%)

Interaction objects	Travel time	Small-sized firms	Medium-sized firms	Large-sized firms	Total
Materials	Within 30 min	100	37.5	50	43.7
	30 min–2 h	0	50	50	43.7
	Over 2 h	0	12.5	0	6.3
	Overseas	0	12.5	0	6.3
Parts	Within 30 min	100	66.7	20	53.8
	30 min–2 h	0	33.3	80	46.2
	Over 2 h	0	0	0	0
	Overseas	0	0	0	0
OEM	Within 30 min	100.00	100.00	100.00	100.00
	30 min–2 h	0	0	0	0
	Over 2 h	0	0	0	0
	Overseas	0	0	0	0
Outsourcing	Within 30 min	0	100	100	100
	30 min–2 h	0	0	0	0
	Over 2 h	0	0	0	0
	Overseas	0	0	0	0
Logistics	Within 30 min	0.00	100	100	100
	30 min–2 h	0	0	0	0
	Over 2 h	0	0	0	0
	Overseas	0	0	0	0
Marketing	Within 30 min	0	100	100	100
	30 min–2 h	0	0	0	0
	Over 2 h	0	0	0	0
	Overseas	0	0	0	0
Legal service	Within 30 min	0	100.00	66.7	80
	30 min–2 h	0	0	33.3	20
	Over 2 h	0	0	0	0
	Overseas	0	0	0	0
Financial service	Within 30 min	100	100	100	100
	30 min–2 h	0	0	0	0
	Over 2 h	0	0	0	0
	Overseas	0	0	0	0
Accounting	Within 30 min	100	100	100	100
	30 min–2 h	0	0	0	0
	Over 2 h	0	0	0	0
	Overseas	0	0	0	0
Partnership	Within 30 min	0	0	50	50
	30 min–2 h	0	0	50	50
	Over 2 h	0	0	0	0
	Overseas	0	0	0	0

Table 11.5 Firm with or without R&D vs. firm size (%)

	Small-sized firms	Medium-sized firms	Large-sized firms	Total
With R&D	0	12.9	16.1	29
Without R&D	22.6	38.7	9.7	71

Table 11.6 Investment of firms in R&D activity (%)

		Small-/medium-sized firms	Large-sized firms	Subtotal
The ratio of R&D expenditure/ sales	0–3%	30	30	60
	Over 3%	20	20	40
Number of R&D personnel	Under 3	11.1	11.2	22.3
	3–5	22.2	22.1	44.3
	Over 5	11.2	22.2	33.4

Table 11.7 Sources of R&D information and R&D purposes vs. firm size (%)

		Small-/medium-sized firms	Large-sized firms	Subtotal
The sources of R&D information	Companies themselves	23.1	38.5	61.6
	Foreign companies	7.7	15.3	23.0
	R&D institutes	0	15.4	15.4
R&D purposes	To decrease production cost	12.5	12.5	25
	To promote the quality of products	6.3	25.0	31.3
	To create new products	12.5	31.2	43.7

Some 43% of factories that do not conduct in-house research also do not cooperate with others in conducting research because they focus on low-technology production using small-scale facilities. This study is concerned on companies with small capital investment and found that the small size of such firms generally prevents them from requiring research. However, the companies would nevertheless like to receive development assistance:

> Actually many small and medium firms know the importance of research and development, although they probably don't have enough capital for it. Therefore, most of these firms need the technical assistance of a central institute to help them develop new technology or products.
>
> —Mr. Liu, chairman of Taiwan Bicycle Export Association

On the other hand, the bicycle industry is considered a traditional industry, and competition for labor from the newly developed Central Science Park created an unwillingness among local workers to enter the bicycle industry:

> Many research and innovation difficulties come from the lack of human resources. Many people see the bicycle industry as a traditional one and ignore its new development trend. Such thoughts reduce the willingness of young and professional workers to join the industry. Furthermore, it decreases our potential for innovation.
>
> —Mr. Tsai, manager

The difficulty in attracting labor is the main concern of most bicycle manufacturers.[2]

11.4.2 "A-Team" Program

The "A-Team" was initially composed of the two largest bicycle companies in Taiwan, Giant and Merida, and 18 bicycle parts factories. The member firms officially announced its establishment during the 2003 Taipei International Bicycle Fair. The A-Team's goals are to increase the performance of associated firms' bicycle production efficiency and quality, supply high-quality products to global markets, research and develop more creative designs and new products, create new brands with high added value, initiate a progressive and advanced new phase for the bicycle industry, and place the industry in a key global market position in terms of creative, innovative, and valuable products. The A-Team proposed resolutions and set out to provide yet unknown needs and creative values for consumers. In short, it made "bicycles made in Taiwan" instead of "Taiwan's bicycles" meaningful.

The requirement for membership in the A-Team is establishing a "just-in-time" (JIT) warehouse in Taiwan and commissioning a Taiwanese company as a producer or establishing a factory in Taiwan. The targets and functions of the A-Team are discussed below.

11.4.2.1 Dedication to Creating International Brands

The A-Team seeks to preserve production capacity for top quality bicycles, maintain a quality differential between Chinese and Taiwanese products, encourage parts upgrading and creative redesign, promote the overall image of the Taiwanese bicycle industry, and increase the global competitiveness of Taiwan. Giant and Merida founded and lead the group. All members are expected to help Taiwan create international brands and ensure the maintenance of a quality differential between China and Taiwan in bicycle production. The efforts of the A-Team are concentrated on meeting the needs of consumers and end retailers.

The establishment of the association has provided members with clear directions and visions; its international reputation grew as a result of this success. As more orders flowed into the A-Team, many European and American customers witnessed an improvement in the quality of the Taiwanese bicycle industry. Taiwan's "dedication to high value-added" impressed buyers, who began to associate Taiwan with excellent bikes. In other words, the A-Team greatly impressed the global bicycle industry.

---

[2] The Taichung Science Park attracts most of the July and August graduates every year. Graduates make Science Park their priority when hunting for jobs after leaving school. Those who choose to join traditional industries such as bicycle production are few and far between.

11.4.2.2 Forming a Mentor Team

The members of the A-Team dedicated themselves to implementing the proposals of the association on a daily basis. They meet regularly to brainstorm, set common goals and plans, and make key strategic decisions. The current operating structure of the A-Team is such that participating teams check in at 3-month intervals. Each member also sends a team to visit the other firms and observe their operations; factories that are visited eagerly share the details of their operations. All members strive to help one another and enjoy doing so. The A-Team project has been supported by Toyota, the Industrial Development Bureau (IDB) of Taiwan's Ministry of Economic Affairs, and the Corporate Synergy Development Center (CSD) of Taiwan. With guidance from the IDB and assistance from the CSD, in coordination with Kuozui Motors Ltd., the world-famous Toyota Production System (TPS) was launched. The mentor team contains two representatives of Giant, two from the CSD, and five from Toyota. The team makes monthly tours and performs a final check every 5 months. The team also does counseling, coaching, training, and factory visits.

11.4.2.3 Factories Exchange Visit Tours

The greatest influence on A-Team members during the process came from exchanges between participants, as most A-Team members were friends who had typically discussed ideas but not visited one another's factories. Giant organized the first high-profile factory visit involving the A-Team, inviting Merida to tour its manufacturing facility. VP Components Co. Ltd subsequently opened its bicycle paddle factory to Wellgo, a producer of the same products. The tours made strong impressions on factory owners: When they saw firsthand the advanced operations of others, it galvanized them to improve their own operations. This virtuous circle gave the mentoring team opportunities for future development.

11.4.2.4 Close Cooperation Between Chain Production Manufacturers ("Just-in-Time")

When executing JIT plans, the A-Team added the strategy of supply-demand chain affect. Recently, world-famous bicycle brands Trek, Specialized, and NBDA (National Bicycle Dealers Association) decided to join the A-Team as supportive members, who will provide valuable assistance and suggestions.

The president of the A-Team, Mr. Lo Shean An, is the manager of Giant. He noted that the Taiwanese bicycle industry had struggled for over a decade to gain 30% of the global market for top-end bicycles. However, the A-Team—Giant, Merida, and 18 parts factories—has recently begun collaborative R&D production with manufacturers of whole bicycles or parts. The A-Team looked around for and purchased foreign parts factories and combined the SBR method through retailers as one of the strategies. These measures helped boost the Taiwanese bicycle industry's production of quality bicycle parts to 60% of total production and export two million first-class bicycles in 2003. Taiwan thus cornered 50% of the global market for quality bicycles and became the center of the high-end bicycle industry.

11.4.2.5 The A-Team Gradually Develops Its Achievements

From January to September 2003, the Giant and Merida bicycle factories (complete bicycle production firms) achieved an 18% climb in export sales, compared to the industry average of 11%. Furthermore, unit prices rose 21% to US$245.00. Eleven other bicycle exporters (bicycle parts factories) achieved 21% growth in export sales during the same period, vis-à-vis the industry average of 12%, to reach US$3 million. The A-Team recently established general goals for all of its members. Whole bikes are to be delivered within 14 days of order placement (compared to a historical lag of 1 month) and bike parts, within 10 days (compared to 21 days historically, with only weekly deliveries). In 2000, the average price of a bicycle was US$111; by 2003, it was USD$146. Giant produces just one third of what it once did in, yet its profits have been growing. Five or six years ago, it was cranking out 1,600,000 bicycles annually for a total profit of US$14 million; now, it just produces 600,000 units and rakes in US$16 million.

11.4.3 Formation of Dajia Bicycle Industry Village

11.4.3.1 Access to Factor Conditions

Dajia, located in central Taiwan, is near central Taichung Harbor, which is very accessible to both importers of raw materials and exporters of finished products. Through its interviews, this study found that businesses in Dajia use central Taichung Harbor instead of Kaohsiung harbor because it saves them 25% in transportation time and costs. All these factors give the businesses a competitive advantage.

11.4.3.2 Clustering with Related and Supportive Industries in Taichung, Plus Ready Access to Plenty of Professional Talent

Central Taiwan is home to a high-precision machine tool industry cluster; 68.1% of cutting machine production facilities in Taiwan are located there. Thus, the main industry in the Taichung area is machinery, and the bicycle industry is part of its chain production. The Taichung machine tool industry propelled the development of the Dajia bicycle industry. This corresponds with our research, which selected the Dajia area for geographical reasons. Our study showed a close relationship between the bicycle factories and the machinery industry in central Taiwan.

Moreover, the core factories in Dajia, such as those of Giant, Apro Tech, Willing Ind. Co. Ltd., and Yulun Bicycle Co. Ltd., produce whole bicycles, and the parts factories clustered around them exist in a vertical production relationship. The factories of whole bicycles are the leaders in this relationship, assisting and guiding the parts makers. This study found that the small parts factories are unable to conduct R&D, and instead attach themselves to the core factories for survival. Therefore, the large factories play a crucial role in the village.

Another important factor is professional investment. With the development of the industry into groups, production requires both labor and professional skills. The support from numerous professionals working in bicycle factories around Dajia means the area has an ideal competitive base in terms of manpower. Other service resources near Dajia offer the professional services required to assist in industry operations, including legal, logistical, and financial services. Most of the supporting industries and services are located within 30-min travel time of Dajia, underscoring the importance of geographic proximity for cooperative relationships.

11.4.3.3 Government Resources and Investment

The most direct form of government assistance for the Dajia bicycle industry village occurred at the end of 1970, when the government set up the Youth Industrial Park there. The complete public facilities and industrial business service network offered in the park prompted bicycle manufacturers to establish production facilities in the area. Bicycle factories mushroomed during the 1980s, demonstrating that government investment bolstered the production environment.

This study also found that factories around the Dajia Youth Industry Park enjoy a competitive advantage over those located south of the Da-an river. The advantage holds true for factory investment capital, the interaction between factories, and academic interaction. Again, this proves that appropriate government investment in a well-regulated and planned production environment helps overall enterprise competitiveness.

11.4.3.4 Tight Industrial Chain, Along with the Establishment of Social Networks and Support from Research Groups

The bicycle industry is based on chain production involving a high level of the machine product business. Developing a close social enterprise network thus has become crucial for survival. Social networks enable factories to share information and benefit from relationships with others. Social networks are indicators of the ease with which producers can organize affiliations during industry development. Affiliations such as the Bicycle Import and Export Association, Industrial and

Commercial Development and Investment Promotion Committee, A-Team, and strategic alliances have greatly benefited the Taiwanese bicycle industry.

Research and development groups and institutes supply the technology and manpower necessary to maintain the advantage and competitiveness of Dajia. Institutions such as the Industrial Technology Research Institute, Central Science Yuan, and Metal Industries Research and Development Centre (MIRDC) transfer technical know-how. Moreover, the Bicycle Research Center specializes in advancing bicycle industry technology and specifically serves central Taiwan.

11.4.3.5 Confirm the Goal of Industrial Investment and Lead in the Innovation System

Porter (1990) reported that when the national market becomes saturated, businesses are forced to develop international markets. Only in this way can they maintain growth or dispose of excess production. Following national market competition, firms must get ready to compete abroad. During the 1960s, the government protected local firms and helped some of the parts factories. In fact, however, the most competitive factories (top 20%) were only established in 1970s and 1980s, when the industry expanded to the European and American markets. Factories receiving American OEM orders gained international competitiveness, as well as production technology and skills. It was these OEM orders that spurred the industry into becoming internationally competitive.

11.4.3.6 Opportunity: The Trend Toward Leisure Bicycles

The above environmental factors and industry efforts placed the Dajia bicycle group in an advantageous position. Other key factors were opportunity and the power of government. During the early 2000s, sports and leisure became a major trend, and developed countries became concerned with ecology and green business. This gave bicycle producers an opportunity to maintain healthy production levels. At the same time, the government mobilized national resources to assist in researching new directions for development, and the results were transferred back to the industry. Thus, government influence has emerged as the prime driving force behind the creativity and growth of the Taiwanese bicycle industry.

11.4.4 Composition of the Dajia Industry Cluster

Dajia bicycle valley has major industrial and socioeconomic aspects (Table 11.8) which stimulate the development of the local bicycle industry. From previous discussions and the company survey, this study has found that all components contribute to and interact in the knowledge generation process; the components

Table 11.8 Composition of Dachia bicycle valley

Socioeconomic aspect	Industrial aspect
Industry association	Convenient transportation access
Government support	Proximity of supportive industries
Public atmosphere	Proximity of energy suppliers
	Human capital
	Technical assistant from research institutes

include the following: convenient transportation access (transportation around the industry valley), proximity of supportive industries (neighboring parts and machinery factories and material suppliers), proximity of energy suppliers (adequate supply of utilities), human capital (human resources employed in the machinery industry), technical assistance from research institutes (such as the ITRI, Central Science Yuan, MIRDC, and Bicycle Research Center), industry associations (social associations such as the Bicycle Import and Export Association, Industrial and Commercial Development and Investment Promotion Committee, A-Team, and strategic alliances), government support (policy support, investment, and pertinent infrastructures), and public reception (popularity of bicycling).

11.4.4.1 Socioeconomic Aspect

Related Social Association

Social associations generally gather companies and focus on how to improve industry development and stimulate investment. They do not merely make technical contributions, but also contribute to the creation of an integrated environment that improves overall industry competence as regards investment, technical cooperation, technological improvements, the direction of development, and strategy development. The A-Team is an exemplar of such an association, since it is intended to improve bicycle production and develop quality production to help establish a good international reputation for Taiwanese brands. The close relationships formed between association members embody an industry social network and stimulate informal interaction between firms and employees.

Prevalence of Bicycles

Bicycling is a popular form of exercise around the world. Also, recent environment protection concerns have indirectly encouraged the use of bicycles for daily transportation. Therefore, the demand for bicycles is increasing, particularly at the high end of the market.

Government Support

Government support, which has been crucial to the development of bicycle valley, includes building infrastructure, policies that encourage investment, and strategies supporting technical research and improvement. Such policies have fostered industry growth; government support has been especially helpful in establishing industry clusters. The government has also channeled funds to research institutes for related technology innovation, boosting the interaction between firms and research institutes. Recently, owing to bicycles being a green means of transportation, the government has poured in additional support to encourage the industry and innovation.

The socioeconomic components of bicycle valley play a pivotal role as well. Associations provide opportunities for firms to communicate and interact in the integration and implementation of development strategies. Numerous informal social activities are derived from the association, and the corporations create informal and social network ties.

11.4.4.2 Industrial Aspect

Supportive Industries

The supportive industries include chain parts factories, material factories, and machinery factories. Their presence and regular cooperation strengthens production relationships. Easy and frequent contact increases face-to-face activities, which consolidate human relations and form networks (Isaksen and Aslesen 2001). Dajia bicycle valley provides an environment in which it is easy to connect and communicate with related firms, reduce production time, and simplify management. Furthermore, technological exchange and experience spread more rapidly and intensively than in sparser areas.

Transportation

Convenient transportation gives bicycle valley a huge edge, as it saves travel cost and time. Furthermore, it offers the benefit of "just-in-time" production, enabling the highly efficient production and transportation of materials and finished goods. Access to a combination of road, rail, air, and sea transportation bolsters interaction with firms and cooperative partners located elsewhere—a key advantage in market expansion.

Human Capital

Human capital is crucial in the knowledge economy. Worker skill and knowledge determines production technology. In bicycle valley, besides the manpower of the bicycle industry, the machinery industry also brings relative human capital and technologies to improve the bicycle industry. The proximity of the machinery and bicycle industries is also advantageous in that their workers can easily interact and improve production.

Energy Suppliers

The relation between energy suppliers and bicycle valley is simple and is exclusively about energy supply. The presence of suppliers guarantees energy sufficiency, giving the area a significant development advantage.

Research Institutes

Numerous technological improvements are readily available from research institutes, which provide technical assistance in R&D and offer the industry new or improved technologies for better production. However, most of the research institutes are located in northern Taiwan, not in Dajia.

11.5 Conclusions and Suggestions

The major findings of this study regarding Dajia bicycle valley are as follows:

1. The strong foundation of the cluster is indispensable in assisting the development and transformation of traditional industry.
2. The technological innovations implemented by the bicycle industry are mostly derived from research institutes, and the industry has applied them to the production process.
3. The nearness of related industries and factories to each other has a significant influence on bicycle industry development. The proximity binds the existing cooperative relationships between firms and matches the close-chain characteristics of the bicycle industry.
4. Most of the research institutes providing technology to the bicycle industry are not located in Dajia. This fact does not negate the influence of proximity to research or academic institutes, but it does highlight the different characteristics of traditional industry compared to high-tech industry.
5. Most of the improvements implemented in Dajia bicycle valley are related to the production process. The interaction between firms contributes to the spread of knowledge. Even for small firms that lack R&D capability, interaction becomes a channel that promotes production technology. This feature of Dajia bicycle valley differentiates it from other high-tech industry clusters, which are more focused on generating original knowledge.

6. The A-Team plays a critical development role. The involvement of the A-Team in cooperation, the extension of partnerships, and strategies reduce the chances of lock-in.

The bicycle valley of Dajia represents a different type of knowledge community, which applies, derives, and transforms to improve production. The experiences and characteristics of Dajia bicycle valley demonstrate the successful transformation of a traditional industry into a potential development industry.

This study has its weaknesses. Whether or not the interactions resemble those between workers in a high-tech cluster, the manner of interaction remains unclear. Additionally, this study does not really determine whether a tightly knit cluster is really good for a traditional industry. This study merely demonstrates the advantages of geographic proximity in the bicycle industry, though these may be due to special chain characteristics. This investigation cannot guarantee the same effectiveness for other traditional industries, particularly with regard to the lock-in phenomenon.

References

Baptista R, Swann P (1998) Do firms in clusters innovate more? Res Policy 27:525–540

Bell ERJ (1993) Some current issues in technology transfer and academic-industrial relations: a review. Technol Anal Strateg Manag 5(3):307–321

Boschma RA (2005) Proximity and innovation: a critical assessment. Reg Stud 39(1):61–74

Breschi S (2000) The geography of innovation: a cross-sector analysis. Reg Stud 34(3):213–229

Camagni RP (1995) The concept of innovative milieu and its relevance for public policies in European lagging regions. Pap Reg Sci 74(4):317–340

Chang HL, Chang HW (2009) Exploring recreational cyclists' environmental preferences and satisfaction: experimental study in Hsinchu technopolis. Environ Plan B Plan Des 36:319–335

Chuang WS (2005) A study on the competitive factors of bicycle industrial cluster in Dajia area. Master's thesis, National Taiwan University

Cowan R, Hultén S (1996) Escaping lock-in: the case of the electric vehicle. Technol Forecast Soc Chang 53:61–79

Feldman MP (1994) Technological infrastructure. In: Feldman MP (ed) The geography of innovation. Kluwer, Dordrecht, pp 51–75

Hu T-S, Lin C-Y, Chang S-L (2005) Role of interaction between technological communities and industrial clustering in innovative activity: a case of Hsinchu district, Taiwan. Urban Stud 42(7):1139–1160

Isaksen A, Aslesen HW (2001) Oslo: in what way an innovative city. Eur Plan Stud 9(7):871–887

Malmberg A (1997) Industrial geography: location and learning. Prog Hum Geogr 21(4):573–582

Malmberg A, Maskell P (2002) The elusive concept of localization economies: towards a knowledge-based theory of spatial clustering. Environ Plan A 34:429–449

Massey D, Quintas P, Wield D (1992) High tech fantasies: science parks in society, science and space. Routledge, London

Monck CSP, Porter RB, Quintas P, Wynarczyk P (1988) Science parks and the growth of high technology firms. Croom Helm, London

Oakey RP (1985) High-technology industry and agglomeration economies. In: Hall P, Markusen A (eds) Silicon landscapes. ALLEN & UMWIN, Boston

Porter ME (1990) The competitive advantage of nations. Macmillan, London/Basingstoke

Ritsila JJ (1999) Regional differences in environments for enterprises. Entrep Reg Dev 11 (3):187–202

Saxenian A (1994) Regional advantage: culture and competition in Silicon Valley and Route 128. Harvard University Press, Boston

Storper M (1995) Competitiveness policy options: the technology-regions connection. Growth Chang 26:285–308

Torre A, Gilly JP (2000) On the analytical dimension of proximity dynamics. Reg Stud 34:169–180

Torre A, Rallet S (2005) Proximity and localization. Reg Stud 39(1):47–59

Westhead P, Batstone S (1998) Independent technology-based firms: the perceived benefits of a science park location. Urban Stud 35(12):2197–2219

Wu JH, Chen HS (2001) A study on innovation diffusion and spatial interaction of firms in the industrial zones firms of Taiwan. Sun Yat-Sen Manag Rev 9(2):179–200 (in Chinese)

Yigitcanlar T, Velibeyoglu K, Martinez-Fernandez C (2008) Rising knowledge cities: the role of knowledge precincts. J Knowl Manag 12(5):8–20

Chapter 12
Acceptability of Personal Mobility Vehicles to Public in Japan: Results of Social Trial in Toyota City

Ryosuke Ando, Ang Li, Yasuhide Nishihori, and Noriyasu Kachi

Abstract Many challenges have to be dealt with to create and maintain a low-carbon transport society. One representative technological innovation that is being considered to assist in achieving this goal is a personal mobility vehicle (PMV). A key factor in successfully introducing PMVs for use in the future is understanding social acceptability, i.e., whether the public will accept the new transport innovation. The goal of this study is to gain an understanding of the basic attitudes of the general public toward PMVs and the significant sociopsychological factors that influence the acceptance of PMVs by society. A brief overview of studies pertaining to PMVs is presented. A trial held in the city of Toyota is summarized, and the sociodemographic characteristics of the respondents are analyzed. The intended uses of PMVs are examined using principal component analysis and type I quantification theory, and the empirical results are discussed. Conclusions are drawn, and suggestions for further study are made.

Keywords Personal mobility vehicles (PMV) • Acceptability • Social trial • Toyota • Citizens' attitudes

---

R. Ando (✉) • N. Kachi
Research Department, Toyota Transportation Research Institute,
Wakamiya-cho 1-1, Toyota, Aichi 471-0026, Japan
e-mail: ando@ttri.or.jp; kachi@ttri.or.jp

A. Li
Urban Transport Center, Ministry of Housing and Urban–Rural Development,
No. 9 San Li He Road, Hai Dian District, Beijing 100037, China
e-mail: anglipeking@gmail.com

Y. Nishihori
Chubu Branch, Chuo Fukken Consultants Co. Ltd, Front Tower Bldg.,
Nishiki 2-3-4, Nagoya 460-0003, Japan
e-mail: nishihori_y@cfk.co.jp

M. Kawakami et al. (eds.), *Spatial Planning and Sustainable Development: Approaches for Achieving Sustainable Urban Form in Asian Cities*, Strategies for Sustainability,
DOI 10.1007/978-94-007-5922-0_12,

12.1 Introduction

Ohta (2008) notes that many challenges have to be dealt with to ensure sustainable transport. The issues, summarized in Table 12.1, can be divided into three categories: environmental, economic, and social issues. In the environmental category, it has been recognized that local air pollution is caused by an increase in vehicular emissions, while global warming is being aggravated by CO_2 emissions. In addition, waste of energy and resources is being caused by the consumption of fossil fuel, e.g., gasoline. Furthermore, another pressing concern is that road accidents are one of the principal culprits posing a danger to human life and health. Moreover, it has been accepted that a lifestyle of overdependence on motor vehicle results in some health problems. Regarding the economic category, traffic congestion decreases personal efficiency, and a decrease in the users of public transportation system, such as railways and buses, and management deterioration can lead to the degradation of public transport services. Moreover, another serious concern is the significant decline in the number of tourists coming to the central city, both local and foreign. Lastly, in the social category, it is important to ensure public transport service for mobility, especially for elderly people.

Technologies, such as intelligent transportation systems (ITS) in the field of traffic management, have played a significant role in solving these current challenges. One representative technological innovation that is being considered to address these challenges is a personal mobility vehicle (PMV). A PMV is defined as a single-person-use vehicle that explores the possibilities of personal mobility. Some advantages of PMVs are summarized in Table 12.2. First, in the environmental category, the use of PMVs instead of motor vehicles with an internal combustion engine can lead to a decrease in exhaust emissions and CO_2 emissions, an increase in fuel efficiency because of compact travel modes, and a reduction in traffic accident casualties. Second, in the economic category, the use of PMVs for commuting or for business could lead to a decrease in the occupied road space per person. Furthermore, the use of PMVs could increase the use of the public

Table 12.1 Current issues in transport planning and policy

Category	Issue
Environmental	Local air pollution caused by vehicular emissions
	Damage to the global environment caused by CO_2 emissions
	Waste of energy and resources caused by the consumption of fossil fuels, such as gasoline
	Traffic accidents posing a danger to human life and health
	Detriment to human health caused by of a lifestyle of overdependence on motor vehicle use
Economic	Cost of traffic congestion to the local economy
	Degradation of public transport services due to decreased use and management
	Urban deterioration and sprawl
Social	Need for public transport services to ensure mobility

Table 12.2 Potential benefits provided by PMVs

Category	Problem	Method and effect	
Environmental	Local air pollution	Method: use of PMVs instead of motor vehicles with internal combustion engines	◎
	Global environment	Effect: reduced exhaust emissions and CO_2 emissions and conservation of fuel	
	Waste of energy and resources		
	Traffic accidents	Method: use of PMVs instead of motor vehicles Effect: reduced number of traffic accident casualties	◎
	Health	Method: use of PMVs in daily life Effect: possibility of increased mobility, participation in social activities and exercise	▲
Economic	Traffic congestion	Method: use of PMVs instead of motor vehicles for commuting and business Effect: decreased occupied road space per person and decreased overall traffic congestion. Overall space use efficiency is improved due to decreased number of occupied parking spaces	◎
	Public transport service	Method: PMVs are used for access to and egress from transit Effect: access to and egress from public transport is improved	▲
	Activation of city	Method: use of PMVs for urban sightseeing tours Effect: tourism is promoted and diversified by reduced travel resistance	◎
Social	Mobility	Method: use of PMVs in daily life Effect: possibility of increased mobility and participation in social activities. Individual mobility is enhanced. Travel modes are diversified	◎

◎, possible; ▲, possible but there are also negative effects

transportation system because PMVs can conveniently access and egress the public transportation system. Moreover, the use of PMVs can increase and diversify sightseeing tours and activities in consideration to reduce travel resistance. Lastly, in the social category, the use of PMVs can meet the potential mobility needs, especially for elderly people, for social activity participations by connecting the public transit and residential locations.

A key success factor in the future introduction of PMV is understanding social acceptability, i.e., whether the public will accept the new transport innovation. Most PMVs can be categorized as one of two types: self-balancing personal transporters, such as the Segway and Winglet, and single-seated personal mobility cars, such as the i-REAL. Some of these PMVs have been introduced as a means of transport in the real world. For example, the Segway was first introduced to the US retail market in December 2001. In Japan, however, PMVs such as the i-REAL can only be used on private roads, e.g., for airport security, because this kind of transport device cannot be used on the public roads according to the Road Traffic Law of Japan. Therefore, it is necessary to measure social understanding that can lead to possible acceptance before PMVs can be introduced meaningfully to Japan. The goal of this study is to gain an understanding of the basic attitudes of the general public toward PMVs and the significant sociopsychological factors that can influence the acceptance of PMV by society.

This chapter is structured as follows. A brief overview of the studies pertaining to PMVs is presented. A trial held in the city of Toyota is summarized, and the sociodemographic characteristics of the respondents are analyzed. The intended uses of PMVs are examined using principal component analysis and type I quantification theory, and the empirical results are discussed. Conclusions are drawn, and suggestions for further study are made.

12.2 Literature Review

Some studies have been conducted to analyze and discuss PMV use, especially for self-balancing personal transporters, such as Segway. Ulrich (2005) pointed out that PMVs offer several potential benefits for users and society, including lower transportation costs, reduced trip times, and low environmental impact. Shaheen et al. (2005) reviewed the safety literature regarding low-speed crashes and offered some suggestions for using the Segway. That review included a summary of the regulatory and legislative history of PMVs in the USA for the purpose of examining how concerns about interaction between PMVs and pedestrians have resulted in legislation that includes specific safety requirements. Sawatzky et al. (2007) studied the use of the Segway as an alternative mobility device for people with disabilities and concluded that subjects with disabilities thought the Segway was easy to use and were excited about its potential to assist them. Miller et al. (2008) analyzed the approach speed and passing clearance that Segway devices exhibit on encountering a variety of obstacles on the sidewalk. Boniface et al. (2011) described a series of emergency department visits for injuries related to Segway use.

Since the introduction of the Segway in Japan in 2006, research into PMV use has steadily increased. For example, Nakaga et al. (2009, 2010) proposed a new PMV that can change between self-balancing mode and bicycle mode and investigated the safety and comfort of the proposed vehicle in pedestrian flows. Okabe et al. (2009) discussed the development of PMVs to support lifestyles for elderly people. Hamada et al. (2009) summarized the technology used in the Winglet, which is a self-balancing personal transporter developed by Toyota, and its possible future applications. Nishihori et al. (2010) summarized the issues related to the current urban transportation system in Japan and discussed issues related to the possible introduction of PMVs into society, by considering the characteristics of the various PMVs, the limitations of the current laws and rules, and the present road conditions in Japan. Li et al. (2011) identified differences in acceptability before and after use and proposed a theory to measure the acceptability of PMV.

Nevertheless, there is still room for further research. There have been few applications of single-seated three-wheeled personal mobility cars, although some new vehicles of this type, such as the i-swing and i-REAL, have been developed. This study focuses on the attitudes and usage intentions related to self-balancing personal transporters, such as the Winglet, and single-seated three-wheeled personal mobility cars, such as the i-REAL. The significant sociopsychological factors that can influence acceptance of PMVs by society are analyzed by applying type I quantification theory.

12.3 Data

12.3.1 *Outline of the Toyota Trial*

The data used in this study are derived from a survey in the Toyota trial. The trial is briefly summarized below. The purpose of the trial was to investigate the use of PMVs, specifically the Winglet and i-REAL, as shown in Fig. 12.1. The trial was held in the city of Toyota for 2 weeks in October 2010. At the beginning of the trial, vehicle demonstrations by type were conducted, based on the assumption that giving demonstrations of the vehicles could have a greater impact than merely giving information about them. During the trial, details were provided regarding the reasons for the development of PMVs, the potential applications of PMVs, and the various types of PMVs. This chapter focuses only on the attitudes of the respondents who did not use the PMVs but rather watched the demonstrations and received the information distributed.

The questionnaire used included questions regarding the possible use of PMVs, but also included specific questions about driving and experience. Nine questionnaire items were used to assess the respondents' judgments about PMVs, which were categorized as useful or useless, pleasant or unpleasant, good or bad, likeable or irritating, practical or impractical, nice or annoying, effective or superfluous, desirable or undesirable, and convenient or inconvenient. Individual item scores ranged from −2 to + 2. These items can effectively measure the public's perception of the usefulness of and satisfaction with PMVs.

12.3.2 *Respondents*

The sociodemographic characteristics of the 66 survey respondents are presented in Table 12.3. A total of 66 people responded to the survey about their attitudes about PMVs. Overall, 60.6% of the respondents were male, and 39.4% were female. The study group consisted of the following: 5 persons (7.6%) with ages from 10 through 19 years old, 8 persons (12.1%) with ages from 20 through 29, 21 persons (31.8%)

Fig. 12.1 PMVs in the Toyota trial (*left* i-REAL, *right* Winglet)

Table 12.3 Sociodemographic characteristics of respondents

Item	Category	Sample size	Percentage (%)
Gender	Male	40	60.6
	Female	26	39.4
Age of individual (in year)	10–19	5	7.6
	20–29	8	12.1
	30–39	21	31.8
	40–49	13	19.7
	50–59	5	7.6
	60 and above	13	19.7
	Missing	1	1.5
Area of residence	Toyota City	44	66.7
	Nagoya City	3	4.5
	Aichi Prefecture	15	22.7
	Other	4	6.1

with ages from 30 through 39, 13 persons (19.7%) with ages from 40 through 49, 5 persons (7.6%) with ages from 50 through 59, and 13 persons (19.7%) with ages 60 or over. Overall, 66.7% of the respondents lived in the city of Toyota, 22.7% of respondents lived in the Aichi Prefecture except the cities of Toyota and Nagoya, and only 4.5% of respondents lived in the city of Nagoya.

Among the respondents, 71.3% reported being aware of the existence of the self-balancing Segway personal transporter before the survey, but only 6.1% reported having used a Segway. Similarly, 73.7% of the respondents knew of the Winglet, but only 6.1% reported having used one. A somewhat smaller percentage of respondents (65.1%) reported knowing of the single-seated three-wheeled personal mobility car, the i-REAL. Only 3% of respondents reported having used the i-REAL. Although only 3–6% of the respondents reported having used a PMV, the percentages of people that knew of PMVs were quite high, considering that PMVs have not yet been introduced for public use in Japan.

In response to a question about whether they thought they could drive a Winglet or not, 45.5% of the respondents said that they believed that they could drive it easily, while 13.6% of respondents said that they thought that the Winglet would be difficult to use. Similarly, 45.5% of the respondents thought that they could easily drive the i-REAL, but 9.1% did not think that they could. There was, however, a higher percentage of individuals who thought it would be difficult to use Winglet compared to individuals who thought it would be difficult to use the i-REAL. This difference is probably because the Winglet is a self-balancing type of PMV.

12.4 Empirical Analysis and Discussion

12.4.1 Usage Intentions of PMV

The summary of usage intentions of PMV by type is shown in Table 12.4. Among the respondents, 32 respondents (15.6%) preferred the Winglet for short-distance trips in downtown area and moving in buildings. Moreover, 31 respondents (15.1%) choose the Winglet as a transport mode for the purpose of tourism and excursions. Furthermore, 22 (10.7%) and 21 (10.2%) respondents preferred the Winglet for the purpose of accessibility between their destination and train station/bus stop as well as between their home and the nearest station for commuting, respectively. On the other hand, only 2 respondents (1%) thought the Winglet was available to extend travel range as a senior car. Additionally, only 6 respondents (2.9%) believed that the Winglet could be used by disabled or elderly people.

Regarding the usage intentions of i-REAL, 44 respondents (15.9%) preferred to use PMVs to extend travel range as a senior car, and 43 respondents (15.6%) thought the vehicles are available for transport support for disabled or elderly people. Moreover, 30 respondents (10.9%) choose i-REAL as a mobility mode to go to the neighborhood hospital. On the other hand, most of the respondents believed that i-REAL is not suitable to use in a building (2.5%), move in short-distance trip in downtown area (4%), and access between home and the nearest station for commuting. Surprisingly, there are only 13 respondents (4.7%) who choose i-REAL as a mobility mode for the purpose of tourism and excursions.

The survey results shown in Table 12.4 indicate significantly different usage intentions for the self-balancing Winglet and the single-seated i-REAL. However, the respondents only watched the demonstrations and received the distributed information. After experiencing the vehicles, these people could change their initial usage intentions concerning the Winglet or i-REAL.

12.4.2 Principal Component Analysis

The nine items used to assess the respondents' judgments about the PMVs were analyzed by principal component analysis. An equamax rotation with Kaiser normalization was performed to define the components in terms of subsets of

Table 12.4 Respondents' usage intentions of the Winglet ($n = 66$) and the i-REAL ($n = 66$)

Item	Winglet	i-REAL	Sample size by PMV
Tourism and excursions*	31 (15.1%)	13 (4.7%)	
Short-distance trip in downtown area*	32 (15.6%)	11 (4.0%)	
Moving within a building*	32 (15.6%)	7 (2.5%)	
Access between home and the nearest station	21 (10.2%)	13 (4.7%)	
Shopping in neighborhood	15 (7.3%)	17 (6.2%)	
Transport support for disabled or elderly people*	6 (2.9%)	43 (15.6%)	
Access between destination and train station/bus stop	22 (10.7%)	16 (5.8%)	
Going to the neighborhood hospital*	7 (3.4%)	30 (10.9%)	
Access between home and Toyota-shi station	10 (4.9%)	19 (6.9%)	
Extended travel range as a senior car*	2 (1.0%)	44 (15.9%)	
Medium-distance commute*	7 (3.4%)	27 (9.8%)	
Daily transport in mountain area*	6 (2.9%)	21 (7.6%)	
Business travel in urban area	14 (6.8%)	15 (5.4%)	

*$p \leq .01$

items, as shown in Table 12.5. The first set of subscales for both the Winglet and the i-REAL, termed useful, practical, effective, and convenient, could be interpreted as the usefulness of a system. The second set of subscales, termed pleasant, good, and desirable, could be interpreted as reflecting the satisfaction with a system.

In Table 12.6, the mean scores for the usefulness and satisfaction subscales are given by different variables. As Fig. 12.2 shows, there are significant differences between the subscales for the two types of PMVs. Among the respondents, those aged between 20 and 59 rated the potential usefulness of and satisfaction with the self-balancing Winglet most highly. The respondents older than 59 rated the potential usefulness of and satisfaction with the self-balancing vehicle lower. For the i-REAL, only the category of respondents with driver's licenses gave positive evaluations for both usefulness and satisfaction. The lowest evaluations for the i-REAL were given by respondents without licenses. The plot also suggests that men and women have different attitudes toward the usefulness of the i-REAL. There are also significantly different attitudes toward the i-REAL among the different age groups. Respondents of ages less than 20 evaluated the i-REAL negatively in terms of both usefulness and satisfaction.

12.4.3 Analysis of Type I Quantification Theory

Type I quantification theory was employed to analyze the significant sociopsychological factors that can influence acceptance of PMV by society.

Table 12.5 Principal component analysis[a] of nine items for the Winglet ($n = 66$) and the i-REAL ($n = 66$)

Item	Winglet		i-REAL	
	Component 1	Component 2	Component 1	Component 2
1. Useful	0.66	0.40	0.64	0.33
2. Pleasant	0.27	0.77	0.09	0.93
3. Good	0.18	0.87	0.55	0.60
4. Likeable	0.55	0.72	0.61	0.59
5. Practical	0.77	0.35	0.88	0.20
6. Nice	0.42	0.76	0.67	0.42
7. Effective	0.70	0.54	0.83	0.15
8. Desirable	0.70	0.36	0.48	0.61
9. Convenient	0.88	0.11	0.67	0.52
Cumulative %	61.98	72.26	59.55	68.99
Component name	Usefulness	Satisfying	Usefulness	Satisfying

[a]Rotation method: equamax with Kaiser normalization, rotation converged in three iterations

Table 12.6 Scores for the usefulness and satisfaction subscales by variable

Variable	Category	Winglet		i-REAL	
		Usefulness	Satisfaction	Usefulness	Satisfaction
Gender	Male	0.00	0.05	0.17	−0.06
	Female	0.00	−0.06	−0.24	0.08
Age of individual (in years)	Less than 20	−0.17	0.13	−0.14	−0.30
	20–59	0.05	0.11	−0.01	0.07
	60 and above	−0.10	−0.55	0.02	−0.25
Driver's license	Yes	−0.01	0.03	0.03	0.03
	None	0.13	−0.32	−0.28	−0.36

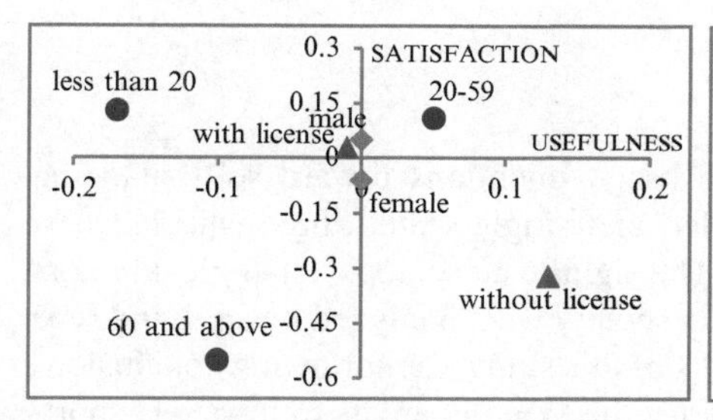

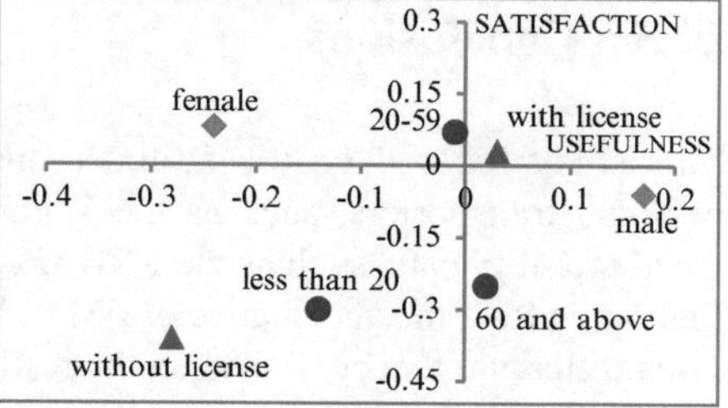

Fig. 12.2 Scores for the usefulness and satisfaction subscales (*left* Winglet, *right* i-REAL)

As mentioned in Sect. 3.1, nine questions were asked to assess the respondents' judgments regarding the Winglet and the i-REAL. Individual item scores ranged from −2 to + 2. The items are analyzed by applying principal component analysis. In consideration of the principal component as an index to represent the total acceptability of the PMV, the extracted principal component is proposed as the dependent variable for analysis of type I quantification theory. The independent variables include the individual attributes, the vehicle attributes, and the attitudes explored in the survey. The analysis results of the type I quantification theory analysis are summarized in Table 12.7 and Table 12.8 for the Winglet and the i-REAL, respectively.

As a result, among the independent variables considered, the variable *design of the Winglet* has the strongest influence on the acceptability of the self-balancing Winglet. The variables *transport only one person*, *usefulness in buildings*, and *whether person can drive the vehicle* also significantly affect acceptability of the Winglet. On the other hand, the variables *gender* and *age*, which were presumed at the beginning of the study to be important factors, are not statistically significant. Turning to the sociopsychological factors of the i-REAL, among the independent variables considered, the variable *design of the i-REAL* affects the acceptability most strongly. The variables *usefulness on roads*, *safety on roads*, and *harmony when used in buildings* also significantly affect the acceptability of the i-REAL. The variables *environmentally friendly* and *gender* have little influence on the acceptability of the i-REAL.

The *design* of the vehicle appears to be important to the respondents' acceptance of the PMVs. There was less difference among the respondents by individual attributes, such as *gender* and *age*. The Winglet and the i-REAL probably have different sociopsychological factors with respect to their acceptability because people have different usage intentions for them, as discussed in Sect. 4.1. However, most of the respondents have no experience driving these vehicles but rather know them only through demonstrations and information provided to them. Therefore, whether their attitudes would be different after experience driving, PMVs should be investigated in future research.

12.5 Conclusions

This study focused on the attitudes and usage intentions toward self-balancing personal transporters, such as the Winglet, and single-seated three-wheeled personal mobility cars, such as the i-REAL. The significant sociopsychological factors that may affect the acceptance of PMVs by society were analyzed by applying type I quantification theory. Based on the results of this study, the following conclusions are drawn. The top three usage intentions for the Winglet are as follows: (1) short-distance trips in the downtown area, (2) moving around in buildings, and (3) tourism and excursions. The top three usage intentions for the i-REAL are as follows: (1) extending travel range using the vehicle as a senior car, (2) transport support for the disabled or elderly, and (3) going to the neighborhood hospital. The design of the Winglet, namely that it transports only one person, and the respondent's attitudes concerning whether the Winglet is useful in buildings significantly affect the acceptability of the Winglet. The design of the i-REAL, whether it is useful, whether it is safe to use on the road, and whether it is harmonious when

Table 12.7 Results of the type I quantification theory analysis for the Winglet[a]

Variable	Category	Sample size	Score	Range	Partial correlation coefficient
Gender	Male	29	0.11	0.24	0.10
	Female	22	−0.14		
Age	Less than 59	41	0.02	0.08	0.02
	60 and above	10	−0.06		
Do you know of the Winglet?	Yes	36	0.10	0.36	0.13
	No	15	−0.25		
Do you think you can drive a Winglet?	Yes	26	0.63	1.68	0.44
	I don't know	18	−0.49		
	No	7	−1.05		
Do you think the Winglet is safe to use on the road?	Yes	13	−0.08	0.55	0.17
	I don't know	10	0.43		
	No	28	−0.12		
Do you think the Winglet is useful on the road?	Yes	31	0.46	1.47	0.47
	I don't know	15	−1.01		
	No	5	0.18		
Do you think the Winglet is harmonious on the road?	Yes	18	−0.05	0.40	0.12
	I don't know	10	−0.25		
	No	23	0.15		
Do you think the Winglet is safe to use in buildings?	Yes	21	−0.09	0.84	0.22
	I don't know	11	−0.47		
	No	19	0.37		
Do you think the Winglet is useful in buildings?	Yes	38	0.45	2.38	0.50
	I don't know	6	−0.62		
	No	7	−1.93		
Do you think the Winglet is harmonious in buildings?	Yes	30	0.18	0.88	0.27
	I don't know	9	−0.71		
	No	12	0.09		
Design of the Winglet	Approve	42	0.29	3.54	0.44
	Average	7	−0.81		
	Disapprove	2	−3.25		
Transport only one person	Approve	41	0.32	2.40	0.48
	Average	9	−1.57		
	Disapprove	1	0.83		

[a]Sample size, 51; constant variable, −0.10; multiple correlation coefficient, 0.89

Table 12.8 Results of the type I quantification theory analysis for the i-REAL[a]

Variable	Category	Sample size	Score	Range	Partial correlation coefficient
Gender	Male	28	0.04	0.09	0.04
	Female	21	−0.05		
Age	Less than 59	39	0.12	0.57	0.21
	60 and above	10	−0.45		
Do you know of the i-REAL?	Yes	32	0.30	0.86	0.31
	No	17	−0.56		
Do you think you can drive the i-REAL?	Yes	26	0.33	0.80	0.31
	I don't know	18	−0.47		
	No	5	0.00		
Do you think the i-REAL is safe to use on the road?	Yes	14	0.61	1.40	0.46
	I don't know	16	−0.79		
	No	19	0.21		
Do you think the i-REAL is useful on the road?	Yes	31	0.65	2.68	0.65
	I don't know	7	0.31		
	No	11	−2.03		
Do you think the i-REAL is harmonious on the road?	Yes	16	0.04	0.42	0.14
	I don't know	10	−0.31		
	No	23	0.11		
Do you think the i-REAL is useful in buildings?	Yes	32	0.26	1.06	0.37
	I don't know	6	0.10		
	No	11	−0.80		
Do you think the i-REAL is harmonious in buildings?	Yes	24	−0.54	1.39	0.44
	I don't know	11	0.10		
	No	14	0.85		
Design of the i-REAL	Approve	39	0.17	3.51	0.58
	Average	7	0.39		
	Disapprove	3	−3.12		
Transport only one person	Approve	34	0.24	1.11	0.29
	Average	11	−0.42		
	Disapprove	4	−0.87		
Environmentally friendly	Approve	38	−0.02	0.07	0.02
	Average	11	0.06		
	Disapprove	0	–		

[a]Sample size, 49; constant variable, −0.24; multiple correlation coefficient, 0.88

used in building were statistically significant factors in the acceptability of the i-REAL. In summary, the self-balancing Winglet and the single-seated three-wheeled i-REAL have different usage intentions and different sociopsychological factors affecting their acceptability.

In future research, several issues need to be addressed before PMVs are introduced for public use in Japan. Among these are safety concerns and legislation issues. The safety of PMVs in interaction with other vehicles and with pedestrians on public roads is a key factor in whether the public supports the introduction of PMVs. Therefore, safety concerns should be further studied based on the usage intentions identified in this study.

References

Boniface K, McKay MP, Lucas R, Shaffer A, Sikka N (2011) Serious injuries related to the Segway® personal transporter: a case series. Ann Emerg Med 57(4):370–374

Hamada H, Takase M, Yamashita K, Oh S, Ohtaka T (2009) Moving vehicle 'Winglet' in the near future. J Inst Electr Eng Jpn 129(11):758–761 (in Japanese)

Li A, Ando R, Nishihori Y, Kachi N (2011) Measuring acceptability of self-balancing two-wheeled personal mobility vehicles. Proc 2011 JSAE Annu Congr (Spring) 41(11):7–12 (in Japanese)

Miller S, Molino JA, Kennedy JF, Emo AK, Do A (2008) Segway rider behavior: speed and clearance distance in passing sidewalk objects. Transp Res Rec J Transp Res Board 2073:125–132

Nakaga C, Suda Y, Nakano K, Nabeshima K (2009) Proposal for personal mobility vehicle. Seisankenkyu 61(1):71–74 (in Japanese)

Nakaga C, Nakano K, Suda Y, Kawarasaki Y, Kosaka Y (2010) Safety and comfort of the personal mobility vehicles in the pedestrian flows. Trans Soc Autom Eng Jpn 41(4):941–961 (in Japanese)

Nishihori Y, Kawai M, Kachi N, Inagaki T, Ando R (2010) A discussion for introducing personal mobility vehicles into real society. Proc 2010 JSAE Annu Congr (Autumn) 142(10):23–26 (in Japanese)

Ohta K (2008) The evolution, challenges to and direction of "transport-based Machizukuri (community development)". IATSS Rev 33(2):136–139 (in Japanese)

Okabe K, Tomokuni N, Shino M, Kamata M (2009) Development of personal mobility to support lifestyles for seniors. J Soc Autom Eng Jpn 63(9):104–105 (in Japanese)

Sawatzky B, Denison I, Langrish S, Richardson S, Hiller K, Slobogean B (2007) The Segway personal transporter as an alternative mobility device for people with disabilities: a pilot study. Arch Phys Med Rehabil 88(11):1423–1428

Shaheen SA, Rodier CJ, Eaken AM (2005) Improving California's Bay Area rapid transit district connectivity and access with Segway human transporter and other low-speed mobility devices. Transp Res Rec J Transp Res Board 1927:189–194

Ulrich KT (2005) Estimating the technology frontier for personal electric vehicles. Transp Res C Emerg Technol 13(5–6):448–462

Chapter 13
Urban Form, Transportation Energy Consumption, and Environment Impact Integrated Simulation: A Multi-agent Model

Ying Long, Qi-zhi Mao, and Zhen-jiang Shen

Abstract More energy is being consumed as urbanization spreads. Extensive research has found that a dominant share of urban energy consumption belongs to transportation energy, which has a strong relationship with urban form in the intracity level. However, little attention has been paid to the relationship between urban form, transportation energy consumption, and its environmental impact in the inner-city level. This chapter aims to investigate the impact of urban form, namely, the land-use pattern, distribution of development density, and the number and distribution of job centers on the residential commuting energy consumption (RCEC). We developed a multi-agent model for the urban form, transportation energy consumption, and environmental impact integrated simulation (FEE-MAS). Numerous distinguishable urban forms were generated using the Monte Carlo approach in the hypothetical city. On the one hand, the RCEC for each urban form was calculated using the proposed FEE-MAS; on the other hand, we selected 14 indicators (e.g., Shape Index, Shannon's Diversity Index, and Euclidean Nearest Neighbor Distance) to evaluate each generated urban form using the tool FRAGSTATS, which is loosely coupled with the FEE-MAS model. Afterward, the quantitative relationship between the urban form and RCEC was identified using the calculated 14 indicators and RCEC of all generated urban forms. Several

Y. Long (✉)
Beijing Institute of City Planning,
No. 60 South Lishi Road, Beijing 100045, China
e-mail: longying1980@gmail.com

Q.-z. Mao
School of Architecture, Tsinghua University,
Haidian District, Beijing 100084, China
e-mail: qizhi@mail.tsinghua.edu.cn

Z.-j. Shen
School of Environment Design, Kanazawa University, Kakuma Machi,
Kanazawa City 920-1192, Japan
e-mail: shenzhe@t.kanazawa-u.ac.jp

M. Kawakami et al. (eds.), *Spatial Planning and Sustainable Development: Approaches for Achieving Sustainable Urban Form in Asian Cities*, Strategies for Sustainability,
DOI 10.1007/978-94-007-5922-0_13,

conclusions were drawn from simulations conducted in the hypothetical city: (1) the RCEC may vary three times for the same space with various urban forms; (2) among the 14 indicators for evaluating urban form, the patch number of job parcels is the most significant variable for the RCEC; (3) the RCECs of all urban forms generated obey a normal distribution; and (4) the shape of an urban form also exerts an influence on the RCEC. In addition, we evaluated several typical urban forms—e.g., compact/sprawl, single center/multicenters, traffic-oriented development, and greenbelt—in terms of the RCEC indicator using our proposed model to quantify those conventional planning theories. We found that not all simulation results obey widely recognized existing theories. The FEE-MAS model can also be used for evaluating plan alternatives in terms of transportation energy consumption and environmental impact in planning practice.

Keywords Land use • Development density • Transportation energy consumption • Environment impact • Multi-agent model (MAS) • Monte Carlo

13.1 Introduction

The global environment is deteriorating with the accelerating consumption of fossil fuel. Human activities related to energy consumption are reported as the dominant driving force of global warming (IPCC 2007). Energy consumption in urban areas comprises 75% of total global consumption and greenhouse gas emissions, 80% of the world total (Shen 2005). With this as the backdrop, the low-carbon society (LCS) has been extensively discussed planetwide. China has been the focus the discussion (Hourcade and Crassous 2008; Remme and Blesl 2008; Shukla et al. 2008) due to the enormous challenges it faces in energy consumption and the corresponding greenhouse gas emissions (Wang and Chen 2008; Zhuang 2008). Mixed land use, compact cities, and smart growth have been recognized as effective tools for solving the urban energy consumption problem, by introducing reasonable spatial organization into urban systems. Fuel for transportation has been proved to have a significant relationship with urban form by many researches, which use the whole city as a sample for intercity comparison. However, not too much attention has been paid to identification of the quantitative relationship between urban form, and transportation energy consumption and environmental impact in the inner city.Conventional land use and integrated transportation models generally have comprehensive structures and a number of modules, requiring large-scale datasets and long-run time (Johnson and McCoy 2006). These models are not suitable for retrieving general rules dominating urban systems, especially the relationship between urban form, and transportation energy consumption and environmental impact. This chapter will thus construct an urban form-transportation energy consumption-environment integrated multi-agent model (FEE-MAS) to identify the quantitative influence of the urban form (e.g., land-use pattern, development density distribution, and the number and distribution of job centers) on transportation energy consumption and environmental impact in the inner-city level. For this analysis and simulation, we derive insightful results based on urban microsamples, such as parcels in the physical space and residents in the social space. In particular, we will focus on the residential commuting energy consumption (RCEC) sector in transportation energy consumption in this chapter as the first stage of the FEE-MAS model.

Three types of factors have been proved to influence the RCEC: urban form (e.g., land-use characteristics), transportation system characteristics (e.g., accessibility, convenience, and service quality), and the socioeconomic attributes of the individual or family (Wang et al. 2008). Urban form, as the important carrier for energy conservation and low-carbon economy, is the basic outcome of spatial plans, and it can guarantee the sustainable development of the urban system from the very beginning of urban development. Many studies have empirically indicated that urban form has strong relationships with energy consumption, especially transportation energy consumption—including passengers and cargo (Owens 1987; Anderson et al. 1996). Generally, the urban form featuring polycentric, high-density, and mixed-use areas corresponds to a lower average transportation energy consumption per capita. For example, Newman and Kenworthy (1989), using many cities as samples, found that the average transportation energy consumption per urban form decreases with population density. Holden and Norland (2005) discovered significant relationships between urban form and household and transportation energy consumption by analyzing eight neighborhoods in the greater Oslo region, which indicates that the compact city policy corresponds with a sustainable urban form. Shim et al. (2006) analyzed the impact of city size, density, and number of centers on transportation energy consumption. Alford and Whiteman (2009) examined the relationships between transportation energy consumption and urban form, as well as the choice of transportation infrastructure, via the evaluation of various urban forms in different subregions of the Melbourne area in Australia.

Urban form, as one of the factors, has significant impact on the traveler's commuting behavior choices and total commuting distance, influenced by the RCEC. The activity-based modeling approach is widely applied using travel diaries as the basic dataset for these researches, in which urban forms at housing and job sites are used as variables for quantitative evaluation. These empirical researches range from the impact of urban form on traveling behavior, mobile traveling behavior, and children's traveling behavior to pedestrian traveling behavior and nonwork traveling behavior (Dieleman et al. 2002; Giuliano and Narayan 2003; Horner 2007; Maat and Timmermans 2009; McMillan 2007; Pan et al. 2009; Schlossberg et al. 2006; Zhang 2005). Moreover, Krizek (2003) also indicated that the traveling behavior of a family could differ from that of their neighborhood.

Multi-agent systems (MAS), as a type of bottom-up approach based on the theory of complex adaptive systems (CAS), can be borrowed to induce the relationship between urban form and the RCEC in the inner-city level. The agent in MAS is the entity with high autonomous ability running in a dynamic physical environment (Zhang et al. 2010). Various researches have been conducted using MAS for simulating residential location choice (Benenson et al. 2001; Brown and Robinson 2006) and commercial facility location choice (Yi et al. 2008). In addition, Kii and Doi (2005) developed an integrated land-use and transportation model based on MAS, incorporating spatial economics to evaluate the compact city policy in terms of quality of life (QOL). Zellner et al. (2008) evaluated the impact of various planning policies on urban form, development density, and air quality in a hypothetical city using a MAS model. Our chapter will also use the MAS approach to develop an explicitly spatial model to evaluate potential total commuting distance, RCEC, and the environmental impact of various types of urban forms to identify their quantitative relationship.

This chapter will investigate the following topics: the varying extent of the RCEC for distinguished urban forms within the same area, the most significant spatial indicator of urban form influencing the RCEC and the weight of each spatial indicator or urban form, the shape of the urban form influencing the RCEC, and the differences between various typical urban forms originating from conventional urban planning theories. Thus, the factors influencing the RCEC, hypothesis of the FEE-MAS model, and simulation procedures will be elaborated in Sect. 13.2. The preliminary results of the model will be introduced in Sect. 13.3. Finally, the concluding remarks and discussion of the FEE-MAS will be presented in Sect. 13.4.

13.2 Approach

13.2.1 Conceptual Model

The residential commuting energy consumption and corresponding environmental impact are the results of various urban activities of residents. The integrated model proposed in this chapter will focus on the RCEC and its environmental impact, as well as their relationship with urban form. Therefore, factors related to the RCEC should be identified from aspects of the socioeconomic attributes of residents, the physical spatial layout of the city, and the characteristics of the urban transportation system (see Fig. 13.1). The RCEC depends on commuting frequency, distance, and mode. Commuting frequency is related to the socioeconomic characteristics of commuters; commuting distance, to land-use characteristics; and commuting mode, to both the socioeconomic characteristics of commuters and land-use characteristics.

13.2.2 Hypothesis of the Model

We propose the following hypotheses for the FEE-MAS model to test planning theories and identify the general rules governing urban systems in a hypothetical city, especially the relationship between the urban form and RCEC:

1. The hypothetical city as a closed system has no transport link with outside regions, and every resident works within the city.
2. Parcels in the city are square and identical in size. The road networks are grids with no subway system.

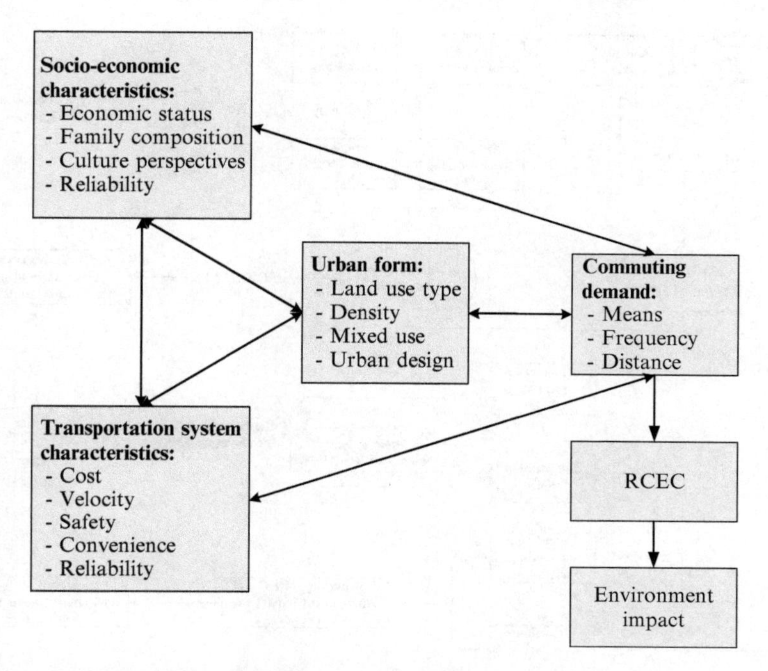

Fig. 13.1 Factors related to the RCEC and environmental impact

3. The hypothetical city is fully developed, and only residential (R) and commercial (C) types of land use are accounted for.
4. A parcel with a floor-area ratio (FAR) of 1 corresponds to one resident living in the parcel.
5. Every resident works and commutes.
6. Only residential commuting energy consumption is counted for residents; household, entertainment, shopping, and other types of energy consumed are excluded.
7. There is no capacity limitation on the job count in each parcel.
8. Residents choose the residing parcel randomly; it is not related to their socio-economic attributes.
9. Three types of commuting modes are considered: car, bus, and biking or walking.
10. The socioeconomic characteristics of the residents remain the same for all generated urban forms and will not vary through simulations.

13.2.3 Simulation Procedures

We developed the FEE-MAS model based on ESRI ArcGIS Geoprocessing, using Python. Residential agents and urban parcels are the two types of primary elements of FEE-MAS, whose simulation procedures are illustrated in Fig. 13.2. On the one

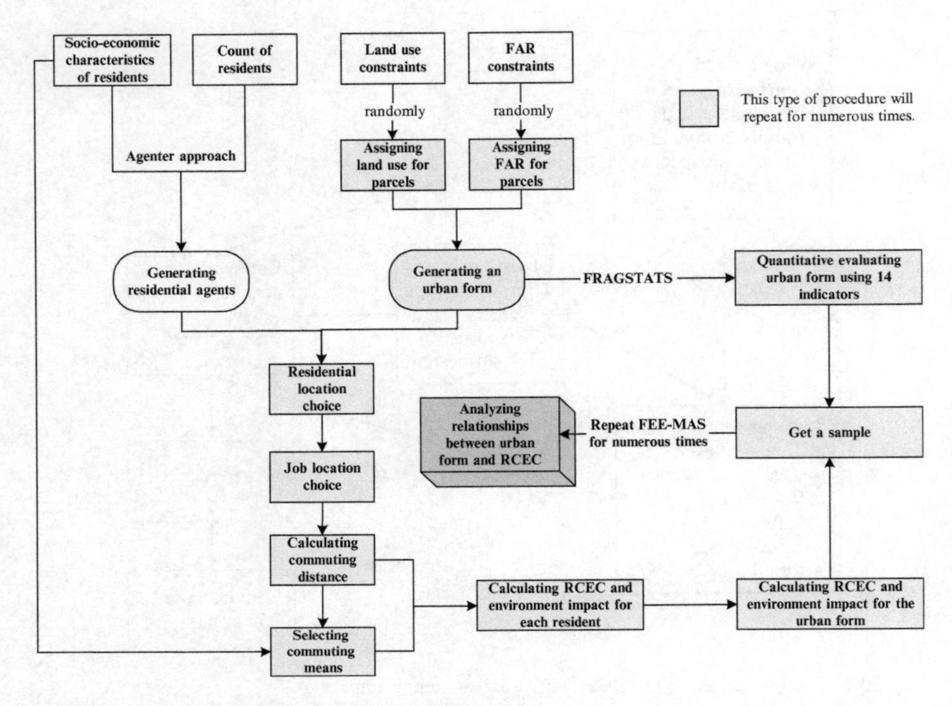

Fig. 13.2 The flow diagram of the FEE-MAS model

hand, each urban form can be generated by setting the land-use pattern and density distribution randomly and measuring them quantitatively using several indicators. On the other hand, residential agents can be generated using the total number of residents and their socioeconomic statistical characteristics. Each residential agent will choose a residential place and a job place in the generated urban form with residential and commercial parcels, then select a commuting mode based on his/her socioeconomic attributes and commuting distance. The RCEC and related environmental impact of each residential agent can thus be calculated, then aggregated in the whole city level by summing all residential agents. At this point, the simulation for an urban form is finished. Finally, we will run the FEE-MAS model for several times to get sufficient samples for analyzing the relationship between the urban form and RCEC.

The detailed simulation procedures are as follows.

13.2.3.1 Generating Residential Agents

Residential agents as the input of FEE-MAS can be generated using the disaggregation approach proposed by Long et al. (2010) from the population census report of Beijing (Beijing Fifth Population Census Office and Beijing Statistical Bureau 2002) and common sense regarding residents' socioeconomic characteristics. Two thousand resident agents were generated for simulations and computing the RCEC in the level of resident. The residential agent can be expressed as A_j, where j is the resident ID. J as 2,000 is the total number of residents. The socioeconomic attributes of residential agents are expressed as $A_j{}^s$, where s is the ID of socioeconomic attributes.

13.2.3.2 Generating Urban Forms and Residential Location Choice

The hypothetical urban form is supposed to have a size of 20 parcel × 20 parcel, composed of 400 parcels (see Fig. 13.3). Two thousand residents live in the hypothetical city. In this chapter, the key components of an urban form include the land-use pattern, spatial distribution of density, as well as the number and distribution of commercial parcels. The urban form generation procedure is made up of two steps. First, the land-use pattern is assigned for an urban form F^i (i is the urban form ID). There are two types of land use in the hypothetical city: the residential type (R) and commercial type (C). The number of commercial parcels obeys the uniform distribution of 10–40 for generated urban forms. For instance, if 25 is chosen as the number of commercial parcels for an urban form, 25 parcels will be randomly selected from the 400 parcels and assigned as land-use type "C"; the rest of the parcels are assigned as "R." The land-use type of parcel m in the urban form i is defined as $TP^i{}_m$. In this regard, the land-use pattern is defined for this urban form, with the number and location of commercial parcels defined. Second, a float number randomly selected from 0 to 10, standing for the FAR value $\mathrm{FAR}^i{}_m$, is assigned for each residential parcel m in the urban form i. $\mathrm{FAR}^i{}_m$ is rescaled to meet the sum of FAR values for all residential parcels equal to the count of all residents (2,000), enabling $\mathrm{FAR}^i{}_m$ to represent the resident count of the residential parcel m in the urban form i. The number of urban forms is generated using the Monte Carlo approach for the hypothetical city.

As to the choice of residential location, the 2,000 residents will randomly select a residential parcel, obeying the FAR value of each parcel. For example, three residents randomly selected from the total of 2,000 will reside in a parcel with the FAR as 3. Then, $\sum_{m=1}^{M} AC_m = 2000$, where AC_m is the residential agent count in the parcel m and is equal to its FAR, and M is the parcel count in each urban form (400 in this chapter).

13.2.3.3 Job Location Choice

Two extreme conditions are considered for the process of choosing a job location, without counting residents' socioeconomic characteristics. For the first condition, each resident with full rationality will select the nearest commercial parcel in which to work. For the second condition, each resident will randomly select a commercial

32	4	4	9	3	4	9	5	7	30	4	2	46	7	4	11	6	4	4	8
7	5	5	3	4	8	27	8	5	8	11	12	55	5	5	5	6	5	5	6
6	2	25	5	4	26	1	5	5	7	4	2	50	9	3	2	4	4	5	4
3	6	6	4	5	32	6	9	7	4	6	7	6	10	8	6	5	29	7	4
43	8	5	10	66	4	2	4	7	5	2	13	6	3	6	7	5	20	4	6
6	5	6	4	4	2	1	5	12	8	44	11	10	48	3	3	8	4	1	2
5	9	9	6	4	95	2	1	6	10	10	9	4	3	5	8	7	5	5	31
7	11	5	3	6	4	5	6	7	6	6	7	2	9	0	34	4	6	137	1
10	4	11	5	5	1	4	5	7	5	2	5	5	6	7	5	2	3	7	27
7	7	8	4	4	7	3	13	2	4	3	6	7	9	5	4	10	8	7	6
6	6	5	4	3	224	6	6	2	10	7	9	5	7	9	7	6	5	7	3
2	7	11	7	4	6	7	8	10	6	6	4	7	1	6	5	4	7	132	1
6	9	7	8	8	3	10	2	5	5	7	8	7	49	4	6	9	4	4	2
8	3	5	11	9	5	4	7	4	7	6	2	6	8	0	4	17	4	6	239
10	3	2	3	8	9	7	3	7	6	6	0	6	2	3	5	4	42	6	4
6	6	8	4	9	5	13	12	4	3	5	4	1	8	7	5	4	8	7	4
9	8	5	5	8	4	4	5	4	8	2	74	5	2	4	6	0	6	5	4
4	101	8	6	10	4	5	4	9	4	2	4	2	9	2	6	8	8	27	7
3	7	4	7	7	5	9	5	6	35	3	6	5	6	7	5	8	2	17	6
4	79	3	4	4	5	3	6	6	3	2	4	6	12	7	19	8	6	2	7

The commercial parcel

The residential parcel

Fig. 13.3 A generated urban form in the hypothetical city (the number in each residential parcel is the FAR value, and the number in each commercial parcel is the number of working persons)

parcel in which to work, with no rationality and regardless of the commuting distance. For residents with limited rationality between the two extreme conditions, they will select commercial parcels as job places obeying a predefined probability. Therefore, we introduce the variable r as the rational degree of a resident. When r is 1, the resident with full rationality will select the nearest place to work. When r is 0, the resident with no rationality will select a place to work randomly. When r is 0.3, the resident will select the nearest place to work with a probability of 30% and randomly select a place to work with a probability of 70%. All residents are supposed to be identical in terms of r ($r = 1$) as for focusing on the relationship between the urban form and RCEC. It should be noted that attributes (e.g., the scale and quality), in addition to the spatial location of a commercial parcel, are not accounted for in the job location choice procedure of our chapter, although they may have an important impact on people's working choice behavior in reality.

13.2.3.4 Commuting Mode Choice

The choice of commuting mode is not only determined by the socioeconomic characteristics of a resident but also by his/her commuting distance. The latter can be elaborated as $\mathrm{COMM\_TYPE_j} = f(A_j, \mathrm{dist}_j)$, where $\mathrm{COMM\_TYPE_j}$ is the commuting mode of resident j, A_j are the socioeconomic attributes of resident j, dist_j is the commuting distance of resident j, and f is the commuting mode choice function, which is used to determine the resident j's commuting mode based on his/her socioeconomic attributes A_j and commuting distance dist_j. We simplify this process by using a decision tree. Suppose the commuting mode COMM_TYPE is related to the resident's monthly INCOME (unit: CNY) and his/her commuting distance *dist* (unit: km), the decision tree in the form of Python is expressed as follows:

```
if INCOME>= 5000 and dist>= 4:
  COMM_TYPE="Car"
elif dist>= 3:
  COMM_TYPE="Bus"
else:
COMM_TYPE="Biking or Walking"
```

Note: This rule is generated from the household travel surveys of Beijing conducted in 2005.

13.2.3.5 Calculating the RCEC and Environmental Impact for Each Generated Urban Form

The commuting distance can be calculated from the results of residential location choice and job location choice. Regarding the confirmed commuting mode of each resident, the RCEC E_i and environmental impact C_i (e.g., pollutants SO_2 and NO_X, as well as house gas CO_2) can be calculated using indicators for various commuting modes (see Table 13.1). The RCEC and environmental impact on the whole city can be then calculated by summing up all residents. Since this chapter mainly aims to identify the quantitative relationship between the urban form and RCEC, the indicators shown in Table 13.1 are not real values and are only used to illustrate the relative relationship among various commuting modes in terms of the RCEC and environmental impact.

Table 13.1 Transportation energy consumption and environmental impact indicators for various commuting modes

ID	Travel mode	Consumed energy per kilometer per capita	Environment impact per kilometer per capita
1	Car	10	10
2	Bus	2	1
3	Bike or walk	0	0

13.2.3.6 Selecting Indicators for Measuring Urban Forms

We selected 14 indicators developed by McGarigal et al. (2002) for measuring generated urban forms using FRAGSTATS. These indicators, initially developed for evaluating ecological landscape, were borrowed to measure urban forms and are divided into two types (detailed descriptions for these indicators are available at: http://www.umass.edu/landeco/research/fragstats/documents/Metrics/MetricsTOC.htm). The first type is for evaluating the land-use pattern (i.e., the spatial distribution of commercial parcels) and includes seven indicators. The second type is for evaluating the spatial distribution of density (i.e., FAR) and includes the other seven indicators. The calculated indicator is expressed as I_i^k, where k is the ID of the indicator, k is the number of indicators (14 in this chapter), and i is the ID of the generated urban form.

- The type for measuring land-use pattern (see red parcels in Fig. 13.7)
  - CLS_CA: Total Area (of commercial parcels)
  - CLS_NP: Number of Patches (A group of adjacent parcels is defined as a patch.)
  - CLS_LPI: Largest Patch Index
  - CLS_ENN_MN: Euclidean Nearest Neighbor Distance
  - Shape indicators
    - CLS_SHAPE_MN: Shape Index
    - CLS_LSI: Landscape Shape Index
    - CLS_PARA_MN: Perimeter-Area Ratio
- The type for measuring FAR (see colored parcels in Fig. 13.6)
  - Diversity indicators
    - LD_SHDI: Shannon's Diversity Index
    - LD_SHEI: Shannon's Evenness Index
  - LD_ENN_MN: Euclidean Nearest Neighbor Distance
  - LD_COHESION: Patch Cohesion Index
  - Contagion-Interspersion
    - LD_CONTAG: Contagion Index
    - LD_DIVISION: Landscape Division Index
    - LD_AI: Aggregation Index

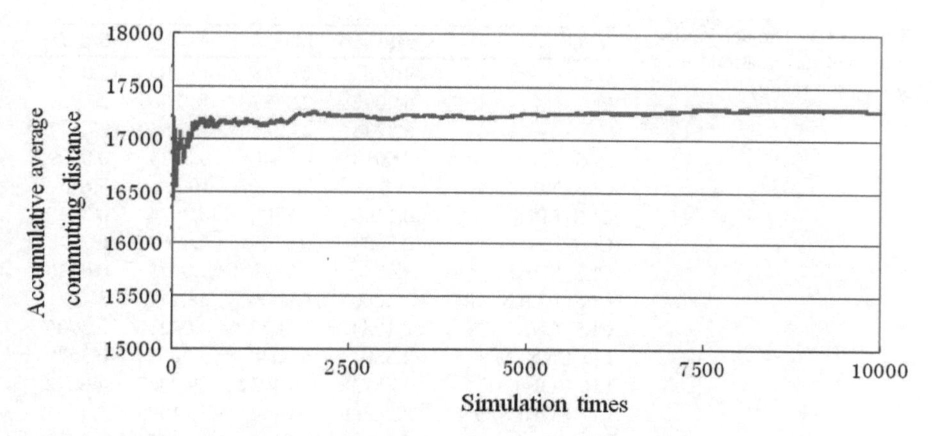

Fig. 13.4 The accumulative average commuting distance for each simulation

13.2.3.7 Identifying the Relationship Between the Urban Form and RCEC

The quantitative relationship between the urban form and RCEC can be identified from the calculated urban form indicators and RCEC. The details are as follows:

1. Conduct correlation analysis among indicators for measuring the urban form to eliminate indicators with high correlation (greater than 0.8 or less than −0.8).
2. Identify the dominant factors influencing the RCEC of the urban form using the global sensitivity analysis approach.
3. Evaluate the influence of the shape of the urban form on the RCEC.
4. Calculate the RCEC for various typical urban forms to test conventional planning theories.

The results of the above tests are shown in Section 3.

13.3 Results

A total of 10,000 parallel simulations, with 10,000 urban forms generated, were conducted using the FEE-MAS model with 112 h consumed. Convergence was reached in terms of the accumulative average commuting distance (see Fig. 13.4), indicating that the 10,000 urban forms generated can represent almost all possible urban forms in the hypothetical city. A big number of simulations were run to guarantee that the identified relationship between the urban form and RCEC is stable and represents the objective characteristics of the urban system.

The descriptive statistical information of 10,000 simulations is shown in Table 13.2, in which *dist* is the total commuting distance, *E* is the total RCEC, and *C* is the total pollutant emission of the whole city.

Table 13.2 The descriptive statistical information for simulation results

Name	Min	Max	Ave	Std. Dev.
dist	9,186	27,848	17,300	2,932
E	64,092	238,378	140,500	27,969
C	62,236	233,844	137,500	27,517
CLS_CA	0.0010	0.0040	0.0025	0.0009
CLS_NP	6	33	19.2	5.6
CLS_LPI	0.2500	3.7500	0.7624	0.3248
CLS_LSI	2.2857	6.0769	4.4267	0.7948
CLS_SHAPE_MN	1.0000	1.2375	1.0423	0.0326
CLS_PARA_MN	34,000	40,000	38,741	877
CLS_ENN_MN	2.1554	6.2072	3.0421	0.5705
LD_ENN_MN	2.3301	5.6048	3.3062	0.3579
LD_CONTAG	24.4728	50.4032	39.2857	4.0672
LD_COHESION	82.6889	96.7945	93.1993	1.4731
LD_DIVISION	0.5431	0.9249	0.6957	0.0553
LD_SHDI	0.7929	1.0989	0.9778	0.0671
LD_SHEI	0.4978	0.7560	0.6090	0.0406
LD_AI	35.8211	54.1918	44.6079	3.0234

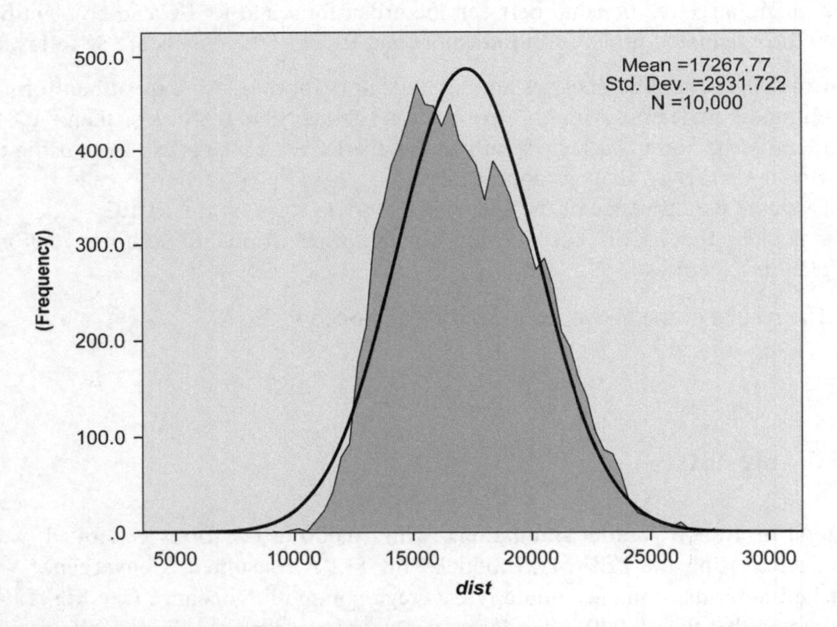

Fig. 13.5 The frequency density distribution of dist compared with the normal distribution

The total commuting distance for each urban form (*dist*) is the core variable in the simulation results. From the frequency distribution of *dist*, the variation of total commuting distance *dist* varies by urban form, ranging from 9,186 to 27,848 among all generated urban forms. Fig. 13.5 shows the probability density distribution of

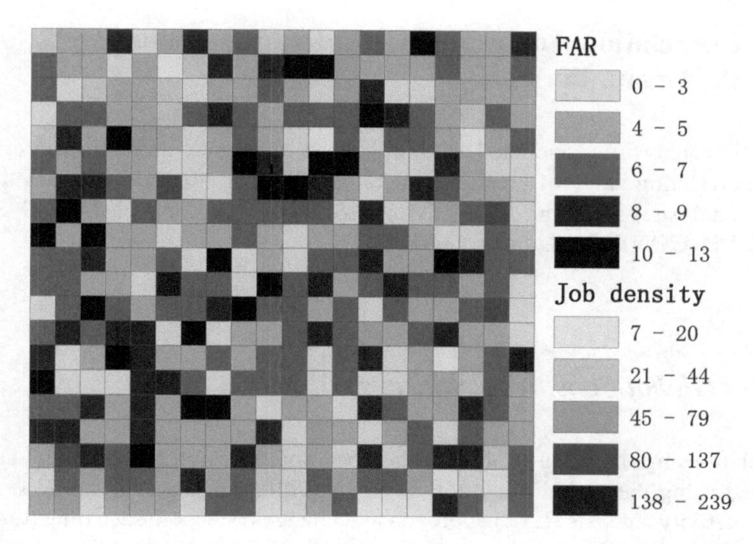

Fig. 13.6 FAR distributions for an exemplified urban form with the results of job location choice

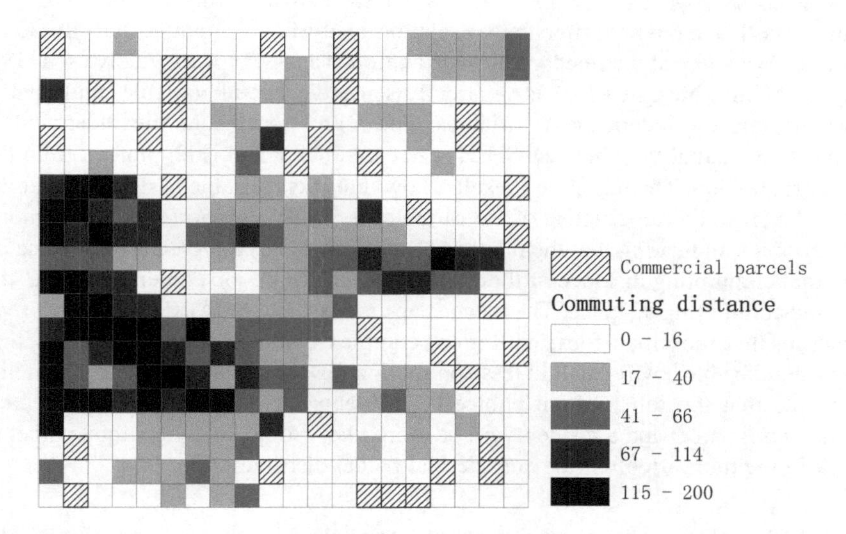

Fig. 13.7 The commuting distance map for an exemplified urban form

dist, which is very similar with the normal distribution curve shown in the figure, with the average value of 17,267.77 and standard deviation of 2,931.722.

To illustrate the general information regarding the 10,000 simulations, Fig. 13.6 shows one generated urban form with the results of job location choice and Fig. 13.7, the spatial distribution of commercial parcels and commuting distance for each residential parcel.

13.4 Correlation Analysis of Indicators for Evaluating the Urban Form

Correlation analysis is conducted on all indicators to measure urban form. Variables with a correlation value of greater than 0.8 or less than −0.8 are eliminated from further analysis, including CLS_CA, CLS_LSI, CLS_ENN_MN, LD_SHEI, LD_AI, LD_CONTAG, and LD_COHESION.

13.4.1 Global Sensitivity Analysis

The relationship between spatial indicators and commuting distance can be identified using the global sensitivity analysis (GSA) approach. In contrast to the local sensitivity analysis (LSA) approach that changes one factor at a time (OAT) to see the effect of factor variation on the output, the GSA can detect the parameters' sensitivity by adjusting all parameters' values in the whole parameters' value space. That is to say, we can see the "tree" via LSA and the "forest" via GSA. The indicators that remained after the correlation analysis are inputted into the GSA process. We adopted the linear regression approach as one type of widely used GSA approach, in which ln (*dist*) is regarded as the dependent variable, and spatial variables are the independent variables. The regression results are illustrated in Table 13.3; spatial variable LD_SHDI is not significant and is eliminated from the linear regression. The regression results show that each variable is significant at the 0.001 level, and the coefficient of the variable CLS_NP is negative and least among all variables, indicating that the number of job centers has the greatest influence on the total commuting distance of the whole city. The more job centers there are, the less the commuting distance. The mean shape index CLS_SHAPE_MN is negative, denoting that the urban form with more complex commercial parcel distribution and irregular shape will result in less commuting distance. The reason may lie in the principle that the job location choice of residents in this chapter is the nearest commercial parcel and a resident will be more likely to find a commercial parcel to work in the more irregular the commercial parcel distribution is.

Table 13.3 Global sensitivity analysis for simulation results

Variable	Standardized coefficient	*t*	Sig.
Constance		198.017	.000
CLS_NP	−.771	−121.363	.000
CLS_SHAPE_MN	−.114	−19.246	.000
CLS_PARA_MN	.076	14.449	.000
LD_ENN_MN	.040	7.359	.000
CLS_LPI	−.030	−4.451	.000
LD_DIVISION	−.026	−3.975	.000

13.5 Identifying the Relationship Between the Shape of the Urban Form and RCEC

The previous results are based on square urban forms with 400 parcels. To test the influence of the shape of the urban form on total commuting distance, we select four types of shapes, as shown in Table 13.4—square, rectangle, circle, and poly-clusters. For each shape, we generated 5,000 different urban forms based on the approach elaborated in Sect. 13.2.3.2, using the FEE-MAS model. The total commuting distance for each urban form with various shapes is calculated as shown in Table 13.4, in which the average value of *dist* denotes the influence of the shape of the urban form on total commuting distance. According to the computation results, the circle has the least commuting distance and the square, the greatest.

Table 13.4 Simulation results of commuting distance for urban forms with various shapes

Shape	Shape	Min	Max	Ave	Std. Dev.
Square	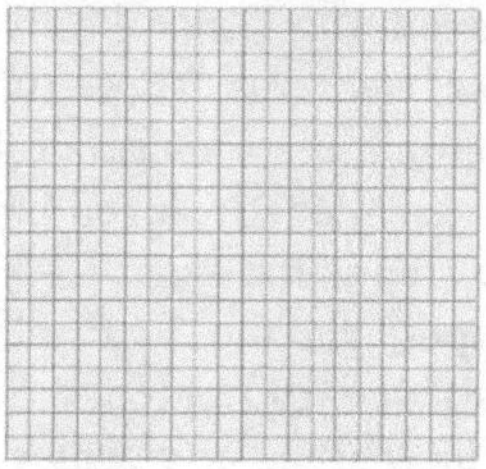	9,186	27,848	17,300	2,932
Rectangle	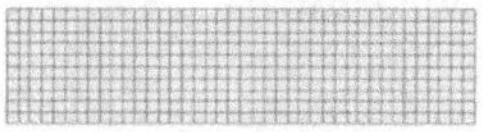	6,943	33,187	16,801	4,904
Circle	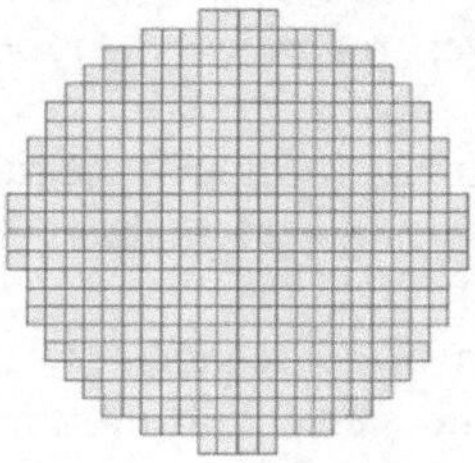	7,419	25,750	13,460	2,906
Poly-clusters	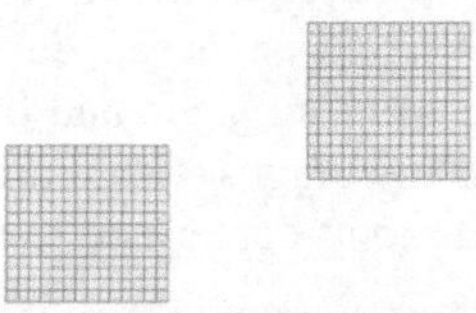	6,583	47,535	14,948	4,910

Note: The parcel in each urban form may differ from the others due to different map scales

The multi-cluster has a shorter commuting distance than the mono-cluster. The three conclusions above are all in accord with conventional planning theories. However, the rectangle has a shorter commuting distance than the square, which is not the same as the conventional planning theory. This may be because the resident with full rationality in our model will choose the nearest commercial parcel to work in. The results may vary in the random selection of the work place, which will be explored in future research.

13.5.1 Evaluating Typical Urban Forms

For evaluating conventional planning theories, such as greenbelts, transit-oriented development (TOD), poly-centers, and the compact city, we generated six typical urban forms (see Fig. 13.8), taking the number of job centers and the distribution of development density into account. The shape of each urban form is the same as the urban form generated in Sect. 13.2.3.2 "Generating Urban Forms and Residential Location Choice," with 400 parcels and 2,000 residents. These typical urban forms are generated according to conventional planning theories rather than the approach in Sect. 13.2.3.2. For the sprawl pattern with low-density developments, the FAR of each residential parcel is 5 and the built-up area is the size of 400 parcels. For the compact pattern with high-density developments, the FAR of each residential parcel is 20 and the built-up area is the size of 100 parcels. In the urban form based on TOD, the FAR decays as the distance to the city center increases. As for the urban form with a greenbelt, parcels within the belt remain undeveloped.

We calculated the total commuting distance for each typical urban form (see results in Table 13.5):

- The total commuting distance of the urban form with a sprawl pattern (e.g., A and E) is double that of a compact pattern (e.g., B and F). This could be due to their differences in total urban built-up area, regardless the mono-center or poly-center urban form.
- In contrast to the mono-center and TOD urban form with the same shape (C), the urban form with mono-center and sprawling pattern has greater commuting distance.
- For mono-center cities, the introduction of a greenbelt (D) will increase the development density of the city with the same built-up amount and slightly increase the total commuting distance compared with the sprawl pattern (A).
- The urban form with poly-centers and compact pattern (F) has the least commuting distance because reducing the size of an urban developed area also reduces the general distance between the working place and living place.
- The urban form with poly-centers (E) has the biggest total commuting distance, which is similar to the urban form with a mono-center and greenbelt (D).

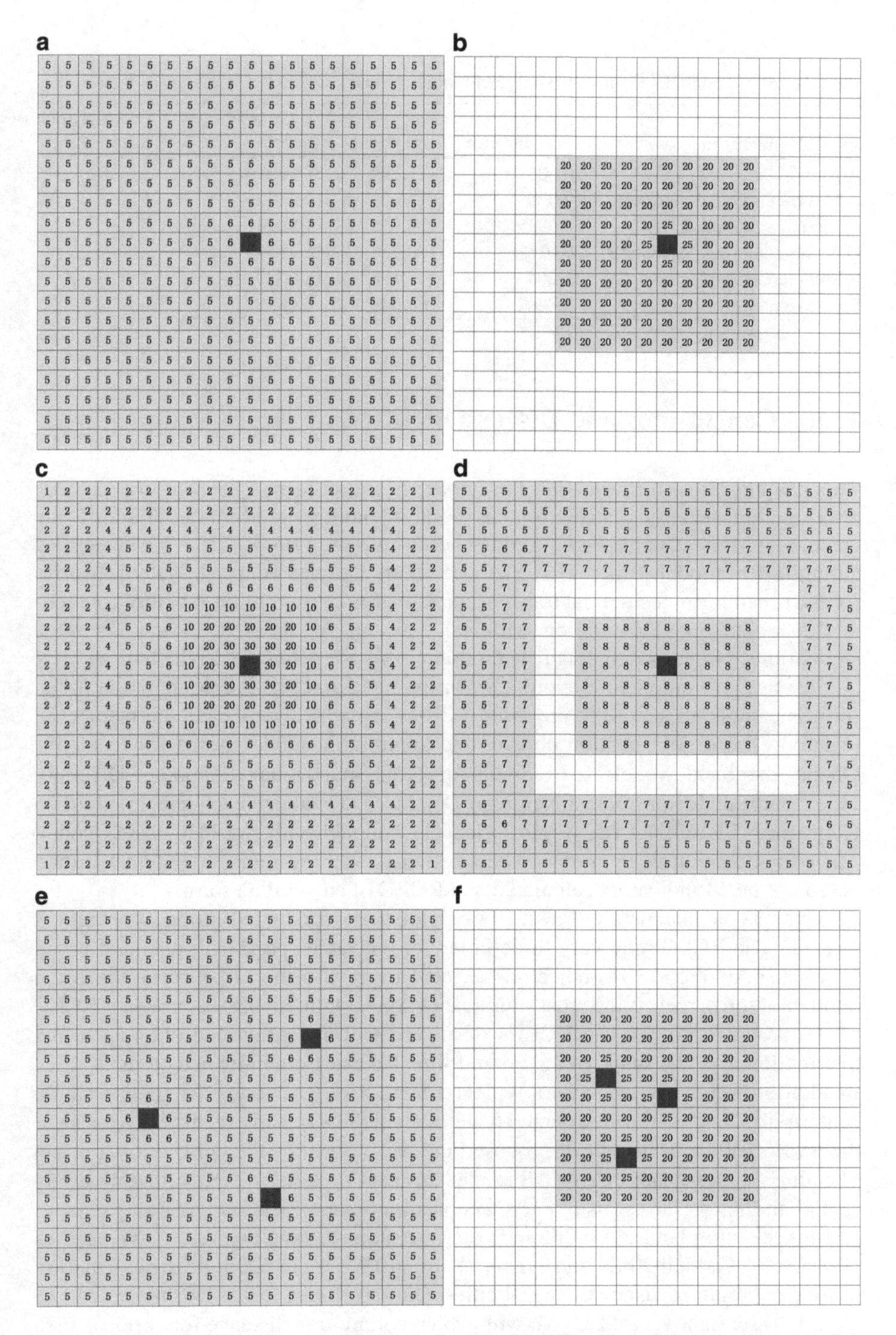

Fig. 13.8 Six typical urban forms in the hypothetical city. Note: Parcels in dark grey are commercial; in light grey, residential; and in white, undeveloped. The figure in each parcel is its FAR. We only use one commercial parcel to illustrate the spatial layout of job centers rather than the real size of job centers

Table 13.5 Simulations results for six typical urban forms

Typical urban form	Total distance	Energy consumed	Environment impact	Order of total distance
A Mono-center and sprawl	20,006	171,792	168,391	3
B Mono-center and compact	10,020	78,528	76,344	5
C Mono-center and TOD	14,092	113,456	110,673	4
D Mono-center and greenbelt	20,026	170,228	166,674	2
E Poly-centers and sprawl	22,264	191,168	187,429	1
F Poly-center and compact	8,860	64,648	62,879	6

13.6 Conclusions and Discussion

This chapter aims to investigate the impact of urban form—the land-use pattern, development density distribution, as well as number and distribution of job centers—on the residential commuting energy consumption (RCEC). We developed a multi-agent model named FEE-MAS based on complex adaptive system theories for the urban form, residential transportation energy consumption, and environmental impact integrated simulation, with residents and parcels as the basic units in the simulations. Numerous distinguished urban forms are generated in the hypothetical city using the Monte Carlo approach. On the one hand, the RCEC for each urban form is calculated using the proposed FEE-MAS, which integrates the residential location choice, job location choice, and commuting mode choice for residents. On the other hand, we selected 14 indicators (e.g., Shape Index, Shannon's Diversity Index, and Euclidean Nearest Neighbor Distance) to evaluate each generated urban form using FRAGSTATS, which is loosely coupled with the FEE-MAS model. Then, the quantitative relationship between the urban form and RCEC is identified based on the 14 indicators calculated and RCEC of each urban form.

Several conclusions are drawn from simulations conducted in the hypothetical city: (1) The RCEC may vary three times for the same space with different urban forms. (2) Among the 14 indicators selected for measuring urban form, the patch number of job parcels is the most significant variable influencing the potential RCEC of the urban form. (3) The RCECs of all urban forms generated obey a normal distribution. (4) The shape of the urban form also exerts an influence on the RCEC. In addition, we evaluated several typical urban forms (e.g., compact/sprawl, single center/poly-centers, traffic-oriented development (TOD), and greenbelt) in terms of the RCEC indicator using our proposed model to quantify the conventional urban planning theories. The FEE-MAS model can also be applied for evaluating urban spatial alternatives in terms of energy consumption and environmental impact.

Most existing land-use and transportation integrated models (LUTMs), such as UrbanSim (Waddell 2002) and Tranus (Putman 1975), can facilitate the calculation of total commuting distance for real cities. Our developed FEE-MAS model can be regarded as a lightweight LUTM, with which commuting distance for various cities can be calculated. The FEE-MAS model is developed to identify the principal rules governing dynamic urban systems, rather than perform empirical applications for real cities of LUTMs. The FEE-MAS model features the identification of the dynamic relationship between urban form and commuting distance in a manner of inner-city analysis, which is not possible for intercity researches.

The spatial-explicit FEE-MAS model can meanwhile be applied for evaluating planning alternatives in real cities in terms of commuting distance, energy consumption, and environmental impact, thus providing a low-carbon alternative to planners and decision makers. This process can guarantee low-carbon city development in the planning stage by embedding it in the comprehensive planning alternative evaluation procedure.

In addition to applying the model to the practical city, we will be conducting further research. (1) The energy consumption of households, entertainment, and industrial sectors will be taken into account besides the residential commuting energy consumption that was analyzed by the current FEE-MAS model. (2) The relationship between the socioeconomic attributes of residents and their rationality will be considered in the form of $r = g(A_j)$, where A_j is the resident j's socioeconomic attribute set and g is the function for calculating the rationality of a resident using his/her socioeconomic attributes. The socioeconomic attributes of the residents will then be introduced into the residential location choice process of the model. (3) The capacity of jobs in commercial parcels will be incorporated in the job location choice process to simulate a more practical result. (4) The transportation network is expected to be included in the model to replace the current Manhattan distance in this chapter for a more precise calculation of residential commuting energy consumption.

References

Alford G, Whiteman J (2009) Macro-urban form and transport energy outcomes: investigations for Melbourne. Road Transp Res 18:53–67

Anderson WP, Kanaroglou PS, Miller EJ (1996) Urban form, energy and the environment: a review of issues, evidence and policy. Urban Stud 33:7–35. doi:10.1080/00420989650012095

Beijing Fifth Population Census Office, Beijing Statistical Bureau (2002) Beijing population census of 2000. Chinese Statistic Press, Beijing

Benenson I, Omer I, Hatna E (2001) Agent-based modeling of householders' migration behavior and its consequences. In: Billari FC, Prskawetz A (eds) Agent based computational demography., Springer, pp 97–115

Brown DG, Robinson DT (2006) Effects of heterogeneity in residential preferences on an agent-based model of urban sprawl. Ecol Soc 11(1). Available at http://www.ecologyandsociety.org/vol11/iss1/art46/

Dieleman FM, Dijst M, Burghouwt G (2002) Urban form and travel behaviour: micro-level household attributes and residential context. Urban Stud 39:507–527. doi:10.1080/00420980220112801

Giuliano G, Narayan D (2003) Another look at travel patterns and urban form: the US and Great Britain. Urban Stud 40:2295–2312. doi:10.1080/0042098032000123303

Holden E, Norland IT (2005) Three challenges for the compact city as a sustainable urban form: household consumption of energy and transport in eight residential areas in the greater Oslo region. Urban Stud 42:2145–2166. doi:10.1080/00420980500332064

Horner MW (2007) A multi-scale analysis of urban form and commuting change in a small metropolitan area (1990–2000). Ann Reg Sci 41:315–332. doi:10.1007/s00168-006-0098-y

Hourcade JC, Crassous R (2008) Low-carbon societies: a challenging transition for an attractive future. Clim Policy 8:607–612. doi:10.3763/cpol.2008.0566

IPCC (2007) IPCC Fourth Assessment Report: Climate Change 2007 (AR4)

Johnson RA, Mccoy MC (2006). Assessment of integrated transportation/land use models. Information Center for the Environment. Department of Environmental Science & Policy, University of California, Davis

Kii M, Doi K (2005) Multi-agent land-use and transport model for the policy evaluation of a compact city. Environ Plan B: Plan Des 32(4):485–504. doi:10.1068/b3081

Krizek KJ (2003) Residential relocation and changes in urban travel—does neighborhood-scale urban form matter? J Am Plan Assoc 69:265–281. doi:10.1080/01944360308978019

Long Y, Shen Z, Mao Q, Dang A (2010) Retrieving individuals' attributes using aggregated dataset for urban micro-simulation: a primary exploration. Proc Geoinform. doi:10.1109/GEOINFORMATICS.2010.5567738

Maat K, Timmermans HJP (2009) A causal model relating urban form with daily travel distance through activity/travel decisions. Transp Plan Technol 32:115–134. doi:10.1080/03081060902861285

McGarigal K, Cushman SA, Neel MC, Ene E (2002) FRAGSTATS: spatial Pattern Analysis Program for Categorical Maps. Computer software program produced by the authors at the University of Massachusetts, Amherst. Available at www.umass.edu/landeco/research/fragstats/fragstats.html

McMillan TE (2007) The relative influence of urban form on a child's travel mode to school. Transp Res Pt A Policy Pract 41:69–79. doi:10.1016/j.tra.2006.05.011

Newman PWG, Kenworthy JR (1989) Gasoline consumption and cities: a comparison of US cities with a global survey. J Am Plan Assoc 55:24–37. doi:10.1080/01944368908975398

Owens S (1987) Energy, planning and urban form. Pion, London

Pan HX, Shen Q, Zhang M (2009) Influence of urban form on travel behaviour in four neighborhoods of Shanghai. Urban Stud 46:275–294. doi:10.1177/0042098008099355

Putman SH (1975) Urban land use and transportation models: a state of the art summary. Transp Res 9:187–202. doi:10.1016/0041-1647(75)90056-8

Remme U, Blesl M (2008) A global perspective to achieve a low-carbon society (LCS): scenario analysis with the ETSAP-TIAM model. Clim Policy 8:60–75. doi:10.3763/cpol.2007.0493

Schlossberg M, Greene J, Phillips PP, Johnson B, Parker B (2006) School trips-effects of urban form and distance on travel mode. J Am Plan Assoc 72:337–346. doi:10.1080/01944360608976755

Shen Q (2005) Study on urban energy sustainable development: a view from urban planning. Urban Plan Forum 6:41–47 (in Chinese)

Shim GE, Rhee SM, Ahn KH, Chung SB (2006) The relationship between the characteristics of transportation energy consumption and urban form. Ann Reg Sci 40:351–367. doi:10.1007/s00168-005-0051-5

Shukla PR, Dhar S, Mahapatra D (2008) Low-carbon society scenarios for India. Clim Policy 8:156–176

Waddell P (2002) Modeling urban development for land use, transportation, and environmental planning. J Am Plan Assoc 68:297–314. doi:10.1080/01944360208976274

Wang ZW, Chen J (2008) Achieving low-carbon economy by disruptive innovation in China. IEEE Int Conf Manag Innov Technol 1–3:687–692. doi:10.1109/ICMIT.2008.4654448

Wang J, Mao Q, Yang D (2008) Energy influences of urban planning: an urban transportation energy consumption model. China Planning Annual Conference, Dalian, China (in Chinese)

Yi C, Li Q, Zheng G (2008) Commercial facility site selection simulating based on MAS. Proc SPIE—Int Soc Opt Eng 7143:71431–71438

Zellner ML, Theis TL, Karunanithi AT, Garmestani AS, Cabezas H (2008) A new framework for urban sustainability assessments: linking complexity, information and policy. Comput Environ Urban Syst 32:474–488. doi:10.1016/j.compenvurbsys.2008.08.003

Zhang M (2005) Exploring the relationship between urban form and network travel through time use analysis. Landsc Urban Plan 73:244–261. doi:10.1016/j.landurbplan.2004.11.008

Zhang H, Zeng Y, Bian L, Yu X (2010) Modelling urban expansion using a multi agent-based model in the city of Changsha. J Geogr Sci 20:540–556. doi:10.1007/s11442-010-0540-z

Zhuang GY (2008) How will China move towards becoming a low carbon economy? China World Econ 16:93–105. doi:10.1111/j.1749-124X.2008.00116.x

Chapter 14
Mapping Walking Accessibility, Bus Availability, and Car Dependence: A Case Study of Xiamen, China

Hui Wang

Abstract As stated in the existing literature, travel behavior and transportation choice are closely related to urban forms and the built environment. The study presented in this chapter attempts to conduct a citywide evaluation of the walking accessibility to urban facilities and the availability of public transport, as well as relevant potential car dependence. By taking the city of Xiamen, China, as a case study, the current study generates some useful information vis-à-vis both public transport and urban facility providers, by illustrating the city's spatial patterns and identifying problematic areas. The methodology developed in this study might also serve as a reference for future studies.

Keywords Sustainable transportation • Walking accessibility • Bus availability • Car dependence • Xiamen

14.1 Introduction

Newman and Kenworthy's (1989) phrase "car dependence/automobile dependence" has been frequently used in the travel behavior, transport planning, traffic management, and urban policy-making literature. Although no common definition for this term has been agreed upon to date, it might be briefly described as characterizing "a high level of physical need or/and mental demand for car transport."

Car dependence is associated with a range of environmental, economic, and social problems. In general, an increase in car dependence can worsen the

H. Wang (✉)
Department of Urban Planning, Xiamen University, Xiamen 361005, China

Cross-Straits Institute of Urban Planning at Xiamen University, Xiamen University, 361005 Xiamen, China
e-mail: wanghui618@yahoo.cn; xmuwanghui@gmail.com

M. Kawakami et al. (eds.), *Spatial Planning and Sustainable Development: Approaches for Achieving Sustainable Urban Form in Asian Cities*, Strategies for Sustainability,
DOI 10.1007/978-94-007-5922-0_14, © Springer Science+Business Media Dordrecht 2013

transportation structure of cities, retard the construction of green transportation systems, intensify oil vulnerability, create a shortage of urban land and infrastructure, aggravate environmental stress, and contribute to greenhouse gas (GHG) emissions. Additionally, it can also result in a loss of street life and community, suburban isolation, access problems for carless people, and health problems among car drivers. Compared to other forms of transportation, car traffic has the highest cost and lowest efficiency and creates the most pollution—including the highest volumes of GHG emissions per passenger. Therefore, it is generally suggested that car transport be ranked the least preferred form in terms of "greening" urban transportation systems (CNU 2000; Huang 2004; Pan, et al. 2008; China Society of Urban Studies 2009). A city cannot be considered "sustainable" if it is car-dependent (Newman and Kenworthy 2000), and researchers use words such as "reducing," "overcoming," "combating," and "abandoning" in their work, to distinctly express their opposition to car dependence (Newman and Kenworthy 1999; Stradling 2002, 2003; Guo 2010). With the growing popularity of the "sustainable development" concept, there has been burgeoning public awareness of the damaging impact of car use on cities. As a result, reductions in car dependence in order to improve transport sustainability now constitute a major goal of transport policy in many cities (Cullinane and Cullinane 2003).

In reviewing a variety of literature, we can say that the factors that determine or influence car dependence are quite numerous, and vary from place to place. However, according to Gorham (2002), car dependence can be categorized thus: (1) *physical/environmental dependence*, which is caused by the built environment, including the urban form, regional structure, the distribution of activities within the structure, and the nature or status of collective transportation modes; (2) *social/psychological dependence*, which is caused by the symbolic meanings or "signs" of cars in society and in a consumption culture and/or the mental desire for the convenience and comfort of car travel; and (3) *circumstantial/technological dependence*, which describes the relationship between the technological capacities of a car (speed, carrying capacity, privacy, etc.) and the particular requirements of an individual life (e.g., a professional musician needs a car to carry a music instrument between home and working place, and his or her instrument itself is large, bulky, and expensive and therefore not suitable for public transit). In China, circumstantial/technological dependence manifests when people need to travel with pets (pets cannot be carried on public transport in China). "The[se] three types of car dependence outlined above interact significantly," says Gorham, and therefore, "it is probably difficult in practice to identify particular behavior with any one of them" (Gorham 2002, p. 110).

Conventional thinking states that an increase in the number of private cars derives from a growth in wealth. However, owning a car does not necessarily imply frequent car use, let alone car dependence. As shown in the studies of Kenworthy and Laube (1999, 2001), based on data gathered from over 100 cities worldwide between 1960 and 2000, patterns of automobile dependence in cities do not significantly relate to differences in wealth among cities, but they do vary in a clear and systematic way in terms of land-use patterns. The most car-dependent cities (i.e., low-density cities in the United States and Australia) are less wealthy than some densely settled and more transit-oriented cities in Europe and Asia (e.g., Singapore, Hong Kong, and Tokyo) that have very low automobile dependence. Therefore, "there are stronger factors determining automobile dependence" (Newman and Kenworthy 2000, p.112). According to a multiple regression analysis undertaken by Newman and Kenworthy, very strong correlations can be found between the level of automobile use (indicated by transport energy use) and certain parameters such as the level of public transport provision, public transport speed to traffic speed ratio, total length of roads, and parking provision; however, the strongest correlation is with urban density. Thus, "achieving a more sustainable urban form inevitably involves the development of densities that can enable transport, walking and cycling to be viable options" (Newman and Kenworthy 2000, p. 113). Urban structure does matter in shaping travel behavior and car dependence (Naess 2006). In *Seven Rules for Sustainable Communities*, Condon (2010) also emphasizes how essential it is to create an urban structure and built environment that is conducive to reductions in car dependence and the promotion of sustainability. New Urbanism, as a dominant planning theory in postmodern society, also advocates a "compact urban form," "mixed land use," "a pedestrian-friendly built environment," and "transit-oriented development" as key elements of sustainable urban development (CNU 2000). Therefore, an increase in car use cannot be attributed merely to a growth in wealth. We need to consider the ways in which we build and organize our cities.

For this reason, urban planning and design is widely recognized as essential to a city's coping with car dependence. In reshaping cities' built environments, planning tools can be used to influence the amount of travel carried out via various modes of conveyance, and the proportions thereof. The "right design" will encourage walking and discourage car use. A citywide evaluation of car dependence and of related walking accessibility and bus availability is necessary. A city-specific analysis of the spatial pattern of car dependence would be even more helpful for local planners and decision-makers in acquiring information on where the problematic areas are and what corresponding solutions there may be. This considered, by taking the city of Xiamen as a case study, with the aid of ArcGIS software and based on an examination of the walking accessibility to urban facilities and the availability of public transport, the current study attempts to evaluate and map potential car dependence. The study results will be useful in identifying the most appropriate locations for new urban facilities, as well as public transport services.

This chapter is organized into six sections. Following this introduction, the background of the research is presented in Sect. 14.2, in order to provide the reader with a "bigger picture" of the issues at play. Section 14.3 discusses the appropriate level of detail needed to measure overall walking accessibility from a given city block to all types of urban facilities. The study starts with the measurement of walking accessibility, because motor traffic demand would be low wherever walking accessibility is good. Then, public transportation availability is measured in Sect. 14.4, because the greater the availability is of public transit lines in a block, the better that area is connected to other areas of the city. In Sect. 14.5, in view of the fact that car dependence is somehow inversely proportional to walking accessibility and the availability of public transport, the potential car dependence across the city of Xiamen is measured and mapped, in the hope of providing useful information for local planners and decision-makers. Finally, the chapter concludes in Sect. 14.6 with a summary.

14.2 Research Background

Since China's undertaking of reforms and general "opening up," Chinese cities have been experiencing rapid growth and remarkable restructuring. As a result, both the macro- and microlandscape of Chinese cities have been changing markedly. While the overall urban form is becoming larger and increasingly fragmented, the spatial patterns of people's daily lives have also changed. People's work and nonwork lives are becoming more and more separated each day, with the commercialization of housing (e.g., many commercial houses are built in newly developed suburban areas) and the disappearance of "work-unit compounds" (i.e., welfare housing or factory dormitories nearby the work place). As a result, more and more people in China's cities face long, frustrating commutes on a daily basis. Meanwhile, urban facilities (such as shops, hospitals, schools, and public transport) have also become completely or partially "commercialized," resulting in service dropouts whenever or wherever the residential density is not sufficiently high to make such service provision profitable, which in turn makes it more difficult to introduce a transit-oriented development (TOD) model. With the inducements created by car advertisements and increases in people's incomes, more and more people are choosing to buy a private car when they can just barely afford to do so. As a result, there has been a surging "car boom" in China.

In 2009, China overtook the United States to become the world's largest market for personal vehicles, with sales of 13 million cars and light trucks in that year. It is still too early to get "the annual sale in 2012". In fact, China has become the world's number one automaker and seller since 2009, with two-digit rates of annual increase in sales figures. Road networks in megacities like Beijing and Shanghai (as well as in many second-tier cities such as Xi'an, Dalian, and Xiamen) have doubled or tripled over the past decade to adapt to the growing popularity of automobiles, but they continue to suffer from ever-worsening congestion and parking problems. As more cars appear on the city streets of China, more and more pedestrians and cyclists are being crowded out. It is reported that between 1995 and 2005, bicycle ownership dropped by 35% in China. In the meantime, the development of public transport has been far from satisfactory. As a result, urban transportation in Chinese cities is tending to be increasingly car-dependent.

Increasing car dependence may be particularly alarming for a city like Xiamen. The city of Xiamen, on the southeast coast of China (117° 53′ E–118° 27′ E; 24° 25′ N–24° 55′ N), is one of the earliest four special economic zones in China and a major city on the west bank of the Taiwan Strait. It has a total population of 3.53

million, about half (52.7%) of which lives on Xiamen Island, i.e., the earliest developed area of the city stretching over a 131-km^2 area. The city enjoys a fine reputation of being "a garden city at sea," and it is a winner of the UN Habitat Award, on account of its beautiful natural scenery and living environment. It has also been designated as one of China's ten low-carbon pilot cities since 2010, and it has set for itself the ambitious target of becoming a Chinese role model for low-carbon cities. However, one of the biggest challenges in reaching this goal has been the rapid increase in motor-vehicle use (especially private cars).

According to the latest data released by the Department of Traffic Police, the number of motor vehicles in Xiamen reached 827,150 at the end of February 2012, of which 438,400 are private cars (53.01%). In each of the three most recent months, an average of 6,628 new motor vehicles has been added to the streets of Xiamen. The number of newly registered motor vehicles in January 2012 alone was 9,324—over 70% of which are private cars. According to the figures of the sixth national population census, the resident population in Xiamen is 3.53 million, and they form up to 1.24 million households. This means that car ownership in the city has reached a ratio of 35.36 per 100 households in Xiamen (i.e., one car per 2.83 households, or 12.42 cars per 100 people). Although these numbers appear modest compared to the US ratio of 94 cars per 100 people, the rate at which automobile use is expanding in Xiamen is alarming. Traffic jams have become so serious that the average vehicle speed has fallen to less than 20 km/h at peak commuter times, over half of all intersections are saturated, and there is a serious shortage of public parking lots. Casualties and economic loss stemming from traffic accidents also continue to grow, not to mention a series of other negative consequences, such as environmental pollution, oil shortages, GHG emissions, the extrusion of public space, access problems for carless people, a worsening of personal mobility, and an erosion of quality of life. These effects not only challenge the city's efforts to become a role model for low-carbon cities, but it is also seriously damaging its image as an "ecological garden city" and its attractiveness for investors and individuals of talent.

The damage of car-oriented transportation has aroused public concern and complaints. Pressures have also been placed on local authorities at all levels since China announced at the Copenhagen Conference its goal of a 40–45% reduction in carbon emissions by 2020, based on 2005 levels. Besides developing new vehicles driven by alternative "clean energy" fuel sources—like those described in China's National Climate Change Program (NDRCC 2007)—more and more policy analysts have come to see the manipulation of the urban form as a tool by which carbon emissions could be cut, by affording reductions in travel demand and car dependence. To undertake such manipulations, citywide analysis and region-specific guidance is much needed by local decision-makers, as they need to know where the problematic areas are (i.e., those with the highest car dependence) and what the corresponding solutions may be.

14.3 Walking Accessibility to Urban Facilities in Xiamen

14.3.1 Relationship Between Walking Accessibility to Urban Facilities and Motor Traffic Demand

It has become common knowledge that the energy consumption and environmental impact per passenger of different transportation modes are in the following ascending order: walking → bicycle → rails and subway → bus → taxi → carpool → single-occupancy car. Thus, to develop an environmentally friendly and low-carbon transport system, a major principle is to promote nonmotorized transport, especially walking.

There is a growing acknowledgement that travel demand and travel-mode choice (motorized vs. nonmotorized, transit vs. car) in urban areas are affected by the urban spatial structure and land-use characteristics. A number of studies have addressed this issue. The general wisdom is that enforced dependence on automobiles and long-distance travel will be mitigated by the establishment of compact cities (as indicated by urban density) with well-distributed urban facilities and job opportunities. Car dependence can be largely reduced through critical design that creates a diversity of urban amenities within a 10–30-min walking area (Newman and Kenworthy 2006). Thus, overcoming car dependence, to a large extent, becomes a question of whether people can access the facilities of a city without the use of a car. Walking accessibility to urban facilities—which refer to the services and activity possibilities that the residents and visitors of a city may use and visit—then becomes an important indicator. In general, motor traffic demand will be low if the walking accessibility to urban facilities is high; as walking is a comfortable mode of transport in accessing needed services, motor transport would be less of a necessity, in such a scenario. The converse is also true: motor traffic demand will be high if walking accessibility is poor.

14.3.2 Measuring Walking Accessibility to Urban Facilities

Accessibility is a measure of the ease of reaching places (Grengs 2001). Measuring the walking accessibility of a certain place to urban services and facilities is one of the key steps in evaluating potential car dependence. Moreover, in planning terms, measuring accessibility to local services and facilities also allows improvements to be made more precisely, to either transport or the ways in which services are provided.

A rich body of literature addresses the methodology of measuring accessibility, and a variety of accessibility measures have been developed. Most published measures of spatial accessibility to urban facilities tends to be one of four types: (1) provider-to-population ratios, (2) travel impedance to nearest provider, (3) average travel impedance to a set of providers, or (4) gravitational models of provider influence (Winter 1992). By taking "path distance" as the indicator of "travel impedance"—and by taking Cooper et al.'s (2009) relevant work as its main reference—the measurement of "walking accessibility" developed in this research combines the second and third aforementioned types.

Measuring the walking accessibility of a city requires a view of a grid of areas at an appropriate level of detail; this grid serves as the basis of all calculations. This research divides the case city into hundreds of "blocks," each of which is a grid defined by the city's second-class roads, and it excludes areas of water, wetland, farmland, forest land, and protected ecological areas.

Suppose that there are in total n blocks in the city, that m types of urban facilities are considered, and that d_{ij} stands for the average shortest path distance from the center of block i to the nearest three facilities of type j. (The nearest three facilities of a certain type are the most frequently visited ones for the people living in a certain block.) Evidence from Reneland's report (2000) suggests that there is a relationship between path distance and crow-fly distance. More specifically, as a common rule in the field of transportation planning, the former is roughly 20–30% greater than the latter (Reneland 2000). The specific number is 31% greater in the case of Xiamen, according to a test sample of that city's road map. This means that the value of d_{ij} can be calculated as being 1.31 times the average crow-fly distance from the geographic centroid of block i to the nearest three facilities of type j.

Those urban facilities that apparently generate or attract travel within a city are investigated:

- *Type 1*: Shopping centers (supermarkets, department stores, shopping malls) with business space exceeding 5,000 m^2 each
- *Type 2*: Kindergartens and primary schools
- *Type 3*: Secondary schools
- *Type 4*: Hospitals/medical centers ranked as class I or class II
- *Type 5*: Major public culture, entertainment, and sports centers

A formula for "the *overall* walking accessibility from a given block (e.g., the block i) to *all types* of facilities" (marked as A_i) is designed as follows:

$$A_i = \sum_{j=1}^{m} W_j a_{ij}$$

where W_j ($0 < W_j < 1$) is the weighting coefficient of a certain type j facility ($j = 1, 2, \ldots., m$; in the case study, $m = 5$), which will be set according to the results of a questionnaire survey about the frequencies that people visit a variety of urban facilities. The index a_{ij} stands for "the *individual* walking accessibility from block i to the type j facility." The value of a_{ij} relates to the value of d_{ij} as well as general opinions about what distance constitutes a "travel impedance." According to investigations conducted in our field survey in Xiamen, "within five minutes" is generally considered a "comfortable distance for walking"; most people are willing to walk if the walking time is less than 10 min, and they are more likely to use motorized transport (public or private) if the time exceeds 15 min. Therefore, a walking time of 15 min can be considered a watershed value that separates "good"

Table 14.1 Correspondence among path distance, walking time, evaluation value of walking accessibility, and status of walking accessibility

Path distance (d_{ij}) (m)	0–300	301–600	601–900	901–1,200	1,201–1,500	>1,500
Walking time (min)	<5	5–10	10–15	15–20	20–25	>25
Evaluation value of walking accessibility (a_{ij})	100	80	60	40	20	0
Status of walking accessibility	Comfortable	Easy	Fine	Tolerable	Barely accessible	Difficult to do so
Overall walking accessibility (A_i)	>90	71–90	51–70	31–50	11–30	0–10
Evaluation level	Very good	Good	Fine	Tolerable	Poor	Bad

walking accessibility from "bad." Table 14.1 shows the correspondence between path distance, walking time, the evaluation score of walking accessibility, and the status of walking accessibility.

Finally, "the overall walking accessibility to urban facilities from block i"—namely, A_i—can be obtained by carrying out the aforementioned processes. A_i ranges from 0 to 100, and the greater its value is, the better the overall walking accessibility from a given block i to urban facilities—and, hence, the lower the potential demand for motorized transport there will be. Conversely, the smaller A_i is, the poorer the overall walking accessibility is, and the higher the potential demand for motorized transport there will be.

All these calculation processes can be facilitated through GIS software. The results thereof can be visualized as a kind of "walking-accessibility map" where the values of A_i can be shown in different colors after being classified into five levels.

14.3.3 Data and Results

Using the aforementioned methodology, this project carries out an empirical case study of Xiamen, China.

First of all, the "Master Plan of Xiamen City (2010–2020): The Overall View Map" is taken as the study's base map, after subjecting it to vectorization processing in ArcGIS. According to the road grid—combined with land-use classifications and terrain maps—the entire city's construction land is divided into 560 "blocks," each of which is a grid defined by secondary-class roads and excludes the areas of water, wetland, farmland, forest land, and ecology protection zones. Then, using the relevant analysis tools in ArcGIS, the "centroids" of those 560 blocks are extracted as the centers of the blocks.

Various data on urban facilities are obtained from the Xiamen Municipal Bureau of Commerce, Bureau of Education, Bureau of Health, Bureau of City Planning, and the like. In total, there are 78 shopping centers, 696 primary schools and kindergartens, 82 secondary schools, 20 hospitals, and 25 public cultural/sports centers. These data are processed in ArcGIS to create a research database, and the facility locations are marked on the base map.

Then, using ArcGIS analysis tools, the average crow-fly distances from each block's centroid to the nearest three facilities of each type are measured. As mentioned, according to a sample analysis based on the road map of Xiamen, the path distance d_{ij} is 1.31 times the crow-fly distance. Furthermore, the value of d_{ij} can be converted to a_{ij}, according to the score assignment standards shown in Table 14.1.

Then, through expert inquiry and discussion among the members of the research team, using the analytic hierarchy process, the weight coefficient W_j of each type of urban facility is established: $W_1 = 0.4174$ (shopping centers), $W_2 = 0.2634$ (kindergartens and primary schools), $W_3 = 0.1605$ (secondary schools), $W_4 = 0.0975$ (hospitals), and $W_5 = 0.0612$ (cultural/sports centers).

Finally, the overall walking accessibility to urban facilities in block i—namely, A_i—can be obtained according to its calculation formula. A_i ranges from 0 to 100. The greater the value of A_i is, the better the overall walking accessibility is from a given block i to urban facilities—and, hence, the lower the potential demand will be for motorized transport. In accordance with the standards shown in the last two rows of Table 14.1, the final results of A_i are assessed as being at one of six levels: very good, good, fine, tolerable, poor, and bad. Using ArcGIS grading tools, the results can be visualized as Fig. 14.1, below. This figure directly reflects the spatial pattern of the overall walking accessibility to urban facilities and indirectly reflects the spatial pattern of the potential demand for vehicle transport.

As Fig. 14.1 shows, no area in the city achieves the level of "very good"; only the old downtown area on Xiamen Island and a small number of blocks in the Haicang and Tong'an districts are "good" or "fine." Most of the remaining areas are "poor" or even "bad," which means it is not convenient to live there, or one is otherwise unlikely to access urban facilities by walking there, and that such areas may exact a high potential demand for motorized transport. Fig. 14.1 also indicates that almost all types of urban facilities are mainly located on Xiamen Island—i.e., the earliest developed area of the city—leading to an obvious imbalance between areas inside and outside Xiamen Island.

14.4 Availability of Public Transport in Xiamen

14.4.1 Connection Between Availability of Public Transport and Car Dependence

As stated, a major principle in developing a sustainable "green" transport system is to promote nonmotorized transport (especially walking). By adding urban facilities to areas where pedestrian accessibility is poor (i.e., bringing facilities closer to

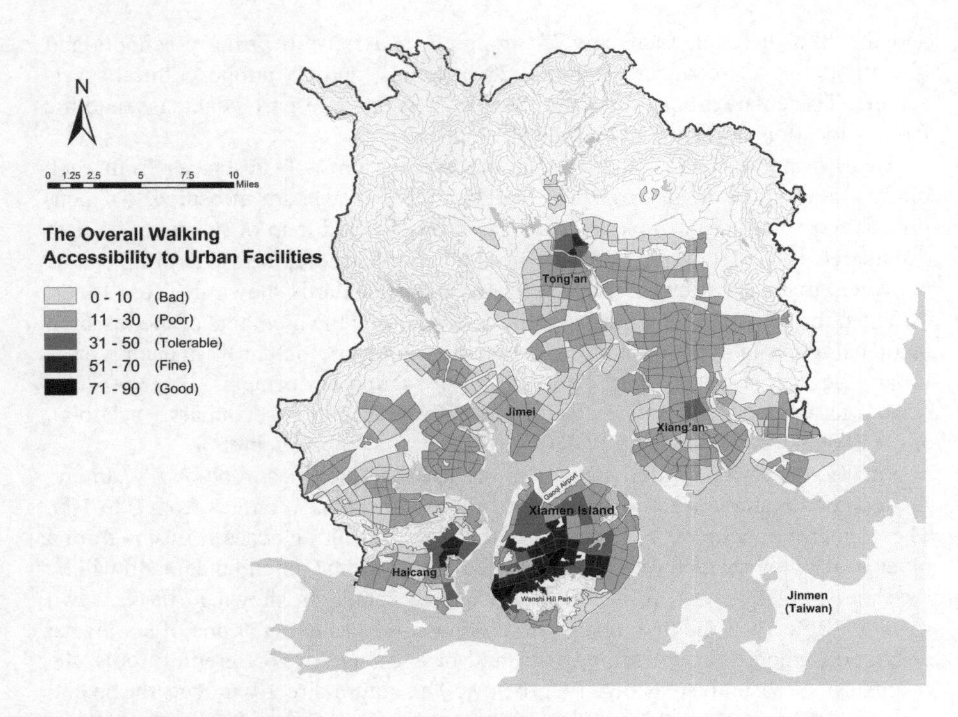

Fig. 14.1 Spatial pattern of overall walking accessibility to urban facilities in Xiamen

people), potential demand for motorized transport could be reduced to some extent. Besides, doing so could also reduce the probability of a need for long-distance travel involving a single destination. However, it is unrealistic to distribute urban facilities to every corner of the city. Besides, high-order services that demand a "threshold radius" that exceeds the walking scale cannot survive if they are too close to each other. Moreover, as the city expands and its structure becomes looser and more fragmented, the distances among residences, work sites, and service facilities will gradually increase, normally far beyond what is walkable. Such changes make the need for motorized transport inevitable, and a "second choice" of mode of transport will be necessary. Comparatively efficient, fair, energy-saving, and low carbon-emitting modes of public transport, rather than private cars, should then be encouraged. Besides encouraging nonmotorized transport (e.g., walking, biking), the promotion of public transport is another main solution to the problem of car dependence.

However, whether or not public transport can become a dominant choice for local residents—and how competitive it can be, compared to private cars—depends heavily on its availability. "Availability" is a concept that connotes "possibility of obtaining it," "ease of access to it," and "utility of using it." The concept is often used to examine whether medical facilities or other resources are fairly allocated, convenient to access, and efficiently used (Winter 1992; Hearn et al. 1998). The greater the availability of public transport is, the less likely individuals will use private cars.

14.4.2 Measuring the Availability of Public Transport

Concerning public transport, the research introduces the concept "availability," to express the possibility and convenience of using public transport, as well as the utility of doing so. The availability of public transport in a certain area is certainly proportional to the number of public transit stations, but it is more determined by the number of transit lines that relate to those stations. The greater the number of public transit lines, the better an area is connected to other areas of the city (i.e., the higher the utility of using public transport will be), and thus the more likely local habitants are to use public transport. In view of this, the current study measures the availability of public transport in a certain block, which is based on the total number of those public transit lines that set up at least one station within the block. Grading tools in ArcGIS are then used to rank the results through a general comparison of all blocks.

14.4.3 Data and Results

Considering that a rail transit system is not yet available in Xiamen and that the Bus Rapid Transport (BRT) is operational in only a small area of the city thus far, the current study focuses only on the city's well-developed mode of conventional public transport: its bus system.

First, according to data collected from the Xiamen Public Transportation Group, Inc., Tuba (www.mapbar.com), and Mabc (code.mapabc.com), there are 208 bus lines, and along with all the bus stops within the city, they are processed into the base map; data items include stop names, stop sites (with latitude and longitude), line numbers, and line paths, among other things.

Then, using the ArcGIS tools for "area + buffer," each of the 560 blocks and their "border streets" are integrated as units. As a result of this process, when we extract a block from the map, we extract not only the internal area of the block but also the streets that surround it and act as boundaries.

Then, using the relevant analysis tools in ArcGIS, all the bus stops are extracted from each block (including its boundary streets), and the total number of bus lines relating to those stops are counted.

In the results garnered in the last step, the minimum value is 0 (i.e., there is no bus stop in that block), and the maximum value is 76 (i.e., there are 76 bus lines

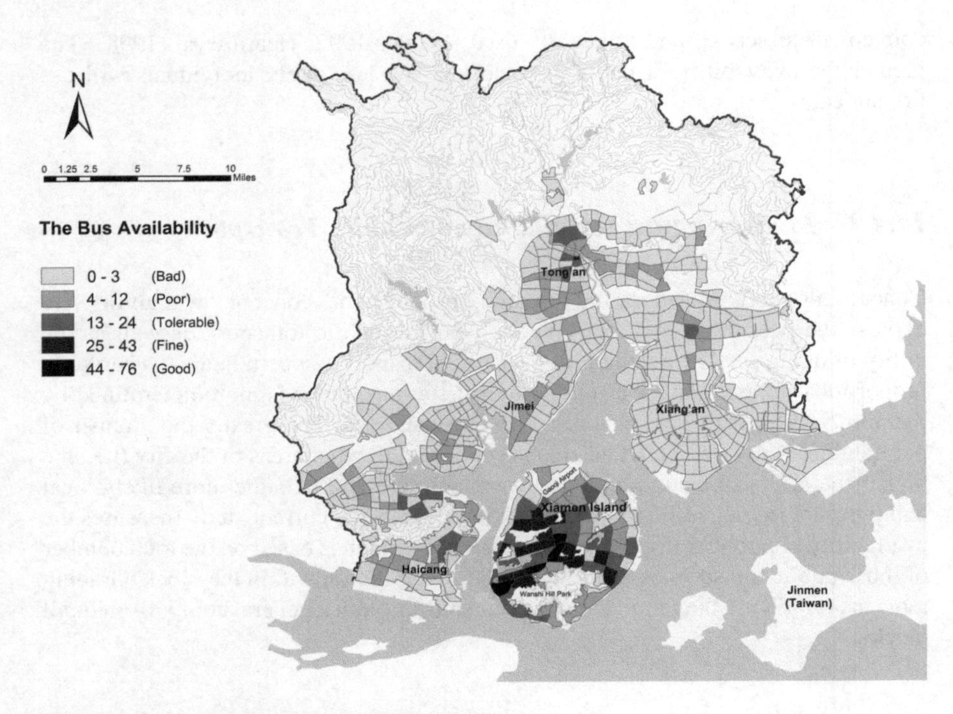

Fig. 14.2 Spatial pattern of bus availability in Xiamen

with stops in that block). By using the classification tool in ArcGIS and the "natural breaks" grading option, all 560 blocks are divided into five levels and differentiated in terms of mapping color. The final result is shown in the Fig. 14.2. Blocks with higher values (i.e., in darker colors) are those with relatively good bus availability, while those with lower values (i.e., in fainter colors) have relatively poor bus availability. Thus, the figure generally reflects the spatial pattern of the availability of public transport across Xiamen.

As shown in Fig. 14.2, the spatial pattern of bus availability in Xiamen closely resembles that of walking accessibility to urban facilities: only the old downtown area on Xiamen Island and a few blocks in the Tong'an district are at the "good" or "fine" level; most areas are only at the "poor" or "bad" level. There is also a significant gap in terms of the availability of public transport between Xiamen Island (i.e., the earliest developed area) and the outer districts on the mainland (i.e., the newly developed areas). This indicates a noticeable lag and insufficiency with regard to the construction of a public transport system in those new districts outside of Xiamen Island; this suggests the ongoing transition of Xiamen from an "island city" to a "bay city," in which the development of those outer districts is part of the "key strategy" of the plan.

14.5 Mapping Car Dependence

14.5.1 Relationship Among Walking Accessibility to Urban Facilities, Availability of Public Transport, and Car Dependence

As stated at the beginning of the chapter, there are three categories of car dependence: physical/environmental dependence, social/psychological dependence, and circumstantial/technological dependence. As Gorham (2002) points out, it is probably difficult in practice to identify behavior endemic to any one of these categories, because they interact significantly. Therefore, it is not easy to measure car dependence, and it is even more difficult to overcome car dependence in practice, especially when social and psychological dependence are at play (Begg 1998; Cullinane and Cullinane 2003; Mackett 2009). Nonetheless, there are still widely recognized interrelations between car dependence and the built environment (i.e., urban form, regional structure, the distribution of activities within those structures, and the nature or status of collective transportation modes). Walking accessibility to urban facilities and the availability of public transport are essential components of a city's built environment. Although there is some debate as to how significantly the urban built environment impacts car dependence (Simmonds and Coombe 2000; Handy and Clifton 2001), it is still safe to say that the level of car dependence is somehow inversely proportional to walking accessibility and the availability of public transport. The better (worse) accessibility and availability are, the lower (higher) car dependence would be: this is why public transport development is still widely regarded as a meaningful and helpful way of optimizing the urban form and built environment and improving public transport, with the endpoint of mitigating car dependence. Despite there being some other factors that urban planning could scarcely affect—e.g., social values, the presence of an automobile culture, personal habits, and psychological needs—the levels of both walking accessibility to urban facilities and the availability of public transport can still reflect the potential extent of car dependence. The analysis of its spatial pattern can provide useful information for urban planning and relevant policy-making.

14.5.2 Evaluation of Potential Car Dependence

The previous sections evaluated the walking accessibility to urban facilities and the bus availability within the 560 blocks of Xiamen; the results of those evaluations were differentiated into five levels: good, fine, tolerable, poor, and bad. Considering that car dependence relates to both walking accessibility to urban facilities and bus availability—and that the level of walking accessibility of a block may different from its level of bus availability—the scoring method detailed below is designed to quantify potential car dependence in each block.

1. Using "five-point scaling," we quantify the five evaluation levels of "good," "fine," "tolerable," "poor," and "bad" into the evaluation scores 5, 4, 3, 2, and 1, respectively.
2. We let "the evaluation score of potential car dependence = the evaluation score of overall walking accessibility to urban facilities + the evaluation score of bus availability." Thus, the minimum evaluation score of potential car dependence is 2 (i.e., both walking accessibility and bus availability are "bad"), while the maximum is 10 (i.e., both walking accessibility and bus availability are "good").
3. We then divide the possible scores of potential car dependence (2–10) into five levels: highest dependence, high dependence, average dependence, low dependence, and lowest dependence. Details are found in Table 14.2.

14.5.3 Data and Results

Applying the aforementioned method, by executing layer accumulation analysis in ArcGIS, all the evaluation scores of potential car dependence in Xiamen's 560 blocks are calculated. Then, according to the five-level division standards listed in Table 14.2, scores are converted to corresponding levels. Finally, a map of potential car dependence is generated (Fig. 14.3), on which different levels are displayed in different colors.

As shown in Fig. 14.3, there are distinct spatial disparities in terms of potential car dependence across Xiamen. Because the best (worst) areas in terms of walking accessibility are very often also the best (worst) areas in terms of bus availability, the spatial pattern of potential car dependence in Xiamen is similar to the patterns of these two factors. The old central area of the city on Xiamen Island is comparatively in the best situation, where the potential car dependence is generally at the level of "low dependence" or "lowest dependence." However, the vast and newly developed areas—including the north and east fringes of Xiamen Island, as well as those outside areas such as Haicang, Jimei, Tong'an, and Xiang'an—are left behind as disadvantaged areas in terms of urban services and public transport. Hence, they have the worst scenario, where the potential car dependence is generally at the level of "high dependence" or "highest dependence." Therefore, there is still an obvious disparity between "on-island" and "off-island" in terms of potential car dependence. Similar to Turcotte's (2001) findings in Canada, areas new and far from the city center and featuring low-density neighborhoods have the highest levels of potential car dependence, suggesting that such areas are problematic and need to be improved. These areas should be deemed a priority for future investment, in terms of both urban facilities and public transport services. Otherwise, an undesirable, highly car-dependent transportation structure could result, in addition to inconvenient living environments. The evaluation and mapping of potential car

Table 14.2 Evaluation scores and ranking levels of "potential car dependence" in Xiamen

Evaluation score of "potential car dependence"	2	3–4	5–6	7–8	9–10
Possible combination of "overall walking accessibility" and "bus availability"	Bad + Bad	Bad + Poor Bad + Bad	Bad + Good Poor + Fine Poor + Tolerable Tolerable + Tolerable	Poor + Good Tolerable + Good Tolerable + Fine Fine + Fine	Fine + Good Good + Good
Ranking level of"potential car dependence"	Highest dependence	High dependence	Average dependence	Low dependence	Lowest dependence

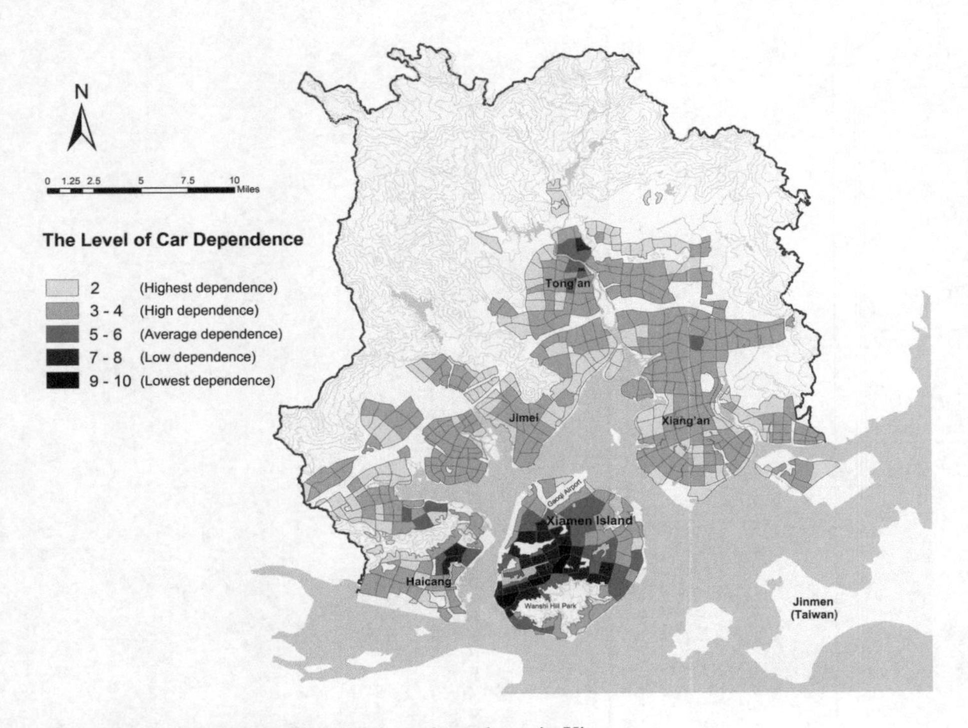

Fig. 14.3 Spatial pattern of potential car dependence in Xiamen

dependence, together with that of walking accessibility and bus availability, can therefore provide very useful information for planners and decision-makers vis-à-vis the ongoing development of areas outside Xiamen Island, as well as the structural optimization of Xiamen Island.

14.6 Summary

Like many other Chinese cities, the case study city of Xiamen has been experiencing a marked spatial transformation in recent decades. Since the strategy of transforming from an "island city" to a "bay city" was initiated several years ago, under the guidance of the "Overall Planning of Xiamen City (2004–2020)," the so-called battlefield of urban development is moving to a much broader space outside Xiamen Island. The spatial pattern of transportation will inevitably present a divergent distribution by taking Xiamen Island as the center. It is foreseeable that travel demand and traffic mileage will thus increase dramatically. The cross-sea, unidirectional traffic flow to and from the island is estimated to amount to 500,000 person-times a day. How to guide and control this unprecedented transformation and enable the city to lay a foundation for future sustainable development—rather than something quite untenable, unsustainable, and undesirable—is now a crucial issue and stern challenge that local decision-makers, researchers, and planners face.

The results of the current study indicate that the construction of urban facilities and public transportation systems has not kept pace with the development of those new areas, leaving them in an obvious state of hysteresis. This leads to the result where the vast and newly developed areas outside Xiamen Island generally suffer from "bad" or "poor" conditions in terms of both walking accessibility to urban facilities and availability of public transport, as shown in the previous sections. The empirical study undertaken in the current study has highlighted that the worst areas of walking accessibility to urban facilities are very often also those with the worst bus availability; these areas are the ones most susceptible to car dependence. More and more people in those areas of Xiamen need to buy a car, despite the fact that they can barely afford one. Such circumstances make these individuals not only "mortgage slaves," but also "slaves to cars." For many cities, an explosive increase in the number of cars is becoming an intractable challenge—and because of the rigidity of their spatial structures and road networks, as well as the mental refractory of car dependence, once a car-dependent transportation system is formed, it is very difficult to change it, due to so-called path dependency and the "locked-in effect." Therefore, it is critical for decision-makers and planners to put sustainable transportation at the tops of their agendas, in order to avoid the trap of car dependence.

By taking the city of Xiamen as a case study, the current study has also generated some useful information for both public transport and urban facility providers, by identifying problem areas and thus allowing them to join forces to promote improvements therein. The methodology developed in this research might also serve as a practical reference for future studies.

Problems in an individual city can also highlight important features associated with wider dynamics. The "car boom" and car-oriented development is now a common challenge that Chinese cities face. Fortunately, Chinese cities still have a chance to escape the trap of car dependence and innovate new ways to live sustainably, but only if they so choose. After all, since most people in China do not yet own cars and have not yet begun to drive, their travel habits and preferences can still be shaped in an environmentally sustainable way (Qian 2010). All in all, cities need to consider more than just road networks and vehicles when attempting to solve transportation problems: it is more fundamentally necessary to examine the urban form and built environment that push people to drive single-passenger cars.

Acknowledgement The research is supported by the National Natural Science Foundation of China (Project Item: No. 41071101), the Research Fellowship Scheme (2011–2012) at Peking University-Lincoln Institute of Land Policy (Project Item: No. FS-20110901), and the Research Start-up Fund for Introduced Talents at Xiamen University.

The author thanks three graduate students, Huang Jiuju, Li Yongling, and Yan Xin, in the Department of Urban Planning, Xiamen University, for their assistance in collecting and processing data.

References

Begg D (1998) Car free cities: reducing traffic in cities: avoiding the transport time bomb. The Third Car-free Cities conference, Edinburgh, June 1998

China Society of Urban Studies (2009) China's low carbon eco-city development strategy. China City Press, Beijing

Condon PM (2010) Seven rules for sustainable communities: design strategies for the post-carbon world. Island Press, Washington, DC

Congress for the New Urbanism (2000) Charter of the new urbanism. McGraw-Hill, New York

Cooper S, Wright P, Ball R (2009) Measuring the accessibility of opportunities and services in dense urban environments: experiences from London. In: Proceedings of European Transport Conference 2009, Leeuwenhorst Conference Center, The Netherlands, 5–7 Oct 2009

Cullinane S, Cullinane K (2003) Car dependence in a public transport dominated city: evidences from Hong Kong. Transp Res Pt D Transp Environ 8(2):129–138

Grengs, J. (2001) Does public transit counteract the segregation of carless households? Measuring spatial patterns of accessibility. Transportation Research Record 1753, paper No.01–3534

Gorham R (2002) Car dependence as a social problem: a critical essay on the existing literature and future needs. In: Black WR, Nijkamp P (eds) Social change and sustainable transport. Indiana University Press, Bloomington, pp 106–115

Guo Z (2010) The value of "abandoning car dependence" for the development of eco-cities in China. Ecol Econ 1:183–185

Handy SL, Clifton KJ (2001) Evaluating neighbourhood accessibility: possibilities and practicalities. J Transp Stat 4:67–78

Hearn MD, Baranowski T, Baranowski J et al (1998) Environmental influence on dietary behavior among children: availability and accessibility of fruits and vegetables enable consumption. J Health Educ 29(1):26–32

Huang F (2004) Towards the future urban planning and design. China Architecture and Building Press, Beijing

Kenworthy J, Laube F (1999) Patterns of automobile dependence in cities: an international overview of key physical and economic dimensions with some implications for urban policy. Transp Res Pt A Policy Pract 33(7-8):691–723

Kenworthy J, Laube F (2001) The millennium cities database for sustainable transport. International Union of Public Transport, Brussels/Institute for Sustainability and Technology Policy, Perth

Mackett RL (2009) Why is it so difficult to reduce car use? In: Proceedings of European Transport Conference 2009, Leeuwenhorst Conference Center, The Netherlands, 5–7 Oct 2009

Naess P (2006) Urban structure matters: residential location, car dependence and travel behavior. Routledge, New York

NDRCC (National Development and Reform Commission of China) (2007) China's National Climate Change Programme, pp 7–9

Newman P, Kenworthy J (1989) Cities and automobile dependence: an international sourcebook. Gower, Aldershot

Newman P, Kenworthy J (1999) Sustainability and cities: overcoming automobile dependence. Island Press, Washington, DC

Newman P, Kenworthy J (2000) Sustainable urban form: the big picture. In: Williams K, Burton E, Jenks M (eds) Achieving sustainable urban form. E & FN Spon, Taylor & Francis Group, London, pp 109–120

Newman P, Kenworthy J (2006) Urban design to reduce automobile dependence. Opolis: Int J Suburb Metrop Stud 2(1):35–52

Pan H, Tang Y, Wu J (2008) China's spatial planning strategies for "low-carbon city". Urban Plan Forum 6:57–64

Qian J (2010) Out of a jam: China can choose sustainable transportation and smart growth instead of car dependence and hyper-motorization. A blog posted 27 September 2010 in Greening China. http://switchboard.nrdc.org/blogs/qian/out_of_a_jam.html

Reneland M (2000) Accessibility in Swedish towns. In: Williams K, Burton E, Jenks M (eds) Achieving sustainable urban form. E & FN Spon, Taylor & Francis Group, London, pp 131–138

Simmonds D, Coombe D (2000) The transport implications of alternative urban forms. In: Williams K, Burton E, Jenks M (eds) Achieving sustainable urban form. E & FN Spon, Taylor & Francis Group, London, pp 121–130

Stradling SG (2002) Combating car dependence. In: Proceedings of the Twelfth Seminar on Behavioral Research in Road Safety, pp 174–187. Dep. of Transport: London, 23 Dec 2002

Stradling SG (2003) Reducing car dependence. In: Hine J, Preston J (eds) Integrated futures and transport choices: UK transport policy beyond 1998 White Paper and Acts. Ashgate, Aldershot, pp 100–115

Turcotte M (2001) Dependence on cars in urban neighborhoods. Statistics Canada (2001), Catalogue-11-008, 20–30

Winter EC (1992) Are we ignoring population density in health planning? The issue of availability and accessibility. Health Policy Plan 7(2):191–193

Part III
Landscape and Ecological System, Sustainable Development: Section Green Design and Landscape

Chapter 15
Effects of Green Curtains to Improve the Living Environment

Masashi Kato, Tsukasa Iwata, Norimitsu Ishii, Kimihiro Hino, Junichiro Tsutsumi, Ryo Nakamatsu, Yoshitaka Nishime, Koji Miyagi, and Masakazu Suzuki

Abstract A green curtain means vertical planting of vines over a building surface. It is expected to improve indoor thermal environment of the building in summer, because it has a number of leaves shading a wall surface and windows of the building. However, it is also supposed to be obstacles to natural ventilation. Some experiments were done to confirm these environmental performances of green curtains. Model green curtains for the experiments were made of bitter gourd vines. A questionnaire survey for green curtain users was also carried out in Hamamatsu, Japan, to collect data about the amount of energy saved by green curtains and about the effect of green curtains on natural ventilation in residential houses. Results of the experiments and the survey indicate obvious effects of green curtains on indoor thermal environment.

Keywords Green curtain • Solar shading • Ventilation • Thermal comfort

M. Kato (✉) • T. Iwata • N. Ishii • K. Hino
Department of Housing and Urban Planning, Building Research Institute,
Tachihara-1, Tsukuba City, Ibaraki Prefecture 305-0802, Japan
e-mail: ms-katou@kenken.go.jp; iwata@kenken.go.jp; ishii@kenken.go.jp; hino@kenken.go.jp

J. Tsutsumi • R. Nakamatsu
Department of Civil Engineering and Architecture, University of Ryukyus,
Senbaru-1, Nishihara-cho, Okinawa Prefecture 903-0213, Japan
e-mail: jzutsumi@tec.u-ryukyu.ac.jp; nkmt_ryo@tec.u-ryukyu.ac.jp

Y. Nishime • K. Miyagi
Ocean Expo Research Center, Ishikawa888, Motobu-cho, Okinawa Prefecture 905-0206, Japan
e-mail: y_nishime@keiayouhaku.or.jp; kj_miyagi@kaiyouhaku.or.jp

M. Suzuki
Faculty of Arts and Design, University of Tsukuba, Tennoudai1-1-1,
Tsukuba City, Ibaraki Prefecture 305-8574, Japan
e-mail: masayan@r9.dion.ne.jp

M. Kawakami et al. (eds.), *Spatial Planning and Sustainable Development: Approaches for Achieving Sustainable Urban Form in Asian Cities*, Strategies for Sustainability,
DOI 10.1007/978-94-007-5922-0_15,

15.1 Introduction

The main purpose of this chapter is to confirm effects of green curtains in domestic lives by a questionnaire survey for actual green curtain users. The questionnaire asks them annual fluctuations of domestic energy consumption and other changes in their lives by green curtains.

Practical greenings on buildings, such as roof and wall greenings, are increasing recently. These building greenings are expected to have various positive effects, for example, improvement of cityscapes, mitigation of heat island phenomena, and carbon dioxide reduction. Such a trend pushes people to start using "green curtains" in particular. A green curtain is a kind of wall greening that is a net or a series of strings covered with lianas or vines, such as morning glory and bitter gourd. Normally it is set separately from the wall surface with a short gap and covering over the windows, the balconies, and the wall. It is not difficult to set green curtains if there is room for planters and the sunshine. A questionnaire survey carried out in Misato, Saitama, Japan, in 2009 indicated that green curtains could reduce domestic energy consumption in August at least by 20%.

It is presumed that the green curtains shaded the houses to reduce operation time of air-conditioning system in the Misato survey. However, shading is different from cooling indoor temperature. Actually, the green curtain users in Misato answered that they kept the windows open in the daytime. That means that the green curtains let the wind pass through the windows as natural ventilation to improve the indoor thermal sensation. A green curtain is expected to be a unique material whose functions are not only solar shading but also natural ventilation and to make better indoor thermal sensation in summer from the Misato survey.

It is easily guessed that higher density of leaves of green curtains should have higher efficiency of solar shading, while it should also become more obstacles to ventilation. A series of experiments was carried out in order to confirm these conflicting characteristics. Model green curtains used in the experiment are made of bitter gourd.

Relations between plants and wind speed and air temperature were already published by Architectural Institute of Japan in 2002 (Architectural Institute of Japan 2002). A paper on characteristics of air flow around plants was published in 1954 by R. F. White (White 1954). Wind load on trees was studied by R. C. Johnson Jr., G. E. Rammey, et al. in 1982 (Johnson et al. 1982). J. Tsutsumi, T. Katayama, et al. studied air flow around a house with a hedge and wind pressure by field experiments and wind tunnel tests in1989 (Tsutsumi et al. 1989). Although these studies cover trees and shrubs as the greening factors, they do not deal with green curtains.

The US Environmental Protection Agency (1992) published the thermal environmental effects including solar shading of building greening. Akbari and Taha (1992) measured the influence of growing trees and plants on air-conditioning heat loads in four cities in Canada. However, these researches did not cover green curtains. A research on green curtains was started by K. Izawa, H. Fujii, et al. in

2006, which evaluated and discussed effects of green curtains on improving the thermal environment but did not discuss relations of ventilation and solar shading (Izawa et al. 2006).

From these points of view, a series of experiments were conducted to confirm the characteristics of both solar shading and ventilation blocking and the relations with density of leaves by the same model green curtains.

Although windows must be opened in order to utilize natural ventilation, it is strongly connected with occupants' lifestyle whether windows are opened or closed. Therefore, it is also necessary to verify how much the natural ventilation is utilized in combination with ventilation blocking effect of green curtains.

This chapter consists of the following contents. Actual research design and method are summarized in Sect. 15.2. Results of experiments on ventilation properties and solar shading effects are discussed in Section3. Results of a questionnaire survey for green curtain users about the effects of green curtains and the changes of their lifestyle by green curtains are described in Sect. 15.4. Outputs of this study are concluded in Sect. 15.5.

15.2 Method for Testing Green Curtain's Effects

15.2.1 Description of Experiment

We carried out the test between the two reinforced concrete walls of Okinawa Commemorative National Government Park in Motobu Town, Okinawa Prefecture. The building wall and a windbreak wall were facing each other with a roof on top, forming a nice wind tunnel.

The green curtain test specimen was created by positioning support poles on planter boxes 75 cm wide, 30 cm high, and 42 cm deep. Netting was spread across the supports, and bitter gourd was then trained across it. The height from the foot of the test specimen to the uppermost part of the nets was 230 cm. Three planters were used; positioned so as to block the opening of the ventilation testing area (refer to Figs. 15.1 and 15.2).

While it would be desirable to cultivate a number of test specimens of different foliage densities and conduct a comparative test, the same foliage density ends up being produced under the same natural conditions. Consequently we decided to prepare a sole test specimen and alter the foliage density by successfully pruning the leaves. The test of ventilation properties and solar shading effects was then repeated for each case. The same test specimen was utilized to keep all other conditions constant but leaf density. These experiments were carried out on August 25 and 26, 2010.

Fig. 15.1 Positioning of the green curtain test specimen

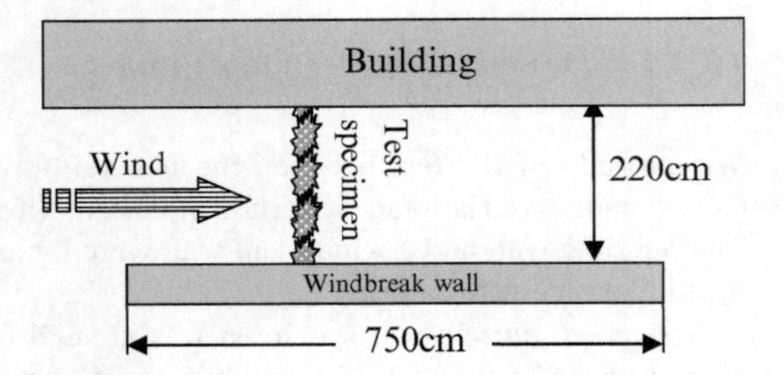

Fig. 15.2 Ground plan of location for testing ventilation properties

15.2.2 Test for Ventilation Properties

As shown in Fig. 15.1, breeze was artificially induced by positioning two commercial fans 5 m away from the test specimen. As shown in Fig. 15.2, the test specimen was set up and anemometers were set up behind and in front of it. To prevent crosswinds from being picked up, air flow transducers (Model 6332D) and probes (Model 0963-00) manufactured by Kanomax that feature wind directionality were used as anemometers. Taking into account the height of the test specimen, the probe sensors were positioned at a height of 120 cm and were placed to take measurements at locations 80 cm away from the test specimen on the upwind side and 50, 100, 150, and 200 cm from the center of the test specimen on the downwind side, respectively. This is because it was assumed that the further away from the test specimen, the more the wind would circulate. Data loggers (midi

LOGGERs manufactured by Graphtec) were set to take measurements at 1-s-intervals, and average figures over a period of 1 min were sought and compared.

We calculated wind speed attenuation (%) as (average wind speed on the upwind side – average wind speed on the downward side)/average wind speed on the upwind side × 100. Draft ratio was calculated as 100 – wind speed attenuation (%).

15.2.3 Testing of Solar Shading Effects

The measuring of the solar shading effects was conducted by moving the test specimen to the sunlit car park after the ventilation properties were tested. Two solar radiation sensors were positioned in front and behind the test specimen, and the differences in the quantity of solar radiation were measured.

However, because the foliage growth of green curtains is not uniform, measurement errors in radiation quantity occur between places where the sunlight penetrates the foliage and places that are in shadow. Consequently a solar radiation sensor platform (91 cm × 23 cm × 3.6 cm) was created, as shown in Fig. 15.3, and the solar radiation sensor was placed on this platform. Screws were inserted on the four corners of the reverse side of this platform in order to ensure it remained level, and as shown in the photograph, two rails were attached to the upper side so that a solar radiation sensor on wheels could be moved horizontally on the rails. Lines were inscribed at 5-cm intervals on the platform so that it was possible to manually move the solar radiation sensor by 5 cm every five seconds and seek an average radiation quantity of a total of 75 s (5 s × 15 intervals). Prede PCM-01 solar radiation sensors were used, and the measurement interval on the data loggers was set at 1 s.

However, a problem surfaced in measuring the shading effect. Because we used sole specimen repeatedly, the time of the measurement was different each case and so caused difference in the elevation of the sun. If it was possible to measure several test specimens at the same time, this problem would not have occurred. But in this approach, it was necessary to make adjustments for the sun's elevation.

If a green curtain was uniform and thin in shape like frosted glass, it would be possible to simply adjust the angle as needed based on the elevation of the sun. In reality however, the various leaves grow at various angles and green curtains do not have a uniform depth. Obviously, it was not enough to simply adjust the angle.

So we decided to measure the shading effects at various time of a day using the same test specimen to secure a grasp of relationship between solar shading effects and sun elevation and use the results as guide for the amounts of adjustments needed.

From September 6 to 7, 2010, we conducted an experiment for the purposes of making the adjustments at a square within Building Research Institute in Tsukuba City, Ibaraki Prefecture. The green curtain test specimen used was formed by positioning support poles on a planter box 75 cm wide, 27 cm high, and 42 cm deep. Netting was spread across the supports, and the height from the ground to the

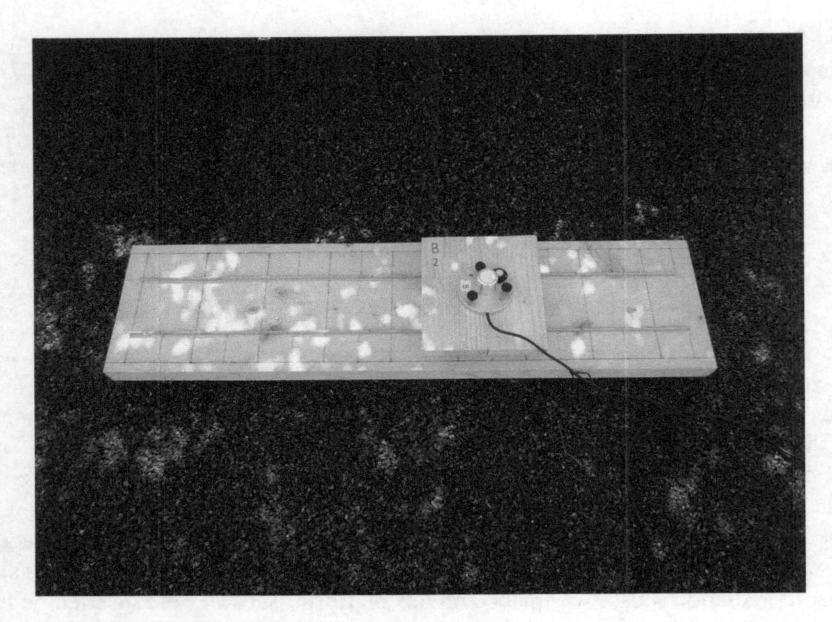

Fig. 15.3 Solar radiation sensor platform

uppermost part of the net was 210 cm. A mark was drawn at a spot 130 cm from the base of this test specimen, and at the location of the mark, a narrow crossbar was established to create a shadow. After the solar radiation sensor platform shown in Fig. 15.3 was placed at the location of that shadow, the crossbar was removed. In other words, by doing this, it became possible to constantly measure the same place on the green curtain.

15.2.4 Method Used to Measure Leaf Area Density and the Result

Green curtain's ventilation and solar shading properties are both related to the density of the foliage, so it was necessary to measure the leaf area density. In this experiment procedure, we employed the green designation method because it was a simple operation and was possible to photograph the test specimen. In the preliminary experiment, the green curtain test specimen was photographed using a thermographic camera so that the differences between the surroundings and the foliage – foliage is lower in temperature to its surroundings – were displayed tonally. Foliage density was then calculated based on these images using image processing software (LIA for WIN32).

In specific terms, a test specimen is placed in front of a white wall or a white cloth is hung behind it, and photos are taken. Having done so, the spaces other than those covered by the plant will show up white. By extracting and calculating the

Table 15.1 Leaf area density at each pruning stage

Pruning stage	Pruned amount (g)	Plant weight (g)	Leaf area density (%)
Initial (before pruning)	0.0	2,380	88.45
First pruning	134.5	2245.5	88.73
Second pruning	160.5	2,085	78.96
Third pruning	204.5	1880.5	68.13
Fourth pruning	426.5	1,454	59.94

green portions of this photographic image, it is possible to determine what the ratio of plant (mainly foliage) is in relation to the flat space, in other words, the leaf area density (%). The relationship between the leaf area measured using this approach and the quantity of pruning is shown in Table 15.1.

15.2.5 Questionnaire to Obtain Quantitative Understanding of the Effects Green Curtains and the Changes in User Behavior Resulting from Green Curtains

Next, in order to verify how much household power consumption declines as a result of the effects green curtains have in improving the thermal environment and whether behavior that improves ventilation (the opening or closing of windows) increases due to the introduction of green curtains, a questionnaire survey of actual green curtain users in Hamamatsu City was conducted.

Japan's power companies employ meter readings to notify households of their monthly power consumption, and these meter readings record a household's power consumption on the month in the year in question. At the same time, they also show the power consumption for the same month of the prior year, and so by comparing these, it is possible to calculate the yearly difference. Yearly differences were obtained for households that introduced green curtains in the present year, not having had them up in the previous year. The survey questions were composed based on the premise that if the power consumption had declined, it could be interpreted as resulting from an improvement in the indoor thermal environment due to installation of the green curtain.

Furthermore, if this was confirmed to be the effect of the green curtains, then in order to determine whether it was simply due to the solar shading effects created by the green curtains or because many users opened windows thanks to the ventilation properties that green curtains possess, we concurrently asked about changes in the conditions with regard to the opening and closing of windows. Additionally, we also asked questions about basic information such as the subjects' attributes and living arrangements.

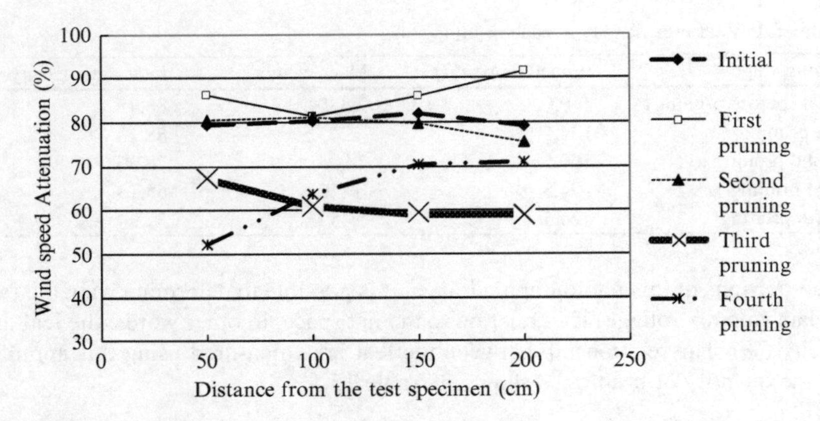

Fig. 15.4 Distance from test specimen and wind speed attenuation (induced breeze)

15.3 Results of Testing of Green Curtain's Effects

15.3.1 Results of Ventilation Properties Test

The outcomes of each test specimen's wind speed attenuation are shown in Fig. 15.4; no fixed pattern was shown in the relationship between the distance from the test specimen and the wind speed attenuation. This is conceivably due to the fact that the wind speed of the induced breeze became slower as the distance from the source of the breeze became greater, and it only had an effect on a limited range. As a result, the relationship between leaf area density and ventilation properties was based on wind speed attenuation data from 50 cm away. The relationship between attenuation and leaf area density is as shown in Fig. 15.6, and as you would expect, the higher the leaf area density, the greater the wind speed attenuation.

15.3.2 Results of Testing Solar Radiation Shading Effects

The solar shading ratios measured are recorded in the solar shading ratios column in Table 15.3, but because each measurement time differed, it was necessary to make adjustments. Table 15.2 shows the results of the solar shading ratios that were derived using the single test specimen on September 6 and 7, 2010, in Tsukuba City, in order to guide the adjustment quantities. The position of Building Research Institute in Tsukuba City is 140.08° east longitude, 36.13° north latitude. However, the location of the experiment site in Okinawa is 127.88° east longitude, 26.69° north latitude. By inputting that position information and the desired time into

Table 15.2 Relationship between sun elevation and solar shading ratios in the same sample

Measurement time (September)	Sun elevation (degrees)	Solar shading ratio (%)	Weather
6th 13:43	48.55	38.44	Slightly cloudy
6th 15:28	29.51	45.83	Fine
7th 16:37	14.57	23.89	Fine
7th 9:12	45.69	33.56	Fine
7th 10:35	56.24	55.75	Fine
7th 10:56	58.80	63.51	Fine
7th 12:10	58.83	69.37	Fine
7th 13:23	50.52	62.54	Slightly cloudy
7th 14:33	40.53	66.82	Slightly cloudy

Table 15.3 Measurement results for test of solar shading effects

Pruning stage	Solar shading ratio (%)	Time of measurement	Sun elevation (degrees)	Adjustment quantity (%)	Adjusted solar shading ratio (%)
Initial (before pruning)	84.55	25th 15:30	44.96	0.03	84.58
First pruning	85.78	26th 8:45	34.67	8.41	94.19
Second pruning	57.74	26th 9:48	47.85	−2.30	55.42
Third pruning	65.24	26th 10:29	57.34	−10.05	55.19
Fourth pruning	56.78	26th 11:22	65.95	−17.06	39.73

"keisαn high-precision accuracy software from HP of © CASIO COMPUTER CO., LTD.," we could determine the sun's elevation at that point in time. The sun elevations determined in this way are recorded in Table 15.2.

A regression equation was used to come up with Fig. 15.5 from Table 15.2. This regression equation was expressed in terms of $f(x)$, and uniform adjusted quantities for sun elevation of 45° were obtained as $f(45) - f(\alpha)$. (α is the sun elevation at the time of measurement that is being sought.) These adjusted quantities are recorded in Table 15.3.

The relationship between leaf area density and the adjusted solar shading ratios in Table 15.3 is shown in Fig. 15.6 in conjunction with ventilation rates. The table shows the tendency for the solar shading effects to increase as the leaf area density increases but also shows that the ventilation properties deteriorate.

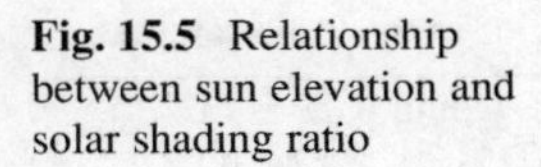
Fig. 15.5 Relationship between sun elevation and solar shading ratio

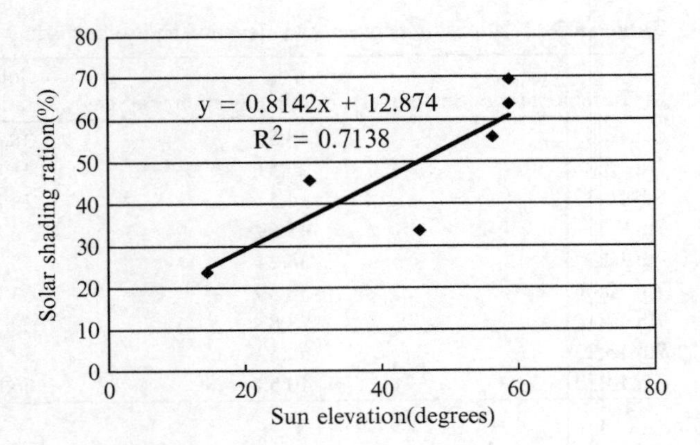

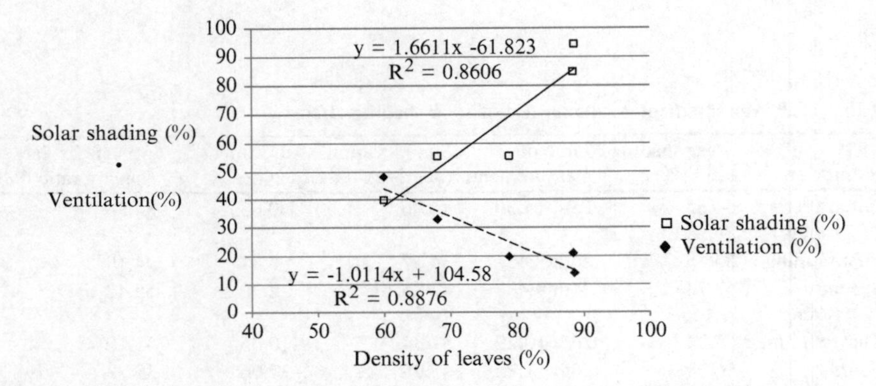

Fig. 15.6 Relationship between solar shading and ventilation properties of green curtains

15.4 Results of Questionnaire Survey

The questionnaire survey was conducted among 150 participants in the green curtain seminar held by Hamamatsu City. The survey was conducted upon explaining the reason behind survey subjects being those who did not install a green curtain in 2009, but will newly install them starting 2010. The items covered in the questionnaire survey are shown in Table 15.4.

15.4.1 Summary of Questionnaire Responses

The questionnaire responses rate was high, at 50.7% (76 responses), but the ratio of valid responses was 23.3% (35 responses) of the questionnaires distributed. The questionnaire survey received a large number of invalid responses, but this was inevitable due to the nature of the survey. Table 15.5 shows the content of the

Table 15.4 Outline of question items in questionnaire survey

Question number	Question items
1	Comparison of power consumption (Power consumption in June–September 2009 and 2010)
2	Changes in the frequency of air-conditioner usage as a result of green curtain
3	Air-conditioner usage situation (usage situation in daytime and evening in 2009 and 2010)
4	Changes with regard to the opening and closing of windows as a result of green curtain (situation in daytime and evening, respectively, in 2009 and 2010)
5	For respondents who answered a larger number of instances in 2010 in response to Q. 4) Reason why the opening and closing of windows changed as a result of green curtain (multiple answers possible)
	1. Because it was cooler inside the room than before
	2. Because inside of the room was no longer visible from outside
	3. Because of increased awareness toward the environment
6	Living arrangement (whether it is a multiunit dwelling or detached dwelling)
7	Total floor area of residence (including second and third floor)
8	Location where green curtain was set up
9	Type of plant used
10	Frequency of watering
11	Attributes of test subjects
12	Other (free responses)

Table 15.5 Breakdown of invalid responses

Reason for invalidity	Qty	Notes
Missing information	18	Data on power consumption were not provided for all four months
Unsuccessful or insufficient growth of green curtains	9	Poor growth status due to the extremely high temperature. Judged based on attached photos and free answer
No air-conditioner	4	Judged based on free answer
Air-conditioner replaced	4	Judged based on free answer
Moved during this year	3	Judged based on free answer
Number of family members increased	2	The number of rooms with air-conditioners increased due to an increased number of family members. Judged based on free answer
Green curtains were used last year as well	1	Green curtains were installed last year as well. Judged based on free answer

invalid responses. To begin with, many questionnaires were not completed adequately due to the fact the respondents were asked to record power consumption figures for all months covering a four-month period. This proved a considerable burden, so many (18 questionnaires) did not fill it out (or filled it out only partially). Next was the problem of poor growth (of the green curtains), which was unavoidable given the nature of a survey dealing with plant material. Additionally, households that had replaced their air-conditioners or did not own one were unsuitable as subjects for this questionnaire. As Saeki and Matsuhara (2000) pointed out, a high response rate does not necessarily lead to "good data," and

Table 15.6 Average monthly household power consumption (Unit: kWh)

Month	2009	2010	Change
June	338.0	316.37	−21.63
July	367.69	378.94	11.26
August	460.37	457.43	−2.94
September	401.31	470.43	69.11

there is also a risk that seeking to lift the valid response rate under any circumstances will lead to distorted survey outcomes. Excluding unsuitable responses as invalid was necessary to preserve the credibility of the questionnaire survey's results.

15.4.2 *Effects of Green Curtains on Improving the Thermal Environment*

The responses to Question 1 are summarized in Table 15.6 and Fig. 15.7. There is no substantial difference between 2009 and 2010, though the green curtains were used only in 2010. It should be noted, however, that the summer of 2010 saw an unprecedented heat wave and the average daily temperature in Shizuoka Prefecture in August 2010 (from data in the website of Japan Weather Association) went up by 1.6° from 2009.

Therefore, average monthly power consumption and average monthly maximum diurnal temperatures in July and August (when the green curtains were fully grown) were plotted in Fig. 15.8 to compare 2009 and 2010. Although the difference in temperatures is significant, the power consumption remained almost the same, which indicates the effects green curtains had.

We also looked at the changes in power consumption regarding the nine green curtain failure cases deemed as invalid in Table 15.5, as shown in Table 15.7. Figure 15.9 compares the failure case results with the successful case results in Table 15.6. The chart reveals that the growth of power consumption in failure cases outpaced that in successful cases, verifying the effects of green curtains in the successful cases.

15.4.3 *Relationship Between Green Curtains and the Opening of Windows*

Our study is based on the assumption that the two features of green curtains, ventilation and solar shading, together will lead to more opening of windows to introduce cool air from outside. To verify this assumption, Question 4 asked about interannual changes in the open/closed status of windows. As shown in Table 15.8, the number of households that kept their windows shut most of the time increased in 2010 compared to 2009, while the number that kept their windows open most of the time also increased in 2010 from 2009, meaning no constant trend was observed overall.

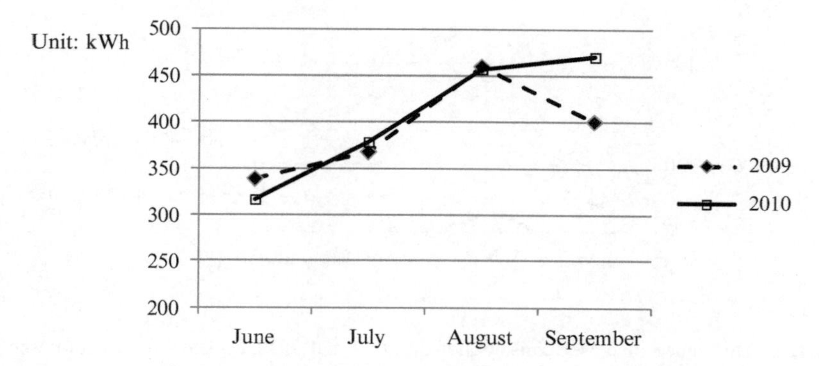

Fig. 15.7 Change in average monthly household power consumption

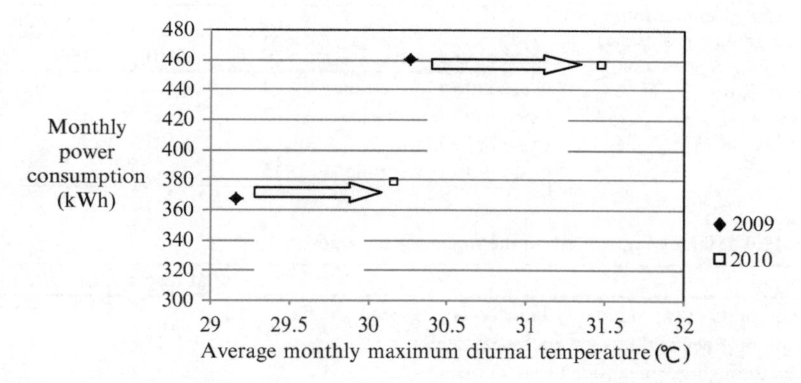

Fig. 15.8 Power consumption and maximum diurnal temperature average in July and August in 2009 and 2010

Table 15.7 Change in power consumption in failure cases (Unit: kWh)

Month	2009	2010	Change	Growth(%)
June	276.22	291.67	15.45	5.6
July	339.56	381.33	41.78	12.3
August	387.44	456.22	68.78	17.8
September	327.67	431.33	103.67	31.6

In the free answer, as summarized in Table 15.9, 42.9% of those who commented noted that it was not very reasonable to measure the effects of green curtains this year, when Japan was hit by the hottest summer ever. It is likely that no marked trend was observed because of the extremely high temperatures.

However, the possibility that green curtains would not influence the opening and closing of windows to begin with cannot be denied either. Because of the possibility that annual climatic conditions will influence the green curtain data, we will consider the possibility of verifying the findings by accumulating data successively the following year also. It will be possible to use this year's study as the fundamental data for this.

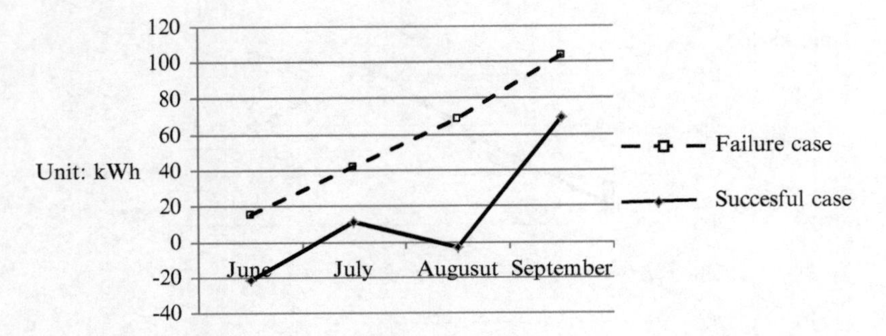

Fig. 15.9 Difference in power consumption in 2009 and 2010 between successful cases and failure cases

Table 15.8 Relationship between green curtains and the open/closed status of windows (Unit: No. of responses)

	Daytime		Nighttime	
Open/closed status	2009	2010	2009	2010
Closed most of the time	2	4	17	16
Open from time to time	7	7	6	10
Open quite often	11	6	8	4
Open most of the time	15	18	4	5

Table 15.9 Major comments from the free answer section

Comments	No. of responses	Ratio (%)
It is too hot this year to identify the effects of green curtains	12	42.9
The look of green curtains gave a breezy feeling	7	25.0
Breezes through green curtains were cool	2	7.1
Others	7	25.0
Total	28	100.0

Incidentally, in Question 5, those who answered in Question 4 that they opened windows more in 2010 than in 2009 were asked their reason for this, and the results are shown in Table 15.10. While there were 14 respondents who answered Question 5, two of them were deemed invalid since in Question 4 they actually answered that they kept windows shut more this year than last year. In addition, Table 15.10 indicates that the screening effect of green curtains is rated highly and that they contribute to raising environmental awareness.

15.5 Conclusion

As a result of this research, we confirmed that green curtains possess contradictory properties in the form of solar shading properties and ventilation properties and gained a quantitative understanding of them. This experiment employed a test

Table 15.10 Reasons for the higher frequency of opening windows

Reason	1 response	1/2 response	Total	Ratio (%)
It is cooler in the room than before	0	2	1	9.1
Green curtains hide the interior of the room (screening effect)	3	4	4	36.4
Green curtains raised awareness of the environment	5	2	6	54.5

specimen, but in the future, we plan to carry out experiments on the solar shading effects and ventilation properties using actual green curtains that are established on buildings.

Additionally, based on the questionnaire survey carried out in this research, we were able to confirm to an extent the effects green curtains have in improving the thermal environment. However, when it came to changes in the opening and closing of windows by users as a result of introducing green curtains, we could not find a clear trend. Conceivably, this may be due to the fact that as a result of the fierce heat, the windows by the green curtains could not be opened adequately, but by carrying out the same survey the following year also, we hope to be able to accumulate data for the purposes of verification. The results of this survey are meaningful in that we were able to obtain basic data that will provide a starting point for this verification.

What became clear from this questionnaire survey was that green curtains help improve internal and external appearances and raise environmental awareness as well as improve the indoor thermal environment, as mentioned above. Green curtain practices could be expected to contribute to crime prevention in the community, given the extra attention paid by users on a daily basis. The effects of green curtains should be evaluated and examined from such diverse perspectives as these, as well as from the standpoint of their functions.

Acknowledgement We thank the officers of the Green Policy Division of Hamamatsu City for their cooperation in conducting the questionnaire survey.

References

Akbari H, Taha H (1992) The impact of trees and white surfaces on residential heating and cooling energy use in four Canadian cities. Energy 17(2):141–149

Architectural Institute of Japan (2002) A greening plan for buildings and cities. Shokokusha, Tokyo, pp 48–53

Izawa K, Fujii H, Yao M, Shukaya M (2006) The passive cooling effect of a large external shading made of creeper. Synopsis of technical presentations at the Architectural Institute of Japan convention, Kanto, pp 651–652

Johnson RC, Ge E, Ohagan DS (1982) Wind induced forces on trees. J Fluid Eng Trans ASME 104:25–30

Saeki Y, Matsuhara M (2000) Statistics as social practice. University of Tokyo Press, Tokyo, pp 47–49

Tsutsumi J, Katayama T et al (1989) Experimental study on wind pressure and Air flow around a dwelling house with arboreal hedges. In: Sun TF (ed) Recent advances in wind engineering. Pergamon Press, New York, pp 568–575

US Environmental Protection Agency (1992) Cooling our communities – a guide book on tree planting and light colored surfacing. Office of Policy Analysis, Climate Change Division, EPA, Washington, DC

White RF (1954) Effects of landscape development on the natural ventilation of buildings and their adjacent areas. Texas engineering experiment station research report no. 45, Texas A.& M. College, College Station, pp 17

Chapter 16
A Comparison of Green Roof Systems with Conventional Roof for the Storm Water Runoff

Sachiko Kikuchi and Hajime Koshimizu

Abstract The overall goal of this chapter is to understand how extensive green roofs influence the storm water runoff reduction for typical Japanese rainfall patterns and to propose a new concept for controlling storm water runoff in densely populated cities. In this study, we calculated the effect on storm water reduction in an urban area and compared the delay time and runoff volume for three roof types: a traditional green roof system, a green roof system with a drainage layer, and a conventional roof. First, to examine how drainage layer thickness affects storm water reduction, runoff was examined for green roof systems with drainage layers of various thicknesses (25, 30, 40, 45, and 55). A typical Japanese rainfall pattern called "local torrential rainfall" with a rate of 150 mm/h was used in an experiment to demonstrate the effect of drainage layer thickness on storm water reduction. The mean storm water reduction was least at 30 mm (0.005 L/min), with a maximum of 45 mm (0.169 L/min), and the mean runoff coefficient was the lowest at 45 mm ($C = 0.63$) and highest at 30 mm ($C = 0.97$). Next, to estimate the effect that roof structure has on storm water reduction, runoff was calculated for the three different roof types in a case study involving 17 flood-prone districts in Chiyoda Ward, Tokyo, Japan. The storm water volume was calculated assuming that traditional green roofs or green roofs with a drainage layer were installed on buildings in a commercial district, on public facilities, and on buildings in an industrial district. Runoff volume was simulated for typical Japanese rainfall events: a local torrential rainfall (July 4, 2000) and a typhoon (August 27, 1993). The calculation results revealed the following: the traditional green roof was able to reduce the peak runoff

S. Kikuchi (✉)
Graduate School of Life Sciences, Tohoku University, 27-1, Kawauchi,
Aoba-ku, Sendai 980-8576, Japan
e-mail: skikuchi@econ.tohoku.ac.jp; kikuchi927b@gmail.com

H. Koshimizu
School of Agriculture, Meiji University, 1-1-1, Higashimita, Tamaku,
Kawasaki 214-8571, Japan
e-mail: koshimiz@isc.meiji.ac.jp

M. Kawakami et al. (eds.), *Spatial Planning and Sustainable Development: Approaches for Achieving Sustainable Urban Form in Asian Cities*, Strategies for Sustainability,
DOI 10.1007/978-94-007-5922-0_16, © Springer Science+Business Media Dordrecht 2013

to 71.7 %, and the extensive green roof with a drainage layer was able to reduce the peak runoff to 57.1 %. Finally, we discuss the significance of green roof systems on new storm water control strategies in urban areas of Japan.

Keywords Traditional green roof • Drainage layer of extensive green roof • Typical Japanese rainfall • Storm water runoff • Runoff coefficient • Runoff delay time

16.1 Introduction

Most highly urbanized areas are covered with impervious surfaces, such as concrete, asphalt, and other artificial structures. In Japan, many local governments have implemented rainwater intrusion prevention measures in order to avoid flooding; as a result, so much rainwater flows through the drains that the processing capacity of the sewerage system is exceeded. Some urban areas face the flood risk posed by an overflowing sewerage system. Most governments aim to install underground reservoirs and permeable pavement to prevent overflow, but these infrastructures require considerable time and money to construct. For these reasons, it is difficult to secure and develop infiltration areas in urban environments.

In the United States and Europe, controlling rainwater runoff has attracted considerable attention. To control runoff, the US government plants trees along the street in urban areas, builds rain gardens and swells, and installs green roofs. Generally, green roof systems act as a substitute for street trees, rain gardens, and swells for capturing rainwater. Green roofs are an effective means of infiltrating storm water in some areas that are unsuitable for rainwater reservoirs.

Many researchers have experimentally investigated the effect of green roof systems on storm water runoff reduction. Ikeda et al. examined the storm water reduction effect of a S*edum spp*. vegetation base and clarified that pearlite and Andosol gave the sedum mat its water-holding capability (Ikeda et al. 2001). Dunnett et al. examined the runoff volume associated with different vegetation and plant structure types and suggested that the grass monoculture vegetation type caused the greatest runoff loss and that plant structure was not important in capturing rainwater (2008). Getter et al. quantified the storm water retention effect for four roof slopes and showed that the green roof slope has an effect on runoff retention (2007). Unfortunately, these findings listed above are difficult to apply to Japanese green roof systems for three reasons: they are case studies in climates that differ from the Japanese climate, the structures of the green roof systems are different than the common Japan structures, and Japanese green roof systems delay storm water runoff in torrential rainfall. Therefore, we resorted to simulating rainfall in an experiment using common Japanese green roof materials to measure the runoff volume. Using the runoff volume, we calculated the runoff coefficient and the runoff delay time. Finally, we simulated storm water runoff to estimate and

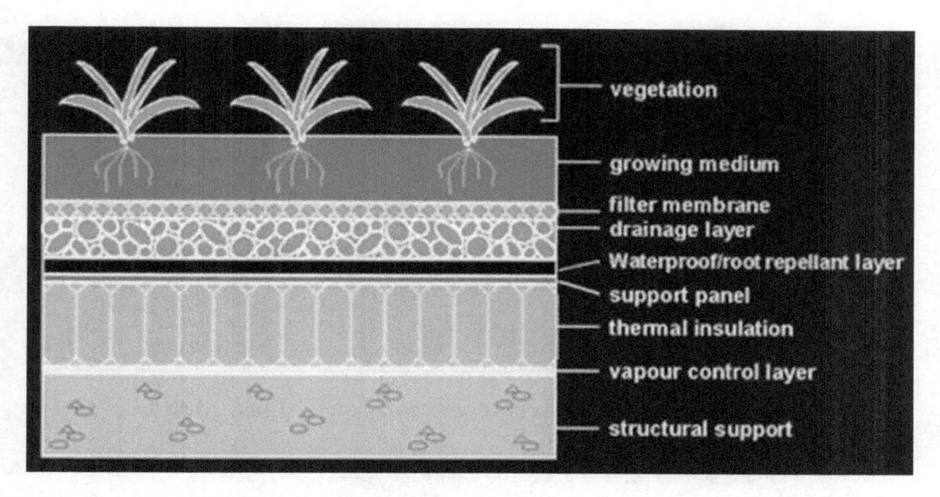

Fig. 16.1 Principal green roof technology components (Green Roofs for Healthy Cities 2009)

discuss the rainwater runoff delay of Murayama and Koshimizu's green roof system (Murayama and Koshimizu 1997).

16.1.1 Green Roof Systems

A green roof system is an addition to an existing roof consisting of a high-quality waterproofing system, a drainage system, a filter cloth, a lightweight growing medium, and plants (Green Roofs for Healthy Cities 2009). Figure 16.1 shows a cross section of a green roof from the Institute for Research in Construction, National Research Council.

The first Japanese green roof was installed in the 1900s and served as a roof garden, as opposed to a green roof in today's sense. After World War II, many famous department stores constructed roof gardens containing trees, flower beds, fountains, and small arbors. The green roof served as a recreational space. Since the 1970s, architects and landscape designers have been concerned with how to introduce greening technology to artificial structures, leading Japanese researchers to analyze green roof functions and effects. Table 16.1 summarizes the effects of green roofs (Organization for Landscape and Urban Green Infrastructure 2004). Green roof technologies not only provide building owners of with a proven return on investment, but also present opportunities for significant social, economic, and environmental benefits, particularly in cities (Green Roofs for Healthy Cities 2009).

In Japan, green roof construction methods are determined not only by building usage but also through budget and the intent of the green space. Table 16.2 lists the characteristics used to select the best green roof method.

Table 16.1 Effects of green roofs (Organization for Landscape and Urban Green Infrastructure 2004)

Familiar environment	Urban environment
1. Control of top floor's temperature rise	1. Reduction of greenhouse effect
2. Creation of relaxation and relief space	2. Reduction of storm water runoff
3. Keeping buildings' surface	3. Creation of biodiversity
4. Protection against dust	4. Air pollution abatement
5. Creation of biodiversity	5. Urban landscape improvement
6. Improvement of asset value	

Table 16.2 Evaluation items to select the best green roof methods (Organization for Landscape and Urban Green Infrastructure 2004)

	Plane surface with green	Elevation surface with green	Biotope
Load capacity	40˜100 kgf/m^2 (4,000˜1,000 N/m^2)	200 kgf/m^2 (2,000 N/m^2)	400˜500 kgf/m^2 (39,000˜4,900 N/m^2)
Effects	1. Reduction of greenhouse effect	1. Reduction of greenhouse effect 2. Reduction of CO_2 3. Creation of relaxation 4. Rest space 5. Creation of biodiversity	1. Reduction of greenhouse effect 2. Reduction of CO_2 3. Creation of relaxation 4. Rest space 5. Creation of biodiversity
Application	Existing buildings, inclined roofs and high-rise buildings	Apartment houses and office buildings	Public facilities, commercial buildings, apartment houses and office buildings
Construction costs	Medium	Medium	High
Maintenance costs	Low	Medium	High

The green roof has attracted considerable attention recently due to increases in local torrential rainfall and is expected to improve the urban environment in several ways. Simulating the effect of the green roof on storm water retention is significant, as the Japanese urban water cycle is especially delayed in comparison with the West.

16.2 Materials and Methods

16.2.1 Examination

16.2.1.1 Storm Water Runoff Testing Experiment

We performed our experiment in a greenhouse in order to maintain the simulated rainfall intensity. We selected five distinct green roof drainage layers with thicknesses between 25.0 and 55.0 mm. The five drainage layers were used to create green roofs for different applications. The characteristics of the various drainage layers, each made of Styrofoam, are summarized in Table 16.3. We selected these particular drainage layers because they represented the latest technological developments in the field, and they were easily worked. The drainage layers of each thickness were set into a plastic tray (500 mm × 500 mm × 75.0 mm), then installed in duralumin case (500 mm × 500 mm × 100 mm, 1:100 scale).

Table 16.3 Characteristics of the five drainage layers (Kikuchi and Koshimizu 2010a)

Name	A	B	C	D	E
Face	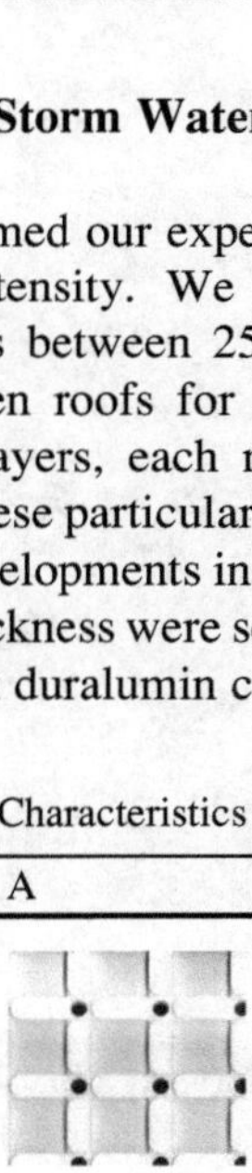	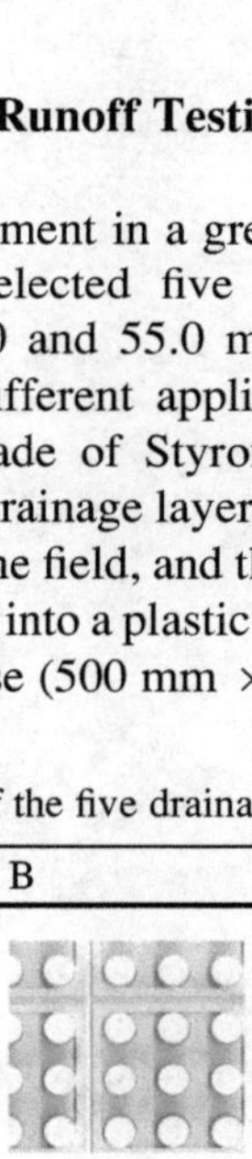	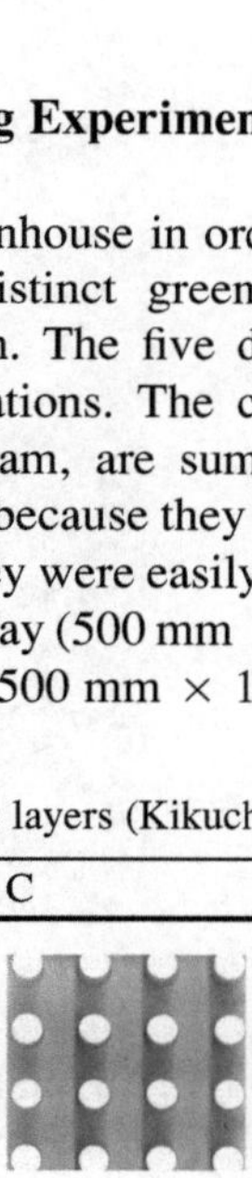	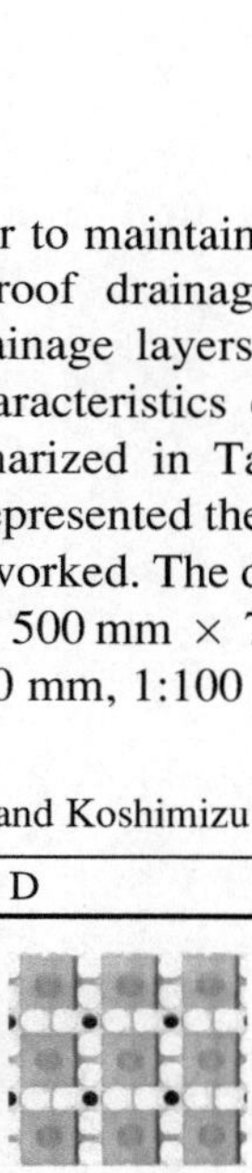	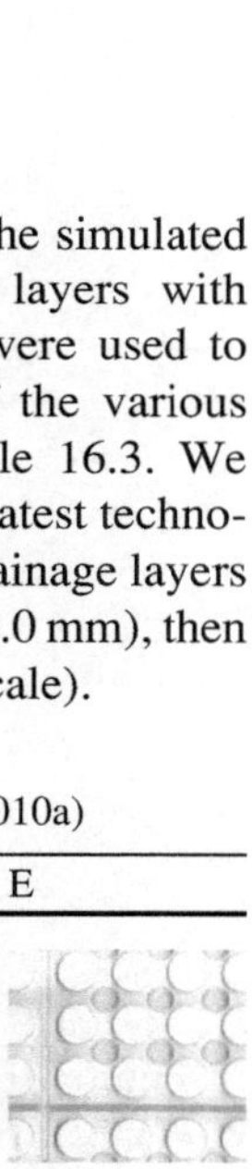
Back	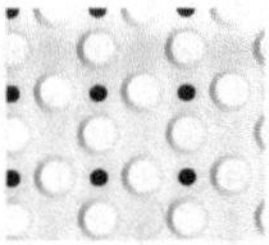	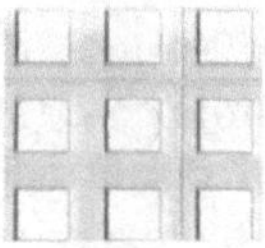	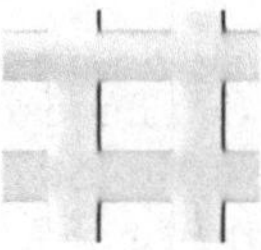	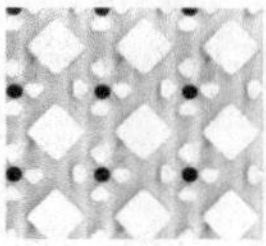	
Side	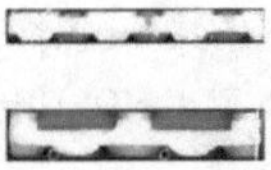				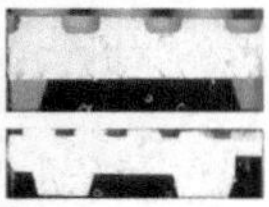
(mm)	25	30	40	45	55
reservoir (L/m^2)	2.2	3.0	10.0	7.1	2.8

Fig. 16.2 Rainfall simulator

Figure 16.2 shows the rainfall simulator (Daiki Rika Kogyo Co., Ltd., Japan 2011) that was used to quantify the exact amount of water that fell onto the drainage layer surface on August 16, 2009.

To collect the runoff water, a rubber hose of 20 mm in diameter was connected to a drainage hole in the side wall of a flat duralumin case (500 mm × 500 mm × 100 mm). We set the rainfall intensity to 150 mm/h during a 5-h interval. The methodology behind this runoff test is the same as in the experiments of the Japan Sewage Works Association (Japan Sewage Works Association 1994) and Kikuchi and Koshimizu (Kikuchi and Koshimizu 2010b). The schematic diagram of the connection of the flow meter is shown in Fig. 16.3a. Fig. 16.3b shows a photograph of the test site. Each rainfall event lasted 1 h.

The water runoff from all drainage flats was measured daily from August 16 to September 11, 2009 (4 times per day for a total of 67 times).

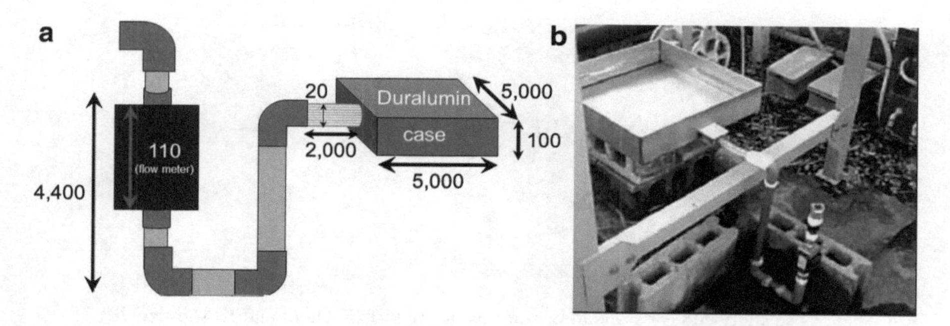

Fig. 16.3 Test setup: (**a**) schematic showing flow meter placement and (**b**) photo of the test site

16.2.1.2 Data Collection

Water runoff (from drainage flat to rubber tube) and time were measured every 0.1 s by a FD-M5AT flow meter (Keyence, Japan). The flow meter was able to measure water runoff regardless of temperature, pressure, density, and viscosity. The measurements were recorded by a TR-V500 data logger (Keyence).

16.2.1.3 Data Analysis

In this study, we modified the analysis in order to simulate the storm water runoff for flood-prone areas. The runoff characteristics were determined by using all of the storm water runoff data, which were categorized by thickness as follows: A (25 mm), B (30 mm), C (40 mm), D (45 mm), and E (55 mm). We first created runoff hydrographs and cumulative hydrographs using all the storm water runoff, compared the hydrographs' shapes, and finally calculated the runoff coefficient and runoff delay time for each drainage layer. The runoff coefficient for a green roof drainage layer was calculated as follows: runoff coefficient = (runoff discharge)/(amount of rainfall).

Analysis of variance (ANOVA) was used to test for significant differences in the amount of water runoff from different drainage layers. When there were significant differences, means were separated by *T* test at the probability level $P < 0.01$. All statistical analysis was performed with Microsoft Excel.

16.2.2 Simulations

16.2.2.1 Case Study Site

The goal of this study was to evaluate the effect of green roofs on storm water runoff in flood-prone areas, so we chose to investigate the most frequently flooded

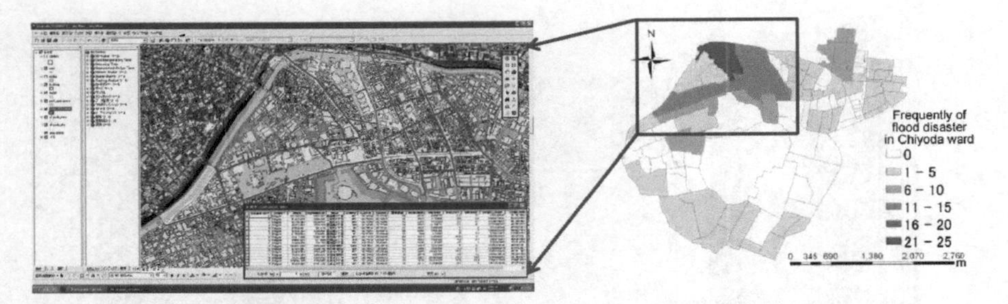

Fig. 16.4 Case study site for simulating storm water runoff (Kikuchi and Koshimizu 2010c)

areas in Chiyoda Ward, Tokyo, Japan (Kikuchi and Koshimizu 2009). Figure 16.4 shows the flooding frequency in Chiyoda Ward from 1974 to 2006; here, red denotes the areas with the highest flooding frequency, and we selected 17 districts as the case study site: Iidabashi 1, 2, 3, 4; Misaki-cho 1, 2, 3; Kudankita 1, 2, 3, 4; Kanda-Jinbocho 1, 2, 3,;and Nishi-Kanda 1, 2, 3 (Kikuchi and Koshimizu 2009).

16.2.2.2 Storm Water Runoff

We used the rational formula to simulate the storm water runoff for an actual location because it permits the peak discharge from the case study area to be determined. The peak discharge was calculated as follows:

$$Q = C * I * A/3600 \tag{16.1}$$

where Q is the maximum runoff volume (m^3/s), C is the runoff coefficient (-), I is the average rainfall intensity (mm/h) within time t of rainfall contact, and A is the land area (ha). The hydrographs were plotted in Microsoft Excel to show the peak discharge per hour.

16.2.2.3 Simulation Input

To gain insight into how runoff retention is affected by a green roof and a green roof with a drainage layer, particularly an extensive green roof, we simulated two typical Japanese rainfall patterns: local torrential rainfall and typhoon rainfall. To obtain local torrential rainfall and typhoon data, we used weather statistics from the Japan Meteorological Agency. We define the July 4, 2000, rainfall event as local torrential rainfall; flooding damage occurred in a 12.72 ha area when this rainfall event caused an inland flood in the case study area. On the other hand, we define the August 27, 1993, rainfall event as typhoon rainfall; flooding damage occurred in a 342 ha area when this rainfall event caused an inland flood in the case study area.

Figure 16.5a shows an example of the local torrential rainfall pattern, and the typhoon pattern is shown in Fig. 16.5b. These figures also show hyetograph patterns, with the gray bar depicting the amount of precipitation per hour; the line graph shows the amount of cumulative rainfall per hour.

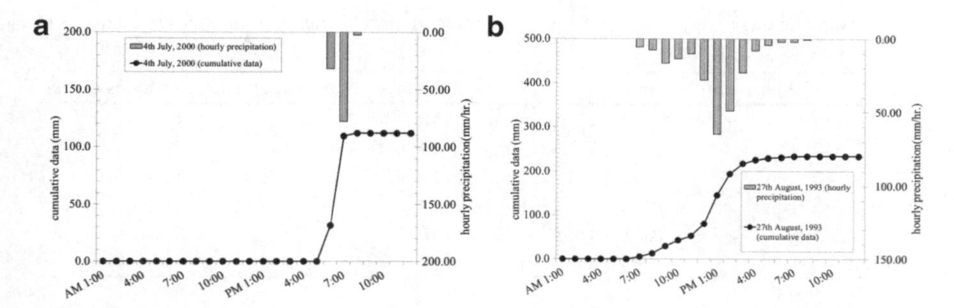

Fig. 16.5 Hydrographs and cumulative hyetographs: (**a**) July 4, 2000 (torrential rainfall event), and (**b**) August 27, 1993 (typhoon rainfall event)

Table 16.4 Runoff coefficient and runoff delay time based on land use (Kikuchi and Koshimizu 2010b)

Land use	Runoff coefficient	Runoff delay time
River, water surface	1.00	–
Commercial and business district	0.95 (conventional roof)	–
Public facilities	0.95 (conventional roof)	–
Industrial district	0.95 (conventional roof)	–
Residential district	0.90	–
Road	0.90	–
Green and open spaces	0.80	–
Traditional green roof system (Murayama and Koshimizu 1997)	0.80	12 min 37 s
Drainage layer D ($d = 45$ mm)	(Experimental data)	(Experimental data)
Development land	0.50	–

As seen in Fig. 16.5a, there is a peak 1 h after the rain starts, after which the amount of rainfall rapidly decreases. An inland flood occurs because a rapid increase in rainfall exceeds the capacity of sewerage system, which is a common feature of Japanese local torrential rainfall. Figure 16.5b shows a slow approach to peak discharge, with the rainfall continuing for an extended period of time. An inland flood occurs because rainfall during a prolonged time exceeds the capacity of the sewerage system, which is a common feature of typhoon rainfall.

Table 16.4 clarifies how the runoff coefficient is associated with land usage and summarizes the runoff coefficient and runoff delay time according to the type of land use. Land use areas were taken from Detailed Digital Information, 10 m Grid Land Use (1994) images from the Japan Map Center. The land use areas were calculated using GIS software (Arc GIS Desktop, version 9.3.1, ESRI, United States).

Table 16.5 Drainage layer thicknesses, experimental period, number of measurements, and analysis objects

Board	Thickness (mm)	Term	Number of measurements	Analysis objects
A	25	23, 24, 25 August	10	4
B	30	4, 5, 6, 7 September	13	7
C	40	8, 9, 10, 11 September	16	5
D	45	19, 20, 21, 22 August	13	3
E	55	16, 17, 18, 19 August	16	5
Control	–	7, 8 August	3	2

The runoff coefficient is required for new building construction, city planning, and redevelopment. Each runoff coefficient was determined from existing research. Increasing the runoff coefficient increases storm runoff to rivers and sewerage systems. The magnitude of the runoff coefficient is affected by the runoff rate of the impervious surface.

According to the survey of the Ministry of Land, Infrastructure, Transport and Tourism, commercial districts, public facilities, and industrial districts have seen more installations of green roofs than other land use types over the last decade. In this simulation, traditional green roofs or green roofs with a drainage layer is assumed to be installed on conventional roofs in a commercial district, on public facilities, and in an industrial district.

For the traditional green roof installation area, the runoff coefficient changed from 0.95 to 0.80. At the installation area of the green roof with a drainage layer, the runoff coefficient changed from 0.95 to the experimental value, and runoff time was delayed accordingly. For the traditional green roof installation area, the runoff time was delayed by 12 min 37 s, and for the installation area of the green roof with a drainage layer, the runoff time was delayed by the experimental value.

16.3 Results and Discussion

16.3.1 Results of Runoff Experiments

Each drainage layer was dribbled out in the frequency range between 10 and 16. We first created hydrographs using all the data and then checked the water runoff trend. We decided to use the average analytical data. Table 16.5 shows the outline of the storm water runoff experiment, along with the thickness of the drainage layer that corresponds to each letter designation.

Figure. 16.6 shows hydrographs and cumulative hydrographs based on the storm water runoff experimental data. The hydrographs show the runoff delay and retention capacity of the drainage layers. When the rainfall started, the runoff starting times were all delayed except for the control case, and the water runoff value showed a small increase. When the retention capacity of each drainage layer

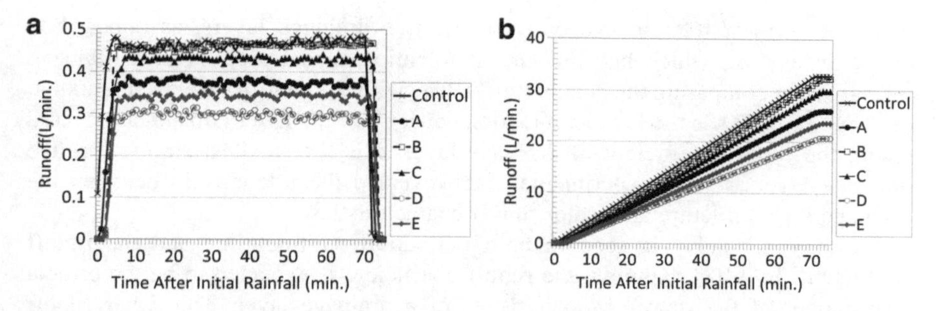

Fig. 16.6 (**a**) Runoff hydrographs and (**b**) cumulative hydrographs: A (25 mm), B (30 mm), C (40 mm), D (45 mm), E (55 mm), and control [9]

Table 16.6 Characteristics of the five drainage layers (Kikuchi and Koshimizu 2010d)

Board	Thickness (mm)	Time of starting of runoff (min.)	Time of arrival of peak runoff (min.)	Time of lasting of runoff (min.) After rainfall stopped.	Peak runoff (L/min.)	Runoff coefficient
A	25	4	7	13	0.374	0.79
B	30	5	7	15	0.462	0.97
C	40	4	9	14	0.431	0.91
D	45	5	7	15	0.298	0.63
E	55	5	8	15	0.341	0.72

was reached, the water runoff value rapidly increased. With the exception of the control case, the highest peak discharge occurred in drainage layer B, which had a thickness of 30 mm, and the peak discharge of drainage layer B is close to the control value. Of the other cases, the peak discharge values in order from highest to lowest were for cases C, A, E, and D. The highest peak discharge was about 1.6 times the smallest peak discharge. The cumulative hydrographs show the drainage layers' runoff velocity, and the cumulative water runoff remains equal to 0 until the retention capacity of each drainage layer is exceeded. The water runoff increased with continuing rainfall, exhibiting a nearly linear increase until reaching the retention capacity of each drainage layer. With the exception of the control case, drainage layer B exhibited the fastest runoff velocity, with C, A, E, and D following in order from fastest to slowest.

Table 16.6 summarizes the results of the runoff experiments and the drainage layer characteristics. The control retained a peak runoff of 0.467 L/min. Table 16.6 shows that the drainage layers have some retention capacity. The time to runoff starting, the time to peak runoff, and the runoff duration were compared with those of the control, and drainage layers A and C were found to have an earlier runoff starting time. Drainage layers A, B, and D were found to have earlier peak runoff arrival times than the control, and drainage layer A was found to have a shorter runoff duration. When comparing the runoff duration, we found drainage layer A to have the shortest retention capacity of the five drainage layers, as opposed to drainage layer E, which had the largest retention capacity of the five drainage layers. When comparing the peak runoff values with the control case, we found that drainage layer D has the highest retention capacity of the five drainage layers; as a result, the runoff coefficient of drainage layer D is the smallest among the five drainage layers. These experimental data revealed the relationship between the drainage layer structure and water runoff characteristics.

Storm water delay is seen in both the runoff starting time and the runoff coefficient. In urban planning, the runoff coefficient is expected to be the critical determinant of the runoff characteristics of a drainage layer. The water runoff trends from each drainage layer were then determined through comparative statistical analysis. In every combination, a significant difference was observed ($P < 0.01$), and the results of the statistical analysis clarified that each drainage layer had a different retention capacity.

The green roof system has been observed to affect storm water delay and retention in the United States and Europe; however, in Japan it is well recognized that it is not necessary to quantify the effects of the green roof on storm water control. In order to prevent flooding in urban areas, it is desirable to quantify the effects of green roofs on storm water delay and retention. Few studies have reported stratified data for green roof structures. The results of our experiment suggest that each drainage layer affects storm water delay and retention. For our next simulation, we selected the lowest runoff coefficient of these drainage layers (which is the runoff coefficient of 0.63 for drainage layer D) and calculated the total volume of storm water runoff in the urban areas of the case study.

16.3.2 Results of Runoff Simulation

16.3.2.1 Runoff for Local Torrential Rainfall

Figure 16.7 shows the calculation results of the simulated local torrential rainfall: the blue bars of the hyetograph indicate the amount of precipitation per hour, the hydrographs indicate the storm water runoff amounts per hour by the three-colored bar graphs, and the cumulative hydrographs indicate the total storm water volume by the three line graphs.

When the maximum rainfall intensity occurs, the storm water runoff volumes were calculated to be 8,331 m^3/h for the conventional roof, 5,976 m^3/h for the traditional green roof system, and 4,761 m^3/h for the green roof with the drainage layer.

Figure 16.7a shows that a large, rapid decrease in runoff volume occurs in the final hours of the rainfall event; after the rainfall ends, the runoff of the green roof systems and the green roof with a drainage layer rebounds due to the decreased runoff coefficients and runoff delay.

Figure 16.8 shows the runoff velocity of local torrential rainfall displayed as a function of rainfall duration. The zero point implies the start of the rainfall event.

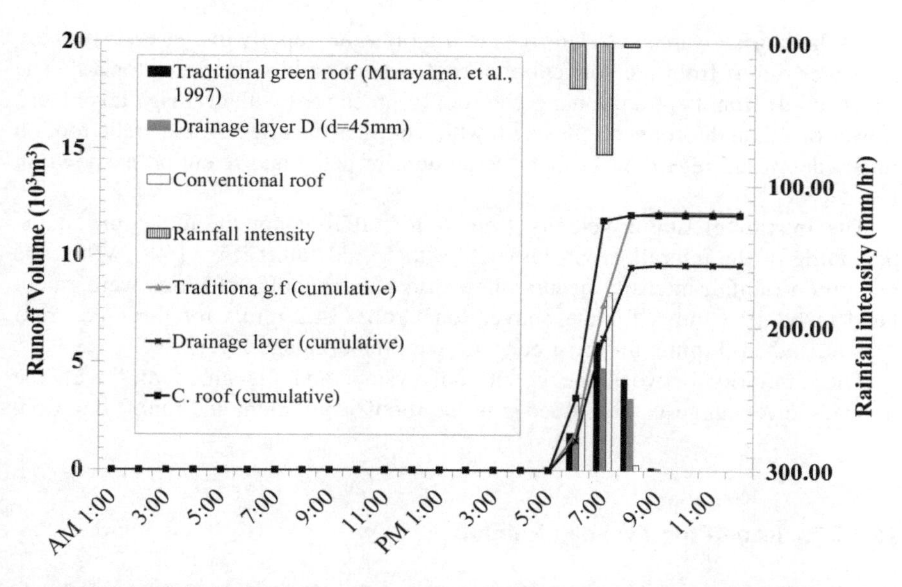

Fig. 16.7 Runoff hydrographs and cumulative hydrographs of selected roof conditions: traditional green roof [12], drainage layer D (d = 45 mm), and conventional roof [7]

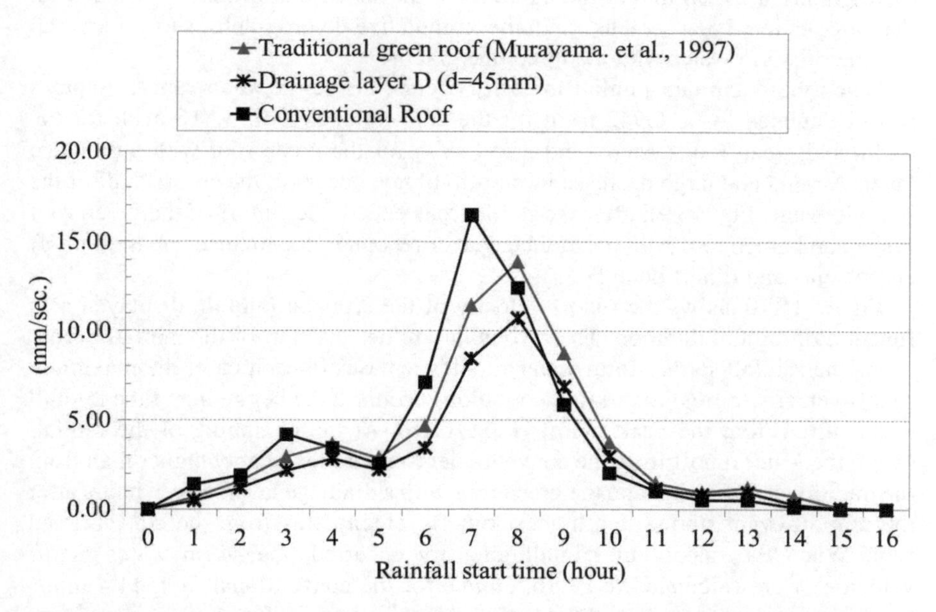

Fig. 16.8 Runoff hydrographs and cumulative hydrographs of selected roof conditions: traditional green roof [12], drainage layer D (d = 45 mm), and conventional roof [7]

As the rainfall starts, each storm water runoff value rapidly increases, especially the water runoff from the conventional roof, which rapidly increases for 3 h. The water runoff from the traditional green roof or green roof with drainage layer has a slower rate than the conventional roof, which implies the effect of the green roof on runoff delay and retention. In fact, the amount of delay and retention is shown in Fig. 16.8.

The maximum runoff velocity (mm/s) for all roof conditions occurs at the beginning of the rainfall event, just before the peak runoff (Fig. 16.8). When the maximum rainfall intensity occurs, the storm water runoff velocities were calculated to be 19.8 mm/s for the conventional roof, 14.2 mm/s for the green roof system, and 11.3 mm/s for the green roof with the drainage layer.

The similarity between the green roof system and the green roof with the drainage layer suggests the influence of the runoff coefficient and runoff duration.

16.3.2.2 Runoff for Typhoon Rainfall

Figure 16.9 shows in the calculation results of the simulated typhoon rainfall: a hyetograph shows amount of precipitation per hour with the blue bar graph, the hydrographs at the bottom of the figure show the storm water runoff per hour with the three-colored bar graphs, and the cumulative hydrographs show the total volume of storm water with the three line graphs.

When the maximum rainfall intensity occurs, the storm water runoff volumes were calculated to be 6,942 m^3/h for the conventional roof, 4,816 m^3/h for the traditional green roof system, and 3,554 m^3/h for the green roof with a drainage layer. A rapid and large decrease in runoff volume occurs in the second half of the rainfall event (Fig. 16.9); after the rainfall has ended, the runoff of the green roof system and green roof with the drainage layer rebounds due to the decreased runoff coefficients and runoff delay.

Figure 16.10 shows the runoff velocity of the typhoon rainfall, displayed as a function of rainfall duration. The zero point implies the start of the rainfall event.

As the rainfall starts, storm water runoff increases in each case; the maximum runoff velocity (mm/s) for all roof conditions occurs at the beginning of the rainfall event, just before the peak runoff (Fig. 16.10). At the beginning of the rainfall event, the water runoff from the conventional roof was noticeably higher than from the traditional green roof and the green roof with a drainage layer. Seven hours after the rainfall event started, the highest runoff velocity was from the conventional roof. When the maximum rainfall intensity occurred, the storm water runoff velocities were calculated to be 16.5 mm/s for the conventional roof, 11.4 mm/s for the green roof system, and 8.44 mm/s for the green roof with a drainage layer for water retention.

The similarity between the green roof system and the green roof with a drainage layer suggests the influence of the runoff coefficient and the runoff duration.

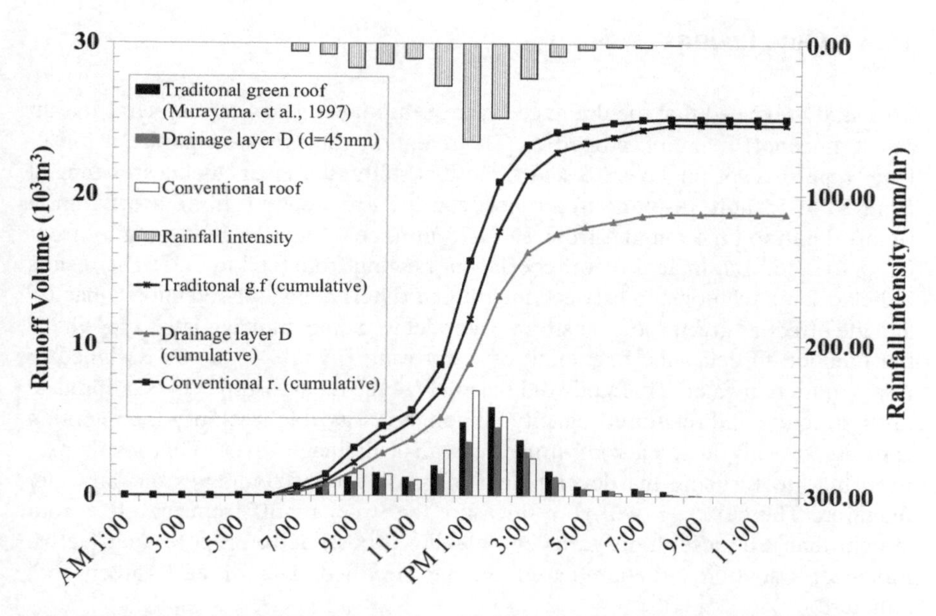

Fig. 16.9 Runoff hydrographs and cumulative hydrographs of selected roof conditions: traditional green roof [12], drainage layer D (d = 45 mm), and conventional roof [7]

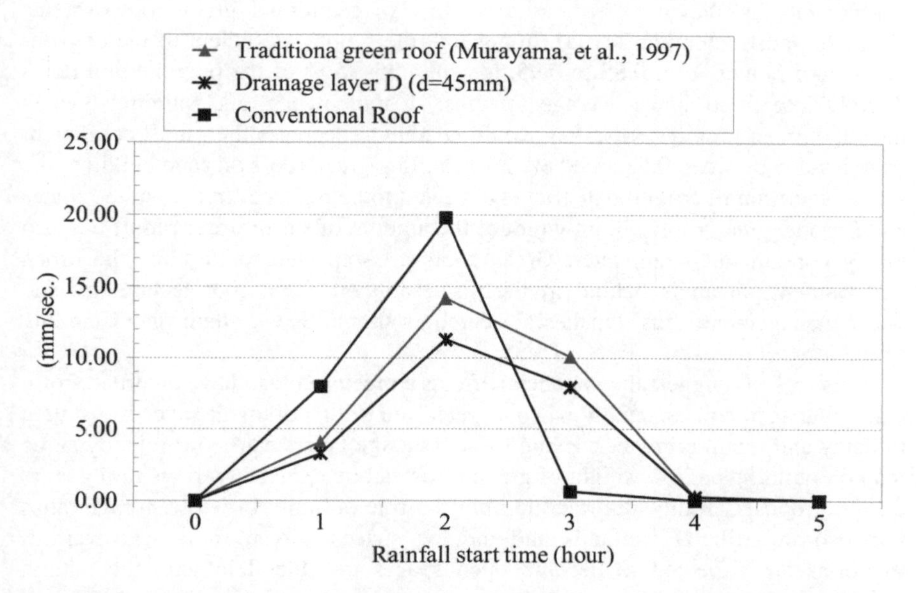

Fig. 16.10 Runoff hydrographs and cumulative hydrographs of selected roof conditions: traditional green roof [12], drainage layer D (d = 45 mm), and conventional roof [7]

16.4 Conclusions

This study revealed that the drainage layer of thin-layer green roof systems has an effect on runoff retention properties. The retention values were significant for all layer types except for layers B and C. In this study, the layer thicknesses ranged from 25 to 55 mm, resulting in a runoff starting time ranging from 4 to 5 min, a runoff duration time ranging from 13 to 15 min, peak runoff values ranging from 0.298 to 0.462 L/min, and runoff coefficient ranging from 0.63 to 0.97. The results indicate some relationship between the time of runoff delay, the retention capacity, and the effect of green roofs on storm water delay. Some drainage layer types have the function to delay the beginning of water runoff, while some are designed to retain more rainwater. This study did not analyze the relationship between drainage layer structure and retention capacity, although it was able to clarify the retention capacity of only a green roof structure with a drainage layer. This result may contribute to focusing the development of green roof technologies on rainwater retention. The next step will be to measure the water runoff from the green roof system that use these drainage layers, making it possible to propose an effective rainwater retention structure based on the stratified data of each green roof structure.

Using the lowest runoff coefficient of 0.63, another study showed that different roof conditions affect the storm water runoff delay. The drainage layer runoff coefficient was shown to be similar to that of traditional green roof systems (Kikuchi and Koshimizu 2010d). In general, the runoff coefficient of the conventional roof ranges from 0.85 to 0.95, meaning 85–95 % of the precipitation flows directly into rivers and sewerage systems. If homeowners and building owners installed green roof systems, they would be able to decrease the runoff coefficient from a value between 0.63 and 0.80. By installing green roofs on each building, the storm water runoff from the district is expected to be reduced. In the United States and Europe, green roofs not only reduce the amount of storm water runoff but also purify contaminated rainwater. Green roofs are expected to improve the urban environment. Japan is behind in the discussion on green roof technology and water management; thus, Japanese researchers should expand their understanding of the issue.

These results suggest the beneficial effects a green roof can have on urban storm water management, especially when a green roof with a drainage layer is installed to delay and retain rainwater. Using these drainage layers can reduce the dynamic load on buildings. If the weight of green roofs can be reduced, various roof usages can be proposed. Rainwater retention plays a role not only in water storage tanks but also in artificial wetlands and biotope spaces. Green roofs with various functions are expected to become open spaces and blend in with the natural environment.

We could not measure the stratified data of each green roof structure to discuss the relationship between the drainage layer characteristics and retention capacity. In the future, we will construct green roof systems with these drainage layers and measure the runoff water from each green roof system. We will then discuss the effective combination of plants, soil, and drainage layer type to prevent storm water runoff. We expect these results are valid for green roof technology development and the promotion of green roof installation in urban areas.

Acknowledgments This study was supported by KAKENHI (Grant-in-Aid for Young Scientists B: 21710187) and a joint study at the Graduate School of Meiji University. Simulation work was supported by collaborative research between the Center of Spatial Information Science, the University of Tokyo (research number: 231).

References

Daiki Rika Kogyo Co. Ltd. (2011) http://www.daiki.co.jp/images/rain/rain0.jpg, Rain simulator

Dunnett N, Nagase A, Booth R, Grime P (2008) Influence of vegetation composition on runoff in two simulated green roof experiments. Urban Ecosyst 11(4):385–398

Getter KL, Rowe DB, Andresen JA (2007) Quantifying the effect of slope on extensive green roof storm water retention. Ecol Eng 31:385–398

Green Roofs for Healthy Cities (2009) http://www.greenroofs.org/index.php/about-green-roofs, About Green Roofs

Ikeda Y, Takami M, Inuyama M, Teraguchi Z (2001) Moderation of rainfall flow by Sedum Mat, the materials for roof top greenery. J Jpn Soc Reveg Technol 27(1):201–204

Japan Sewage Works Association (1994) Planning and designing sewage facilities: guidelines and explanations. Japan Sewage Works Association, Tokyo

Kikuchi S, Koshimizu H (2009) Assessment of a rooftop greening suitable site map by analysis of rainfall pattern and inundation disaster. J Jpn Inst Landsc Archit 72(5):871–874

Kikuchi S, Koshimizu H (2010a) Evaluations of reservoir characteristics of some type of storage-drain boards of green roof forward torrential rain. J Jpn Inst Landsc Archit 73(5):693–696

Kikuchi S, Koshimizu H (2010b) Storm water runoff simulation of water retainage green roof systems for sustainable hydrology in urban area. URBIO 2010. CD-ROM

Kikuchi S, Koshimizu H (2010c) Storm wated runoff simulation of water retainage green roof systems for sustainable hydrology in urban area. Urban Biodivers Des, 18–22, Nagoya, Japan

Kikuchi S, Koshimizu H (2010d) Possibility of green roof storage-drain board of storm water control facilities in flood-prone area. Architectural Institute of Japan, Summary of Technical Lectures, F-1, CD-ROM

Murayama T, Koshimizu H (1997) Storm water runoff depression effect of roof replanting. J Jpn Soc Reveg Technol 28:319–322

Organization for Landscape and Urban Green Infrastructure (2004) Green space design manual 4. NEO-GREENSPACE DESIGN. Seibundo Shinkosha, Tokyo

Chapter 17
Evaluation and Regulation of Ecological Security When Implementing Urban Planning: Review and Suggestions for Spatial Planning and Sustainable Development in China

Lin-yu Xu and Zhi-feng Yang

Abstract It is important to carry out planning evaluation and regulation of ecological security (PERES) after implementing urban planning, which is the most urgent demand in China's national-level urban planning according to China's national strategic target for intensive urbanization. In this chapter, the importance of research on PERES is discussed after related literature review, and a method for PERES is designed in order to integrate ecological planning and sustainability into urban planning process in China. The suggested PERES system should include (1) determining the key bottlenecks in the implementation of urban planning, (2) identifying the basic elements of the evaluation and regulation of urban ecological security, (3) developing criteria to evaluate ecological security, (4) ecological safety monitoring, and (5) building regulations and a system of spatial regulations.

Keywords Urban planning • Ecological security • Evaluation • Regulation

17.1 Introduction

With the rapid development of its social economy, China has witnessed rapid urbanization. While enjoying the positive effects of urbanization, people in urban areas also have to face a series of problems and threats such as severe air pollution, water pollution, waste pollution, ground deformation, floods, and other disasters. Many cities are therefore giving much more importance to ecological planning in their development plans so as to ensure ecological safety during the implementation period. However, some of the potential hazards and risks caused by ecological problems may be avoided when the urban planning programs are prepared. In order

L.-y. Xu • Z.-f. Yang (✉)
State Key Joint Laboratory of Environment Simulation and Pollution Control,
School of Environment, Beijing Normal University, No. 19, Xinjiekouwai Street,
Haidian District, Beijing, China
e-mail: xly@mail.bnu.edu.cn; zfyang@bnu.edu.cn

M. Kawakami et al. (eds.), *Spatial Planning and Sustainable Development: Approaches for Achieving Sustainable Urban Form in Asian Cities*, Strategies for Sustainability,
DOI 10.1007/978-94-007-5922-0_17, © Springer Science+Business Media Dordrecht 2013

to achieve sustainable urban development in China, it is crucial that practical steps are taken to assess, regulate, and safeguard ecological security and to amend urban planning in real time.

Urban security plays a fundamental role in national or regional ecological security. Currently, most Chinese cities are facing problems such as traffic congestion, housing shortages, environmental pollution, and water scarcity, which will restrict their sustainable development and may pose a great threat to social stability and national security. It is therefore essential to identify the key elements threatening urban ecological security, in order to assess and regulate the level of ecological security of urban planning, which will play an increasingly important role in improving urban planning and achieving sustainable urban development. Better insights into techniques and methodology to assess ecological security in cities, or how to control it dynamically, will provide not only China but also the rest of the world with the ecological theories and standards which are needed. The techniques and methodology which can meet the requirements for protecting the safety of the urban ecology during China's rapid urbanization can also meet the requirements for promoting the healthy development of cities and providing people with a good ecological environment. The results of this research will bring considerable benefits to China's urban planning programs.

On the basis of the planning and post-implementation stages, this chapter proposes systematic and comprehensive techniques and methods to either assess ecological security in cities or to control it dynamically. This will provide us with a scientific basis for the ecological safety of urban planning and its implementation.

17.2 Security Assessment Techniques and Analysis for Ecological Security of Cities

With the worsening of the ecological environment, the issue of ecological security, as part of national security, has gained public attention. China's "eco-environmental protection program" explicitly states the national goals for ecological protection, which stipulate that the quality of people's living environment will not deteriorate, or be threatened by, the process of national economic and social development. Scholars in China and elsewhere have noted that there are limitations to ecological safety. Some of them have recognized the ecological risks rather than the vulnerability of the ecology. However, others have only recognized that ecological security is a static state rather than a dynamic state (Cui et al. 2005). According to the variety of concepts about ecological security that have been suggested by academics, ecological safety can be summarized as follows. (1) The health, integrity, and sustainability of the ecosystem are essential. (2) The ecosystem can provide human beings with perfect ecosystem services or a safe living environment (Chen and Zhou 2005). Therefore, the essence of ecological security will be reflected in the following aspects: ecological risk and ecological vulnerability (Cui et al. 2005).

As an important aspect of ecological security, urban ecological security refers to the fact that the required urban ecological environment can meet the current and future needs of urban development (Shi et al. 2005). Before the year 2000, there were few studies on urban ecological security. At the start of the twenty-first century, and with the worsening urban environment, more and more scholars are paying increasing attention to urban ecological security.

17.2.1 Evaluation of System of Targets

At present, the most popular way to assess urban ecological security is by evaluating the targets that have been set. A variety of frameworks have been proposed. The most popular model is the pressure–state–response (PSR) model (Zhou and Wang 2005). Starting from the interaction between human beings and environmental systems, the PSR framework has systematically classified the environmental indicators. Many scholars have taken advantages of the PSR framework to conduct ecological security assessments and have studied a number of regional ecological security assessments. However, despite the fact that there has been sufficient research into ecological security evaluation systems based on the PSR framework, the following problems remain (Wang et al. 2007). (1) As there have been fewer studies on the security trends, the research cannot objectively reflect the dynamic nature of ecological safety. (2) As there have been fewer studies on the differences from place to place, it is very difficult to compare the spatial differences. For example, we cannot determine where it is unsafe when the spatial differences in a certain place are given. (3) The PSR model, based on the "response" to reflect the ecological security caused by human activities, pays more attention to the phenomenon of ecological security than the nature and essence of ecological security. It cannot explain problems such as how upstream has influenced on the ecological security of basins downstream. (4) Although there has been sufficient theoretical research into ecological security, there has been less practical guidance, thus the results of the evaluation often lag significantly behind the actual changes in ecological security. The evaluation has no practical significance for regional development and decision-making. Therefore, it is essential to rethink the evaluation system for ecological safety.

17.2.2 Land Use/Cover Change Model

Some scholars use the land use/cover change (LUCC) model to study urban ecological security. They point out that the negative influence of irrational land use on nature includes biophysical effects caused by changes in hydrology, climate, geology, and other environmental factors at different scales. It also includes pollution stresses caused by waste water, waste gas, solid waste, etc., which include metabolic activity produced by urban areas. Conversely, damage to the natural ecosystems will result in the loss or reduction of integrated ecosystem services and then will endanger the urban ecology on which the existence and development of cities will depend (Zhang et al. 2007). At the international level, the systematic study of the LUCC model and ecological security is a new topic. Currently, there are some issues worth in-depth exploration due to the universality and complexity of the topics under study as well as immature theoretical systems and research methods. For example, the current research programs only focus on the complete, quantitative urban ecological security assessment index system and the corresponding model in one city or more in a certain region. However, there are fewer case studies which provide very systematic reviews of urban ecological security (Xie and Zhang 2004a, b). The current research only focuses on deductions from the LUCC model, the number of changes, and biophysical feedback. However, there are relatively limited studies on the ecological impacts perceived by the LUCC model and only a few case studies in China and elsewhere. What is more, most research has been carried out in a single city or town. Obviously, it cannot accurately describe the relationship between ecological security and LUCC, which is driven by the urban socioeconomic activities of the current process of globalization.

17.2.3 Other Methodologies

In addition to the indicator methods and LUCC models, some assessment methods for urban ecological security are based on the theory of dissipative structures, and the ecosystem analysis is based on complex urban evolutions (Fang et al. 2003; Fang and Chen 2004; Zhao et al. 2007). Here the dissipative structure refers to an organized nonequilibrium state of matter created and maintained due to dissipative processes in nonequilibrium thermodynamics. And urban ecosystem is a typical dissipative system (Dilip and Ilya 1998; Lee 1972). Some scholars have studied urban ecological security from the point of view of the components of the urban ecological system, such as the influence of urbanization on urban ecosystem biodiversity (Savard et al. 2000; Pauchard et al. 2006); the impact on birds in urban areas (Catterall et al. 1998; Crooks et al. 2004; Lim and Sodhi 2004); the impact of urbanization on rivers, groundwater, and marine water quality (Foster 2001; Ren et al. 2003); the study of the influence of urbanization on aquatic creatures (Preston et al. 2003); and the influence of urbanization on human health (Vogele 2000; McDade and Adair 2001).

17.2.4 Analysis of Assessment and Technologies for Urban Ecological Security

With exception of the PSR model, the LUCC method, complex urban ecological systems, and the theory of dissipative structures, in general, we should continue to study the following aspects if we wish to analyze the assessments and technologies for urban ecological security. (1) A single indicator should be adopted to conduct the sensitivity analysis of the overall ecological safety of a city or to conduct a correlation analysis among the indicators. In this context, a sensitivity analysis refers to the analysis of its contribution and the influence of a single indicator, while the correlation analysis refers to the analysis of the interaction between target indicators. (2) The analysis of the trends in the evolution of the ecological security of a city should also be conducted. Urban ecological security will not be achieved instantaneously. The ecological security analysis refers to the description of the state of the system in a given period or at a certain stage, which is a dynamic process or a continuous time function. Therefore, the ultimate goal of our study will not be to establish an urban ecological security index but to find the trends in the evolution of urban ecological security. If the ecological security tends to be insecure, people can make good use of early warning theory to control the direction of the evolution so that it can be regulated away from an unsafe state. However, at present, the index system based on the PSR model cannot do this. Thus, we must continue to study this field in combination with other methods.

17.2.5 Studies on Urban Ecological Risks

After analyzing the above-mentioned trends and the sensitivity indicators, we find that some indicators exert a considerable influence on the overall ecological safety of a city. However, these indicators do not help us to understand the relationships between the indicators or provide guidance about regulating the urban ecological security. Therefore, we must analyze the relationship between the indicators and the potential ecological risks. It may be a one-to-one or a many-to-one relationship. We should analyze how the potential ecological risks affect the ecological safety so as to finalize the regulations and controls.

17.2.6 Analysis of Spatial Differences in Urban Ecological Security

There are regional differences in ecological security risk factors and also the harmful effects of ecological risks. However, although we cannot find the differences between various regions of the city, we can establish the consistency between the affected region and the place where the risk occurs if the PSR framework evaluation and index system are applied. Therefore, an urban ecological security assessment based on its risk factors can help us to find the regional differences in ecological safety. The evaluation may result in ecological safety regulation and management in line with the local conditions.

17.3 Analysis of Control Technologies for Urban Ecological Security

17.3.1 Regulation Technologies Based on Landscape Ecological Security

During the process of urbanization, developed Western countries have experienced a variety of environmental problems. To get rid of urban sprawl and ecological decline, European and American countries have attempted to regulate ecosystems in order to promote the integration of urban and natural ecosystems, and to put urban planning into practice. The early slogans or campaigns such as "Garden City," the construction of the London greenbelt, the "Green City" idea, the green movement corridors, and the green infrastructure have had a profound impact (Zhang et al. 2007) on the world's urban development. As for maximizing the efficiency of land use and minimizing ecological risks, McHarg in the 1960s further developed the concept of ecological planning and developed the classical framework of ecological planning. Since then, the philosophy of ecological planning has aroused much interest from land management workers, urban planners, and environment conservationists.

In recent years, with the in-depth understanding of urban ecological regulation and control, people have gradually recognized the importance of the landscape ecological security pattern for harmony between the city and nature. Xiao et al. (2002) recognized that ecological security should be based on this pattern. As a newly coined term, the landscape ecological security pattern refers to the composition of the key parts, location, and spatial links in the landscape, which will be important for the maintenance and control of certain ecological processes. Although, at an earlier stage, this term was mainly used in the field of biodiversity conservation, it has also been applied to the field of urban ecology planning. In recent years, China has begun to place much importance on ecological security in the practice of urban ecological planning. Research has been carried out in Guangzhou, Shanghai, Lanzhou, and other cities. Starting from the theory of landscape ecological security patterns, Li and Tang (2006) have made suggestions to prevent disasters such as debris flow in some towns in Dongchuan District, in the Kunming prefecture. Li and Tang point out that, firstly, we should take land resources protection and utilization and the strategic layout of the landscape into consideration. Then we should carry out the following engineering measures: disaster system forecasting and layout, lifeline protection system planning, disaster relief facilities and their layout, the construction of disaster prevention measures, evacuation planning, and evacuation channel planning. In addition, we should activate disaster prevention and mitigation so as to achieve the fundamental goal of an ecologically safe city with the support of the following nonengineering measures: the command system construction of a disaster relief organization, disaster relief team-building, the development of disaster contingency plans, and disaster relief propaganda as well as disaster policies and building regulations.

General speaking, urban ecological security control measures and theories are not well developed in China and elsewhere. In promoting the optimal use of land and reducing the ecological risks, the concepts of ecological planning and the landscape ecological security pattern concept have gradually been applied in China. With its immature theoretical system and lack of implementation, China's urban planning process has not yet obtained legal status, making it difficult to implement it in urban planning and construction (Zhang et al. 2007).

17.3.2 Modulation Technologies for Elements of Urban Ecological Systems

Starting from a single element of the urban ecosystem, some scholars have conducted research on modulation technologies for urban ecological systems. After studying the lakes in the east and west of Wuhan city, Wang and Cheng (2003) pointed out that the lakes in the urban area should help to improve the city's environmental quality and to beautify it. They also called our attention to some key issues in its reconstruction. After discussing the disadvantages of the waterfront embankment of urban rivers, Xia et al. (2005) suggested that the urban riverside banks should play an important role. Taking the ecological restoration of the riverside banks of the Grand Canal in Zhenjiang City as an example, he recommended ecological restoration technologies, their implementation, and other options such as concrete slope protection technology and vegetation selection, in the hope of giving full play to the ecological performance of urban rivers. After conducting a thorough investigation of the meaning of a healthy life, Bai et al. (2006) analyzed the current status of urban rivers and suggested some methods for restoring river channels, the riverbed, bed slope, waters, and banks. After evaluating the ecological security status of Beijing, Zhou and Wang (2005) also proposed a series of ecological security measures in urban areas such as adjusting the energy consumption structure, improving the road traffic system, strengthening the control of dust pollution and site management, adjusting the industrial structure, strengthening the treatment of soil erosion and forest conservation, popularizing urban living, introducing garbage collection and disposal, optimizing the aquaculture layout, intensifying the construction of green barriers, and enhancing the construction of the environmental infrastructure. Although some of these measures have been adopted, a systematic urban ecological security control system and specific, executable program measures have not been drawn up. Yang et al. (2007) proposed that regulation and control of the urban ecological system should take the limited capacity of the ecological system into consideration. In order to regulate and control the ecology in urban areas, we should finally deal with the imbalance in supply and demand of the bottlenecks factor. We should be given the level, methods, and blueprint of the urban ecological regulations and controls. That is to say, we should take measures at the level of nature, function, and culture in order to widen the bottlenecks. Depending on the circumstances, the measures for supply or demand can be used to regulate the ecological controls. Martin-Duque et al. (2006) have recognized the close relationship between urban spatial development and ecological safety. It is crucial to regulate urban spatial development in order to achieve ecological safety. The importance of a city's own system to its ecological safety is not further discussed in this chapter.

17.3.3 Problem Analysis for Regulation of Urban Ecological Security

From the above research, we can observe that the regulations and technologies for urban ecological security are still not that well developed. The methods and technologies mentioned above are still relatively crude and will neither be implemented nor get obvious achievements. One reason for this is that the stress-response mechanism has not been fully studied; for another, the control programs and technologies do not meet the requirements of ecological laws and the practical situation. In the future, scholars should therefore focus their attention on the regulation of urban ecological security, in particular, its operability and feasibility. Spatial regulation is needed, as well as systematic regulation.

17.4 Suggestion: Assessment and Regulation Technology System of Ecological Security When Implementing Urban Planning

According to the above analysis of assessment and regulation technology methods for ecological security, an integrated assessment and regulation system for integrated ecological security can be established when implementing urban planning (see Fig. 17.1). The system includes the identification of the bottlenecks affecting urban ecological security, the extraction of key factors for urban ecological security monitoring and assessment, the formulation of standards and critical factors' values for urban ecological security, and the establishment of a dynamic monitoring and evaluation model for urban ecological security using GIS technical support. It also includes the development of dynamic, computer-based simulation of the urban landscape in order to simulate the spatial regulation optimization for urban ecological security at different levels and areas and to formulate the regulatory guidelines. In response to interactions with the urban systems, the supply and demand of the natural ecosystems and socioeconomic system should be coordinated, regulations for the urban ecological security

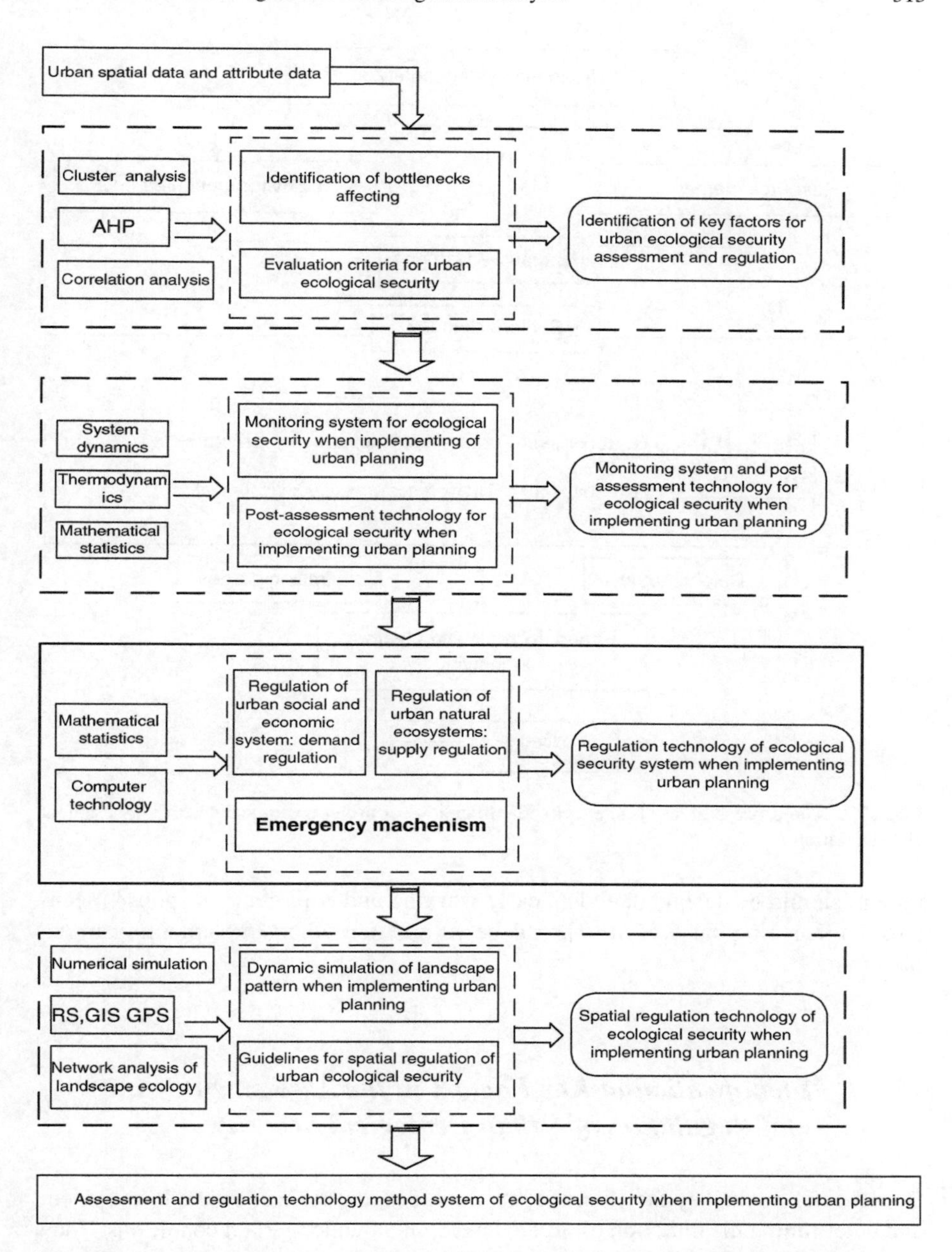

Fig. 17.1 System for assessment and regulation for ecological security when implementing urban planning

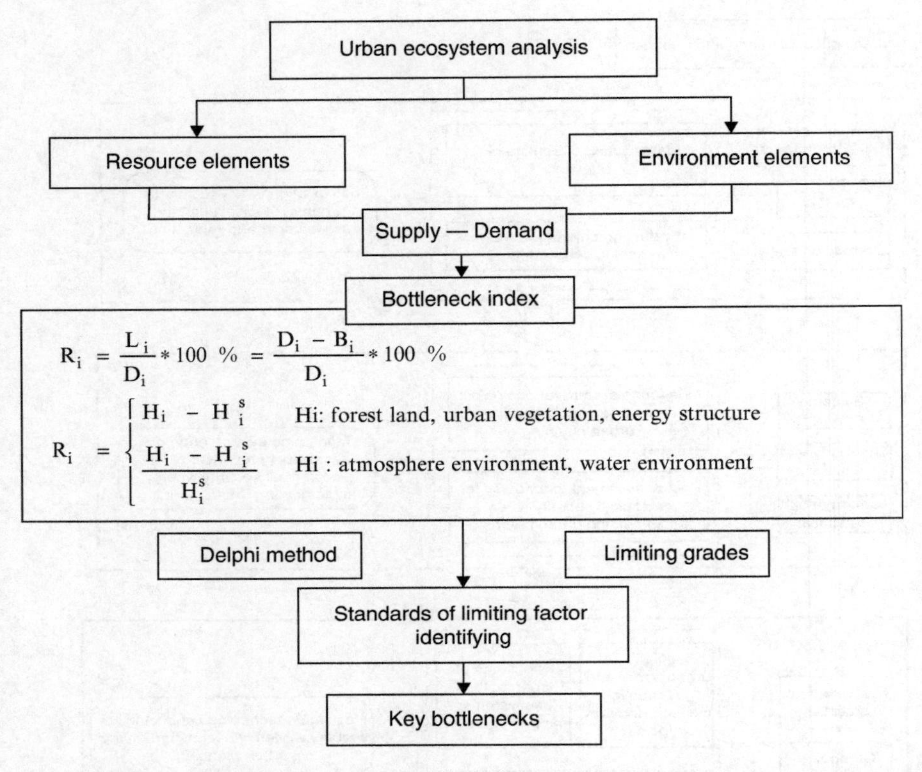

Fig. 17.2 The process of key bottlenecks identification for urban ecological security assessment and regulation

system should be designed, and an early warning and emergency response mechanism for ecological security should be formulated when implementing urban planning.

17.4.1 Identification of Key Factors in the Assessment and Regulation of Urban Ecological Security

Bottlenecks are the key factors impeding the speed and scale of urban development and determining the direction of urban development under certain conditions. They determine the bottom line of urban ecological security. Extracting the key factors for urban ecological security assessment and regulation from the bottlenecks can help to implement urban ecological security monitoring and build an assessment model for urban ecological security.

The process of bottlenecks identification in assessment and regulation of urban eco-security is shown by Fig. 17.2. The obstacles and restrictions of urban development can be selected by bottleneck index calculation.

17.4.2 Monitoring and Assessment Techniques for Ecological Security When Implementing Urban Planning

17.4.2.1 Monitoring of Ecological Security When Implementing Urban Planning

Depending on the different levels of urban ecological security, an appropriate monitoring system can be established. Such a system includes environment-based GIS, online monitoring of the environment, environmental management systems, dynamic management systems for pollution sources, emergency disposal systems for environmental incidents, emergency response systems for environmental accidents, and government information management systems for environmental protection. The key factors for ecological security assessment of medium-sized and large-scale projects (including online resources, social and economic aspects) are monitored online during the implementation of urban planning. Thus, dynamic monitoring can be achieved when implementing urban planning and can be the basis for an early warning and emergency response mechanism for urban ecological security.

17.4.2.2 Assessment Technology for Ecological Security When Implementing Urban Planning

Computer technology can be used to simulate the state of the ecosystem when demonstrating the implementation of planning for a city. An ecological security assessment model can be built and incorporated into the monitoring system by using system dynamics, a thermodynamic model, and statistical techniques. Thus, seamless integration, online monitoring, and dynamic assessment are achieved and can be updated on a rolling basis.

And the process of monitoring and assessment techniques for eco-security when implementing urban planning is illustrated in Fig. 17.3. It can be used for the whole urban planning, especially suitable for urban energy-use planning, industrial planning, and land-use planning.

17.4.3 Regulation Technology for Ecological Security System When Implementing Urban Planning

Integrated regulations and technologies for the ecological processes need to be established on the basis of an urban ecological security system. The allocation of resources and the regulation model for socioeconomic optimization should also be established. The coordination of the regulations and techniques of the natural and socioeconomic components at the different scales should be developed during the process of urban construction. Insecurity in the urban ecosystem is mainly due to the imbalance between supply and demand, which causes bottlenecks, so the

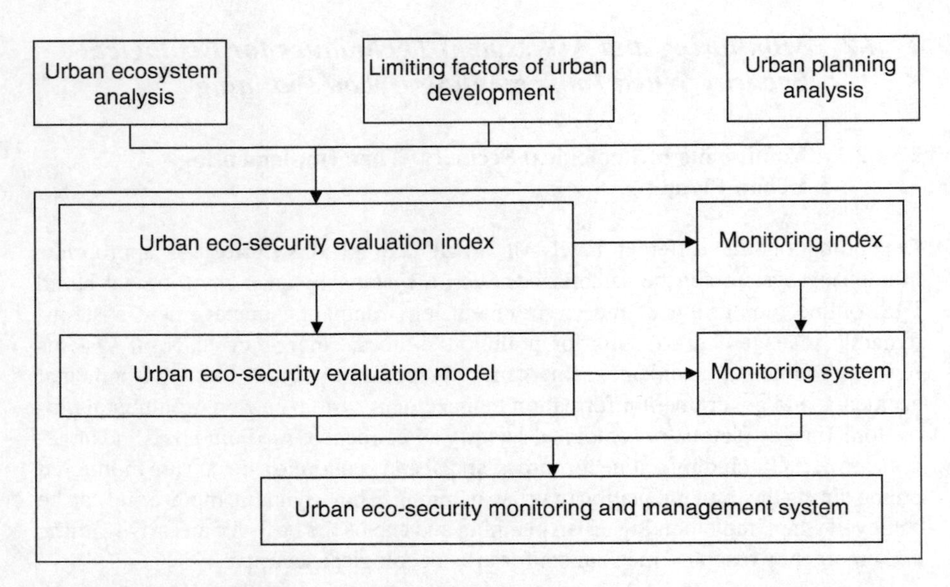

Fig. 17.3 The process of monitoring and assessment for eco-security when implementing urban planning

ecological regulation of an urban system should usually start with the supply and demand (see Fig. 17.4).

17.4.3.1 Regulation: Adjustment of the Needs of the Urban Social and Economic System

Timely feedback is needed from the eco-security monitoring and assessments when implementing urban planning. Regulation of population size, economic scale, and the intensity of human and economic activities will ensure ecological security while implementing urban planning. By controlling the size and speed of socioeconomic development, the pressure on the bottlenecks and the pressure from the human social and economic activities on natural ecosystems will be reduced.

17.4.3.2 Regulation: Adjustment of Supply of Urban Natural Ecosystems

By realigning and configuring the resources, the supply capacity of the bottlenecks can be enhanced and the capacity of natural ecosystems to support social and economic activities can be improved. Based on the quality and quantity of natural ecosystems and their space–time differentiation, the present authors suggest a rational utilization method and an optimal deployment of the main resources, such as water resources, land resources, and energy. Sustainable solutions and environmental protection measures are also presented, in order to lay the foundations for an appropriate scale of urban development. When implementing

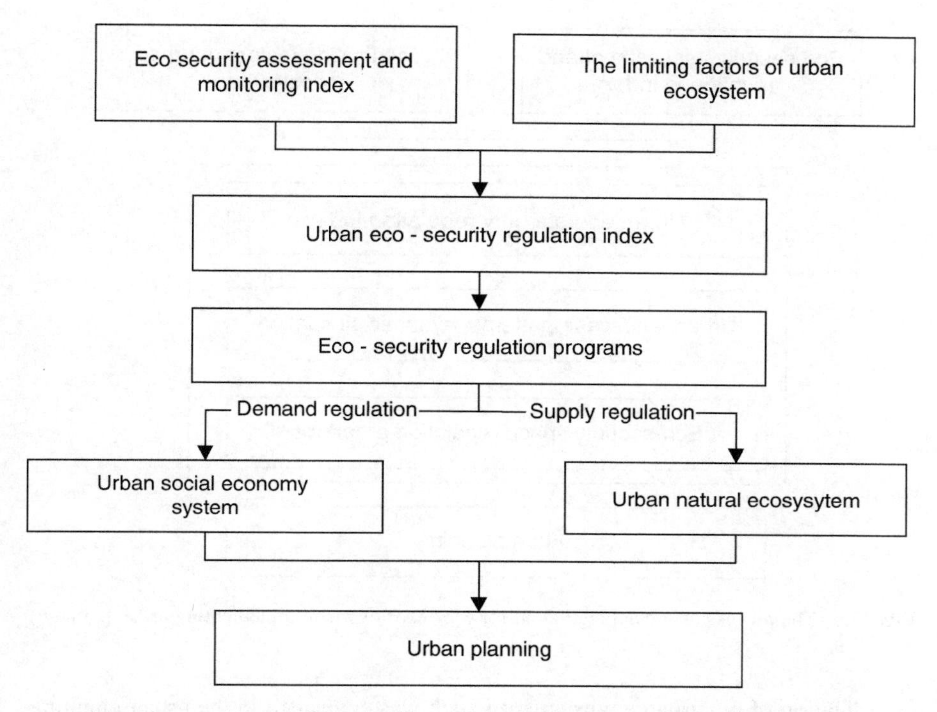

Fig. 17.4 The process of urban ecosystem regulation for eco-security when implementing urban planning

urban planning, the supply of resources and control of environmental pollution must interact with the needs of the socioeconomic system. On the other hand, regulations for the long-term operation of urban ecological security should be established so as to guarantee and provide a reference for ecological protection and restoration.

17.4.3.3 Urban System Coordination and Interactive Regulations

Quantitative accounting for substances and energy flows among urban social, economic, and natural systems and the recognition of the overall optimization of the system must be conducted so as to develop the technology for the long-term regulation of the urban ecological security system.

17.4.3.4 Early Warning and Emergency Response Mechanisms

Urban development may cause large-scale changes in the structure and function of the urban ecosystem. These changes may bring about the remodeling or

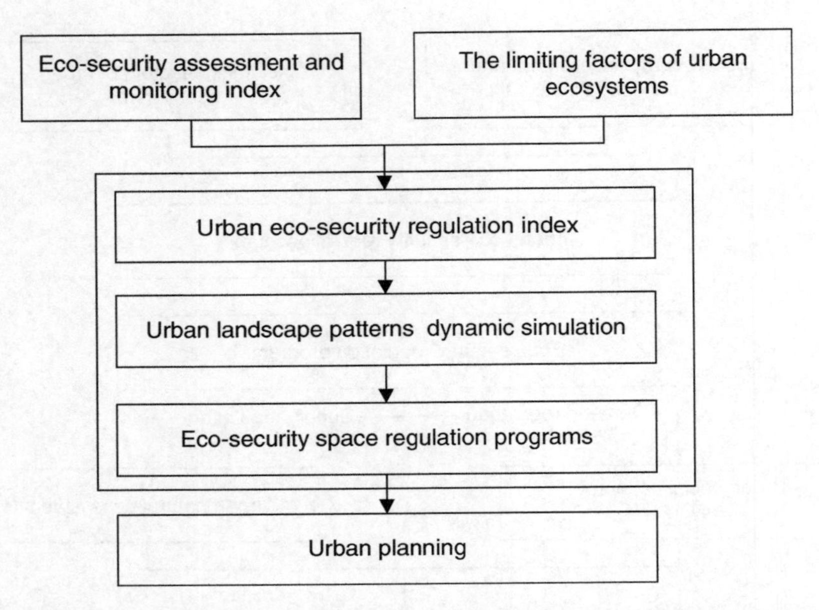

Fig. 17.5 The process of special regulation for eco-security when implementing urban planning

optimization of the urban environment which was envisaged in the urban planning or may cause unexpected disasters and pollution. Therefore, based on the monitoring and assessment of the ecological security in the implementation of urban planning, it is very important to establish an early warning and emergency response mechanism for accidents (including natural disasters and man-made pollution). Based on dynamic monitoring and evaluation, the ecological security alarm systems of computer-based technology, immediate processing systems for disasters, etc. should be established to deal with the potential risks from medium-sized and large-scale urban planning projects.

17.4.4 Spatial Regulation Technology for Ecological Security When Implementing Urban Planning

17.4.4.1 Dynamic Simulation of the Landscape Pattern When Implementing Urban Planning

By using GIS and RS (remote sensing) technologies and the boundary conditions for a particular project, the spatial simulation for different blocks can be conducted for the dynamic changes in the urban landscape pattern, so as to obtain the best spatial regulation for ecological safety (see Fig. 17.5).

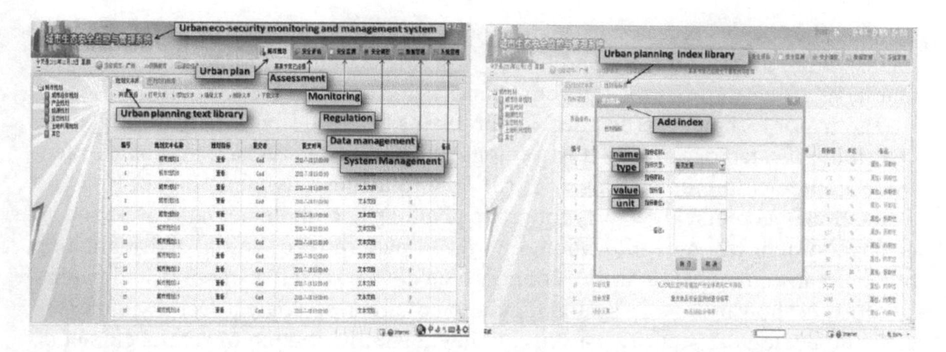

Fig. 17.6 The interface of the urban eco-security monitoring and management system

17.4.4.2 Guidelines for the Spatial Regulation of Urban Ecological Security

The conflicts and contradictions between the resources and the environment and urban development can be analyzed on the basis of the spatial differentiation of key elements of the urban ecosystem. Based on the theory of landscape ecology and the landscape security pattern method, ecological security models for urban structure and layout at different spatial scales can be studied by integrated applications of the numerical simulation, remote sensing, and GIS technology. By analyzing the spatial distribution of a city's main resources, the environmental capacity of the main pollutants, and the structural features of the urban ecological space, a series of guidelines for the spatial regulation of ecological security can be developed. Regulations and technologies for the inspection, evaluation, feedback, and revision of the urban space structure and layout can also be established, so as to provide technical support to ensure ecological security during urban development.

An integrated software package, urban eco-security monitoring and management system (UEMMS), has been developed in order to achieve the process of eco-security assessment and regulation when implementing the urban planning. The system contains six modes: urban plan, security assessment, monitoring, regulation, data management, and system management. In these, the module of urban plan includes the urban planning text library and the urban planning index library (see Fig. 17.6). This UEMMS will work perfectly with the environmental pollution online monitoring system. It will provide direct and instant access to the data from the online database and the information from the experts. The system will simplify the assessment model operation into background and provide the user with the clearest results of eco-security assessment and regulation.

17.5 Conclusion

The existing literatures related to assessment and control technology of ecological security have been reviewed in this chapter. Thereby, a system framework for assessment and regulation of ecological security when implementing urban planning has been suggested in order to achieve China's national strategic goals of intensive urban development and to deal with the serious environmental problems that have arisen during urban development.

The major technical difficulties and problems of the systems include (1) how to determine the factors involved in ecological security monitoring and assessment when implementing urban planning; (2) how to achieve dynamic monitoring and evaluation of ecological security for urban planning and implementation; (3) how to achieve system resource allocation, optimized pattern recognition, and regulation of the urban ecological security system, which covers social, economic, and natural systems; (4) how to achieve spatial regulation of urban ecological security under the different ecological security patterns aiming at the distribution of spatial elements; and (5) how to implement a security assessment and urban ecological systems.

Neither the PSR method nor the LUCC model has realize a dynamic assessment of urban ecological safety. This chapter proposes the use of computer simulation and online monitoring technologies and the simulation of the space–time dynamic changes of urban ecological security in order to implement the process of urban planning and provide a dynamic assessment of the process. In addition, regulating the ecological security of urban planning and implementation from the perspectives of system and space, respectively, can ensure overall coordinated development of the urban ecosystem. It can also establish the regulations and technologies for inspection, evaluation, feedback, and revision of the urban spatial structure and layout, in order to provide a model for the ecological security of the urban structure and layout at different spatial scales. It provides the possibility of sustainable urban development and ecological safety.

References

Bai SQ, Zhang CM, Lu SG (2006) Study on urban channel ecological rehabilitation based on keeping the health of river. People's Yellow River 28(8):3–4

Catterall CP, Kingston MB, Park K et al (1998) Deforestation, urbanization and seasonality: interaction effects on a regional bird assemblage. Biol Conserv 84(1):65–81

Chen X, Zhou CH (2005) Review of the studies on ecological security. Prog Geogr 24(6):8–17

Crooks KR, Suarez AV, Douglas T et al (2004) Avian assemblages along a gradient of urbanization in a highly fragmented landscape. Biol Conserv 115:451–462

Cui SH, Hong HS, Huang YF et al (2005) Progress of the ecological security research. Ecology 25 (4):861–868

Dilip K, Ilya P (1998) Dissipative structures (ch. 19). In: Modern thermodynamics—from heat engines to dissipative structures. Wiley, New York, pp 427–457

Fang YP, Chen GJ (2004) Appraisal of ecosystem security and ecological appropriation in Xichang. Water Res Environ Yangtze River 13(3):212–217

Fang YP, Chen GJ, Su CJ (2003) Appraisal of ecological compensation and ecosystem security in Xichang. J Mt Agric 22(6):537–543

Foster SSD (2001) The interdependence of groundwater and urbanization in rapidly developing cities. Urban Water 3(3):185–192

Lee AS (1972) Dissipative structure: an explanation and an ecological example. J Theor Biol 37 (3):545–559

Li H, Tang C (2006) Establishment of disaster prevention and relief system in the town with mud-rock flow frequently occurring base on landscape ecological security pattern. Disaster Relief in Urban Area 13(1):18–29
Lim HC, Sodhi NS (2004) Responses of avian guilds to urbanisation in a tropical city. Landsc Urban Plan 66(4):199–215
Martin-Duque, JF, Brebbia CA, Emmanouloudis DE et al (2006) Geo-environment and landscape evolution II: evolution, monitoring, simulation, management and remediation of the geological environment and landscape. WIT Trans Ecol Environ 89:305–315
McDade TW, Adair LS (2001) Defining the "urban" in urbanization and health: a factor analysis approach. Soc Sci Med 53(1):55–70
Pauchard A, Aguayo M, Pena E et al (2006) Multiple effects of urbanization on the biodiversity of developing countries: the case of a fast-growing metropolitan area (conception, Chile). Biol Conserv 127(3):272–281
Preston CD, Sheail J, Armitage P et al (2003) The long-term impact of urbanisation on aquatic plants: Cambridge and the River Cam. Sci Total Environ 314–316:67–87
Ren WW, Zhong Y, Meligrana J et al (2003) Urbanization, land use, and water quality in Shanghai 1947–1996. Environ Int 29(5):649–659
Savard JL, P C, Mennechez G (2000) Biodiversity concepts and urban ecosystems. Landsc Urban Plan 48(3–4):131–142
Shi XQ, Zhao JZ, Ouyang ZY (2005) Urban eco-security and its dynamic assessment method. J Ecol 25(12):3237–3243
Vogele J (2000) Urbanization and the urban mortality change in imperial Germany. Health & Place 6(1):41–55
Wang JJ, Cheng WL (2003) Original discussing of research on ecological rehabilitation and exploitation and utilization of the group of lakes: case study in Dongxihu District of Wuhan. Environ Sci 16(5):33–36
Wang G, Wang L, Wu W (2007) Recognition on regional ecological security definition and assessment system. J Ecol 27(4):1627–1637
Xia ZY, Xu WN, Dai FX (2005) Approach to techniques of ecological rehabilitation at urban riverside banks. Chin Soil Water Conserv 10:30–38
Xiao DN, Chen WB, Guo FL (2002) On the basic concepts and contents of ecological security. J Appl Ecol 13(3):354–358
Xie HL, Zhang XS (2004a) Research on measuring and countermeasure of ecological security in suburb—a case study on Haidian district, Beijing city. China's Popul Res Environ 14(3):23–26
Xie HL, Zhang XS (2004b) Study on model of matter-element evaluation of city eco-security. Geogr Geogr Inf Sci 20(2):87–90
Yang ZF, Hu TL, Su MR (2007) Urban ecological regulation based on ecological carrying capacity. Ecology 27(8):3224–3231
Zhang H, Ma WC, Ho HH (2007) Recent advances in research on LUCC: based urban ecological security. Ecology 27(5):2109–2117
Zhao XG, Wei LH, Ma CH et al (2007) Dynamics of ecological footprint and ecological safety in Xi'an. Arid Land Res Environ 21(1):1–5
Zhou WH, Wang RS (2005) Methodology assessment of urban ecological security—a case study of Beijing. Ecol Mag 24(7):848–852

Part IV
Landscape and Ecological System, Sustainable Development: Section Agriculture and Ecological System

Chapter 18
An Investigation of Changes in the Urban Shadow of Beijing Metropolis Under Agricultural Structural Adjustment in China

Dai Wang, Yue-fang Si, Wen-zhong Zhang, and Wei Sun

Abstract The purpose of this study is to clarify changes in agricultural production, especially medicinal crop production, in Zhengzhang Village, which is located in the urban shadow of Beijing Metropolis under agricultural structure adjustment. In addition, the reasons behind the changes were analyzed. This study discusses the prospects of the developmental trends of agricultural production and farm management in farm households in the urban shadow of Beijing Metropolis. The People's Republic of China (hereafter, China) started to implement agricultural structure adjustment in 1999. In particular, to prepare for China's entry in the World Trade Organization (WTO), the government strengthened the quality control on some agricultural products that constitute part of China's agricultural structure adjustment. Zhengzhang has a tradition of medicinal crop production, and from the time the Reform and Opening Policy was introduced, farm households have developed the business of cultivating and processing medicinal crops in order to increase their income. However, since 1999, due to the lack of capital to improve processing facilities, some farm households have had to abandon the business. At the same time, other farm households that were short of farmhands abandoned medicinal crops and shifted to food crops, which provided lower but stable income. As a result, medicinal crop production in Zhengzhang Village is dwindling under the agricultural structure adjustment, and the income polarization among farm households is becoming more defined. These changes can be mainly attributed to government policy adjustments. Other factors, such as market forces, the social

---

D. Wang (✉) • W.-z. Zhang • W. Sun
Institute of Geographic Sciences and Natural Resources Research, Chinese Academy of Sciences, 11A, Datun Road, Chaoyang District, 100101 Beijing, China
e-mail: wangdai@igsnrr.ac.cn; zhangwz@igsnrr.ac.cn; sunw@igsnrr.ac.cn

Y.-f. Si
Department of Geography, Economic Geography, Justus-Liebig-University Giessen, 1, Senckenbergstrasse, 35390 Giessen, Germany
e-mail: yuefang.si@geogr.uni-giessen.de

M. Kawakami et al. (eds.), *Spatial Planning and Sustainable Development: Approaches for Achieving Sustainable Urban Form in Asian Cities*, Strategies for Sustainability,
DOI 10.1007/978-94-007-5922-0_18,

situation, and globalization, also played major roles. This study carried out a detailed field survey of land use and farm management in farm households.

Keywords Agricultural structure adjustment • Urban shadow of Beijing Metropolis • Profitability of agriculture • Medicinal crops • China

18.1 Introduction

In this chapter, we attempt to clarify what caused the changes in the developmental trends of agricultural production, including medicinal crop production, in the urban shadow of Beijing Metropolis. Zhengzhang Village is taken as a case study area for illustrating changes in medicinal crop production, including cultivation and processing. Because it is very difficult to collect statistical data at the local level in China, we analyzed the changes in Zhengzhang through a field survey.

In 1978, China adopted its Reform and Opening Policy, thus spurring economic development and social progress in some areas. However, peasants' incomes were still low, and there was a substantial gap between the incomes of urban and rural residents (Liu et al. 2002). After China's entry in the World Trade Organization (WTO), the increasing importance of quality and low-priced agricultural products affected the domestic agricultural industry to some extent (Motoki 2005).

Many studies have been conducted on agriculture and rural areas in China (Yang 2010), but from the standpoint of the national administration, which focused on industrial structure and gross agricultural production. Thus, the government's sensitivity was weaker in the areas that were much more relevant to the peasants' lives, such as the form and profitability of agricultural management.

In northern China, where Beijing and Tianjin are located, the central government's policy was accomplished quickly. As for research on changes in urban fringe areas and urban shadow areas, reports are available as well. Wang proved that changes in the inner fringe areas of Beijing City were due not only to external factors, such as policy adjustment, but also to internal reasons (Wang 2000). He illustrated the relationship between resource-use patterns and government policies in the rural area around Beijing City from the political ecology viewpoint in China's economic growth (Wang 2006). However, the two studies analyzed the external affects brought to rural areas by modifications in government policies. Other important factors, such as the market economy and globalization process, were not examined. Using secondhand data and statistical analysis, Gale described in detail the actual situations caused by the diversification of agricultural production and agricultural structure adjustment (Gale 2005). It is difficult, though, to find detailed information or proper analysis from the standpoint of individual farm households for this research.

In 1999, in order to establish a competitive agricultural industry and increase peasants' incomes, the Chinese government implemented its agricultural structure adjustment. Therefore, structural adjustments in the industrialization of agriculture and the commercialization of agricultural products were strongly implemented in the urban shadow of Beijing Metropolis. Furthermore, after China's entry in the WTO in 2001, rural areas were increasingly affected by the globalization process.

The structure of this chapter is as follows: Section 18.2—we conduct a literature review in order to clarify policy changes in the Chinese agricultural structure adjustment and medicinal crop production; Sect. 18.3—we take Zhengzhang Village, which is located between Beijing and Tianjin, as a case study area; Sect. 18.4—the field survey and analysis regarding agricultural production and farm management in Zhengzhang are presented; in Sect. 18.5—we reach a conclusion after a discussion of the development trends in agricultural production in the urban shadow of Beijing Metropolitan Area under the policy changes on agricultural structure adjustment.

18.2 Literature Review on Agricultural Structure Adjustment and Medicinal Crop Production in China

Peasants used to be compelled by the government to undertake agricultural production. Since the Reform and Opening Policy of 1978, the Production Responsibility System, which paid more attention to productivity instead, was introduced to rural areas in China. On the precondition that land was still owned collectively, the purpose of the system was to divide and allocate land to peasants. Because agricultural reforms gave peasants the right to decide for themselves, it fired up their enthusiasm to engage agricultural production. Bolstered by irrigation technology and fertilization, agricultural productivity was greatly improved, which resulted in stable increases in the incomes of rural residents. However, development was much slower in rural areas than in urban areas. After 1992, the Market Economic Systems were applied in China and the reforms expanded the market economy. The income difference quickly widened between urbanites, who could enjoy economic development, and the rural folk (Ministry of Economy, Trade and Industry, Japan 2003). Problems in agriculture, rural areas, and peasants have become the Chinese government's most important hurdles.

One of the reasons peasants' incomes were restrained was the change in the supply of and demand for agricultural products. For a long time before 1999, the Chinese government kept urging peasants to produce food crops such as rice, wheat, and corn. Therefore, of the country's total planted area, the majority was devoted to food crops. The China Agriculture Yearbook 2000 reported that annual food crop production during the period 1996–1999 exceeded 0.5 billion tons. In other words, the average production per person was more than 400 kg. Due to the oversupply of agricultural products, such as food crops, the market became favorable for buyers. To cope, the Conference of Central Committee Chinese Communist Party on Rural Affairs was held in 1998, and new guidelines were adopted for strategic adjustments to the agricultural structure. The key policy of taking yield increase as the first priority since the establishment of China in 1949 was adjusted. Targets became multiple: stable production was pursued (food crops included) and the rational distribution of crops and type choices was effected by using regional advantages.

The agricultural structure adjustment adopted after 1999 concentrated on increasing peasants' incomes. Thus, the reorganization of regional agriculture was introduced to improve profitability. In urban inner fringe areas, food crops were replaced by horticultural products, such as vegetables and flowers, which were more profitable (Yang and Akiyama 2002). In urban shadow areas, it was strictly prohibited by the Regulation on Protection to Basic Farmland to use farmland for commercial or industrial purposes. The central government could, through the stabilization of the producer price of food crops, protect the peasants' income and maintain the stability of food crop production. In addition, peasants were allowed to cultivate high-profit oil crops, fruits, and medicinal crops that were considered suitable for the region's natural and geographical conditions, history, and technical advantages (Wang 2010).

As raw materials for crude drugs, medicinal crops have a long cultivation history in China. From 1954 to 1984, sales of medicinal crops were monopolized by the government, and the cultivation and trade of medicinal crops were under its strict control (Wang 2008). The monopoly system was revoked in 1984, when the cultivation of medicinal crops as a type of commercial crop was greatly expanded in rural areas. The area planted to medicinal crops in China in 2003 was 1,284,000 hectares (ha), an increase of more than 50% since 2000 (Ministry of Agriculture of China 2004). The increase can be mainly attributed to the higher profits earned by medicinal crop cultivation compared to food crop production. Since 1984, when the control in the trade of medicinal crops was eased, the supply-demand factor has dictated the prices in the crude drug market. After the Market Economic System was applied in 1992, the price of medicinal crops used to have frequent significant fluctuations; unavoidably, so did gross profits from their cultivation. In any case, even the smallest gross profit from medicinal crops is still 2–3 times as big as the usual earnings from food crops. On good seasons, it can be as high as ten times (Wang 2007).

In its effort to join the WTO, China was challenged to produce crops of higher quality and safety. It became an important task in the agricultural structure adjustment to improve the quality of agricultural products and to establish a globally competitive agricultural industry. The Chinese government promulgated laws and regulations that covered cultivation, processing, and the trade of agricultural products. Some laws and regulations enacted after 1999 are related to medicinal crops: Good Manufacture Practices for Drugs (GMP), Good Supply Practice for Drugs (GSP), and Good Manufacture Practices for Chinese Crude Drugs Production (GAP).

The case study area, Zhengzhang Village, is located in Anguo City, Hebei Province. Anguo City possesses the longest cultivation history and largest planted area for medicinal crops in the urban shadow of Beijing Metropolis, while Zhengzhang has one of the highest yields of medicinal crops in Anguo City.

In addition, with stricter quality control by the government, some GAP cultivation zones were designated in Zhengzhang, with the application of fertilizer or pesticide heavily restricted. Therefore, it is possible to reflect the developmental trends of medicinal crop production in the urban shadow of Beijing Metropolis by illustrating the changes in medicinal crop production in Zhengzhang.

18.3 The Study Area

Zhengzhang Village lies 250 km from both Beijing and Tianjin (Fig. 18.1). In April 2005, there were 3,990 people in the village, grouped into 997 farm households. Of the total area of the village of around 347.4 ha, 90% or 312.7 ha is arable land. Farm households are dense in the southeastern part (densely settled area), where the Villagers' Committee, school, and markets are located.

In the southern part of the densely settled area, there still exist some private plots, which were historically called Zi Liu Di. From the 1950s to the mid-1980s, influenced by the People's Commune Movement, the agricultural system imposed collective production. Zhengzhang, as a member of the Zhengzhang Production Brigade, converted all its arable land into collectively owned land. From 1959 to 1961, due to natural disasters, China was seriously affected by food shortages. Therefore, in 1962, a very small portion of arable land, still owned by the collective, was allocated to farm households as private plots. Peasants were permitted to cultivate what they needed on the plots (Chen et al. 1991).

In 1979, with the adoption of the Production Responsibility System in Zhengzhang, the right to use most collectively owned arable land was transferred to peasants regardless of sex or age. As usage rights had been transferred to peasants, the land was now considered individual land instead of rented arable land. On such individual land, farm households could make their own choices concerning production management. By the end of 2005, each peasant obtained an average of 0.087 ha of arable land.

18.4 Agricultural Production and Farm Management in Zhengzhang Village

18.4.1 Agricultural Production and Land Use

At present, agricultural production in Zhengzhang Village comprises food and medicinal crops. Other crops, such as vegetables or peanuts, are planted in small amounts merely for household consumption. In early April, the village cultivates about 40 types of medicinal crops, such as Danshen, Ziwan, Shanyao, Shashen, Baizhu, and Fangfeng—the so-called spring-planted medicinal crops. Corn is sown

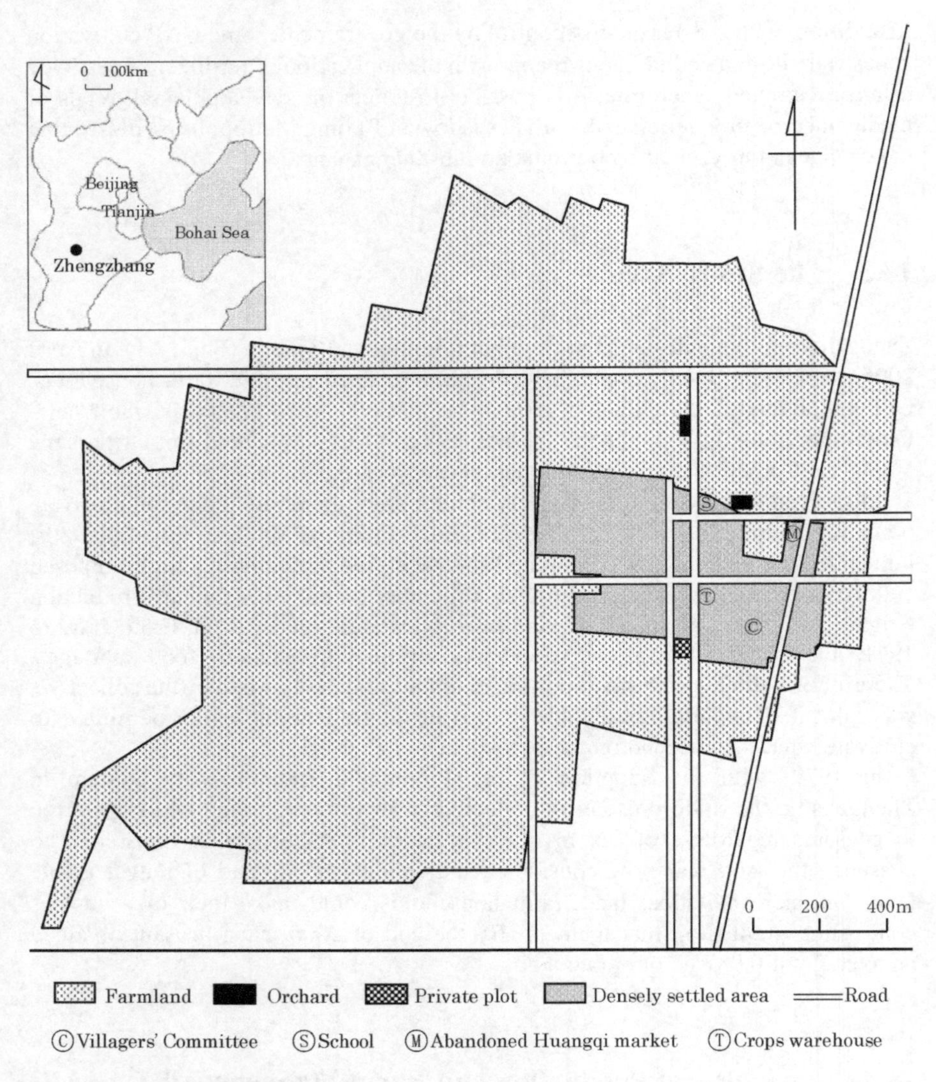

Fig. 18.1 Map of Zhengzhang Village

1 to 2 months later, followed by "summer-planted medicinal crops," such as Niuxi and Dihuang, in June. Corn is harvested throughout an entire month, starting on September 20; this is also the sowing season for winter wheat.

The field survey on the land-use situation in Zhengzhang was carried out in early May 2005, and a corresponding land-use map was prepared (Fig. 18.2). Most of the arable land (83%) is cultivated with wheat; there is plenty of wheat in the northern part of the densely settled area. In the past, the cultivation of wheat in Zhengzhang

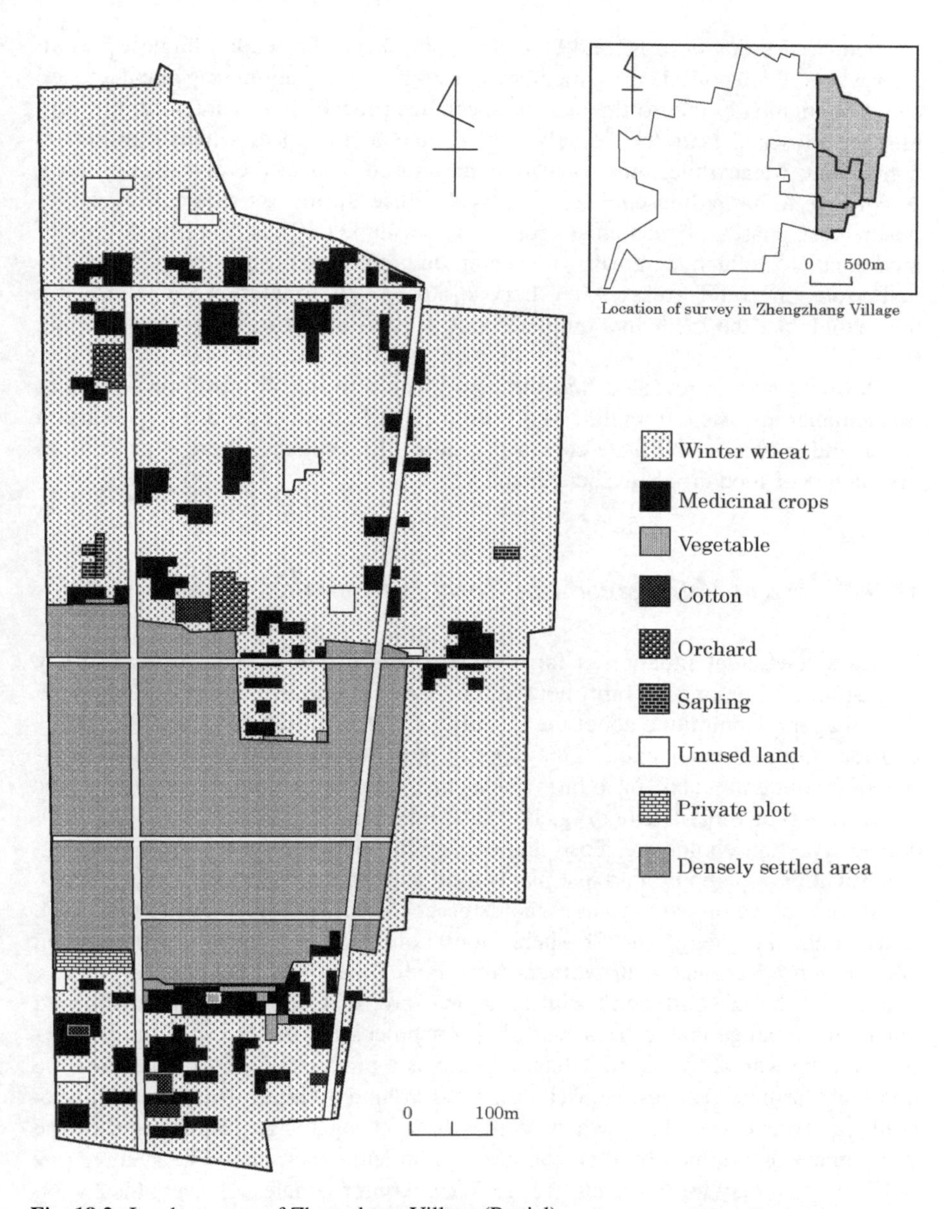

Fig. 18.2 Land-use map of Zhengzhang Village (Partial)

was done manually, with the help of only a few implements. Recently, to increase efficiency, some farm households have started to use combines to sow and harvest. Farm households have voluntarily merged blocks of wheatland to make better use of mechanization. Thus, working efficiency was greatly improved.

Medicinal crops occupy 12.4% of the arable land, the second biggest portion after wheat. Unlike food crops, medicinal crops need frequent management and are cared for manually due to the lack of specialized machinery. Therefore, they are mostly cultivated near the densely settled area or the road, where transport is convenient. Meanwhile, there is also some unused land in the investigated area. According to farm households, it is used either during the pre-cultivation or postharvest phases of medicinal crops. As the market prices of certain types of medicinal crops fluctuate greatly, it is common practice to postpone sowing medicinal crops in some areas. After harvest, to satisfy to market demand, farm households put the medicinal crops through specialized processing before selling them.

The field survey revealed that the area planted to wheat is 6.7 times that of medicinal crops. Also, from mid-May until the last 10 days of June, the postharvest wheatland is used to cultivate corn and peanuts. In summary, there is an obvious prevalence of food crops in Zhengzhang.

18.4.2 Farm Management in Farm Households

To know the actual situation of farm management in Zhengzhang, a field survey was conducted among 25 farm households. Using the information obtained from the Villagers' Committee about the situation of farm households, the field survey considered the diversification of the types of farm management and the differentiation of production scales. As to farm management, the households were classified as cultivation type (nos. 1–20 in Fig. 18.3) and mixed type (nos. 21–25 in Fig. 18.3). Mixed-type households are those that engage in both cultivation and processing. A cultivation-type household just plants food and medicinal crops.

All of the five mixed-type households process traditional medicinal crops, such as Huangqi, Fangfeng, and Shashen. Household 22 has a long-term cultivation cooperation agreement with partners from both Gansu and Shandong provinces, where 33.5 ha are cultivated with medicinal crops. It also rents the abandoned Huangqi exchange market as a workshop for processing the crops. Household 21 rents a crop warehouse in the village for use as a processing workshop during the non-busy farming seasons. Household 21 has around 40 hired workers and household 22, around 60, all of whom come from Zhengzhang. The salaries of the workers are determined by their job description and experience. The average pay is 15–20 Yuan/day for males and 12–15 Yuan/day for females. Households 23, 24, and 25 do the processing at home, and the workers come from the households themselves.

In terms of agricultural labor resources, all of the 25 farm households face a shortage of young male workers. Since 1985, the family registry system that used to restrict population flow has been relaxed in order to allow the economic development of rural areas and increase peasants' incomes. Thus, it opened the opportunity for agricultural laborers to flow into urban industry. This agricultural labor drain is

Types of Management	No.	Farm household members: Headcount	Farm household members: Agricultural laborer	Cultivated land: Total (are)	Cultivated land: Rent	Farm land use situation(are) (Wheat / Medicinal crops / Vegetable / Other crops)	Household incomes (10,000RMB): Cultivation	Household incomes (10,000RMB): Processing
Cultivation type	1	9	4(○○▲△)	44.7	−33.3		<1	0
	2	10	4(○○▲△)	53.3	−33.3		<1	0
	3	5	2(●○)	43.3	0		<1	0
	4	5	3(○▲△)	43.3	0		<1	0
	5	7	4(○○▲△)	60.7	0		<1	0
	6	5	2(▲△)	43.3	0		<1	0
	7	6	2(▲△)	52.0	0		<1	0
	8	6	3(○▲△)	52.2	0		<1	0
	9	8	4(●○▲△)	69.3	0		<1	0
	10	6	2(●○)	52.0	0		<1	0
	11	7	3(○▲△)	60.7	0		<1	0
	12	7	3(●○△)	60.7	0		≒1	0
	13	8	3(●○▲)	69.3	0		≒1	0
	14	9	5(●●○▲△)	78.0	0		≒1	0
	15	8	3(○▲△)	69.3	0		≒1	0
	16	11	5(●○○▲△)	148.7	+53.3		≒1	0
	17	10	4(●●○△)	86.7	0		≒1	0
	18	7	4(○○▲△)	60.7	0		≒1	0
	19	9	5(●●○▲△)	104.7	+26.7		≒3	0
	20	9	5(●●○▲△)	118.0	+40.0		≒5	0
Mixture type (Cultivation + Processing)	21	3	2(●○)	6.0	−20.0	50 100 150	0	≒40
	22	4	2(●○)	34.7	0		0	≒50
	23	9	5(●●○▲△)	78.0	0		≒1	≒9
	24	11	6(●●○○○△)	95.3	0		≒2	≒15
	25	13	7(●●●○○○△)	179.3	+66.7		≒6	≒7

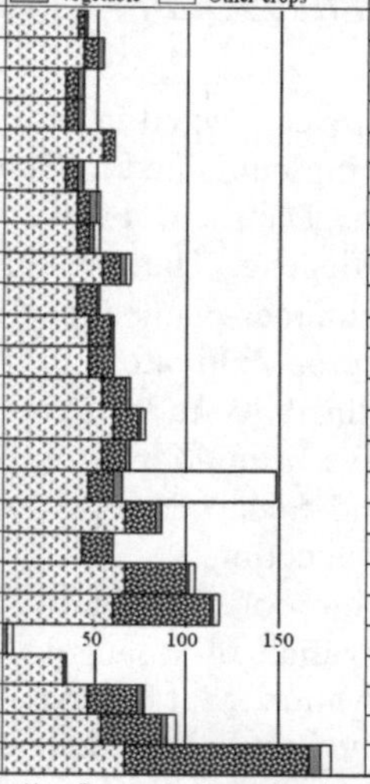

Agricultural laborer: ● Young male laborer; ○ Young female laborer; ▲·△ Older than 60 years old. Rent: − Land lent to others; + Land rented from others.

Fig. 18.3 Current situation of Zhengzhang Village in connection with farm management (May 2005)

evident in Zhengzhang. The 25 farm households surveyed have regular agricultural workers (a total of 92) and hire nonagricultural laborers, who are mostly young males. Only households 21, 22, and 25 do not have nonagricultural laborers. Of the regular agricultural laborers, 69 were either female or more than 60 years old.

The shortage of young male laborers in cultivation-type households is more serious than in the mixed-type households. Of course, some mixed-type households, such as nos. 21 and 22, also have an agricultural labor force shortage, but hiring workers solves this problem. Some households with a large labor force have the potential to expand the area planted to medicinal crops. For instance, Households 16, 19, 20, and 25 are doing crop cultivation both on their individual land and land rented from other households in the village. To be specific, no. 16 is managing an orchard, while the other three households are cultivating medicinal crops. Normally, renting land is done through oral contract, and the lessee is liable for the payment of the agricultural tax for the lessor. Currently, the agriculture tax in Zhengzhang Village is about 630 yuan per hectare of land. Households with a serious shortage of young male laborers can only plant food crops. Households 1, 2, and 21 even used to rent out their individual land to others, as they lacked agricultural labor.

There exist big income differences among the 25 farm households, especially between the two classifications. From the above analysis, it can be concluded that medicinal crop processing can bring a high income to farm households. Due to the higher profits brought by medicinal crop cultivation compared to food crop cultivation, the larger the area planted to medicinal crops, the higher the household's income.

18.4.3 Changes in Medicinal Crop Production in Zhengzhang Village

The field survey conducted in May 2005 showed that about 50 types of medicinal crops were being cultivated in Zhengzhang. The top 10 in terms of planted area were Shashen, Danshen, Fangfeng, Shanyao, Baizhu, Ziwan, Niuxi, Baizhi, Gualou, and Jiegeng. The area of "spring-planted medicinal crops" was larger than that of "summer-planted medicinal crops."

Medicinal crop cultivation in Zhengzhang started before1949. After 1958, it gradually declined due to the People's Commune Movement. Since 1978, a series of reforms have been adopted to loosen government controls on the cultivation, processing, and trade of crops and to encourage farm households towards free agricultural production.

The cultivation of medicinal crops in Zhengzhang resumed in the early 1980s with the cultivation of Huangqi. Most of the land in the village is suitable for Huangqi cultivation, as it is sandy loam with good drainage and abundant organic material. Many farm households own accumulated experience in the cultivation of this crop. After all, Anguo City, where Zhengzhang is located, is well known for Huangqi production; in fact, 25% of the national total is grown here. As Huangqi is widely used as a crude drug, its market price is high and stable.

Starting in 1980, Huangqi became the priority crop of Zhengzhang. In 1984, the biggest Huangqi exchange market in Hebei Province was constructed, and it attracted producers and brokers from Anguo City in northern China and other areas of the country. In 1990, the area planted to Huangqi in Zhengzhang was about 133.3 ha, more than 90% of the total area of medicinal crops (148.1 ha).

By 1995, though, replant failure had set in. The condition is usually attributed to the following: (1) the accumulation of a pathogenic fungus, which ultimately causes the deterioration in soil fertility and (2) the lack of trace elements, which reduces Huangqi's resistance to disease. To solve the twin problems, the rotation of Huangqi with other crops, as well as the application of soil antiseptics, were introduced. However, since it takes at least 8 years for the soil to recover from replant failure, it was too costly to keep sterilizing the soil with antiseptics. Thus, in 2000, Huangqi cultivation was stopped in Zhengzhang.

In 1999, with the introduction of the agricultural structure adjustment concept, medicinal crop cultivation had become the preferred method of increasing income in Zhengzhang. Therefore, when the replant failure of Huangqi became a serious problem, the government encouraged farm households to cultivate other medicinal crops instead. That they did, and in greater variety. In 2000, the area for medicinal crops was more than 130 ha, but by 2005 (when this study's field survey was conducted), it had shrunk to less than 50 ha. The survey revealed two reasons for the reduction.

The first reason is related to the agricultural labor force. The cultivation of medicinal crops needs more agricultural labor than the cultivation of food crops. Figure 18.4 shows the labor consumption of each 0.1 ha of land over 10 days. The

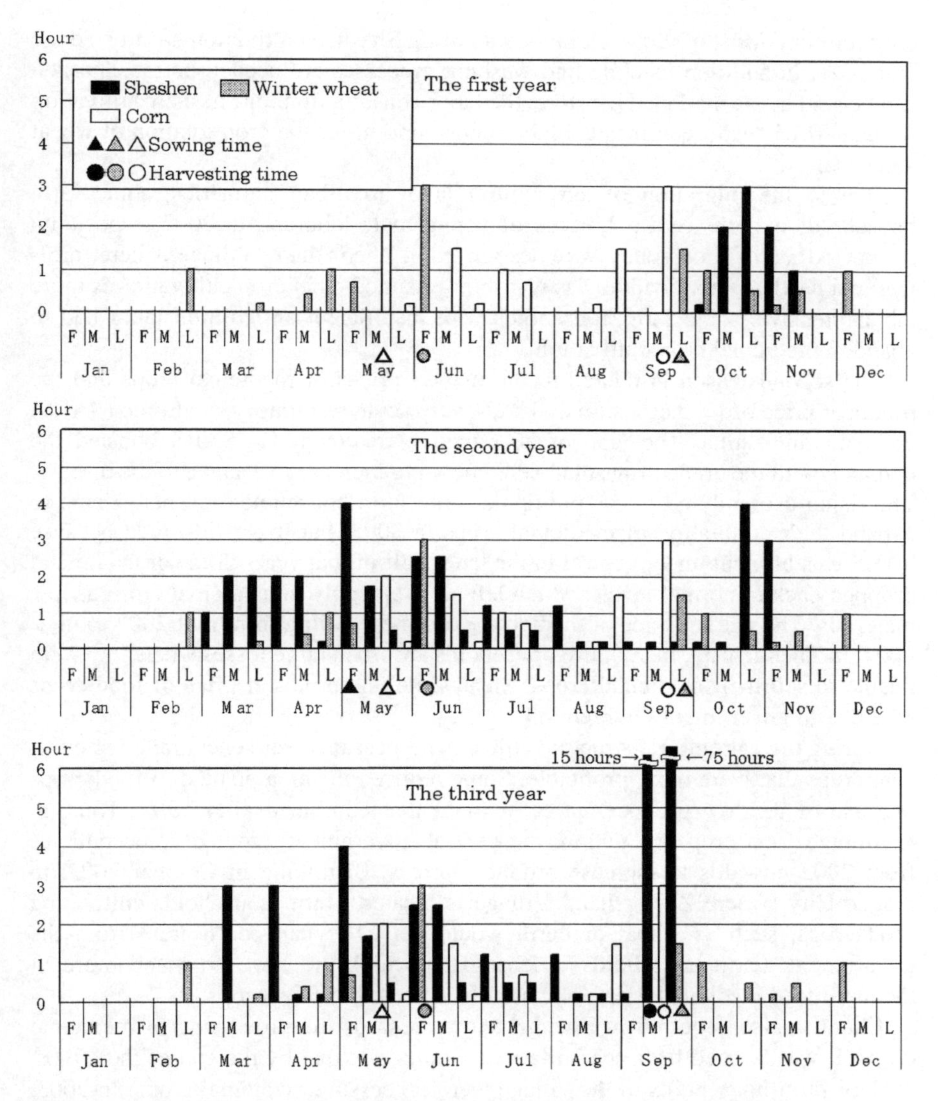

Fig. 18.4 Comparison of labor time for Shashen, winter wheat, and corn (0.1 ha)

land is planted to Shashen, winter wheat, and corn. Usually, May 1–10 is the sowing time of Shashen, but preparations—land choice, tillage, and fertilization—have to be started in October of the previous year. In November (6 months after the sowing in May), the Shashen saplings are dug out and buried in sand. In March of the following year, the buried saplings are replanted and will be harvested 6 months later. In total, the Shashen crop takes 23–24 months from preparation to final harvest. Within the same period, wheat and corn can be harvested twice through crop rotation. Most of the work involved during Shashen cultivation has to be done manually. In addition, peeling and washing processes are needed before Shashen can be sold in the market. Therefore, the labor consumption for Shashen cultivation for each 0.1 ha is three times the labor consumption for the crop rotation of wheat and corn.

Due to the migration of agricultural labor to urban industries, some farm households experienced a shortage of young male laborers. In 2002, especially, some GAP cultivation zones were designated in Zhengzhang Village, where fertilizer and pesticide application was restricted. As medicinal crop cultivation is more labor-intensive, some farm households with insufficient agricultural labor had to abandon medicinal crop cultivation.

The second reason is related to the market prices of medicinal crops and the producer price of food crops. In 2002, a severe acute respiratory syndrome (SARS) epidemic hit China. The lack of an effective treatment for SARS boosted the demand of all the crude drugs that were known to increase resistance to the disease. The rising demand and prices prompted some farm households in Zhengzhang to expand their cultivation of medicinal crops in 2003, but it was ill-timed because SARS was brought under control in the latter half of that year. Crude drug demand dropped back to normal levels, which left an oversupply in medicinal crops as raw materials. The market price of medicinal crops kept falling from mid-2003 to mid-2004. In Zhengzhang, many farm households absorbed large losses when they were unable to sell their medicinal crops. Meanwhile, the producer price of food crops rose due to government intervention.

During the agricultural structure adjustment, peasants were encouraged to cultivate crops that were more profitable. Some reduced the area planted to food crops because of the low producer price, a trend that continued after 1999. Thus, to encourage food crop production, the central government adopted some policies from 2002 onwards to increase producer prices. Beginning in October 2003, in Anguo City (where Zhengzhang Village is located), farm households cultivating food crops, such as wheat or corn, would get 210 yuan per hectare from the government. Farm households in Zhengzhang took the cue: more and more of them reduced medicinal crop cultivation and turned to food crops.

Changes also occurred in Zhengzhang as regards the processing of medicinal crops. It was learned from the Villagers' Committee that by the end of the 1990s, 70% of farm households in the village were processing medicinal crops. In 2003, based on international standards, the Administration of Anguo City adopted an approval and examination system to regulate medicinal crop processing. The target of this system was to maintain global competence in locally processed medicinal crops. All farm households that processed medicinal crops had to meet the requirements set by this system, which is referred to as the GMP of Anguo City. The GMP of Anguo City is very strict in all aspects. For instance, producers are required to have a university education and be able to pass a health examination. Processing workshops are also required pass certain specifications regarding ventilation, humidity control, sterilization of processing equipment, etc. The problem in Zhengzhang is that most farm households that are into medicinal crop processing do not have enough capital to improve production conditions or to hire professional labor. Since 2003, therefore, many farm households have had to give up the medicinal crop processing business.

Of the 20 cultivation-type farm households in the Zhengzhang survey, 17 were processing medicinal crops prior to 2003 (only nos. 3, 10, and 17 were not). All but two had to give up the business because they could not meet the standards set by the GMP of Anguo City. Households 21 and 22, which had sufficient capital, stayed in business by improving production conditions. As of 2005, around 40 farm households had managed to maintain their processing businesses, although the total is less than 10% of the number in 2000. Due to globalization, the government is strengthening quality control of medicinal crop processing, making it very difficult for households with limited capital to squeeze into the processing market.

18.5 Discussion and Conclusion

Since the Reform and Opening Policy, agricultural policies focusing on food crop production were loosened. In the urban fringe of Beijing Metropolis, expanding urbanization and commercialization often ruined farmland or used it for other purposes. At the same time, commercial crops were gradually replacing food crops. In urban shadow areas, such as Zhengzhang Village, there was large-scale cultivation of profitable commercial crops such as medicinal crops. At the same time, the processing of these crops was becoming a high-value-added business. In Zhengzhang, there appeared to be a large number of mixed-type households, which were involved in both the cultivation and processing of medicinal crops.

The government's agricultural structural adjustment of 1999, which strengthened quality control, forced most farm households in Zhengzhang to abandon medicinal crop processing. Only a few households, which had sufficient capital, managed to continue the processing business. The result has been a huge increase in cultivation-type households and very few remaining mixed-type households. Simultaneously, the number of households involved in medicinal crop cultivation also contracted. A polarization of the production scale also appeared, mainly between a few farm households with abundant agricultural labor and the majority, which did not have enough.

Due to the big difference between urban and rural wage levels, there has been an exodus of agricultural manpower to urban industries. This has depleted the agricultural labor force so that more and more farm households have had to give up the cultivation of profitable medicinal crops. Therefore, the agricultural labor shortage became an obstacle to higher household incomes in Zhengzhang. This study strongly challenges the widely accepted notion in China that a big rural population is a barrier to China's agricultural development.

Besides maintaining for the competitiveness of Chinese agricultural products in the world market, the other main target of the agricultural structure adjustment in China is to increase peasants' incomes. From the country's overall perspective, agricultural incomes, which had been stagnant for some time, gradually rose after 2000. However, the picture is not the same for all rural areas. In Zhengzhang Village, for example, only a small proportion of farm households have been able to continuously increase their income. These households are either into large-scale medicinal crop cultivation with the support of a big labor force or are running a processing business because they have sufficient capital. Under the agricultural structure adjustment, the potential for even greater income polarization is becoming more obvious.

References

Chen D, Xie LF, Yao CH (1991) The rural economy. In: Xu GH, Peel LJ (eds) The agriculture of China. Oxford University Press, New York, pp 179–234

Gale F (2005) Regions in China one market or many. In: Gale F (ed) China's food and agriculture: issues for the 21st century. University Press of the Pacific, Washington, DC, pp 20–23

Liu YS, Wu CJ, Lu Q (2002) Orientation and tactics for 21st century sustainable agriculture and rural development in China. Sci Geogr Sin 22(4):385–389

Ministry of Agriculture of China (2004) China agricultural development report. China Agriculture Publisher, Beijing

Ministry of Economy, Trade and Industry, Japan (2003) White paper on international trade 2003. JETRO, Tokyo

Motoki Y (2005) China's restructuring of agriculture in line with global trends: views on food production problems. Spec Publ 40:69–82

Wang PF (2000) The characteristics of rural change in the inner urban fringe of Beijing City, China: a case study of Rutang Village. Geogr Rev Jpn Ser B 73:46–61

Wang PF (2006) Changes of resource use patterns and state policies since 1978: a political ecology of the rural Beijing City, China. Geogr Rev Jpn Ser B 79(5):198–215

Wang D (2007) Changes in medicinal crop production in the urban shadow of Beijing Metropolis under agricultural structure adjustment: a case study of Zhengzhang Village, Hebei Province, China. Geogr Rev Jpn 80(12):758–776

Wang D (2008) Changes in crude drug distribution industry and business evolution by distribution entities of farm household in China: a case study of Anguo City in Hebei Province. Ann Jpn Assoc Econ Geogr 54(3):28–48

Wang D (2010) Development of large scale tenant farming in China: a case study of Gaoyang County, Hebei Province. Jpn J Hum Geog 62(4):1–19

Yang PY (2010) The responsibility in extending function of modern agriculture of government. Territ Nat Res Study 4:54–55

Yang DQ, Akiyama K (2002) The existing conditions of the collective farms in Shunyi District, Peking, China: an investigation of the most recent situation. Bull Fac Agric Kagoshima Univ 52:27–56

Chapter 19
The Spatial Planning of Agricultural Production in Beijing Toward Producing Comfortable and Beautiful Living Environment

Feng-rong Zhang and Hua-fu Zhao

Abstract Since 1984, Beijing's rapid economic growth and urbanization have been heading into resource and environmental problems, such as cropland decrease, environmental deterioration, and traffic jams. Therefore, more space for glades and trees is needed to improve the urban environment. However, the central government's decision to maintain 227,000 ha of cropland for national food security created a dilemma for Beijing: the need to choose between keeping the cropland or planting more trees and enlarging the glades. Actually, people can use the croplands to serve both purposes—food production and eco-environmental service—because croplands are multifunctional anyway. Besides food production, croplands also have ecological, landscape, cultural, and tourism roles, especially in megacities. This chapter analyzed land use and crop structure based on the land-use survey and agricultural statistic data while discussing the existing state of agriculture, including economic profit, labor cost, water resource depletion, and environmental impact. This study then proposed a four-ring agricultural pattern that radiates outward from the city center and exploits the multifunctional nature of croplands: an inner-city ring providing urban beautification with flowers and grass; a city-skirt ring with fruit trees, grain crops, and grass, from which intensive agriculture should be prohibited; a suburban plain ring for efficient, intensive agriculture; and lastly, the mountainous regions, for ecological conservation and unique agricultural products. This way, cultivation is combined with landscaping and ecological protection to give Beijing a more comfortable and beautiful living environment. The proposed spatial planning of agriculture is different from the traditional agricultural ring pattern identified by Thünen in 1826, but it inherits and

F.-r. Zhang (✉)
College of Resources and Environmental Sciences, China Agricultural University,
Yuanmingyuan West Road 2, Haidian District, Beijing 100193, China
e-mail: frzhang@cau.edu.cn

H.-f. Zhao
College of Land Science and Technology, China Geosciences University,
Xueyuan Road 29, Haidian District, Beijing 100083, China
e-mail: huafuzhao@163.com

M. Kawakami et al. (eds.), *Spatial Planning and Sustainable Development: Approaches for Achieving Sustainable Urban Form in Asian Cities*, Strategies for Sustainability,
DOI 10.1007/978-94-007-5922-0_19,

develops the thought illustrated in *Garden Cities of Tomorrow*, a book written by Howard in 1945.

Keywords Agriculture production • Spatial planning • Environmental management • Megacity • Beijing • China

19.1 Introduction

Green space is an important component of the complex urban ecosystem. Parks, forests, and farmlands are the three main types of urban green spaces and have key ecological, social, and economic functions (Bradley 1995; Tyrväinen 2001; Lütz and Bastian 2002). Green space has significant ecosystem advantages, which are defined as "the benefits human population derives, directly or indirectly, from ecosystem functions" (Costanza et al. 1997). People are eager to access green spaces for recreation and to experience nature (de Groot and van den Born 2003; Lynn and Brown 2003).

Multifunctionality is the essential characteristic common to all types or systems of agricultural production (Hall 1999). Besides food, crop production in China also has ecological, landscape, cultural, and tourism applications (Cheng 1999; Zhang 2001; Guo et al. 2004). The ecological functions and their value to various farmland ecosystems in Beijing and China have been previously reported (Yang et al. 2005; Sun et al. 2007).

China's economy has been growing rapidly since 1984, bringing with it increased urbanization. Therefore, large areas of cropland have been replaced by buildings, glades, and trees. To most people, only glades and trees are green space. Instead of building on cropland, city officials also favor converting cropland to glades and trees to beautify the urban scenery. Urbanization has thus been spilling over into cropland. The land-use survey data of the Ministry of Land and Resources show that from 1996 to 2006, cropland was reduced from 130 to 128 million ha in the whole of China and from 344,000 to 233,000 ha in Beijing.

China has the world's largest population (1.32 billion in 2007), and the National Bureau of Statistics expects it to reach 1.47 billion by 2032. With population growth and economic development, cropland is fast disappearing, which may endanger food security. The central government has been stressing cropland protection and has issued the Land Administration Law of the People's Republic of China. The law orders provincial governments to prevent further cropland losses. Thus, if urbanization or other construction activities encroach on a piece of cropland, new cropland must be created elsewhere to replace the loss. Also, every provincial government has to retain a certain amount of cropland during certain periods; it is not allowed to replace cropland with glades and trees during the planning period.

While the city is expanding, its environment is deteriorating. Although the Beijing People's Government and its citizens want to have more space for glades and trees to improve the environment, the central government has asked Beijing to keep 22,700 km^2 for food security purposes. Thus, the city government has a dilemma: keep the cropland or make the environment more comfortable and beautiful. Cropland protection could be a difficult task for the city, as it had to accept the concept that cropland is vital and must be incorporated into its green space system (Zhang et al. 2005).

Urban and modern agriculture in Beijing need to increase the farmers' income through tourism, recreation, and agricultural products; more importantly, urban agriculture must be ecologically and environmentally friendly in order to create a beautiful, refreshing, and rural agricultural landscape for all citizens. Agricultural development in Beijing must integrate rural and urban development and invest vigorously in developing resource-saving and environmentally friendly crop production to meet the requirements of building a livable city. However, both the Beijing Urban Master Plan for 2004–2020 (Beijing People's Government 2003) and Integrated Land Use Planning of Beijing for 2006–2020 (Beijing People's Government 2008) lack detail on the spatial distribution of the various components of agricultural production in Beijing. To better exploit agriculture's multiple functions and develop environmentally friendly agriculture, our study investigates and proposes a layout of cropland usage in the Beijing metropolitan area by arranging the different kinds of agricultural production in different regions of Beijing.

19.2 Current Space for Agricultural Production in Beijing

19.2.1 Essential Information of the Study Area

The municipality of Beijing is located in northern China (39.38–41.05 N) in a temperate zone. It has a total area of 16,807.8 km^2 and a population of approximately 18 million, with an average population density of 888 persons/km^2 in 2004 (Beijing Municipal Bureau of Statistics 2005). The topography consists of mountains, a piedmont, and a plain. The mountains and piedmont account for 62% of Beijing's surface area and are located in the northwest, and the "Beijing plain" is in the southeast, where the elevation is less than 100 m above sea level. The plain makes up 38% of Beijing, and it is the main base for urban extension. Beijing has a characteristic monsoon climate, with a mean annual precipitation of 640 mm. Most of its rivers are seasonal, and Beijing has a serious water shortage problem.

19.2.2 Current Space for Agricultural Production

Agricultural production space is defined as the area for agriculture production activities, including cropland, orchards, pastures, and land used for greenhouses and aquiculture. This chapter calls these kinds of land "farmlands" for "agricultural

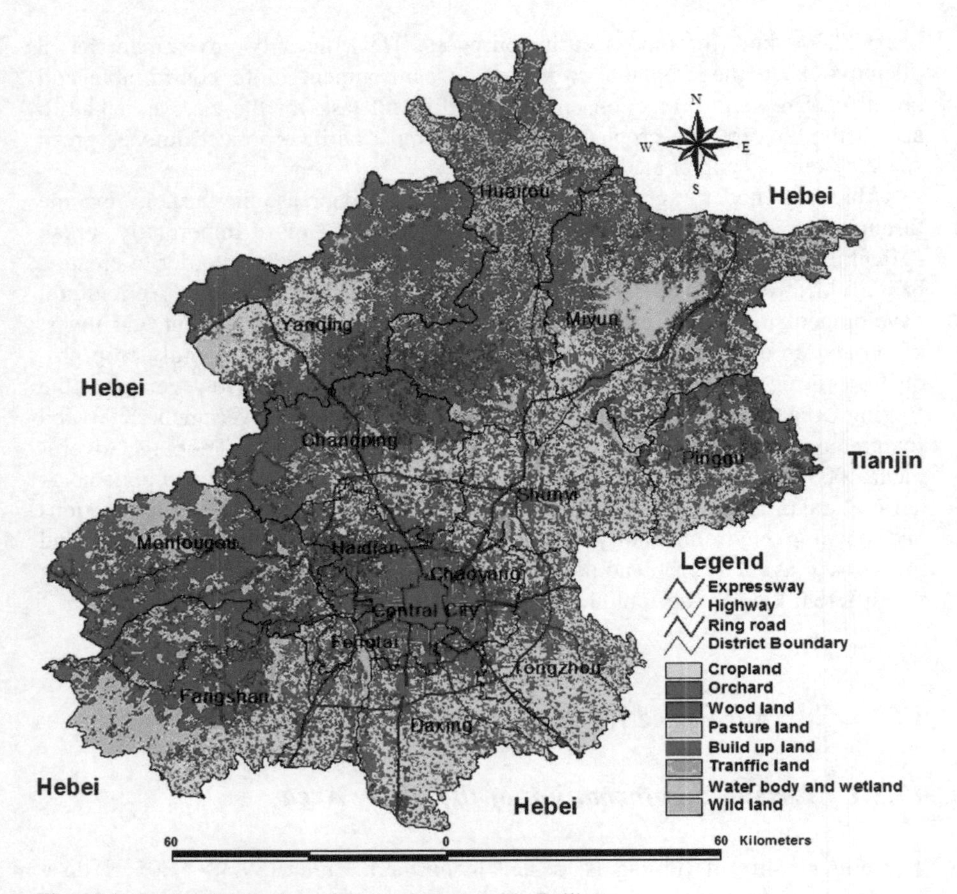

Fig. 19.1 Spatial distribution of the land-use types in Beijing

industry" to differentiate them from "croplands" in classifying land use. Croplands include land for growing grain, vegetables, oil seed plants, cotton, and other crops. Figure 19.1 shows the spatial distribution of land-use types in Beijing.

According to the investigation report on change of land use of Beijing in 2004 (Beijing Municipal Bureau of Statistics 2005), cropland in the municipality of Beijing in 2004 had an area of 2,364.37 km^2 (14.41% of the total land area); orchards, 1,239.40 km^2 (7.75%); pastures, 20.42 km^2 (0.12%); and other agricultural land, 5,510.04 km^2 (3.36%). "Other agricultural land" refers to land for greenhouses, rural roads, poultry and livestock breeding areas, aquiculture farms, field banks, grain-sunning grounds, and irrigation facilities. Thus, the total area used for agriculture was 4,175.23 km^2, accounting for only 25.44% of the total land area of Beijing. The 25.44% designated agricultural land is higher than the 19.48% that comprises the city's built-up area, but much lower than the 42.07% for woodlands. Thus, the proportion of land for agriculture is not very large (see Fig. 19.2), so that the industry can only supply part of greater Beijing's needs.

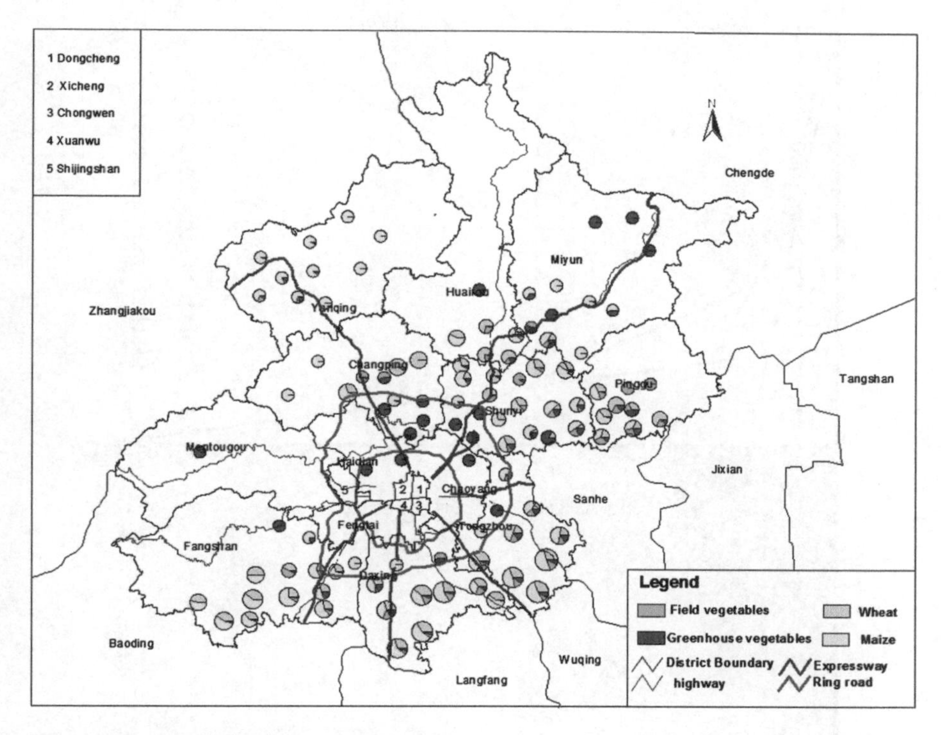

Fig. 19.2 Spatial distribution of crops in Beijing

The proportion of cropland in the suburban plain area is higher. Single pieces of cropland are relatively large and regularly shaped. The city-skirt area is second to suburban plain areas in the proportion of its cropland area. The average patch of cropland in city-skirt areas is bigger than that of urban and mountain areas, but smaller than that of suburban plain areas. However, be it in the city-skirt or suburban areas, cropland is interlaced with other landscapes, such as residential, commercial, and industrial, so the human impact is strong. Cropland in urban areas is, to a great extent, distributed among constructed urban lands, while cropland in the mountain areas is distributed among woodlands and consists of small fragments.

19.2.3 The State of Agriculture in Beijing

Today, although transportation technology has improved, the agriculture of Beijing still looks like the concentric ring feature (Fig. 19.3) described by Thünen (1826). There are three different agricultural zones surrounding the center of Beijing City (Fig. 19.2). The first ring is the plain zone of interlacing city-skirt towns and counties and an area of concentrated vegetable production, including field and

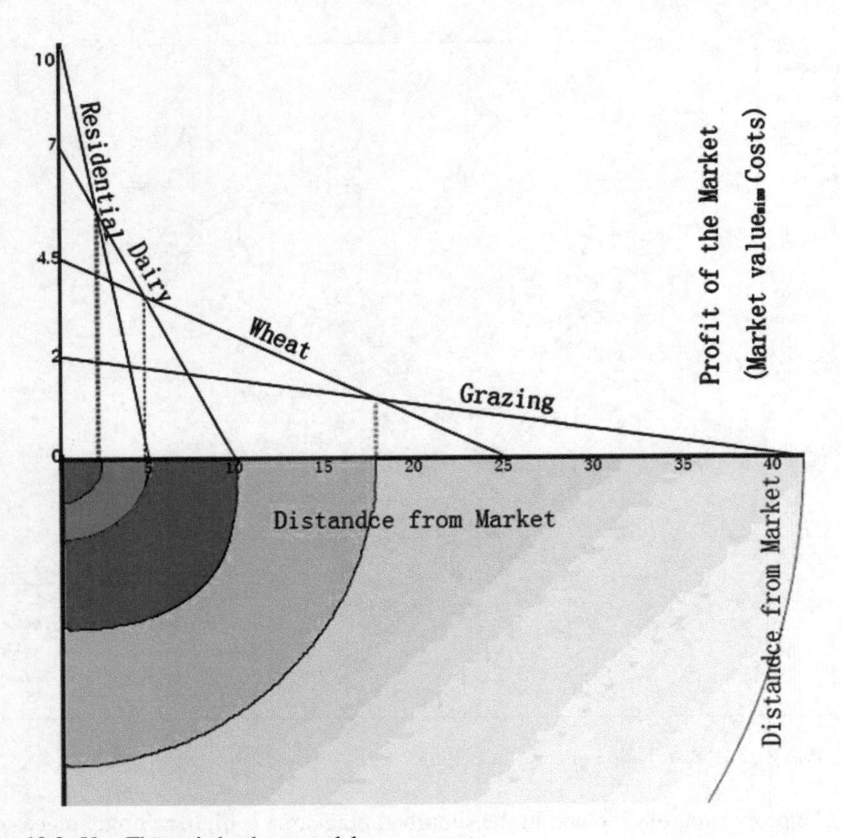

Fig. 19.3 Von Thünen's land-use model

greenhouse vegetables. Most of the greenhouses are located in this area. The second ring is the suburban plain zone and the main production area of grain crops and animal products. The third ring zone (actually a half ring) is the mountainous zone and the main production area of Beijing's forest and fruit products.

The first ring of concentrated vegetable production was formed against the backdrop of the "vegetable basket project" in the 1980s. The project was launched by the city government to resolve the shortage problem and provide the citizens with enough vegetables. It is close to the agricultural product market and supplies fresh vegetables to city dwellers to their convenience. However, the water, soil, and air pollution of this ring is serious, so it is very difficult to guarantee the safety of the vegetables produced. The high use of fertilizers for vegetable production leads to terrible nitrate contamination of groundwater, especially in the city-skirt area (Liu et al. 2006). Soil lead pollution is also a significant threat to food safety (Chen et al. 2006). In addition, irrigation mainly comes from groundwater, further depleting a water resource that is already in short supply and lowering the groundwater table (He et al. 2005). Greenhouse agriculture may bring relatively high economic returns, but its white structures are at odds with the green rural scenery required for the citizens. Labor-intensive vegetable production has created many employment opportunities, but it is not favored by local farmers, who can get better-paying nonagricultural jobs. They thus leave the agricultural work and attendant hardships to migrant farmers. The crude houses, corrals, and barns built randomly by migrant farmers contribute to a dirty, disorderly, and unsightly landscape and environment.

In the second ring of the suburban plain zone, crop production mainly involves grain crops. Whether in labor, water, or fertilizer use, it is not as intensive as the vegetable production in the first ring. Although grain production is well suited to the availability of extensive cropland, it is not well adapted to the region's economic and social conditions. Firstly, the suburban plain areas have an abundant supply of agricultural labor but fewer nonagricultural employment opportunities than in the city-skirt areas, so that the suburban plain areas would be a better choice for labor-intensive vegetable production. Secondly, grain crops are more visually and aesthetically appealing to city dwellers and are therefore better suited to the city-skirt areas. Lastly, since grain crops use less water and fertilizer than vegetables, they are environmentally ideal for areas close to the city center.

As to the above problems of the agricultural situation in Beijing, great books such as *Cities in Evolution* and *Garden Cities of Tomorrow* may help us find harmony in utilizing farmlands' versatility—food production; ecological, landscape, cultural, tourism functions; etc.—and improving the urban environment.

19.3 Proposed Layout for Concentric Rings of Agriculture

The existing agriculture in Beijing is the interim historical result of a combination of natural conditions and economic and social development, but it is not necessarily the best solution for making Beijing a "livable city." The economic and social development of China, especially Beijing, is far removed from the situation described by Thünen in 1826. Advanced transportation networks have shortened the distance between the agricultural production base and the demand market. Free trade has set up a uniform market across the country and even around the world, making the circulation of agricultural products very convenient between regions and countries. This lays a good foundation for Beijing to adjust its existing distribution of agriculture and establish a new layout for modern urban agriculture. It must be more consistent with the current economic and social development level of Beijing; exploit the ecological, landscape, cultural, and production services provided by the farmlands; and make Beijing "a livable city." We thus propose a four-ring agricultural pattern (Fig. 19.4) radiating outward from the city center, based on a ring road structure.

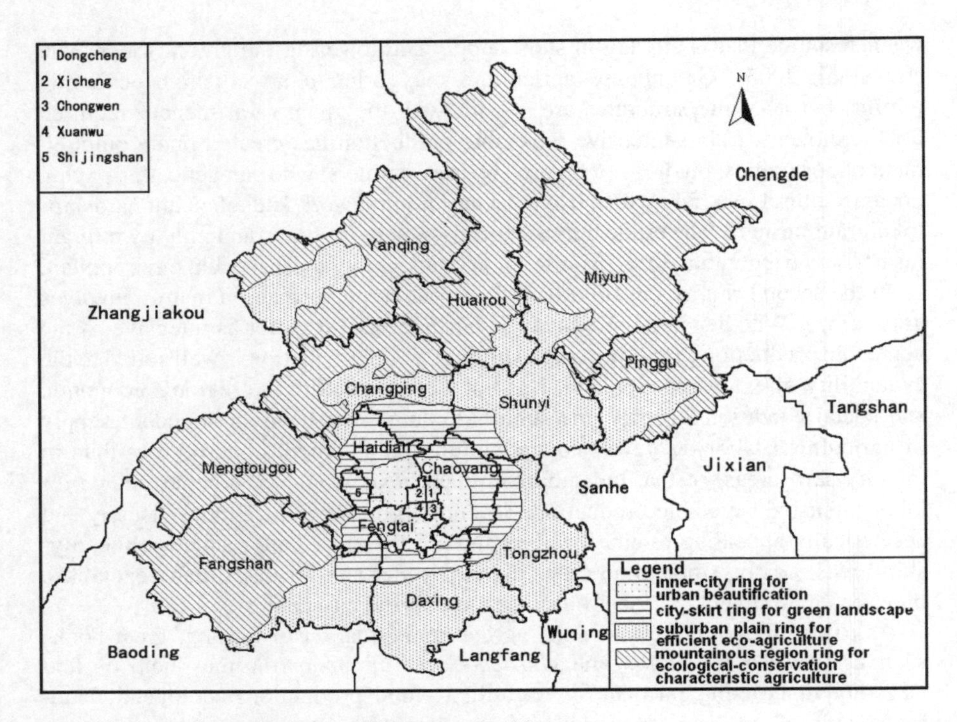

Fig. 19.4 Layout of the agricultural industry of Beijing

19.3.1 Inner-City Ring for Urban Beautification

19.3.1.1 Scope

The city center and the four city-skirt areas of Chaoyang, Haidian, Fengtai, and Shijingshan districts, basically within the fifth ring road (Fig. 19.4).

19.3.1.2 Landscape Characteristic

The total area of the region is about 670 km^2, of which 24 km^2 are cropland; 11 km^2, orchards; 41 km^2, woodland; 21 km^2, other agricultural land; and 585 km^2, built-up land. This area is basically urban, and a great majority of it is cement-enclosed and asphalt-surfaced, with some land for urban gardens and the first noncontiguous greenbelt (Li et al. 2005). There is very little cropland, and it is scattered discretely among built-up land, the land for urban gardens, and the greenbelt. At present, the cropland grows mainly vegetables, mostly in greenhouses.

19.3.1.3 Agricultural Function

The main function should not be agricultural production but city beautification.

19.3.1.4 Layout of Agriculture

This area is at the urban edge of Beijing and changes quickly in land use, so it is likely to be encroached upon by further construction. Croplands in this area should be put into the basic cropland protection zone, protected by the Land Administration Law of the People's Republic of China, to keep them from being developed and prevent urban sprawl. The area should stop producing edible farm products such as vegetables because the serious water, soil, and air pollution make its products unsafe for consumption. Even if pollution were not a concern, some greenhouses should be moved out gradually because greenhouse agriculture is not in harmony with the surrounding high-rise buildings. The most appropriate land use is to grow flowers, grass, and other suitable crops such as cotton for urban beautification.

19.3.2 City-Skirt Ring for Green Space

19.3.2.1 Scope

This is also in the plain area, mainly, the city-skirt area outside of the fifth ring road and within the sixth ring road (Fig. 19.4).

19.3.2.2 Landscape Characteristic

The total area of the region is around 1,606 km^2, of which 327 km^2 are cropland; 84 km^2, orchards; 235 km^2, woodland; 152 km^2, other agricultural land; 758 km^2, built-up land; and 54 km^2, water bodies and wetlands. This area belongs to the region where urban and rural areas interlace, and farmlands and built-up lands are of comparable proportions. Built-up areas, a second greenbelt, and cropland interlock with each other. Cropland is highly fragmented by roads and building blocks and currently produces mainly vegetables and grains.

19.3.2.3 Agricultural Function

Croplands in this ring are a vital component of urban green space; they lower the temperature, dissolve and absorb harmful materials, and have a direct effect on urban environmental quality. The farmlands' principal function should not be agricultural production but the provision of green space and ecological services.

19.3.2.4 Agricultural Layout

This area also belongs to the urban extension area. All the important satellite cities of Beijing—Shunyi, Tongzhou, Yizhuang, Mentougou, and Huangcun—are distributed in this region. The land use/coverage changes quickly and is likely to be encroached upon by new development. To prevent further sprawl, part of the croplands should be put into the basic cropland protection zone to create a boundary of green space between the central city and the various satellite cities. Considering the need to create a pleasant environment, coordinate urban and rural landscapes, guarantee the safety of farm products, and sustain the development of the agricultural industry, it is necessary to phase out greenhouses from this area to reduce shining white plastic pollution and make more green space. Crop production should be open-field agriculture to increase soil permeability and create a rural landscape and scenery around urban clusters or residential developments. Integrating agricultural space with the recreational space of city dwellers not only achieves sustainable agricultural development but also provides citizens with a green recreational and living space. Among grain crops, winter wheat prevents wind erosion from exposed cropland in winter and spring and provides a green landscape, so a certain area of it should be retained, especially in the northwest, which is the windward area.

19.3.3 Suburban Plain Ring for Efficient Eco-Agriculture

19.3.3.1 Scope

The suburban plain outside of the sixth ring road and Yanqing basin areas (Fig. 19.4).

19.3.3.2 Landscape Characteristic

The total area of the region is about 4,272 km^2, of which 1,648 km^2 are cropland; 510 km^2, orchards; 307 km^2, woodland; 356 km^2, other agricultural lands; 1,119 km^2, built-up land; and 127 km^2, water bodies and wetlands. Land in this area is flat, and farmland takes the leading position, making it the matrix of the landscape. Land that has been constructed upon consists of patches embedded in the farmland matrix.

19.3.3.3 Agricultural Function

The farmland of Beijing is chiefly distributed in this ring. The cropland is large, concentrated, linked in parcels, and suitable for developing large-scale farms. It is an important production base of farm products and is agriculture's main function in this region. This region should develop efficient crops, strengthen the rural economy, and promote increases in farmers' income while creating a positive urban and rural environment.

19.3.3.4 Agricultural Layout

This area is relatively far from the city; the land use/coverage changes slowly; water, soil, and air quality are good; environmental protection restrictions are few; and agricultural labor is abundant. Therefore, given the area's function, the layout of the agricultural industry here should be such that it would maximize agricultural production, meet the food demand of citizens, and increase farmers' incomes. Vegetables and other cash crops, even greenhouses, are encouraged here. Certainly, large grain fields are also suitable, especially with better seed breeding, which will give farmers more opportunities for work and extra income.

19.3.4 Mountainous Region for Conservation and Specialized Agriculture

19.3.4.1 Scope

The mountain areas north and west of Beijing City (Figs. 19.1 and 19.4), surrounding Beijing City as a half ring.

19.3.4.2 Landscape Characteristic

The region's total area is around 9,859 km^2, of which 363 km^2 are cropland; 635 km^2, orchards; 6,320 km^2, woodland; 500 km^2, built-up land; 389 km^2, water bodies and wetlands; and 1,618 km^2, barren land that has not been developed in any way, such as mountaintops and rocks. Woodland accounts for 64% of the total area and is thus the matrix of the landscape. Built-up land, cropland, and orchards are interfering patches scattered in the woodland matrix and are connected by infrequent roads.

19.3.4.3 Agricultural Function

This area lies at the windward and upriver belts and is the ecological shield of Beijing. Reservoirs, which provide one third of Beijing's water requirements, are located in this region. Therefore, this region is crucial for supplying Beijing's water and providing ecological conservation. Due to its high altitude, diverse landforms, rocks, soils, and unique ecological conditions exist in this region, which can support the cultivation of specialized crops.

19.3.4.4 Agricultural Layout

The most advantageous locations for producing unique agricultural products, such as fruits, Chinese medicinal plants, and small grains, should be selected. Meanwhile, scenic spots—e.g., Jinhai, Miyun, and Huairou reservoirs and Chaobai, Yongding, and Juma rivers—could be combined with the development of environmentally friendly agriculture to create unique ecological agriculture, recreational agriculture, agricultural sightseeing, ethnic tourism, and ecotourism opportunities which will supplement the region's economy.

19.4 Conclusion and Discussion

19.4.1 Conclusion

Beijing is a large city with a population of 18 million, and its growth is constrained by limited water and land resources. The city's expansion is eating into croplands and degrading the environment. Therefore, agricultural land should not be limited to its production function alone; its landscape, recreational, and ecological functions should also be exploited. The "four-ring" agricultural layout fully demonstrates agriculture's multifunctional characteristic and serves the public's benefit. The spatial planning of agriculture for the municipality of Beijing breaks away from the traditional pattern of agricultural rings identified by Thünen in 1826. It balances the specific land resources, environmental situation, and the city's spatial structure with socioeconomic development and is favorable to recreating Beijing as "a livable city."

19.4.2 Discussion

The idea of agricultural ring patterns in this chapter is meant to resolve the conflicts between food production and environmental improvement in a special background;

i.e., land resources are scarce in Beijing. If there were plenty of land, Beijing could plant a large area to trees and glades. Thus, this concept has some differences from what Howard (1945) proposed in 1945. In Europe, there are relatively rich land resources that allow people to have a large area of forest in the city to solve urban environmental problems.

The proposed "four-ring" agriculture layout is only a model, which is shaped by the unique criteria of Beijing. In fact, there are many other satellite cities in the Beijing district, such as Shunyi, Tongzhou, Yizhuang, and Huangcun, whose agricultural production may also be suitable for such a ringlike structure. Due to the research scale limit, though, these cities were included in the suburban plain ring.

However, to realize such a layout, it is not adequate to rely only on market strength. It is necessary to use the macroeconomic control power of the government and employ various kinds of transfer payments and other means to accelerate the spatial redistribution of the existing agriculture. These means include (1) granting ecological and resource subsidies for crop production that has significant positive ecological and water-saving effects, (2) stepping up the construction of agricultural infrastructure and land consolidation for creating scenic rural areas, and (3) strengthening the extension service of all kinds of agricultural technologies that are suitable for environment protection. The Beijing government has the capacity to reach the goal of the proposed agricultural ring pattern, something that might not be achieved in free market economy countries.

It should be noted that the Beijing People's Government already issued a policy document in which our spatial plan for agriculture was adopted and subsidy policies for environmentally friendly agriculture were announced in 2008. For example, the Beijing People's Government was to give a subsidy of 1,350 CNY/ha of winter wheat grown to beautify the city and protect it from wind erosion.

Beijing's demand for agricultural products is vigorous, but due to high production costs, the shortage of water and land resources, and environmental protection policies, some traditional agricultural industries have to be phased out. Therefore, from the perspective of overall regional planning, an agricultural ring outside the administrative jurisdiction of the municipality of Beijing that cooperates with Beijing and its surrounding provinces should also be created. The hinterland of this wide rural region has abundant farmland, cheap and abundant labor, and a healthy environment. It will become a necessary supplement to the agricultural development space of Beijing. Combined with the capital and technical resources of Beijing, this area could produce large amounts of agricultural products and make up for Beijing's agricultural production deficit.

Acknowledgement Funding for this study was provided by the Natural Science Foundation of China (project number 70673104).and Beijing Agricultural Bureau. Prof. Jijun Liu, Prof. Zhonghong Wu, Dr. Daquan Huang, and Dr. Heng Tang took part in this research. We sincerely thank Dr. Xiying Hao and Dr. Stephen Streng for their technical editing in this manuscript.

References

Beijing Municipal Bureau of Land and Resources (2005) Report 2004 on changes of land use of Beijing. Beijing Municipal Bureau of Land and Resources Press, Beijing (in Chinese)

Beijing People's Government (2003) The urban master plan of Beijing. Beijing People's Government Press, Beijing (in Chinese)

Beijing People's Government (2008) The integrated land use planning of Beijing. Beijing People's Government Press, Beijing (in Chinese)

Bradley GA (1995) Urban forestry landscapes: integrating multidisciplinary perspectives. University of Washington Press, Seattle

Chen TB, Song B, Zheng YM, Huang ZC, Lei M, Liao XY (2006) A survey of lead concentrations in vegetables and soils in Beijing and their health risks. Sci Agric Sin 39(8):1589–1597 (in Chinese)

Cheng T (1999) China's urban agricultural development—taking Beijing as an example. Resour Sci 21(4):39–42 (in Chinese)

Costanza R, d'Arge R, de Groot R, Farber S, Grasso M, Hannon B, Laskin R, Sutton P, van den Belt M (1997) The value of the world's ecosystem services and natural capital. Nature 387 (15):253–260

De Groot WT, van den Born RJG (2003) Visions of nature and landscape type preferences: an exploration in The Netherlands. Landscape Urban Plan 63:127–138

Guo SM, Cheng X, Shi YJ (2004) Restriction of resource and environment and developing countermeasures of ecological friendly type Urban Agriculture in Beijing. Res Agric Mod 25(3):194–197 (in Chinese)

Hall RC (1999) The multi-functional character of agriculture on land: bio-energy. In: Contributed paper in conference on the multifunctional character of agriculture and land, Maastricht

He GP, Zhou D, Yang ZS, Zhao HY, Li CJ (2005) Current status and evaluation of groundwater resources exploitation in the plain area of Beijing. J Hydro-Eng Geol 2:45–48 (in Chinese)

Howard E (1945) The Revenue of Garden City, and how it is obtained The Agricultural Estate. In: Osborn FJ (ed) Garden cities of tomorrow. Faber & Faber Ltd., London

Li F, Wang RS, Paulussen J, Liu XS (2005) Comprehensive concept planning of urban greening based on ecological principles: a case study in Beijing, China. Landscape Urban Plan 72:325–336

Liu HB, Li ZH, Zhang YG, Zhang WL, Lin B (2006) Nitrate contamination of groundwater and its affecting factors in rural areas of Beijing plain. Acta Pedol Sin 43(3):405–413 (in Chinese)

Lütz M, Bastian O (2002) Implementation of landscape planning and nature conservation in the agricultural landscape—a case study from Saxony. Agric Ecosyst Environ 92:159–170

Lynn NA, Brown RD (2003) Effects of recreational use impacts on hiking experiences in natural areas. Landscape Urban Plan 64:77–87

Sun XZ, Zhou HL, Xie GD (2007) Ecological functions and their values in Chinese cropland ecosystem. China Popul Resour Environ 17(4):50–60

Tyrväinen L (2001) Economic valuation of urban forest benefits in Finland. J Environ Manage 62:75–92

Von Thünen JH (1997) Der isolierte Staat in Beziehung auf Nationalökonomie und Landwirtschaft. Akademie-Verlag, Berlin. Chinese edition: Von Thünen JH (1826) (trans: Hengkang W). The Commercial Press, Beijing

Yang ZX, Zheng DW, Wen H (2005) Studies on service value evaluation on agricultural ecosystem in Beijing Region. J Nat Resour 4:564–571 (in Chinese)

Zhang LS (2001) Development of urban agriculture in metropolis and city planning. Urban Plann Forum 3:68–70 (in Chinese)

Zhang FR, An PL, Kong XB (2005) Use and protection of cultivated land and prime farmland in comprehensive land use planning of Beijing. China Land Sci 19(1):10–16 (in Chinese)

Chapter 20
Simplified Ecological Planning Method for Sustainable Landscape Management by *Humantope Index*: Patterns of Land-Use Continuity, Historical Land Use and Landownership

Misato Uehara

Abstract The adaptability of the McHarg planning method seems high. However, separated multiple ecological factors must be evaluated for every land-use proposal. The *LPPV* method needs only two factors (layers): potential natural vegetation and land form. Nevertheless, the artificial environment such as a city area cannot be evaluated using the vegetation division pattern. An actual adaptation of these two methods is limited for reasons above. Also, these divisions by physical and biological index require agreements at many landowners and local administrations. Hence, the purpose of this chapter is to propose a shortened planning process that incorporates the advantages of McHarg and *LPPV* methods using a smaller number of indices in a society's environmental division for environmental management guidelines. We set up the hypothesis that there is the possibility that past land-use patterns could be treated as a result of land evaluation of people. We focused on three important human-induced land patterns to develop the *Humantope index*: (1) continuity and stability of land use, (2) traditional land-use-adapted local environment and (3) landownership as spatial management result. Hence, our *Humantope index* is a highly informative synthesis and corresponding to natural environment and artificial environment. Moreover, *Humantope index* would be a simplified procedures compared to those of complex one like the rank evaluation of McHarg and potential vegetation estimates of *LPPV*. It is convenient to link concerns of landowners and local administrations because the *Humantope* division could be easily acceptable for landowners and local administration unit.

Keywords Landscape strategy • City planning • Natural resources • Environmental assessment • Japan earthquake and tsunami • Comprehensive design science

---

M. Uehara (✉)
Department of Forest Science, Faculty of Agriculture, Shinshu University,
8304 Minamiminowa, Nagano 399-4598, Japan
e-mail: ueharam@shinshu-u.ac.jp

M. Kawakami et al. (eds.), *Spatial Planning and Sustainable Development: Approaches for Achieving Sustainable Urban Form in Asian Cities*, Strategies for Sustainability,
DOI 10.1007/978-94-007-5922-0_20,

20.1 Introduction

McHarg (1969) proposed a method of ecological planning in *Design with Nature*. A distinguishing feature of this method is that it summarises multiple environmental evaluations into one set of visual information. This method of unifying environmental evaluations is defined and known as the overlay method.

This method adaptability is high between new development and nature conservation. The McHarg method uses not only the *Biotope* and *Physiotope indices* but also an artificial environmental evaluation. However, several aspects have been criticised. (1) Multiple index layers require a complicated ex ante analysis. (2) This method requires a considerable amount of time and effort to develop environmental management guidelines assuming land-use continuity. Achievement of continued land use is more important than new land development (diversion) for aging society with fewer children.

In Japan, on the other hand, Kameyama (1973), Takeuchi (1976) and Ide Hisato proposed a method of environmental classification that uses two indices: potential natural vegetation and geography. The potential natural vegetation is defined as the estimated climax vegetation that appears when human influences cease. This method is defined as a landscape planning method based on potential natural vegetation (*LPPV*). Its key feature is that it enables environmental classification using a small number of indices. However, the *LPPV* method cannot assess an artificial environment (town or city). Because the main environmental assessment index is the potential natural vegetation, this index requires a detailed investigation of local flora.

Steiner (1983) compared the following three methods—the US Soil Conservation Service capability classification, the McHarg method and a Dutch approach—and identified the problems of existing methods in analysing resources. Two problems were identified: (1) relevant data must be simplified in terms of saving of the time and money for planning, and (2) it is disadvantageous to invest the effort required to create multilayered data. However, the research above did not propose a simplified method.

Hence, the purpose of this chapter is to propose a shortened planning process that incorporates the advantages of the McHarg and *LPPV* methods using a small number of indices of a society's environmental division for sustainable landscape management. These indices are defined as the *Humantope index* in this chapter. We set up the hypothesis that there is the possibility that past land-use patterns could be treated as a result of land evaluation of people. We focused on three important human-induced land patterns to develop the Humantope index: (1) continuity and stability of land use, (2) traditional land-use-adapted local environment and (3) landownership as spatial management result. *Humantope index* is a pattern unit of environmental division by people (society).

20.2 Methodology and Research Area

Miller et al. (1998) tried to reduce the number of criteria for development of suitability analysis using weighted values for greenway planning. Comins et al. (1993), H.M. Skanes (1997) and Uehara and Shigematsu (2001) overlaid land-use maps of different year. Willemen et al. (2010) appended ecological planning method to assess the correlation of each result from evaluation of suitability in land use (e.g. recreation, tourism and arable areas). Wang et al. (2008) tried to append ecological planning method to compare land evaluation result between different periods.

These results of review imply a tendency in current research to pay attention to consolidating the environmental evaluation index and a historical land-use pattern. However, these researches did not propose a simplified method for different scale and characteristic in some regions.

Hence, the purpose of this chapter is to propose a shortened planning process using a smaller number of indices of environmental division. Human-induced land division pattern indices are defined as the *Humantope index* in this chapter. This index should have a highly informative synthesis and be applicable to both natural and artificial environment. Moreover, this index should be a simplified procedure compared to those of complex one like the rank evaluation of McHarg or potential vegetation estimates of *LPPV*. Therefore, it is necessary to test of *Humantope index* in some environmental scales and each region. McHarg said that the first step is to define the region considered in ecological planning.

Thus, the simplified environmental classification methods in two scales based on *Humantope index* are presented in the main discourse. The first is a case of applied *Humantope index* to local council scale. Roughly categories of Land-use patterns in different years as land-use continuity were used for the index of environmental evaluation to extrapolate to a larger area. For example, forest category includes conifer plantation and broadleaf forest. The second is a case of *Humantope index* in district (old village unit) where direct relationships between the residents and the landowner still exist. It is the smallest unit of regions. This scale is suitable for conservation planning in a physical region because the inhabitants of district own the resources comprising the regional environment directly. The evaluation by two *Humantope indices* was tried to reflect the quality and problem of a more detailed environment: (a) traditional land-use division and (b) ownership division of land.

As a result, Shingu town was selected as the study area for the following reasons. This town area is 1,888 ha including an island approximately 122 ha called Aino-shima. The Shingu town has all typical environmental elements such as mountain, plain, sea and isolated islands. Its natural environment consists of a seaside pine grove, a rare primeval camphor tree forest and farmlands and secondary forests. On the other hand, since this area adjoins a big city, its population growth rate in recent years has been the highest in northern Kyushu. Therefore, balancing urbanisation and nature conservation is important in this city planning.

20.3 Patterns of Land-Use Continuity as *Humantope Index*

20.3.1 Outline of Analysis: Environmental Classification Method Based on a Time Series of Land-Use Changes at a Local Council Scale

First, aerial photographs of Shingu town taken in 1947, 1974 and 2000 were analysed. A land-use map for 3 different years was created using TNTmips (MicroImages, Inc.) to determine changes in land use of local council scale except for the island of Aino-shima, which is analysed in detail in Sect. 20.4. An environmental classification method that focuses on patterns of land-use continuity by overlaying three patterns occurring at 25-year intervals was proposed.

20.3.2 Results

20.3.2.1 Changes in Land Use at a Local Council Scale and the Proposed Environmental Classification Method Based on Time Series of Land Use Serving as the *Humantope Index*

Changing Land-Use Areas at a Local Council Scale

Figure 20.1 shows the land use in Shingu for each 25-year interval on the basis of analysis of the aerial photographs. It was necessary to define roughly land-use categories for the time series analysis. Forest category includes broadleaf forest, bamboo groves and artificial conifer plantations, and farmland includes fields, paddy fields and orchards. Farmland and forests occupied most of the town's area in 1947, and each residential area was small. However, the residential and artificial environments were later expanded on either side of the railroad. Hence, many of the western towns began to be used for housing, commerce and industry between 1974 and 2000, and the farmlands and forest became fragmented.

From the land-use maps, the size of each land-use pattern was calculated at 25-year intervals (Table 20.1). The area covered by farmlands decreased sharply from 772.4 to 333.8 ha between 1947 and 2000. Conversely, the area used for artificial environments (residential, commercial and industrial areas) in 2000 was about 20 times that in 1947. The amount of forested area decreased between 1947 and 1974 and increased between 1974 and 2000. Hence, unlike the area of farmlands, the total forested area did not decrease.

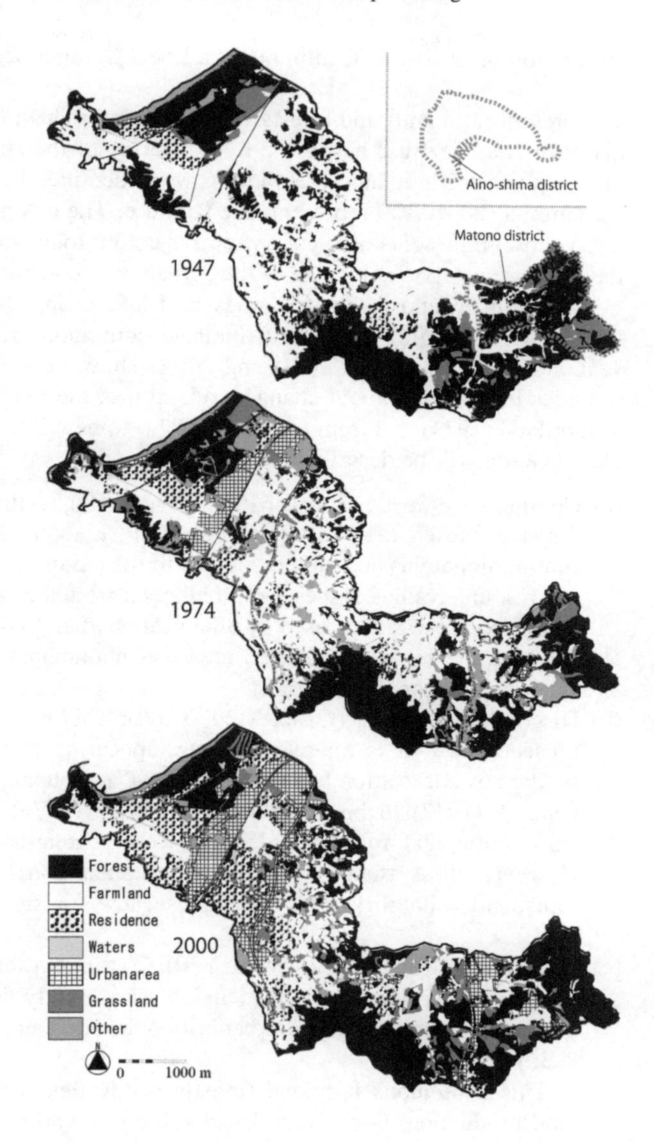

Fig. 20.1 Land use in land council scale of Shingu at 25-year intervals

Table 20.1 Areal extent of each type of land use

	1947	1974	2000
Forest	718.7 ha	578.4 ha	669.6 ha
Farmland	772.4	666.0	333.8
Residence	85.1	109.1	238.1
Urban area	13.2	116.6	277.9
Grassland	120.6	85.6	135.9
Water	11.8	15.0	20.1
Other	41.6	192.3	87.6

Evaluation of Land-Use Continuity at a Local Council Scale

The present farmlands and forests were classified as having continuous land use or discontinuous land use according to an overlay of the land-use patterns for three different years. Three land-use patterns were identified by colour tone conversion; the chroma (20–100%) represents the land use. The colour system (cyan, magenta and yellow) represents the year. After this colour tone conversion, information on the land-use categories in each of the 3 years was collated (Fig. 20.2).

In the superimposed image, areas of 100% cyan, 100% magenta and 100% yellow (C100, M100 and Y100) indicate continuous forest cover for these 50 years. Conversely, C100, M20 and Y100 show new forests; thus, the colour provides information about changes in land use: the transition from forest (1947) to farmland (1974) and from farmland (1974) to new forest (2000). Results of land classification will be described below:

(a) Continuous forest (e.g. cyan100, magenta100, yellow100): The continuous forest consisted of a seashore shelterbelt, a southern primeval forest in the central mountains and an old coppice in the eastern region. The environmental and resource values of these mature forests are stable and high. This continuous mature forest was decreased by urbanisation after 1974. However, it seems that the overall size of the forested area was maintained by an increase in newly appearing discontinuous forest.

(b) Discontinuous forest (C100, M20, Y100): This new forest was 47.4% of the present forest. It is relatively young, appearing after 1947. These areas are typified by a transition from meadow (1947) to plantation forest (1974) or from coppice (1947) to bush (2000) via orchard (1974). In Japan, the domestic agricultural and forestry industries were standardised after World War II. However, these standard discontinuous plantation forest areas did not have forestland suitability. Therefore, these new forests are brownfields (unused land).

(c) Continuous farmland (C20, M20, Y20): Continuous farmland covered 62.4% of the present farmland area. It included many paddy fields from 1947 to 2000. These paddy fields escaped urbanisation despite being near the city because of their poor drainage.

    This continuous farmland (mostly paddy fields) has higher social values: food production, flood control and soil conservation. These are necessary for the city. Thus, such continuous farmland is environmentally important.

(d) Discontinuous farmland (C100, M20, Y100): The area occupied by the new farmland was 37.6% of the current farmland area. This new farmland area has increased over the last 50 years, and the majority of the places that are orchards were forests in 1947. Although this area began rapidly expanding in about the 1970s, it is now being abandoned and will become discontinuous forest (bamboo forest and bush). Compared to the continuous farmland, its functionality and stability with respect to environmental preservation are low.

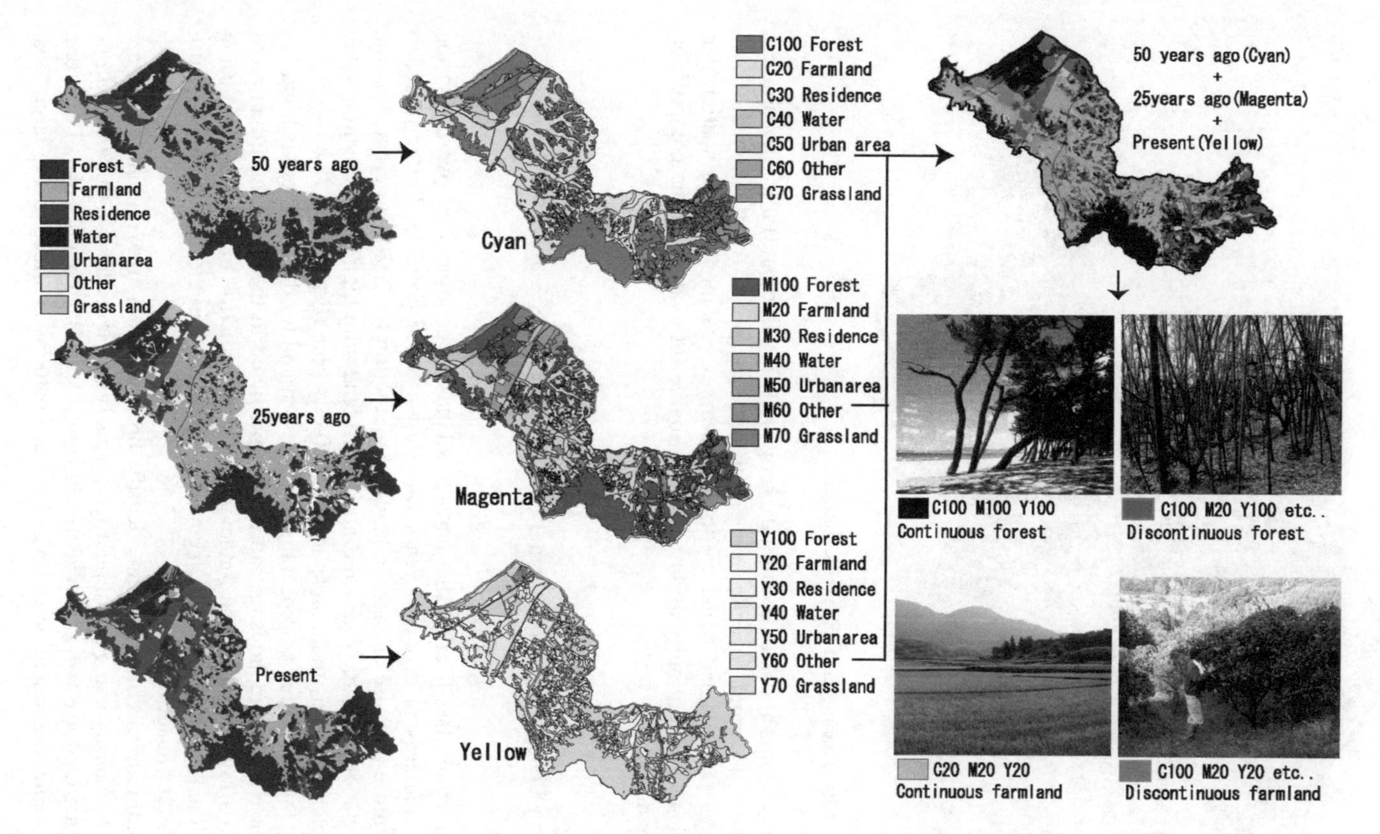

Fig. 20.2 Environment classification using overlay of land use patterns of three different years as land use continuity

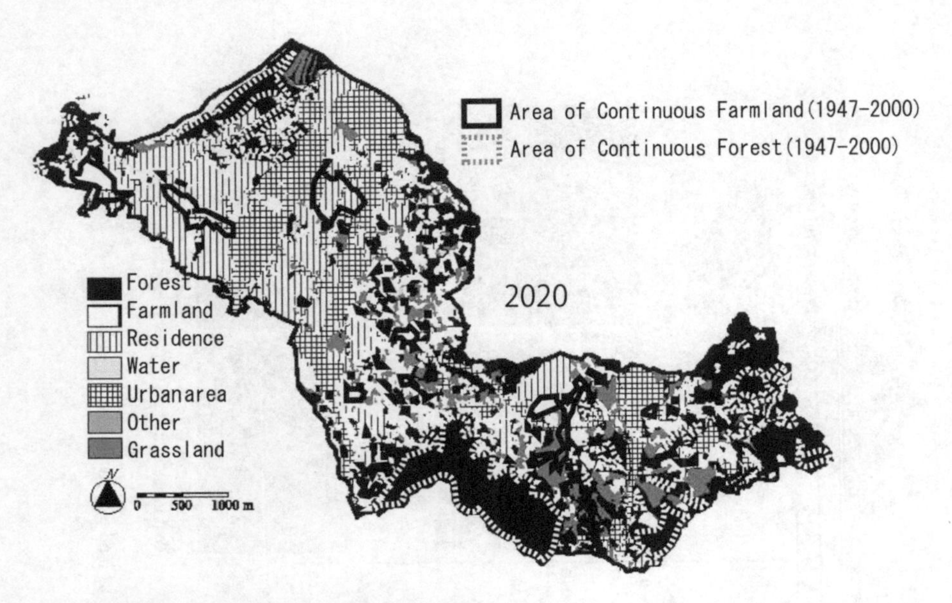

Fig. 20.3 City planning land-use prediction for 2020

Moreover, it is at risk of becoming a brownfield (unused land) if it is abandoned, after which it cannot be called either farmland or forest. As mentioned above, on a large local council scale, we think that an environmental classification that applies land-use continuity as an index is suitable for evaluating the composition of the land.

Predicting Change in Environmental Assessment Results Using Land-Use Continuity

Environmental classification that considers land-use continuity by overlaying the land-use patterns in different years was shown to be possible. Furthermore, to predict future land-use changes, a land-use prediction for 2020 was created on the basis of the city master plan. Continuous farmland with high environmental value clearly changes to artificial areas in the 2020 city plan (Fig. 20.3).

Since the study area adjoins a big city, the demand for development is high, but an increase in surplus housing can also be predicted in the adjoining urban areas owing to a low birth rate and longevity. Such a prediction cannot be made using conventional city planning, which considers the change in the size of each land-use area (e.g. farmland, forest).

In addition, this method can identify brownfields in the region, which are difficult to classify using aerial photograph analysis. Nevertheless, this *Humantope index* (past land-use patterns) was made from old aerial photographs. Therefore, environmental classification of serial land-use patterns is effective at a large local

Photo 20.1 Discontinuous residence receives liquefaction damage by 2011 Tohoku earthquake in Urayasu City

council scale. Moreover, this method is also applicable to artificial environments (residential sections).

Photo 20.1 is another case in point of assessment with availability of land-use continuity. This discontinuous residence based on past mudflat had big damage by land liquefaction by 2011/3/11 earthquake.

20.4 Patterns of Traditional Land Use as *Humantope Index*

Recall that method of this chapter targets regions at two scales: a local council scale and a district scale. This and the next section examine the district scale.

A method of environmental classification by traditional land-use patterns, the *Humantope index*, is proposed at the district scale. We focused the possibility that past land-use patterns could be treated as a result of people's land evaluation.

20.4.1 Outline of Analysis

20.4.1.1 Study Area

Aino-shima in Shingu was selected as the study area for the following reasons: First, its known history dates back to the 1700s, and continuous exploitation of the island's natural resources continued until the 1960s. Second, since the land area is limited, it is suitable for studying the relationship between population and land use.

20.4.1.2 Analysis of Changes in Land Use

To determine the changes in land use in this area, the historical land use from the 1700s to the 1900s was first analysed with pictorial or land register maps. Maps of land use in 1966, 1974 and 1998 were created from analysis of serial aerial photographs taken after World War II.

20.4.1.3 Proposed Environmental Classification Method Using Traditional Land-Use Patterns

On the basis of the above-mentioned analysis, we propose a method for classification in environmental assessment by overlapping traditional land-use patterns (*a Humantope index*) and present land use. The traditional pattern is defined as the division of land use adapted to the regional environment over a long time period.

20.4.2 Results

20.4.2.1 Analysis of Changes in Land Use

Changes from the 1700s to World War II

Since aerial photographs taken before World War II do not exist, other sources of information on land use or local landscape were collected. The result was a pictorial map (Figs. 20.4 and 20.5). By digitisation of a land register map created in 1888, an image showing land use around 100 years ago could be created (Fig. 20.6).

Figure 20.4 shows landscape of Ainoshima in Edo period at Shingu town record. This residence enlarged figure shows pine trees shelter belt on the edge of island.

Figure 20.5 shows landscape in 1895. This old graphic shows that the main vegetation was pine trees. The pine groves were located on sloping ground that surrounded inland farmland. Figures 20.4 and 20.5 show a similar pattern of forest on the edge of the island. This island edge with forest is influenced by salt air, and these forests surround farms and paddy fields. The old town records indicated that the forests provided not only a shelterbelt against the wind but also a forest preserve for nurse tree production during the Meiji era. The above analysis indicates that there is no change landscape from the 1700s to the beginning of the 1900s.

Changes from World War II to the Present

Figure 20.7 shows a map of recent land use on the basis of analysis of aerial photographs taken in 1966, 1974 and 1998. Each area was computed by calculating the number of pixels of each land-use pattern in a colour histogram. The analysis revealed the following results:

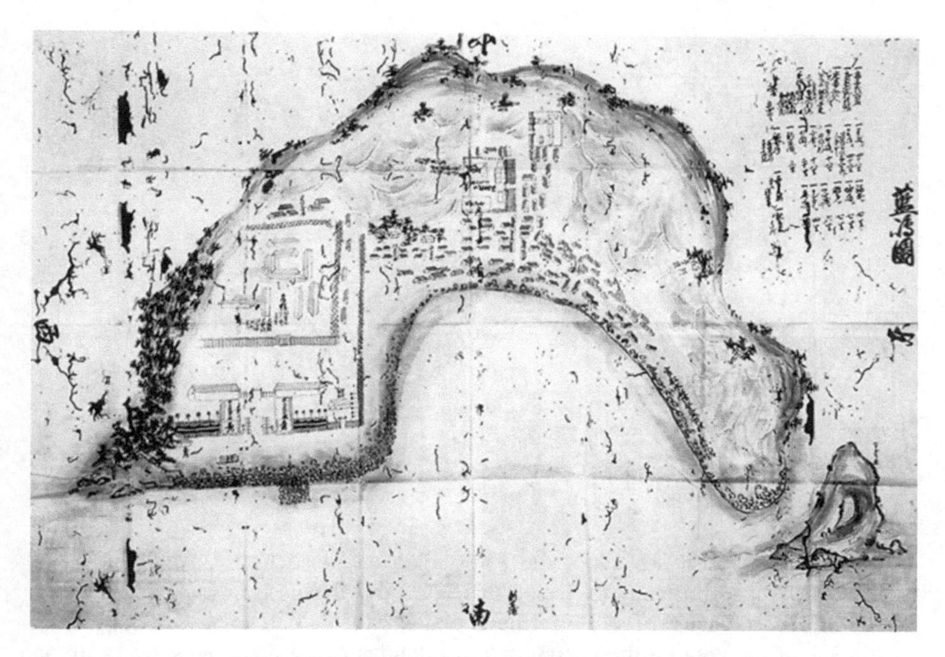

Fig. 20.4 Landscape in the 1748s of Ainoshima at Shingu town record (Iwakuni choko kann)

Fig. 20.5 Landscape in 1895 of Ainoshima at Wakamiya Shrine's old graphic

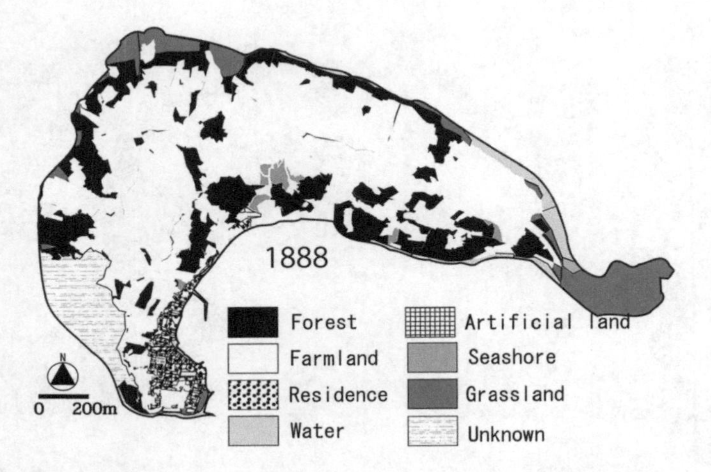

Fig. 20.6 Land use in 1888

(a) Changes in forest: In 1966, the forest also enclosed houses and fields that blocked salty air from the sea (25.5% of this island). It was similar with traditional land use. By 1998, the broadleaf forest had increased rapidly to 63.3% of the area on the abandoned farmland. In contrast, pine grove with no management was mostly lost, owing to pine weevil damage.
(b) Changes in farmland: Fields, paddy fields and orchards were about 41% of the area in 1966. However, the farmland area decreased sharply to about 9% in 1974. Farmland then decreased to its current relative size of 1.2% of the area.

20.4.2.2 Proposed Environmental Classification Using Traditional and Current Land-Use Patterns

The traditional land-use patterns continued over a 100 years have adaptation for local salty wind. Thus, the environment was classified by an overlay with two layers (Fig. 20.8): the traditional land-use pattern (*the Humantope index*) and the present land-use pattern.

The chroma of each colour (20–100%) indicates land use. A chroma of 100% shows mature forest (large pine grove in 1966 and large broadleaf forest in 1998). The cyan indicates traditional land use, and the magenta indicates present year's land use. After this colour tone conversion, the information for two land-use categories was overlaid. For every environmental classification, the characteristics of that environment were analysed.

Changes from Farmland to Forest (Cyan60, Magenta90)

Since 1966, 35.7 ha of farmland changed to forest, and another 11.8 ha became bush. The total abandoned farmland is 39% of the entire area, and a concrete utilisation plan is needed. Traditional farmland is fundamentally suitable for

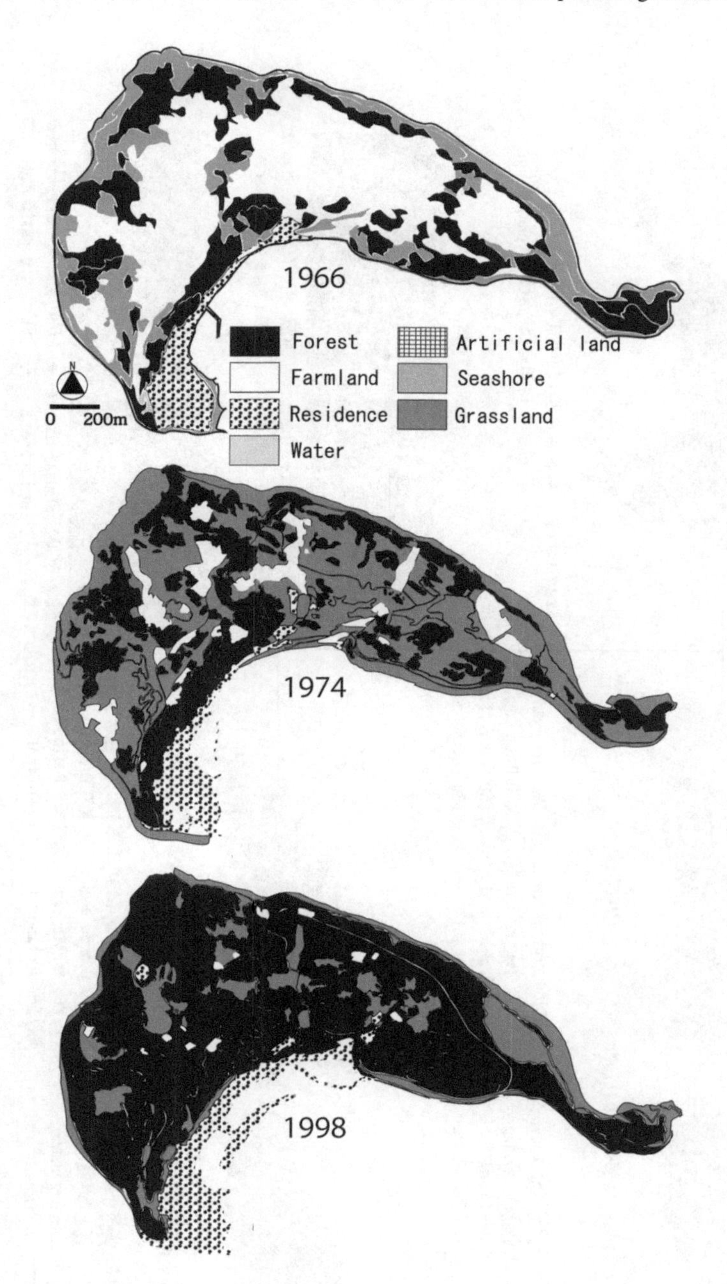

Fig. 20.7 Land use after 1966

re-farming since it is unaffected by salty air. However, places that have already shifted to mature forest because of farmland abandonment must be excluded. This forest expansion by farmland abandonment can be interpreted as a decline in the ability to produce food in the area. On the other hand, it could be regarded as an

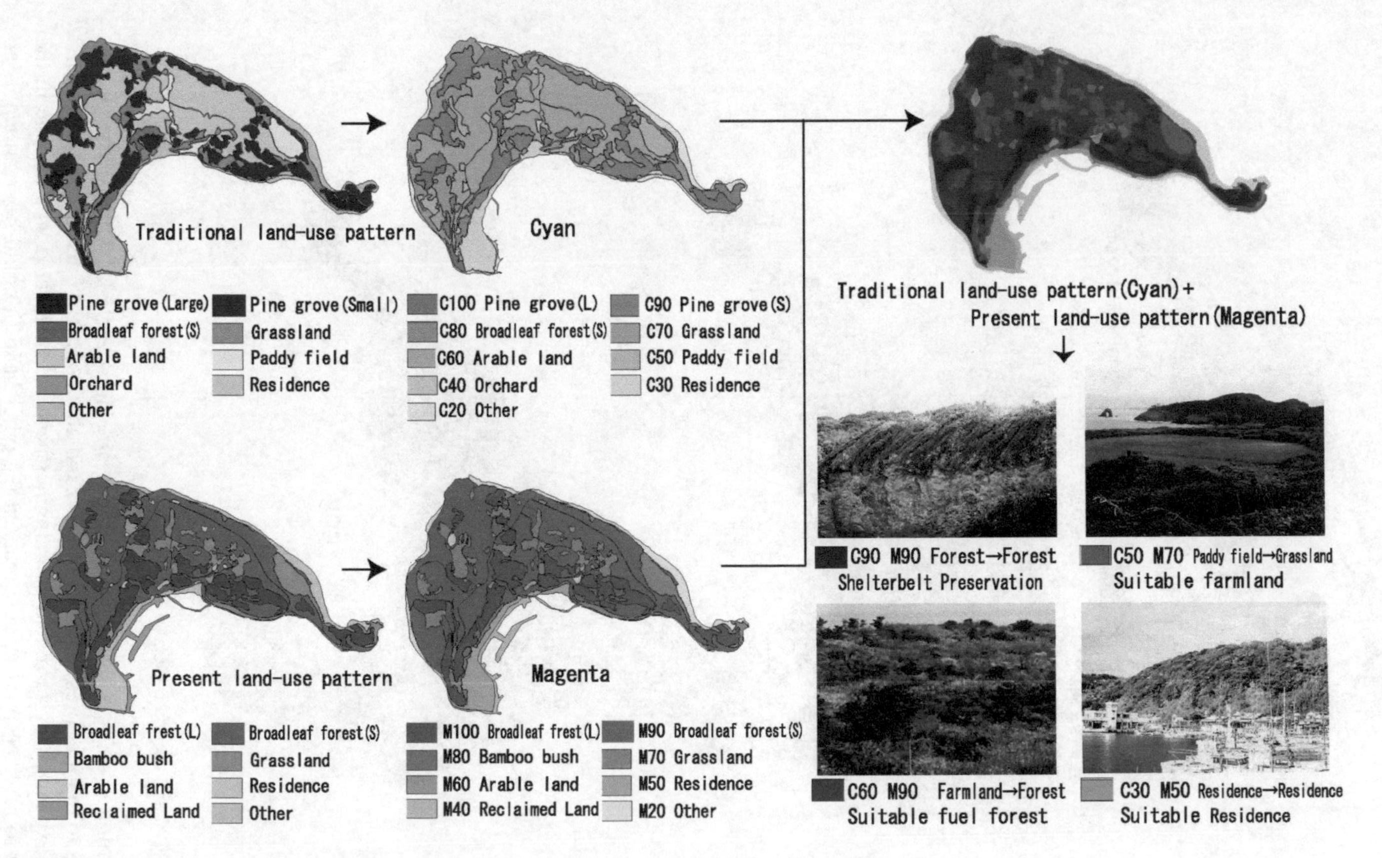

Fig. 20.8 Environment classification using traditional land use patterns and present land use patterns

increase in the amount of biomass resources. The management of corpses for fuel is expected in those forests. Logically, woody biomass does not emit carbon dioxide.

Changes from Pine Grove to Broadleaf Forest (C90, M90)

Between 1966 and 1998, 26.22 ha of area changed from pine grove to broadleaf forest. This area is changing to broadleaf forest in places where the influence of salty wind is significant, particularly along the seashore. Therefore, we think it is necessary to preserve this forest for protecting the residence against storm surges and for nursing the reproductive farmland and forest biomass within this area.

Residential Area (C30, M50)

This residential location has not changed since the Edo era. This location connects to the southern bay, which provides summer breeze and can protect against cold winds in winter. In residential areas located in a region having such a suitable climate, a comfortable life can be expected even without using significant energy for air conditioning and heating.

20.4.2.3 Validity of Environmental Classification by Traditional and Present Land-Use Patterns

In this area, traditional land-use patterns that have persisted for hundreds of years were relinquished in recent years. However, we think that the reuse of farmland and forest based on traditional land-use patterns has been effective in this district, which has escaped the simplification in land use that has occurred nationally. This simplification entails the intensive changes in land use described in Sect. 20.3.

In this example, a forest that traditionally provides protection against storm surges and is still forested is made into a preservation zone (buffer forest). Moreover, the abandoned farmland that changed to grassland and bush can be used for a food production zone. Abandoned farmland that has changed to mature forest can be expected to produce energy from fuel. It is rational to limit the future residential development in the area facing the bay.

This method can achieve both sustainable land use and preservation of nature using only two indices: traditional and present land-use patterns. Figure 20.9 shows a draft plan for land use based on the overlay produced by this method. The planned forest preserve zone can be located in steeply sloping areas unsuitable for other uses, without a land elevation index. Furthermore, farmland or fuel forest has been located on internal land, which is both easy to use and highly productive. The result of this land classification and regional planning provides evidence that traditional land-use patterns are useful as a *Humantope index*. In particular, this planning method does not require information layers about land elevation, such as the *Physiotope index*, which is necessary for both the McHarg and *LPPV* methods.

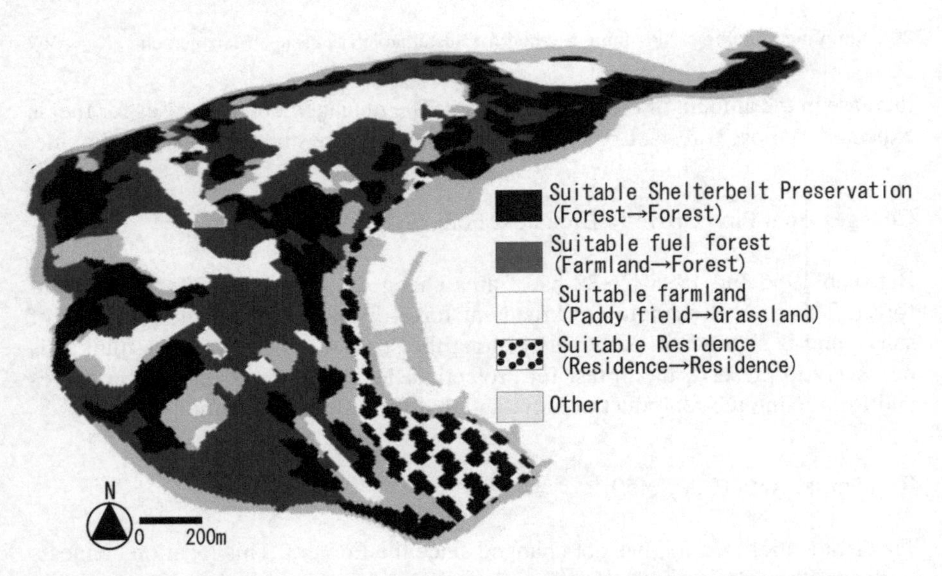

Fig. 20.9 Draft land-use plan created by superimposing traditional and present land-use

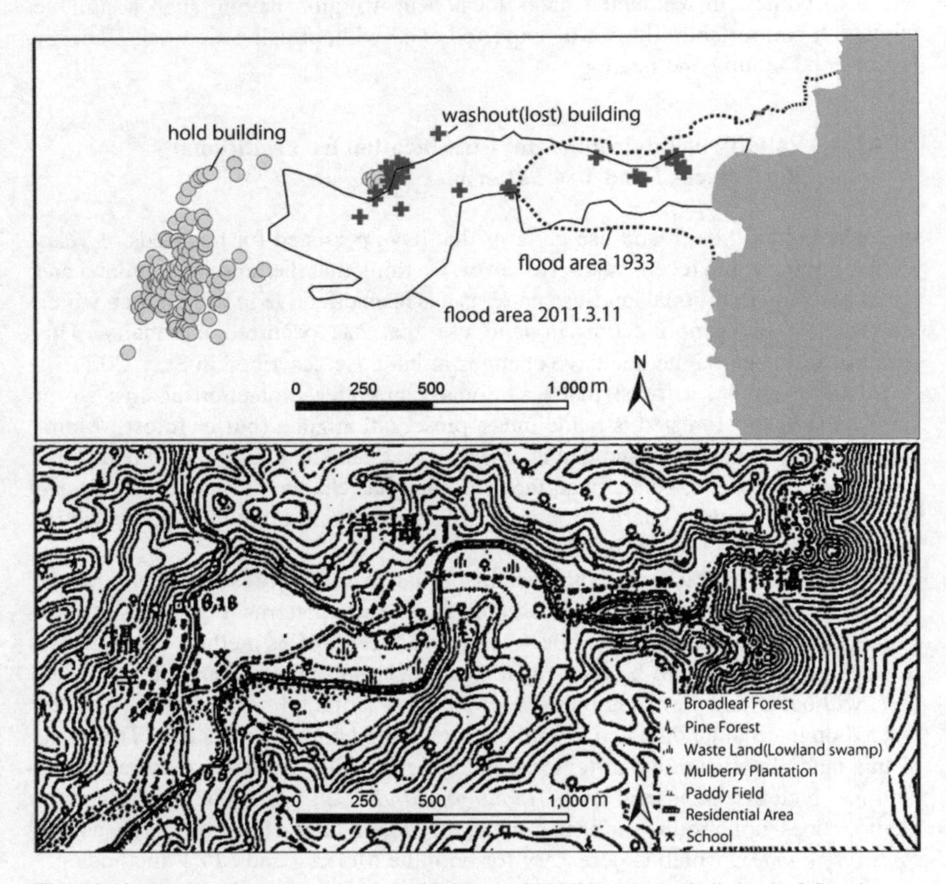

Fig. 20.10 Relationship between Iwate Prefecture 2011/3/11 tsunami disaster of Settai town building and traditional land use 100 years ago (past tsunami disaster area)

In addition, Fig. 20.10 is another region case in point of assessment with traditional land-use pattern's availability. Settai traditional residence had not big damage by Tohoku earthquake and tsunami in 2011.

20.5 Patterns of Landownership as *Humantope* Index

This section focuses on a suburban farm's district scale (the Matono area) because primary industries were transformed from traditional land use after World War II in Japan; monocultural farmland use and expansion of forest plantations occurred. Hence, this section proposes a new ecological planning method at a district scale where traditional land-use patterns cannot be used.

20.5.1 Outline of Analysis

20.5.1.1 Study Area

The Matono area is a suburban farm village with 86 people (24 households). In this section, environmental classification at the district scale is proposed; this village area was covered with orchards and conifer plantations after World War II.

20.5.1.2 Changes in Land Use and Proposed Environmental Classification Method Using Landownership Classification

To understand the process of changes in farmland and forest, which are the main natural environment at the district scale, a traditional land-use map was constructed from the land register map of 1888. Maps of land use after World War II were constructed from aerial photographs. On the basis of the above results, we suggest a new overlay method for the suburban district, which developed from traditional land use to uniform land use. This overlay method uses landownership patterns (*a Humantope index*) and the present land-use pattern.

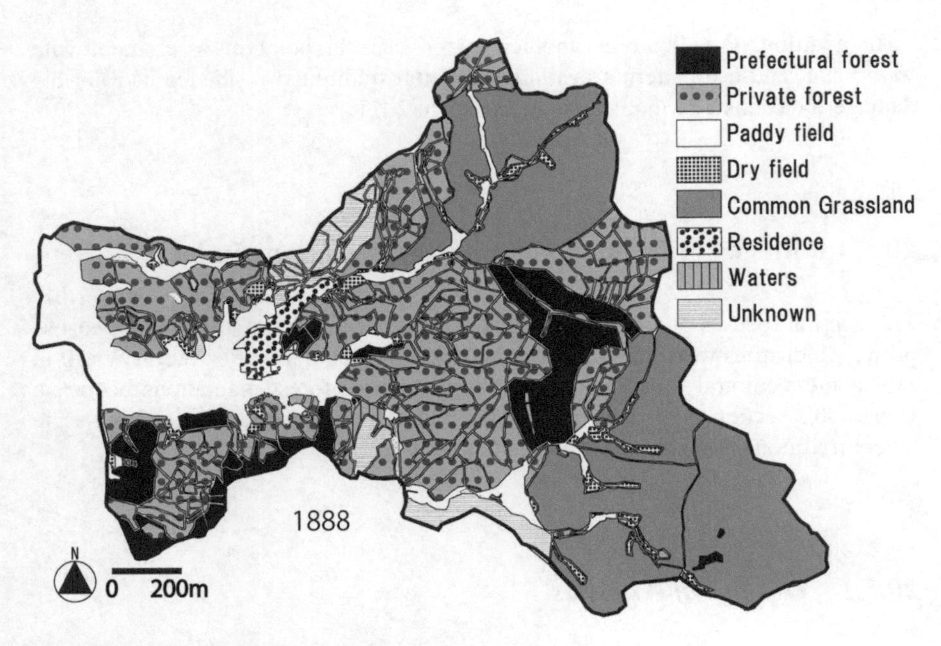

Fig. 20.11 Land use in Matono in 1888

20.5.2 Results

20.5.2.1 Analysis of Land-Use Changes

Changes from 1888 to World War II

Figure 20.11 shows the land-use pattern of the Matono area in 1888 based on the old land register map. Clearly, the paddy field and residential area are arranged so that they are surrounded by the north-south hill range. Furthermore, common grassland was distributed over the slope of the hill, where soil conditions were poor and conveyance was difficult. This land-use pattern is similar to the traditional land-use forms in farm villages in Japan in the past.

Changes After World War II

Figure 20.12 shows the land-use pattern in this area based on analysis of a series of aerial photographs taken after World War II in 1947. This pattern had many features in common with traditional land-use patterns in 1888. However, forestation by needleleaf forest had already begun in some common grassland. Furthermore, the traditional land-use pattern was transformed into single-pattern land use of farm and forest.

Later, when crops produced under this single pattern remained unsold, these monocultural farmlands and forests were abandoned (Fig. 20.12).

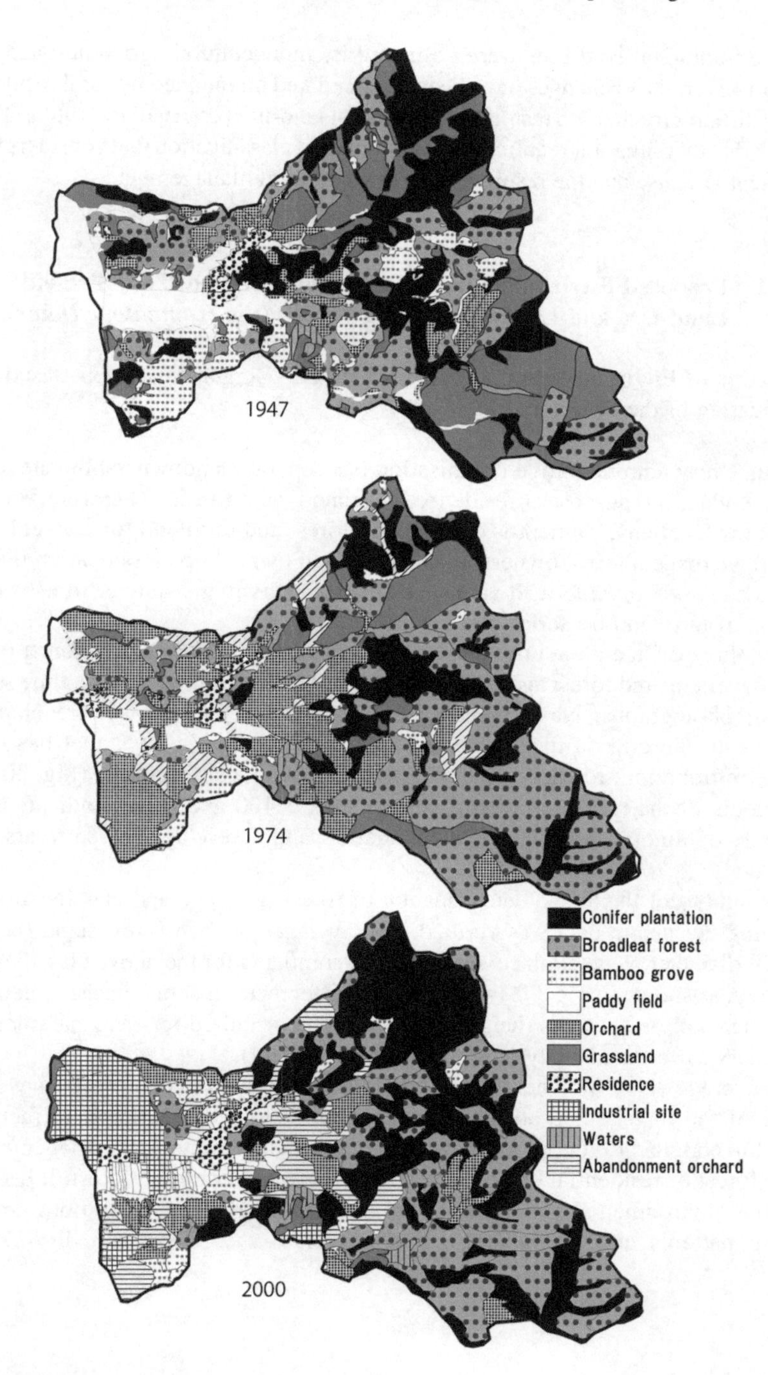

Fig. 20.12 Land use in Matono after 1947

These traditional land uses were destroyed by monocultural agriculture and forestry. However, these changes have been enforced and maintained by rural residents, and it is difficult for them to restore the traditional land-use pattern of the 1950s. Thus, we think that this situation required an environmental classification that considers both the present land use and the results of this monocultural management.

20.5.2.2 Proposed Environmental Classification Method Using Present Land-Use and Landownership Patterns (the *Humantope Index*)

Verification of Environmental Assessment Method Using a Possession-Based Classification of the Forest and Its Validity

In Japan, every administrative organisation has separate landownership categories for tax collection purposes: residence, farmland and forest. Therefore, in this section, the local environment is divided into forest and farmland for convenience.

First, we made a forest ownership and farmland ownership classification map to serve as a *Humantope index*. It is more difficult to survey the state of forest stands than that of farmland by aerial photographs.

Now, three different pattern maps were prepared: present land use (forest type), forest ownership and forest age. The forest age was also determined by a time series of aerial photographs. Next, the colours of these three patterns were changed according to the criteria in Fig. 20.13. Then, we obtained a layer that has three types of information: forest type (land use), ownership and forest age (Fig. 20.13). The colours of the resulting overlay map (C100, M100 and Y100) indicate three attributes: common land, needleleaf plantation and forest age 61–85 years old, respectively.

The validity of the above determination of forest type, age and classification by overlaying ownership data was verified. The averages of each forest stand (height, tree shift diameter, vegetation cover rate) were compared for the above classification separately, as shown in Fig. 20.14. Significant differences in actual measurements of the averages of forest stands (height, diameter) were acquired for each classification using one-way factorial analysis of variance (Fig. 20.14).

The average value (tree height, shift diameter) is related to ownership as well as the age of the forest stand and the management history (Fig. 20.14). In fact, the ownership classification is related to management status: individual management of private forest or residential organisation management of common forest. It became clear that environmental classification by the overlay method using both present land-use patterns and *Humantope patterns* (forest age, ownership division) is useful.

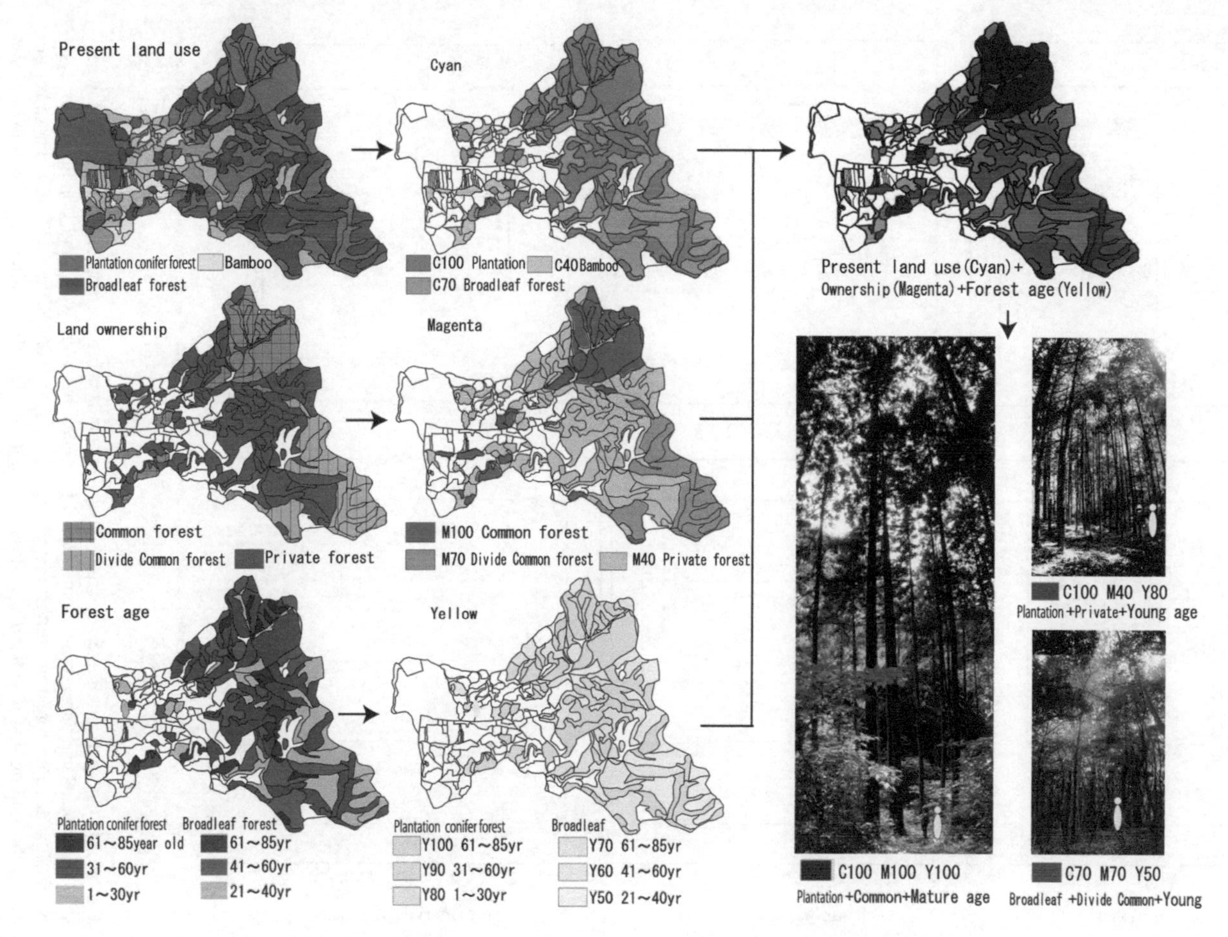

Fig. 20.13 Environment classification of forest using present land use patterns and land ownership patterns

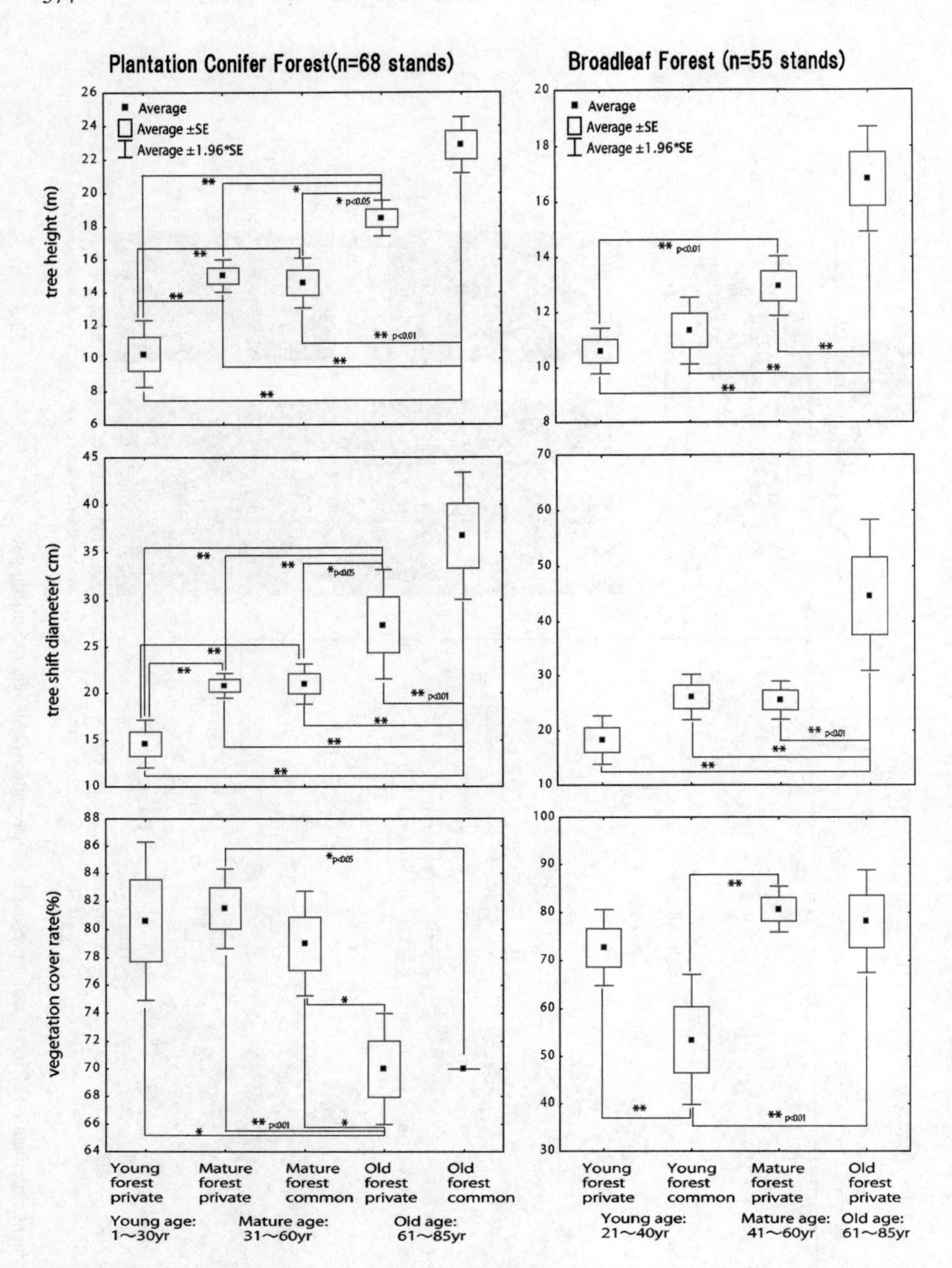

Fig. 20.14 Comparison of averages for each forest stands using present land-use and landownership patterns. There are some significant differences between woodland conditions of each ownership in same forest age division (**$p < 0.01$, *$p < 0.05$: Fisher's PLSD). These results reflect the finding that woodland conditions without detailed observations. (1) Common conifer forest's tree shaft values are larger than private conifer forest's tree shaft values in old-age forest

Classification of Farmland and Verification of Classification Method's Validity

To predict the expansion of abandoned and cultivated farmland in the Matono area, a map showing possession of farmland was created (Fig. 20.15). Furthermore, the landownership pattern was superimposed on the land-use pattern (farmland) to analyse the relationship between type of ownership and management state (Fig. 20.15). Both resident and nonresident owners have already abandoned about 40% of their respective properties. However, farmland abandoned by nonresident owners tends to have already changed to forest, owing to long-term abandonment.

Thus, superimposing the landowner division as a *Humantope index* and the present land-use pattern yielded a classification based on present land use and the results of monocultural agriculture and forestry management. This land classification that reflects the management history cannot be obtained with only an aerial photograph.

20.5.2.3 Validity of Environmental Classification Method Using Both Landownership and Present Land-Use Patterns

Figure 20.16 shows the proposed plan of conservation and use for the environment classified above on the basis of landownership and present land-use patterns. Note that this method can classify local environments that are difficult to classify by analysing aerial photographs.

20.6 Conclusion

Our proposed environmental classification method superimposes two index layers: the *Humantope index* and present land-use patterns. This *Humantope pattern* could be chosen from among three types of index which depends on the size of region and

---

Fig. 20.14 (continued) division. These results reflect the finding that old community's forest management is sufficient than private forest in old-age forest division. (2) There were no significant differences between tree shaft diameter of mature-age broadleaf forests (private) and tree shaft diameter of young-age broadleaf forests (common). These results reflect the finding that logged forest's tree shafts are narrower than seed tree shaft in different-age forest division. Young-age broadleaf forests (common) are seed tree thru plantation pine tree death. (3) Crown's cover rates of young broadleaf common forests are lower than young broadleaf private forests. These results reflect the finding that young-age broadleaf forests (common) have tree crown's gap thru plantation pine tree death

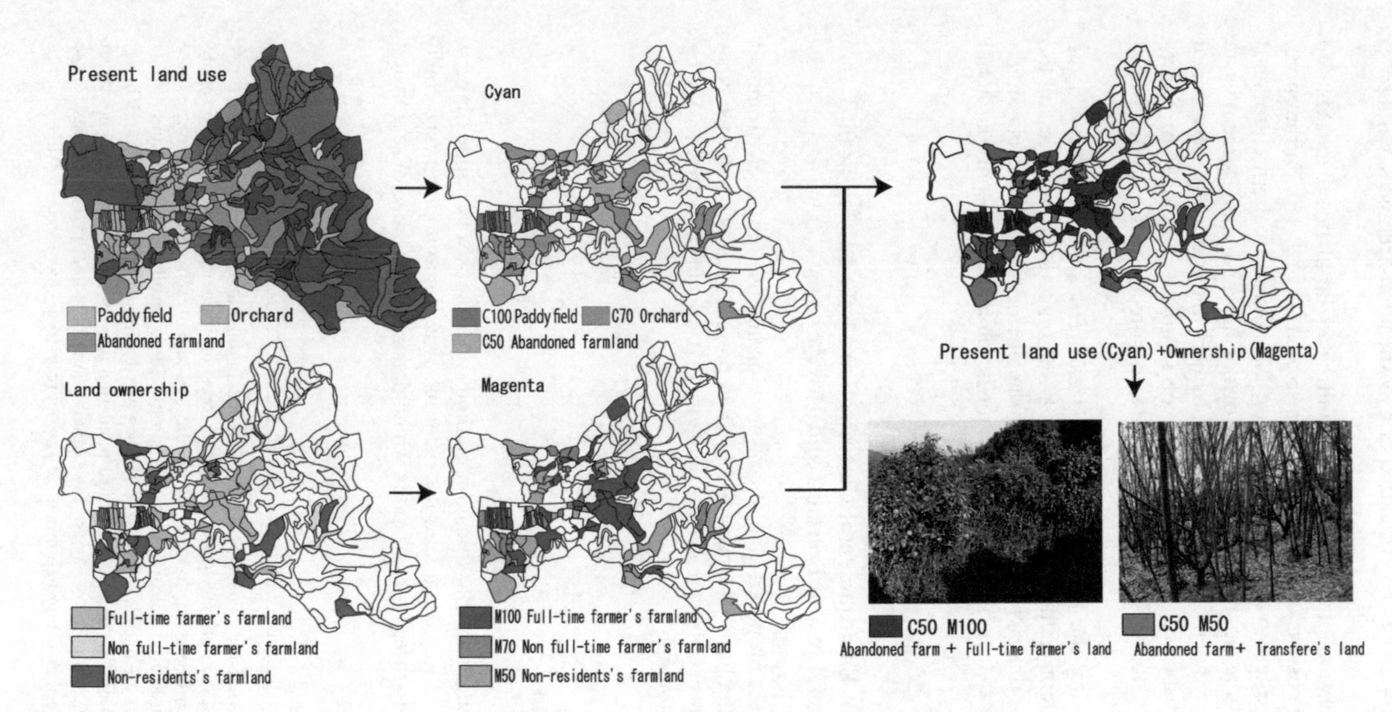

Fig. 20.15 Environment classification of farmland using present land use patterns and land ownership patterns

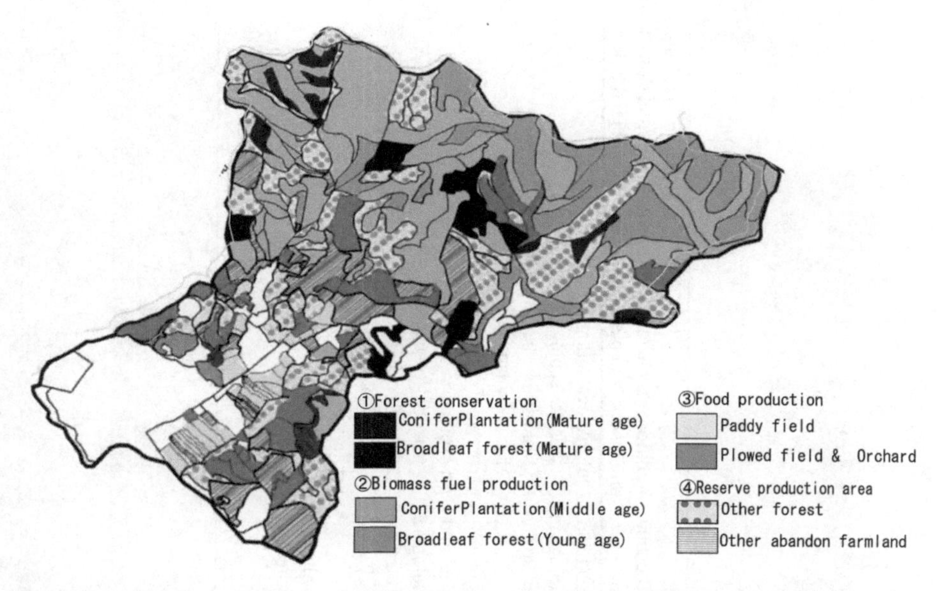

Fig. 20.16 Draft land-use plan superimposing landownership and present land-use patterns

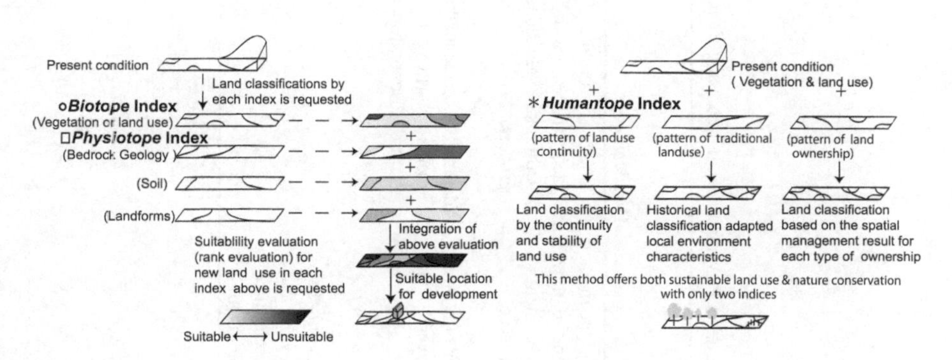

Fig. 20.17 Comparison between McHarg and methods using *Humantope* divisions

environmental characteristics: patterns of past serial land use, traditional land use or landownership.

The proposed method using both a *Humantope pattern* and present land-use patterns is easier to apply than the McHarg and *LPPV* methods (Fig. 20.17, Table 20.2). Indeed, in the McHarg method, the interpretation of multiple evaluations of separate ecological factors must be checked for every land-use

Table 20.2 Comparison between McHarg and LPPV methods and the using *Humantope*

	Advantage	Disadvantage	Land classification criteria	Additional evaluation process for land planning	Coincident of planning unit with actual district and council land unit
McHarg method (Ecological planning)	Adaptability of the method seems high (i.e. new town development to Nature Reserve)	Method needs a lot of factors (layers) with additional evaluations process	Physical and biological pattern	Multiple evaluations of separate ecological factors must be checked for every land use proposal	×
LPPV (Landscape planning based on potential method natural vegetation)	Method needs only two factors (layers)	Adaptability of the method seems low (i.g. only Nature Reserve, forest, farm)	Physical and biological pattern	An estimation of the potential vegetation through a detailed investigation of the local flora is required	×
This chapter's method	Method needs only two factors (layers) and adaptability of the method seems high	–	Humantope pattern	–	○

proposal. The *LPPV* method requires an estimation of the potential vegetation through a detailed investigation of the local flora. Also, these two methods determine which areas are intrinsically suitable for new land-use development. These data are of little use until they are interpreted and evaluated against land-use proposals. In addition, these divisions by physical and biological indices require adjustments between many owners and local administration.

On the other hand, our method treats land use based on the environmental evaluation of human inhabitants over a long period. The proposed method offers both sustainable land use and conservation of nature with only two indices. This method does not require geographical data as the *Physiotope index*, which is necessary in both the McHarg and *LPPV* methods. In addition, the *Humantope pattern* agrees with those of landowners and local administration. Hence, it is convenient to adjust with landowners and local administration. With the *Humantope index pattern*, environmental divisions that cannot be proposed with only an aerial photo can be realised, yet the main *Humantope index* pattern can be made from old aerial photographs.

Acknowledgement This work was supported by a Grant-in-Aid for Young Scientists (B): KAKENHI(22760455) and the "Shinshu University Mountainous Areas Research Project," supported by the Ministry of Education, Culture, Sports, Science and Technology of Japan in fiscal 2011.

References

Comins JS, Sendra JB, Sanz FM (1993) Crisis and permanence of the traditional Mediterranean landscape in the central region of Spain. Landscape Urban Plan 23(3–4):155–166

Kameyama Akira (1973) Phytosociological studies for rural land use planning. Jpn Soc Appl Phytosociol 2:1–52

McHarg IL (1969) Design with nature. The Natural History Press, Garden City

Miller W, Collins MG, Steiner FR, Cook E (1998) An approach for greenway suitability analysis. Landscape Urban Plan 42(2):91–105

Shingu Town (1997) Shingu-choshi (shingu town record). Shingu Town, Fukuoka. Japan

Skanes HM (1997) Towards an integrated ecological-geographical landscape perspective. A review of principal concepts and methods. Norwegian J Geogr 51(3):145–171

Steiner F (1983) Resource suitability: methods for analyses. Environ Manage 7(5):401–420

Takeuchi Kazuhiko (1976) Methode der landschaftsokologischen Beewertung fur die Planung der Landschaft. Jpn Soc Appl Phytosociol 5:1–61

Uehara Misato, Shigematsu Toshinori (2001) Fundamental study on the conservation and utilization of local resources with organic recycle system in AINOSIMA Isle. J Jpn Inst Landscape Arch 64(5):831–834

Wang SY, Liu JS, Yang CJ (2008) Eco-environmental vulnerability evaluation in the Yellow River Basin China. Pedosphere 18(2):171–182

Willemen L, Hein L, Martinus EF, van Mensvoort MEF, Verburg PH (2010) Space for people, plants, and livestock? Quantifying interactions among multiple landscape functions in a Dutch rural region. Ecol Indic 10(1):62–73

Chapter 21
Land Cover Analysis with High-Resolution Multispectral Satellite Imagery and Its Application for the CO_2 Flux Estimation

Jung-Rack Kim, Shih-Yuan Lin, Eun-Mi Chang, In-Hee Lee, and He-Won Yun

Abstract The global warming caused by CO_2 greenhouse effect becomes a world wide critical issue nowadays. In order to make suitable and sustainable policies for fully managing and monitoring the CO_2 emission and reduction, the identification of CO_2 volume and spatial distribution is highly crucial for the local environmental authorities. Therefore, an indirect measurement method using land cover classes and their CO_2 emission estimation per unit area driven from the sample survey or statistical analysis was proposed in this chapter. To prove the feasibility, Boryeong and Yoengi cities in South Korea were selected as the two test sites, and CO_2 emission maps covering these areas were produced using the proposed method. One of the key factors for a successful mapping of CO_2 emission was to produce an accurate land cover map over the test site. To achieve this, the Rapideye satellite imagery with characteristics of high-resolution and multispectral was chosen as the main data source for land cover analysis. Together with the SPOT satellite images and GIS land-use data, the accurate land cover map was created. The emission ratio

J.-R. Kim • H.-W. Yun
Department of Geoinformatics, University of Seoul, Jeonnong 2-dong, Dongdaemun-gu, Seoul 130-743, South Korea
e-mail: kjrr001@gmail.com; hwyun@uos.ac.kr

S.-Y. Lin (✉)
Department of Land Economics, National Chengchi University, No. 64, Sec.2, ZhiNan Rd., Wenshan District, Taipei, Taiwan
e-mail: syl@nccu.edu.tw

E.-M. Chang
Ziin Consulting Inc., #325 Yongbieachunga Naesu 75, Jongno, Seoul 110-070, South Korea
e-mail: emchang21@gmail.com

I.-H. Lee
Department of Environment and Ecology Research, Chungnam Development Institute, 73-26 Yonesuwon-gil, Gong-Ju, Chungcheonnamdo, South Korea
e-mail: inhee@cdi.re.kr

M. Kawakami et al. (eds.), *Spatial Planning and Sustainable Development: Approaches for Achieving Sustainable Urban Form in Asian Cities*, Strategies for Sustainability, DOI 10.1007/978-94-007-5922-0_21,

of each land cover type was subsequently applied on the land cover map to produce CO_2 emission map and also estimate the total CO_2 flux.

Keywords Land cover • CO_2 emission ratio • Multispectral image • Image classification

21.1 Introduction

The global warming caused by CO_2 greenhouse effect becomes a worldwide critical issue nowadays. In order to control the greenhouse gas (GHG) emissions, the CO_2 reduction obligation for developed countries was outlined in the Kyoto Protocol adopted in 1997. Since then, the allocation of emission reduction was discussed continuously in the subsequent Conference of the Parties arranged by the United Nations Framework Convention on Climate Change. It is believed that the pledges for countries to submit emission reduction targets will be introduced worldwide in the near future. In order to make suitable and sustainable policies for comprehensively managing and monitoring the reduction, the identification of CO_2 volume and spatial distribution over the administrative area is highly crucial for the local environmental authorities.

Janssens et al. (2003) reported a study of estimation of regional CO_2 flux, in which two approaches were summarized. One was the land-based analysis through the forestry, agricultural, and soil interpretation. The other method was carried out by atmospheric analysis. In addition to these two methods, a number of research projects dealing with CO_2 emission have been progressed, employing various approaches, such as space-borne sensors for direct gas tracing (Buchwitz et al. 2005), the indirect biomass analysis (Hese et al. 2005), and the direct measurement over some specific test areas. Among them, the direct measurement using sensor-equipped tower, which was called eddy covariance analysis (Baldocchi 2003), was widely accepted for CO_2 flux estimation. Through the analysis, estimation of CO_2 fluxes was estimated using the measurements derived from sample sites. However, the cost for the system construction and data collection was considerable. Moreover, for the cases of relatively large spatial target areas, the reliability of the estimation of total CO_2 fluxes over such areas still remained uncertain by only applying direct measurements obtained from selected sample sites. To address the issues, a method integrating maps recording comprehensive land cover over the target area was applied. Through the analysis of spatially referenced GIS land cover map, the types and changes of land cover were determined and used to estimate GHG emission, especially CO_2 flux (Schimel et al. 2000; Quaife et al. 2008). Together with high-quality and high spatial resolution of land-use information, reliable CO_2 estimation using the indirect method could be achievable.

In order to promote a feasible method executable by local environmental authorities, also considering the large administrative area they dealt with, an indirect method of CO_2 emission measurement was proposed on two cities in

Korea in this chapter. To this end, multiple spectral resolution remote sensed image data was used. An advanced classification scheme using the image data was developed for producing land cover map. Together with the CO_2 emission ratio, the land cover map was then applied as the base data for estimating the CO_2 flux over the test area. At last, the authorities could perform CO_2 flux monitoring and related policy making using the sequential CO_2 emission maps. To describe the overall processing procedure, review of the indirect measurement methods of CO_2 emission, study areas and data applied, data processing, results, and validation are introduced in the following sections.

21.2 Indirect CO_2 Emission Estimation

Normally, the methods applying land cover information for estimating GHG emission could be classified into two categories. In the first group, patches of different land cover types, especially vegetation type, were treated as an interactive function such as plant functional types (PFTs). Following the principle, Bonan et al. (2002) represented the landscape as patches of PFTs and demonstrated its linkage with climate and ecosystem model. While in the other category, average GHG emission ratios were given based on various land cover types. The overall estimation of GHG emission over the target site was then computed accordingly. The latter case was more straightforward and could be easily delivered to authorities in charge of environmental issues and related policy making. Most importantly, the approaches employing land cover map and average emission ratio in each land cover types would be greatly useful to estimate the total CO_2 fluxes over the suburban and urban areas where the analysis with vegetation and atmospheric interaction function was very complicated. A successful example adopting this approach was demonstrated in Soegaard and Moller-Jensen (2003), in which AVHRR and Landsat TM data over Copenhagen City were used to compare with mobile and station measurements. In the study, the texture-based classification for classifying the metropolitan areas by their contexts and homogeneities was firstly performed. Combining with Normalized Difference Vegetation Index (NDVI) showing the vegetation density, CO_2 contributions were calculated in each land cover types.

Two issues have to be considered when using land cover type to estimate CO_2 flux. First of all, for such approaches, normally the satellite image could be the most reliable data sources due to the high temporal availability and the relatively low cost. However, various satellite image sources were of different spatial resolutions, including low-resolution images such as MODIS; medium-resolution images, for example, Landsat TM and ETM; and high-resolution data like ALOS ANVIR and SPOT images. Moreover, the recent modern commercial satellites, such as IKONOS, GeoEye, and Worldview, have extremely high spatial resolution from one meter to submeter. Together with capability of collecting near-IR band and even more spectral bands outside of visual wavelength range, such products become valuable tools for CO_2 flux calculation. The evaluation of the effects employing satellite images with different resolution for land cover extraction was performed by Turner et al. (2000). Through the comparison of net primary production and net ecosystem production, it was proven that error ratios highly depended on the spatial resolution of satellite imagery.

The second issue was the determination of CO_2 emission ratio in each land cover type. As each land cover type interacted with external environment such as wind, temperature, daytime light, and surrounding land covers, the definition of a fixed CO_2 flux value in a specific land cover type was not straightforward. Especially the anthropogenic CO_2 sources involved with many social factors so that the specific CO_2 emission ratio was not readily available. In addition, the appropriate evaluation method for mobile CO_2 fluxes of traffic vehicles and remote sources such as power plants was also not fully achievable. Owing to these limitations, the CO_2 flux estimation employing land cover extracted from satellite imagery rarely exists.

Even so, considering the efficiency of the method, also to explore the potential of evaluation of CO_2 fluxes by satellites imagery and average emission ratio in each land cover types, indirect method was performed in this study under the conditions and assumptions described below. Firstly, we employed unprecedented high-resolution satellite imagery, in which the individual object possessing homogenous land cover type could be clearly discriminated. Consequently, the problems regarding the homogeneity within classified land cover patch raised in Turner et al. (2000) and Soegaard and Moller-Jensen (2003) could be avoided. Secondly, it was realized that only four multispectral bands of the normal high-resolution product were not sufficient to extract overall land covers. For example, the discrimination between artificial bare land and the building was impossible with images of four multispectral bands (red, green, blue infrared (IR)) even if it was clearly understood that there was a huge difference in the CO_2 emission per unit between the two land cover types. Hence, the number of spectral bands of the imagery was also considered. Thirdly, the CO_2 emission ratio in individual land cover was determined based on the results derived from project conducted by Koerner and Klopatek (2002) and survey works performed by government agencies. With the given CO_2 emission ratio and land cover map, the overview of the CO_2 emission amount over the study area could be realized. Moreover, in order to evaluate the potential of the land cover analysis using satellite imagery of different spatial resolution, a comparison of the total CO_2 fluxes over target area computed using the high-resolution and multispectral satellite imagery and the low-resolution land cover maps by MODIS was demonstrated.

21.3 Study Areas and Datasets

The study area is located at the surrounding area of Boryeong City and Yoengi City, Chungnam, Korea (outlined in Fig. 21.1). It included various land cover types, for example, vegetation canopy, sandy beach, water, cultivated area mainly rice paddy, and dense urban environments. Therefore, it was not readily possible to estimate the

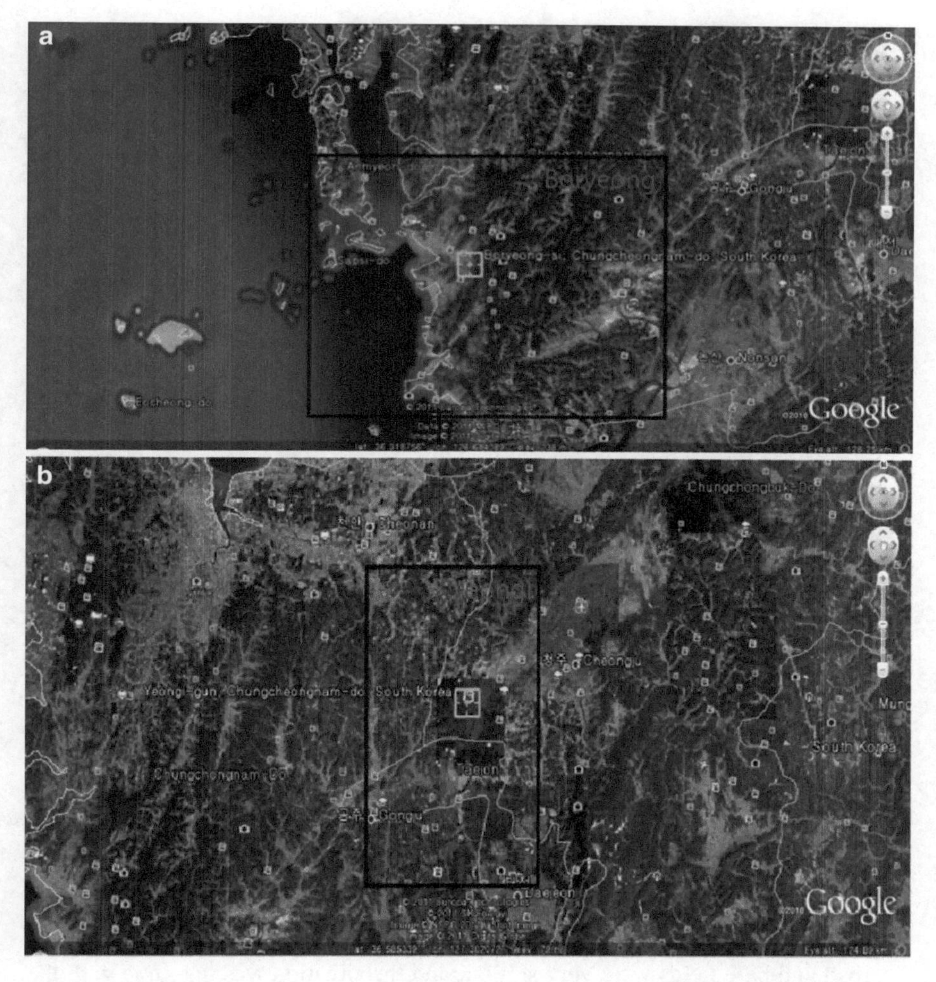

Fig. 21.1 The location of study areas and the coverage of applied satellite images are outlined in the *black rectangles*

CO_2 flux, employing numerical model of vegetation. It was also found that different land cover types were mixed together in satellite image of tens of meter in resolution. Hence, it was not possible to model each possible land cover mixtures using the traditional approach with medium- and low-resolution satellite images. For example, in the identified forest area from the land cover classification analysis using MODIS image, the lowest-level patches failed to discriminate the heterogeneous portions including the small towns, bare lands, and crops. To improve the classification performance, Turner et al. (2000) used the secondary factors, employing IR band pixel values to measure the heterogeneous portion.

The same problems occurred with existing land-use GIS data sets. An example of such homogeneity issues of data resolution is demonstrated in Fig. 21.2. Many land cover types in low-resolution map are not segregated or missing because of the insufficient raw data resolution to trace the boundaries in manual work or the limitation of the automated or semiautomated classification algorithms with low-resolution images. Therefore, if the total CO_2 flux was calculated with such low-resolution data sets, the inhomogenous patches would contribute with the wrong weighting values.

To address the problem of mixed land cover patch, the best possible resolution image must be employed for the classification processor. However, it was also found that the application of higher-resolution multispectral image could not always produce better error-free land cover patch for the CO_2 flux calculation. As shown in Fig. 21.2, the parts of land cover analysis employing very high-resolution satellite imagery are classified with wrong levels and include many small erroneous patches.

A set of multispectral imagery was tested in this chapter to find the best data sources for the reliable land cover extraction. The crucial factors of suitable multispectral satellite for the land cover analysis were identified as below:

(a) Suitable image resolution to extract the refined homogenous land covers boundaries: Even though the finer resolution was not a necessary condition for this purpose, the few meter spatial resolution was essential to delineate the homogenous land cover boundaries. Then it was possible to extract reliable land cover patches for CO_2 fluxes.
(b) Radiometric qualities: In some cases, the radiometric quality was more important than the spatial resolution for the correct land cover analysis. Compared with output from the other meter resolution images, it was clear to observe that the results from SPOT in Fig. 21.*3*c are of the best signal to noise ratio.
(c) Enough multispectral bands: The sufficient number of multispectral channels was a highly important factor to discriminate the similar signatures between different CO_2 sources. For examples, the spectral signatures between concrete road and bare fields were very similar, and the situation was the same with the case of water and rice paddy. The contributions of these land covers for total CO_2 fluxes were very different, although their spectral characteristics shown in the images were similar. It was also found that commercial satellite imagery with four bands (R, G, B, IR) or three bands (R, G, B) was sometimes insufficient.
(d) Photogrammetric quality: Other important factor was the photogrammetric quality of sensor information. Well-established photogrammetric quality made it possible to fuse with other data sources such as GIS or panchromatic image with multispectral channels. For areas where more than single satellite image coverage was required, the photogrammetric quality was essential for creating seamless land cover mosaic from a number of satellite image sources.

Figure 21.3 shows the effects of employment of multispectral image on the qualities of land cover extraction output. It was observed that the clearest boundaries

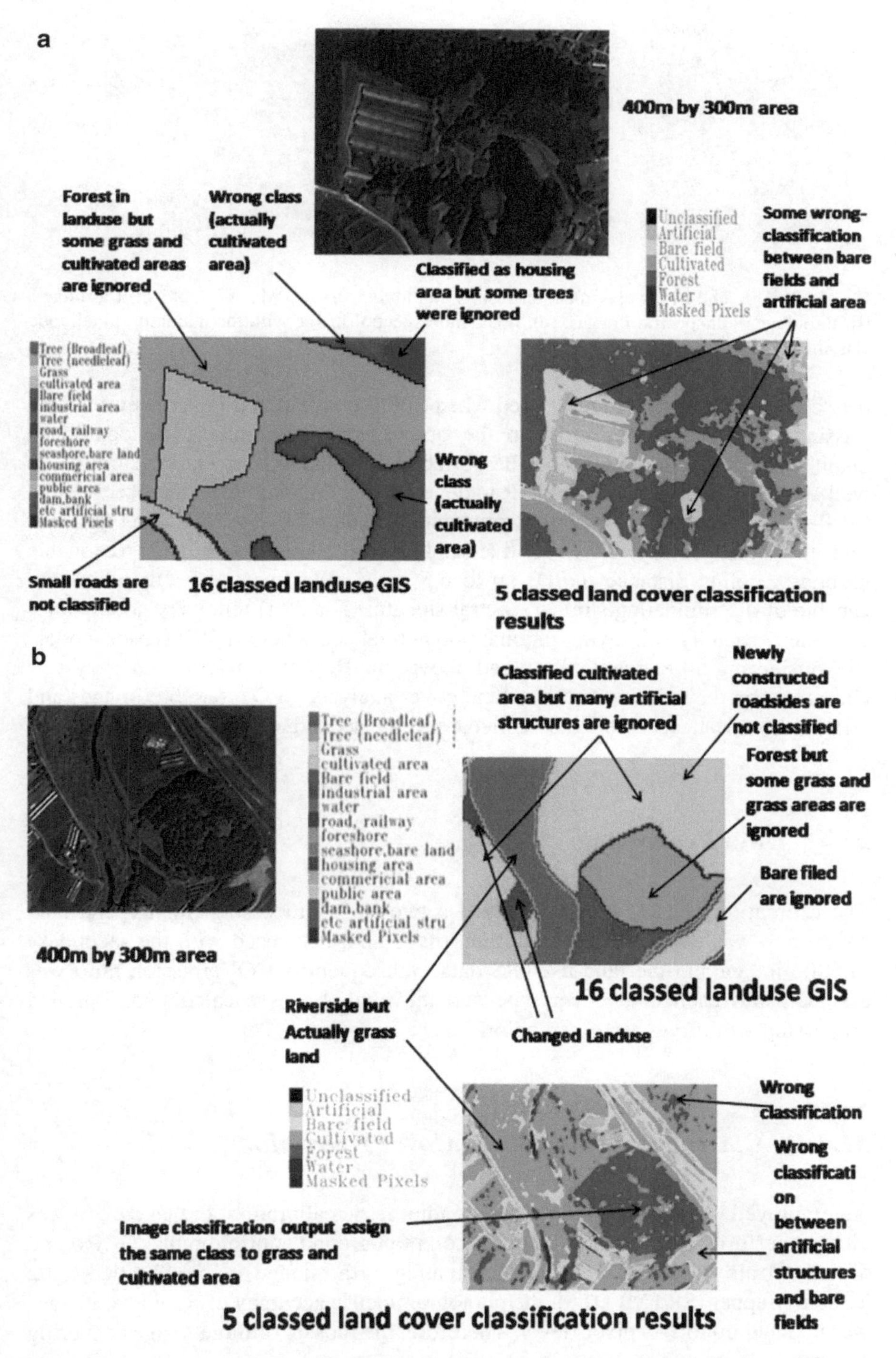

Fig. 21.2 Two cases show that the classification errors occur respectively in land-use GIS map and land cover classification map generated from satellite imagery (1 m pan-sharpened KOMPSAT-2 image)

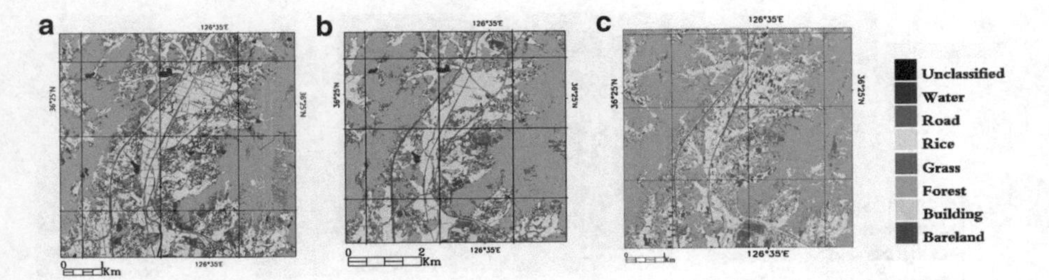

Fig. 21.3 The land cover classification results employing (**a**) KOMPSAT multispectral image, (**b**) Rapideye multispectral images, (**c**) SPOT multispectral image with the maximum likelihood classification

between land cover types are derived when SPOT image was used. However, many missing classifications occurred in the output land cover map. The comprised qualities of clear boundaries and the correctness of land cover classification were well shown in the output applying Rapideye images. In addition to the observation, the Rapideye satellite has highly recommendable characteristics, including (1) short revisiting time ($<$1 day) due to the five-satellite constellation in orbit, (2) reasonable ground sampling distance (GSD) up to 6.5 m, (3) five spectral bands which are capable of discriminating similar spectral signatures, and (4) relatively good photogrammetric quality employing rational polynomial coefficient (RPC) sensor model.

Considering all elements discussed above, the Rapideye satellite imagery was chosen as the main data source for land cover analysis. SPOT satellite images and GIS land-use data sets were also collected as the secondary preferable data.

21.4 Data Processing

The estimation of CO_2 fluxes followed a two-stage processing. Firstly, the land cover map was extracted from satellite image and then fused with the secondary satellite image and the land-use GIS data. Subsequently, CO_2 emission ratio was assigned into each land cover type and the total flux was calculated. Detailed processing workflows are given below.

21.4.1 Classification and Land Cover Extraction

All employed images were only applied radiometric calibration. In that case, it was ideal to perform data registration using common ground control points (GCPs) and digital terrain model (DTM). After creating ortho-image using Shuttle Radar Terrain Mapper (SRTM) DTM, the photogrammetric accuracy of Rapideye images was updated up to 1–2 pixels level. Therefore, the mosaic of three images covering

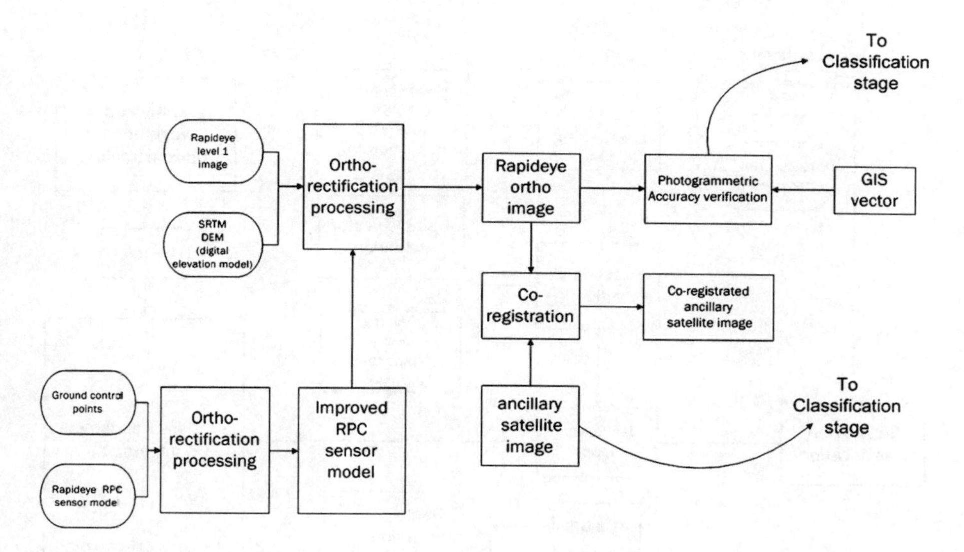

Fig. 21.4 Preprocessing stage of satellite images

whole target area was readily possible together with data fusion with GIS map and secondary SPOT images. Figure 21.4 shows the detail procedure of preprocessing including geo-rectification from the raw image data.

The data delivered for classification is shown in Fig. 21.5. At the first stage, available classification algorithms as shown in Fig. 21.6 were tested. It was found that the support vector machine (SVM) and maximum likelihood (ML) showed equally good qualities. However, SVM appears to be more dependent on spectral signatures rather than training vector definitions. For example, in Fig. 21.6d a large portion of rice paddies was classified as the water body in spite of very broad crop training vector definition over rice paddy. Classical ML algorithm is much faster but showed erroneous results in intermediate portions between different land cover types. Therefore, both the SVM and ML were applied together to classify eight land cover types at first – that is, water, forest, grass, crop, concrete, building, road, and bare land. Together with Enhanced Vegetation Index (EVI), the land cover classes were refined and three more land cover type (rice, mud, and sand which were crucial for the calculation of CO_2) classes were introduced. As EVI values were not easily saturated and very sensitive to the material properties, EVI was very useful to discriminate a close spectral signature such as water/rice paddy, mud/sand, concrete/asphalt, and bare land/sand (shown in Fig. 21.6f). Rather than the introduction of EVI as decision tree condition, the better result was produced when the sixth spectral band of EVI was employed. The SPOT image with the same classification scheme and NDVI values were introduced to fill the gap by the clouds over Rapideye images.

As CO_2 fluxes per unit over road and building land cover types were high, the accuracy of land cover classifications for such types was therefore very important

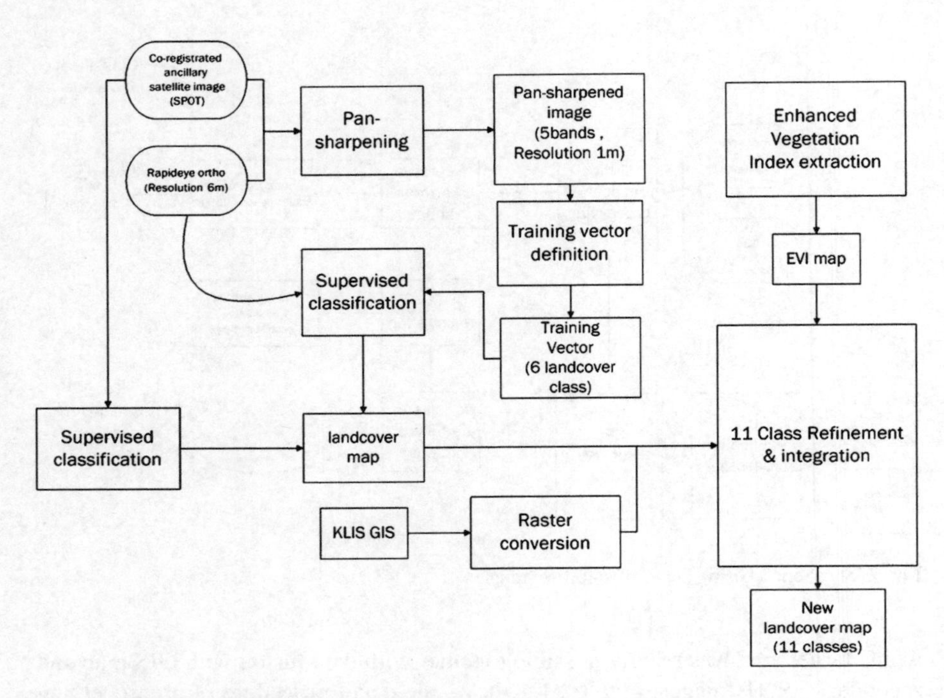

Fig. 21.5 Classification scheme for land cover map extraction

(Koerner and Klopatek 2002). However, it was found that the land cover classification with the satellite image frequently produced wrong land cover patches. To improve the classification performance, land-use GIS map derived from Korea Land Information System (KLIS) was employed at this stage. It was supposed that the concrete structures over residential area or commercial district assigned in KLIS were buildings as long as their morphological elongation was not high (<3.0). Otherwise, they were assigned as the concrete structure such as the road with concrete pavement, bridges, and dam. In addition to these criteria, several heuristic decision rules based on the information of KLIS land-use data set were introduced. As a result, the accuracy of land cover was enhanced and resolution of constructed land cover map was improved.

21.4.2 Estimation of CO_2 Emission Constants in Land Cover Classes and CO_2 Flux Calculation

As described in Sect. 21.2, the emission ratio constants employed in this chapter were based on the previous researches and some surveys. The values are listed in Table 21.1. Time dependence of emission ratio was replaced by introducing the constants averaging the daily CO_2 functions. Especially the contributions from moving vehicles were indirectly measured by the contribution of the road networks.

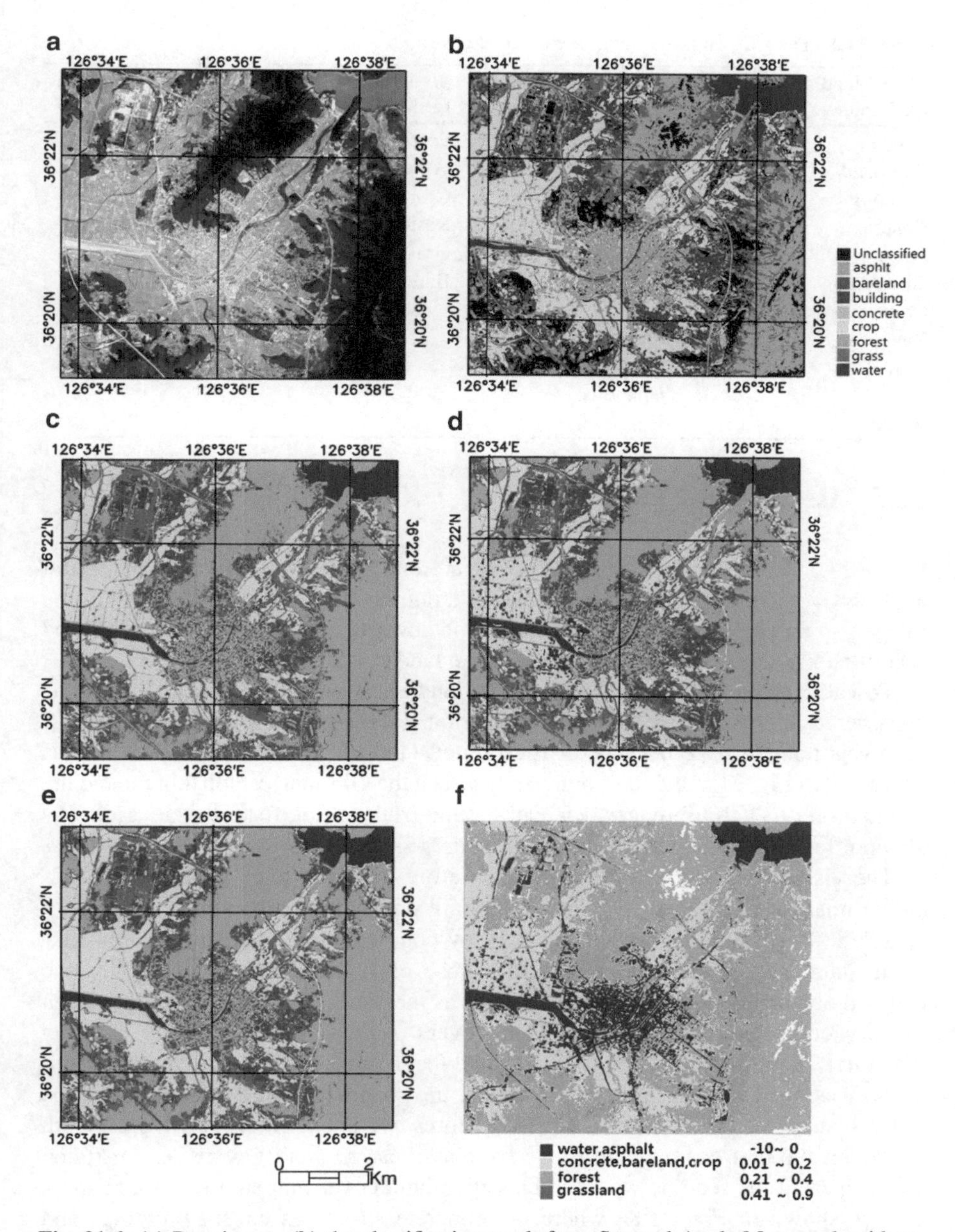

Fig. 21.6 (**a**) Raw image, (**b**) the classification result from Spectral Angle Mapper algorithms, (**c**) the classification result from maximum likelihood, (**d**) the classification result from support vector machine, (**e**) the classification result from neural network, (**f**) the land cover threshold with EVI

Table 21.1 The CO_2 emission ratio in each land cover

Eleven land cover types for Rapideye image	CO_2 emission ratio (kg/m^2/year)	Six land cover types for MODIS image	CO_2 emission ratio (kg/m^2/year)
Asphalt	30.2	Forest	−0.73
Bare land	0.64	Water	Unknown
Building	62.5	Crop	6.3
Concrete	3.1	Residence	42.5
Forest	−0.73	Grassland	−0.53
Grassland	−0.53	Bareland	0.64
Crop	6.1		
Rice paddy	6.6		
Sand	0.6		
Mud	Unknown		
Water	Unknown		

21.5 Results and Validation

Figure 21.7 shows the final land cover outputs from the high-resolution multispectral analysis over two target test sites. In spite of a number of technical issues such as the image gap by clouds and the complicated land cover types (especially in coastal and rebuilt areas), the maps show highly accurate land cover distributions.

To validate the accuracy of the resultant land cover map, the KLIS GIS map was considered as the ground truth for evaluation. However, the land-use KLIS GIS map was not effective because of the limited land-use resolution of GIS map as described in Fig. 21.1 and temporal gap between the GIS map establishment and the acquisition of satellite images. Manual ground truth setting for the evaluation was also not feasible, owing to the large target area size.

The visual observation method was therefore proposed to evaluate the classification quality. Land cover patches possessing precise shape were introduced as a standard. The examples are shown in many concrete structures such as embankment, dam, and bridges in Fig. 21.7a. Those objects accurately maintained the original object shapes and clearly classified as the concrete in land cover type even in very complicated surrounding land covers. Sometimes, large-scale building structures were observed in the classified land covers even though the detail shape was not clear. In both Fig. 21.7a, b, the shapes of some natural landscape objects such as ponds, lakes, and foreshores were extremely well preserved. However, the best examples to prove the classification quality were demonstrated by comparing the road networks in classified land cover map and KLIS GIS map. This is shown in Fig. 21.8a where the concrete structures except buildings and asphalts from land cover map formed the road networks. Figure 21.8b including only the road networks from KLIS is clearly similar with the classified road networks. From the observations, the high accuracy of land cover classification was clearly proved, although the direct numerical comparison was unavailable as some unpaved roads were not included in KLIS GIS map.

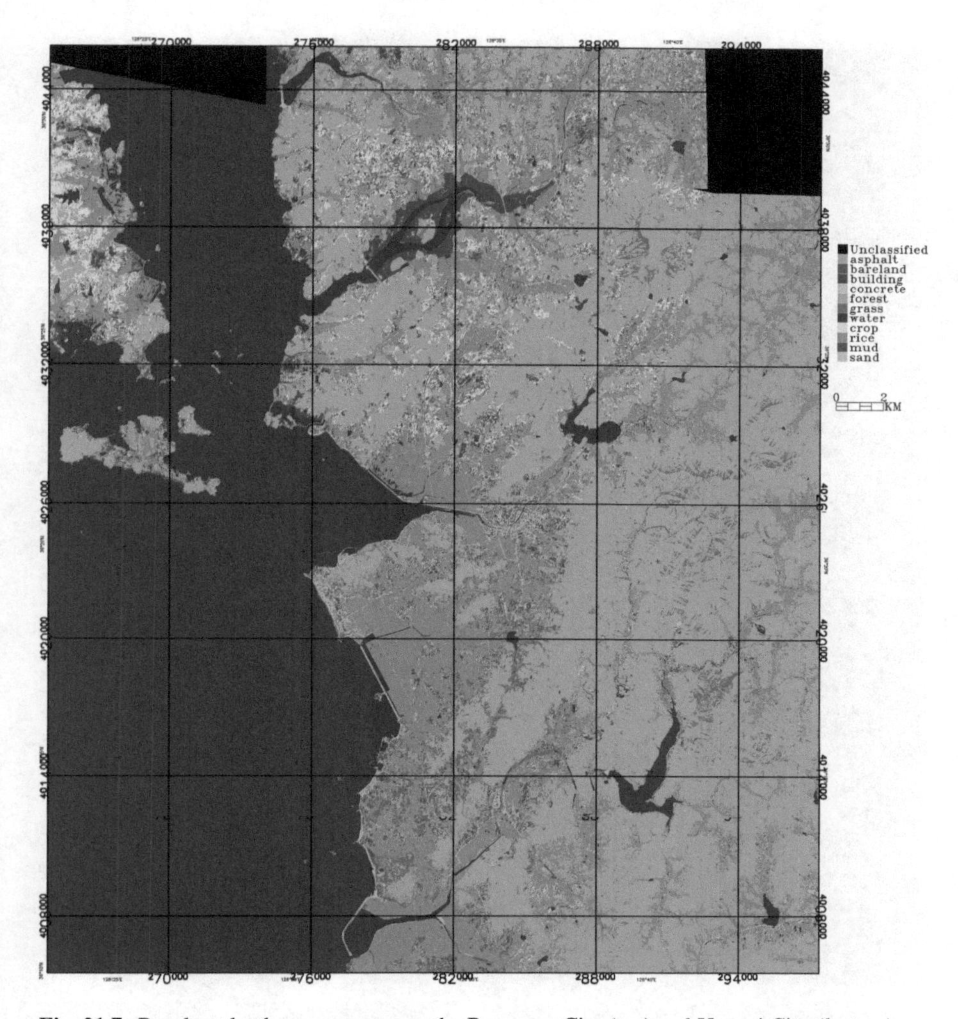

Fig. 21.7 Resultant land cover maps over the Boryeong City (*top*) and Yoengi City (*bottom*)

For the application of CO_2 fluxes, the CO_2 emission maps are created from land cover analysis and shown in Fig. 21.9a, b. The high-density CO_2 flux areas could be easily distinguished in the products. The validation of total CO_2 flux distributions was implemented by comparing with land cover and CO_2 flux values derived from the MODIS which was widely used for environmental analysis. The land cover analysis with MODIS has been published as MODIS level 3 products annually (Freidl et al. 2010). However, the spatial resolution was 500 m so it was not useful to extract the test area's CO_2 fluxes using MODIS level 3 global land cover products. Alternatively, two MODIS subset images including seven multispectral bands for the land surface analysis were prepared on the dates of Rapideye image acquisitions.

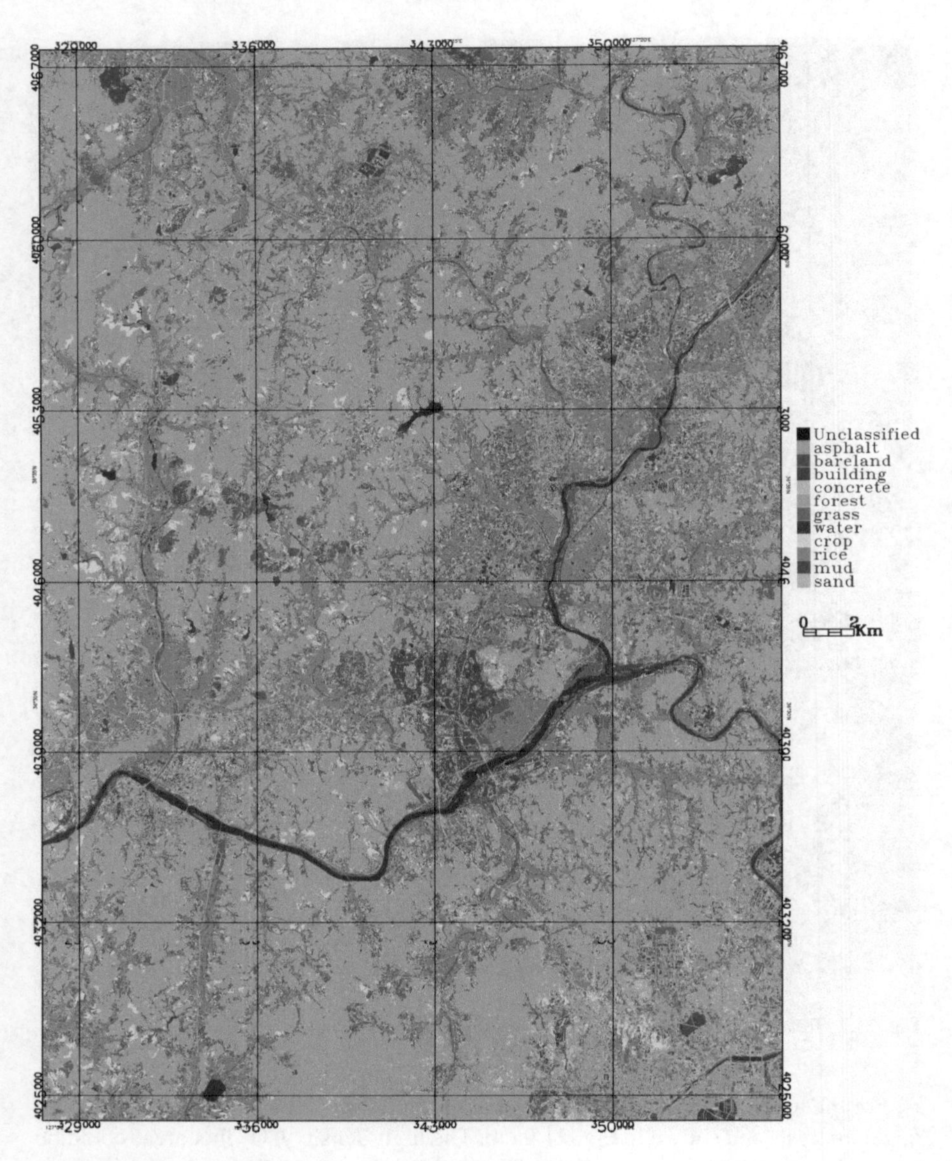

Fig. 21.7 (continued)

Five classes, including water, urban, cultivation, grass, and forest, were classified with 250 m resolution pan-sharpened MODIS image. The strip-wise noises of MODIS prevented the full exploitation of the classification scheme using land-use GIS map and ML/SVM algorithms together. Hence, the neural network classifications were performed and CO_2 flux maps were derived and are shown in Fig. 21.10.

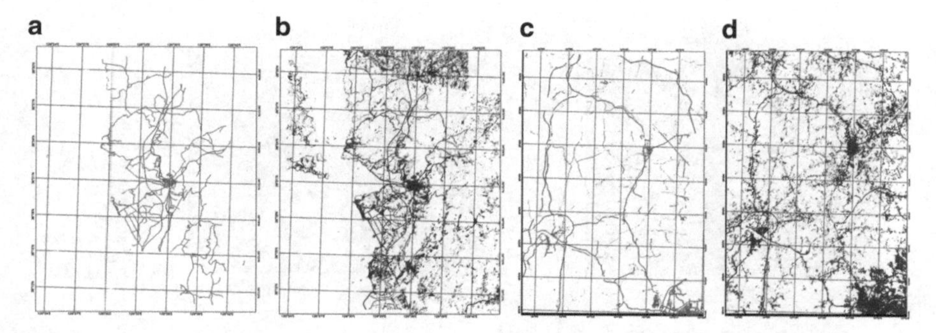

Fig. 21.8 The road network comparison between the land-use GIS data (*red*) and the land cover map (*blue*) generated from multispectral image (*left* two: Boryeong City; *right* two: Yoengi City)

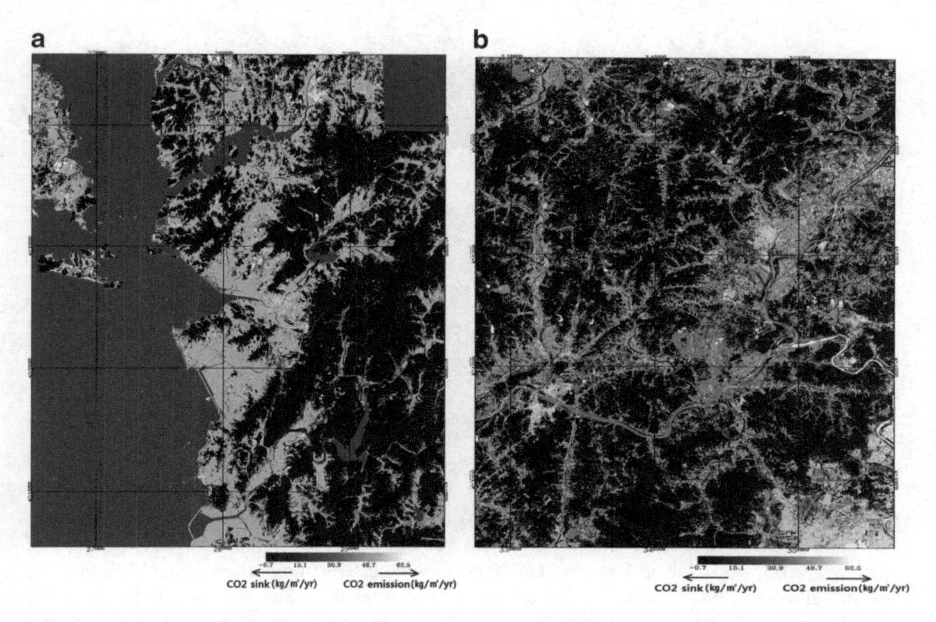

Fig. 21.9 The CO_2 emission maps of Boryeong City (*left*) and Yoengi City (*right*). The total CO_2 flux amount is estimated as 3.609e + 9 kg/year and 7.387e + 9 kg/year, respectively

It was observed that the CO_2 flux distributions from MODIS analysis were similar with high-resolution Rapideye images. However, the calculated total CO_2 fluxes from MODIS and Rapideye were quite different. It should be noted that the CO_2 flux dependence on the spatial resolution of basis satellite image for land cover analysis was already proven in Turner et al. (2002).

According to the analysis, it was summarized that the high-resolution CO_2 products gave more reasonable estimations. The land cover analysis using low-resolution products like AVHRR or MODIS and medium-resolution such as Landsat TM or ETM was demonstrated as a relatively weak basis for GHG estimation.

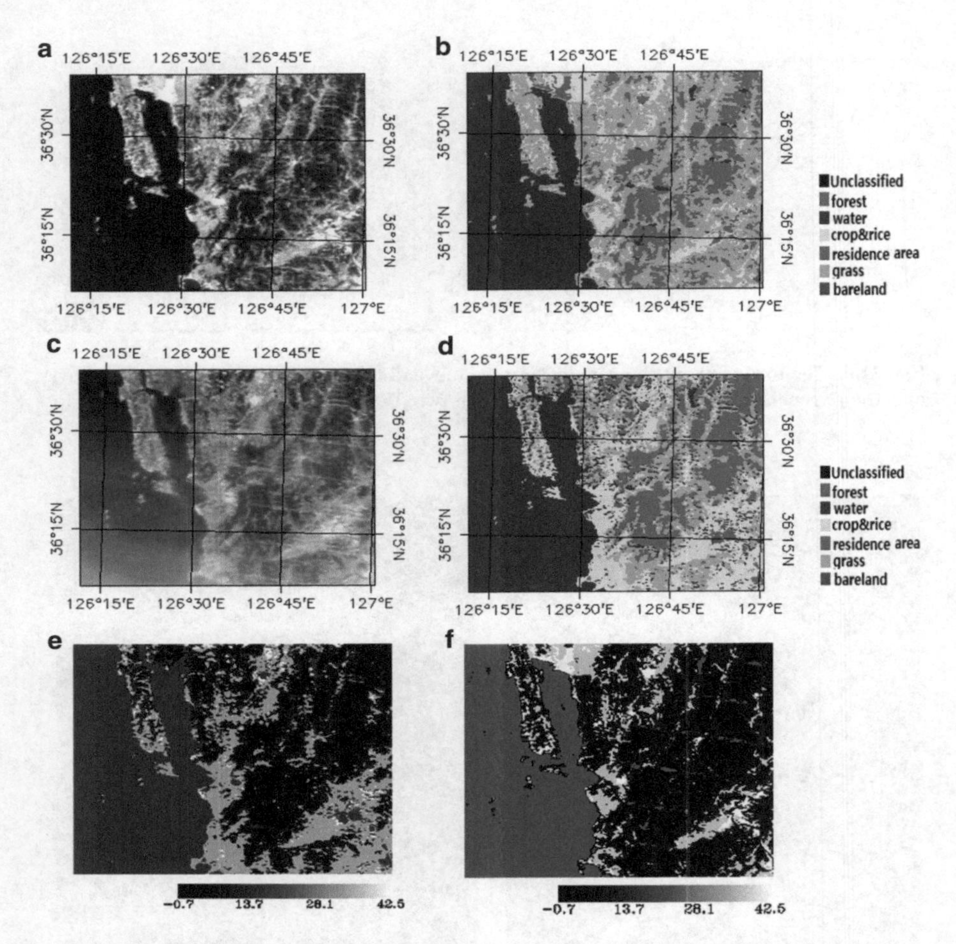

Fig. 21.10 The land cover extraction examples using MODIS data and the resultant CO_2 emission maps. The MODIS data were collected on the 18th of June 2011 (*top left*) and the 16th of September 2011 (top right). The resultant land cover classification maps and CO_2 emission maps are shown in the middle and bottom rows. The total CO_2 flux amount produced in the 18th of June and the 16th of September 2011 is estimated as 6.587e + 9 kg/year and 5.281e + 9 kg/year, respectively

21.6 Conclusions

In this study, we tried the unprecedented wide area land cover analysis with high-resolution multispectral images. Technically, it was a challenge to interpret the land cover up to a few hundred km^2 area with the acceptable accuracy to be applicable for the practical usages such as policy decision. Moreover, available high-resolution multispectral satellite images were tested with the same classification algorithms. As the result, Rapideye multispectral imagery was demonstrated as the best solution for the extensive land cover mapping. The photogrammetric and radiometric quality with 5-bands data were identified as suitable properties for such purpose.

The approaches employed to tackle the technical issues were successful. The classification scheme employing SVM and ML algorithms was applied firstly to classify roughly six land cover types. Subsequently, the refinement scheme employing existing land-use GIS map as the decision criteria together with additional EVI channel was performed. Therefore, the erroneous classification by the similar multispectral signature could be minimized. Based on the high spatial resolution products, individual landscape objects such as bridges, embankments, and buildings were shown in the classification map. After applying the unit CO_2 emission ratio on the products with detailed classification, the CO_2 emission maps were produced. The results were then compared with the land cover map and CO_2 emission from MODIS analysis. As incorrect classifications might occur, the total CO_2 fluxes estimated from low resolution were clearly different from the high-resolution analysis, revealing the importance of spatial resolution.

Conclusively, this work showed the importance of the spatial resolution of multispectral satellite image and the algorithms employed for the land cover analysis. For the application of CO_2 fluxes, the accuracy of land cover classification was sufficient. Once the robust CO_2 emission ratio accounting for the temporal and environmental variation is derived in the near future, an advanced CO_2 emission map can be produced based on the resultant land cover map. With the final product, local areas with high and low CO_2 emission volumes can be observed. This provides useful information for locating areas of serious CO_2 emission. After examining the background of high emission and effects might cause around the area, local environmental authorities are able to make suitable and reasonable policies to ask the CO_2 contributors to reduce the CO_2 emission to the requested volume. As the CO_2 emission can be determined clearly on the maps, the CO_2 reduction progress can be reviewed in the specific time interval. In addition, the CO_2 emission map also plays critical role for land-use planning. As the CO_2 emission ratios associated with specific land cover/use types are given, also the CO_2 flux over the target area can be computed, and the local authorities can use this information as one of the factors for deciding overall land-use plan or single land development permission. At last, the sequential CO_2 reduction can be monitored using multiple temporal CO_2 emission maps. The CO_2 emission reduction can be reviewed and the policies can then be modified accordingly.

References

Baldocchi DD (2003) Assessing the eddy covariance technique for evaluating carbon dioxide exchange rates of ecosystems: past, present and future. Glob Change Biol 9(4):479–492

Bonan GB, Levis S, Kergoat L, Oleson KW (2002) Landscapes as patches of plant functional types: an integrating concept for climate and ecosystem models. Global Biogeochem Cycles 16 (2):1021–1044

Buchwitz M, de Beek R, Noel S, Burrows JP, Bovensmann H, Bremer H, Bergamaschi P, Korner S, Heimann M (2005) Carbon monoxide, methane and carbon dioxide columns retrieved from SCIAMACHY by WFM-DOAS: year 2003 initial data set. Atmos Chem Phys 5(12):3313–3329

Friedl MA, Sulla-Menashe D, Tan B, Schneider A, Ramankutty N, Sibley A, Huang X (2010) MODIS collection 5 global land cover: algorithm refinements and characterization of new datasets. Remote Sens Environ 114(2010):168–182

Hese S, Lucht W, Schmullius C, Barnsley M, Dubayah R, Knorr D, Neumann K, Riedel T, Schröter K (2005) Global biomass mapping for an improved understanding of the CO_2 balance – the Earth observation mission Carbon-3D. Remote Sens Environ 94(1):94–104

Janssens IA, Freibauer A, Ciais P, Smith P, Nabuurs GJ, Folberth G, Schlamadinger B, Hutjes RW, Ceulemans R, Schulze ED, Valentini R, Dolman AJ (2003) Europe's terrestrial biosphere absorbs 7 to 12 % of European anthropogenic CO_2 emissions. Science 300(5625):1538–1542

Koerner B, Klopatek J (2002) Anthropogenic and natural CO_2 emission sources in an arid urban environment. Environ Pollut 116(1):45–51

Quaife T, Quegan S, Disney M, Lewis P, Lomas M, Woodward FI (2008) Impact of land cover uncertainties on estimates of biospheric carbon fluxes. Global Biogeochem Cycles 22:12

Schimel D, Melillo J, Tian H, McGuire AD, Kicklighter D, Kittel T, Rosenbloom N, Running S, Thornton P, Ojima D, Parton W, Kelly R, Sykes M, Neilson R, Rizzo B (2000) Contribution of increasing CO_2 and climate to carbon storage by ecosystems of the United States. Science 287 (5460):2004–2006

Soegaard H, Moller-Jensen L (2003) Towards a spatial CO_2 budget of a metropolitan region based on a textural image classification and flux measurements. Remote Sens Environ 87 (2–3):283–294

Turner DP, Cohen WB, Kennedy RE (2000) Alternative spatial resolutions and estimation of carbon flux over a managed forest landscape in western Oregon. Landsc Ecol 15(5):441–452

Turner DP, Gower ST, Gregory WB, Maiersperger TK (2002) Effects of spatial variability in light use efficiency on satellite-based NPP monitoring. Remote Sens Environ 80(2002):397–405

Part V
Landscape and Ecological System, Sustainable Development: Section Vulnerability of Urban System

Chapter 22
Taiwan's Five Major Metropolitan Areas of Taiwan Vulnerability Assessment of Flood Disaster Comparison Study

Jen-te Pai

Abstract This study explores the connotations of vulnerability, the proposed indicators of vulnerability using the Fuzzy Delphi expert panel, and the establishment of an indicators system. Afterward, a multi-criteria assessment of the Analytic Hierarchy Process and Analytic Network Process was applied to be able to identify the indicators of the vulnerability of urban class relations and the weight of the metropolitan area, in order to finalize the vulnerability and resilience of the indicators system. Finally, the proposed assessment framework was applied to the five major metropolitan areas in Taiwan as an empirical analysis. Research results satisfied all relevant vulnerability assessments, including exposure, sensitivity, and resilience; Fuzzy Delphi survey was utilized to adjust the indicators system for the metropolitan areas of Taiwan to establish the vulnerability of the flood disaster assessment framework. This study also found that hardware and software infrastructure, preparedness, and response capabilities and the ability to reconstruct and rehabilitate the assessment of vulnerability get more than half the weight. The study considers the opinions of different decision-making groups to assess the typhoon hazard vulnerability of each metropolitan area. The overall typhoon hazard vulnerability is Tainan> Taichung> Taoyuan Chungli> Kaohsiung> Taipei. Comparisons were made of the different aspects of vulnerability indicators for each area, and some policy suggestions were proposed, such as the following: (1) through hazard information, integrate and establish a complete disaster-prevention database; (2) strengthen preparation, emergency responsiveness, and resilience and adaptation ability; (3) balance the needs of disaster prevention among all regions; and (4) undertake preparation and disaster-prevention planning in different scales.

Keywords Vulnerability • Analytic Hierarchy Process • Analytic Network Process

---

J.-t. Pai (✉)
Department of Land Economics, National Chengchi University, No.64, Sec.2, ZhiNan Rd., Wenshan District, Taipei City 11605, Taiwan
e-mail: brianpai@nccu.edu.tw

M. Kawakami et al. (eds.), *Spatial Planning and Sustainable Development: Approaches for Achieving Sustainable Urban Form in Asian Cities*, Strategies for Sustainability, DOI 10.1007/978-94-007-5922-0_22,

22.1 Introduction

Global climate change has increased the occurrence of severe natural disasters. In 2007, the assessment report by the Intergovernmental Panel on Climate Change (IPCC) warned of the clear evidence of global climate change, as the frequency of floods and droughts has increased dramatically, and coastal regions worldwide have come under imminent danger. If no immediate measures are taken, there will be dire consequences. Taiwan will not be safe from the disasters wrought by global climate change. The World Bank indicated in a report in 2005 that Taiwan ranks among the top in the risk analysis for several natural disasters. Also, floods are getting worse because the original hydrological environment is destroyed when people continuously compete for land that has rivers in the pursuit of economic development and better life quality. Thus, in the Flood-Prone Area Management Plan by Water Resources Agency, MOEA (Ministry of Economic Affairs) clearly indicates that an 8-year flood-control budget worth 80 billion NTD has been increased to 110 billion NTD. Therefore, the flood-control plan needs to be continuously executed.

Birkmann (2007) said that vulnerability study involves a complicated and multidisciplinary area, including development and poverty, public health, climate study, security study, engineering, geography, political ecology, and disaster risk management.

Many papers indicate that dense population and urban development increase vulnerability. The study by Chen and Chen (2007) pointed out that over time, a typhoon-flooded area inclined to be newly developed areas, but not the downstream, low-lying areas. A similar trend is found in urban areas. It is clear that land development and urban expansion have an interesting correlation with urban vulnerability—a phenomenon that cannot be overlooked. Urbanization in Taiwan is still on the rise. Fortunately, domestic emphasis on disaster and environment vulnerability has increased, and in recent years, many scholars have committed themselves to vulnerability study.

In summary, it is very important to assess and resolve flood problems in disaster-prone, densely populated urban areas. Before a disaster-prevention solution can be sought, a pragmatic overall index to increase the reference value for disaster assessment must first be established, and disaster-prevention, spatial planning, and land use must be reviewed. Those are the primary goals of this study.

22.2 Literature Review

Recently, the vulnerability to natural disasters has been increasingly emphasized internationally. There are many similar disaster-prone areas like Taiwan in the world, including island countries that have a high disaster risk. From the Declaration of Barbados and action plan and the 1997 UN Economic and Social Affairs call for island-country expert groups to study environmental and economic vulnerability indexes, to the reports of natural disaster risk assessment from the World Bank in

2005 and the UN Intergovernmental Panel on Climate Change (IPCC) in 2007, the problems have become increasingly severe and imminent.

US vulnerability studies before 1975 were almost exclusively based on natural science or engineering. In 1945, American geographer Gilbert F. White stated that political, economic, and social conditions would affect the effectiveness of engineering disaster prevention in the book titled *Environment as Hazards*, which was based on flood disaster research. The assessment of research nature hazard by White and Haas in 1975 started the multidisciplinary subject that involves the economy, society, and politics to disaster study (Hsiao 2008). The simple interpretation of vulnerability is an inherent potential or extent subject to disaster. Blaikie et al. (1994) believed that vulnerability should not be limited to inherent disaster potential and expanded the concept to the potential and ability of a specific individual or group to prepare, handle and respond, resist, and recover from natural disaster impact. For Chen (2003), vulnerability is the degree of tendency or susceptibility to harm or damage. Bankoff (2004) believed that when vulnerability is generally applied to hazards and disasters, it refers to a concept that people and the environment sustain and compete for social power, organization, and humanistic value.

Vulnerability involves many concepts. Turner II et al. (2003) proposed two basic models for vulnerability: RH (Risk-Hazard) and PAR (Pressure-and-Release). The basic RH model is used to understand the impact of hazards and has an intrinsic responsiveness to hazards in the mechanism. The model is quantitatively applied to assessing the impact and susceptibility to climate and general stress exposure. It can be seen that vulnerability is defined by a series of processes, not always formally, by a model. Different approaches of investigation manifest the unsuitability of the RH model in some cases. Due to previous recognition, the PAR model was built, in which risk is clearly defined as a vulnerability and stress unit for disturbance, stress, or exposure. It focuses directly on exposure to unsecured states, vulnerability, and the causes of the situations. Its application is mainly focus on social groups facing disasters. The model emphasizes the distinction of vulnerability among different exposures.

In general, social and natural vulnerability are the two primary factors of vulnerability. Social vulnerability is intrinsically determined by "which human system is exposed to which hazard." Therefore, although social vulnerability is not a dependent variable for hazards, regional groups may become more vulnerable to a specific hazard due to some system attributes. Thus, it is necessary to clarify "whose vulnerability or whose vulnerability to what." Only through social vulnerability can the most vulnerable group be identified as well as whether different vulnerabilities can exist within a region or among different regions. The ideal vulnerability assessment should include physical and social vulnerability. Physical vulnerability assessment provides the foundation of natural environment variables, and social vulnerability assessment, the foundation of human attribute variables. Both would make studies on environmental impact more complete (Lin 2004).

Turner II et al. (2003) thought that it would be impractical to consider the system's integrity by understanding the conceptual analysis of vulnerability because the restriction by the data and variability in a real world would reduce the effectiveness of our vulnerability assessment. Despite this, it is still necessary to underscore the attempt to analyze. Being the assessment framework and regional

Fig. 22.1 A diagrammatic summary of the conceptual relations between vulnerability, resilience, and adaptive capacity (Source of Data: Gallopin (2006))

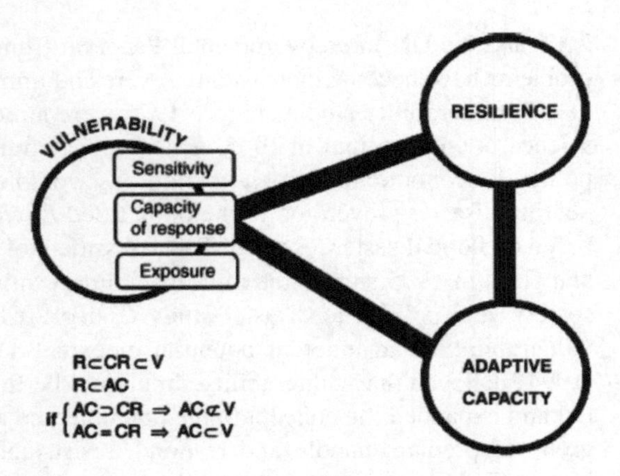

exposure, sensitivity, and resilience for vulnerability in Figs. 22.1 and 22.2, it provides a vulnerability framework that is an appropriate format to simplify analysis, including large-scope issues. This framework is not a complete explanation but offers a broad and extensive scope and constitutes the links to the composite system of hazard vulnerability.

The basic structure comprises the following: (1) in broad definition, linking the states and operation routes between humans and the natural environment in a composite system; (2) the routes in which disturbance and stress appear; and (3) where the vulnerability involving humans and the natural environment is, including exposure, stress, and adjustment. These components are interactive and sometimes interdependent. Such analysis is affected by the composite system and provides conceptualization and the boundary of the study.

Recently, some scholars have included resilience in vulnerability. For example, Gallopin in the study of hazard vulnerability and resilience (2006) believed that vulnerability, resilience, and adaptability are similar in certain situations and different in others. The components form a triangle (as in Fig. 22.1), in which vulnerability includes sensitivity, exposure, and responsiveness, while responsiveness is related to resilience and adaptability. Thus, resilience is a subset of responsiveness, responsiveness is a subset of vulnerability, and resilience is also a subset of adaptability. In short, resilience is a part of vulnerability. This study adopts the opinions of Gallopin (2006) on vulnerability study and also investigates resilience. Nevertheless, the framework is still on the scope of vulnerability.

22.3 Establishment of an Indicators System

Based on literature, discussion, and review on the vulnerability assessment process and concept, the vulnerability framework is proposed, with flow and screening processes. The proposed vulnerability assessment indicators are adjusted and revised after the analysis of the expert questionnaire survey using the Fuzzy Delphi Method.

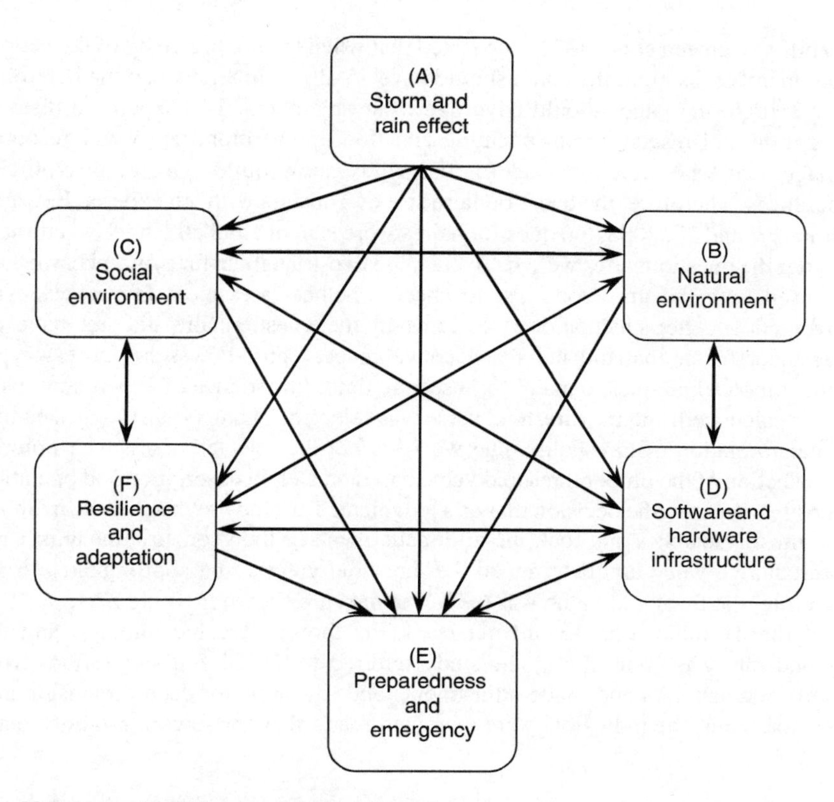

Fig. 22.2 Network relationships between dimensions

The main methodology includes the Fuzzy Delphi Method, Analytic Hierarchy Process (AHP), and Analytic Network Process (ANP), which are described below.

The Fuzzy Delphi Method used in the study was proposed by Cheng in 2001. It mainly has reference to the Fuzzy Delphi Method by Chen (2001) and is partially modified to reduce the number of repeated investigations. The ANP is an extension of the AHP. With the development of the feedback and dependence function, the ANP successfully overcomes the limiting conditions above.

First, the study defines urban typhoon flood hazard vulnerability as the "susceptibility of life, wealth, and public environment damage to typhoon hazards" and uses Turner II's (2003) framework as the blueprint of the assessment framework, which covers exposure, sensitivity, and resilience. With reference to domestic and foreign literature and data, the initial criteria and indices are established to build the framework system for the urban typhoon flood hazard assessment. The framework is used to conduct the Fuzzy Delphi expert questionnaire survey, which concentrates on disaster prevention, planning, and water resource management, and covers public services and academia. As for the number of people, Dalkey (1969) thought a group of at least ten could minimize the error and maximize the reliability. Delbecq et al. (1975) suggested that when the homogeneity of the Delphi panel members is high, the panel should have 15–30 members; when the heterogeneity is high, the panel should have five to ten members. The experts in disaster prevention and rescue, urban planning and flood prevention, and water resource management who were selected for this study show quite a lot of disciplinary differences. Therefore, the questionnaire survey consults with ten experts. Between February 9 and 27, 2008, ten questionnaires were sent out and all ten were returned.

After the questionnaires were recovered, the two-triangular-fuzzy-number method was used with the gray zone test to check whether assessment factors achieved convergence. After calculation, the items in the questionnaire did not have an assessment factor that did not reach convergence. Thus, it was not necessary to conduct the Delphi questionnaire for a second time. Importance of expert consensus Gi was calculated, and the threshold value was selected. Chang (1998) suggested that the determination of threshold value would affect the selected assessment criteria. The selection of the proper threshold value based on Delphi description and operation is mostly subject to the decision maker's judgment. The study referred to the methods from most researches and took the arithmetic mean of the geometric mean of each representative value and determined the threshold value to be approximately 6.31. Therefore, the threshold value was 6.31. The data are shown in Table 22.1.

Of the 41 indicators, 15 did not reach the threshold value initially, and the removal rate was about 36%. The study referred to the Gi and suggestions from experts and scholars and made adjustments and revisions for each dimension and indicator. Thus, the indicators were extracted for further framework establishment.

22.4 Assessment Framework and Comparative Area Analysis

This section focuses on the network relationship assessment framework, weight analysis, and comparative area analysis. The study defines urban typhoon hazard vulnerability as the "susceptibility of life, wealth, and public environment to typhoon hazards" and uses Turner II's (2003) framework as the blueprint to establish assessment indicators for typhoon hazard vulnerability. There are six dimensions in the assessment framework: (A) "storm and rain effect," (B) "natural environment," (C) "social environment," (D) "software and hardware infrastructure," (E) "preparedness and emergency responsiveness," and (F) "resilience and adaptation."

The "storm and rain effect" dimension is similar to "exposure," in an urban area, to an active and unpredictable natural hazard. Thus, this dimension affects the other five. The interaction between the dimensions of "natural environment," "social environment," and human environment's "software and hardware infrastructure" is very clear in the framework and literature. These basic conditions will also affect response ability and reconstruction resilience. The "preparedness and emergency responsiveness" dimension is the short-term measure during disaster and affects "resilience and adaptation." The long-term "resilience and adaptation" dimension affects the other five. The network relationships of each dimension were summarized in Fig. 22.2.

Table 22.1 Fuzzy Delphi expert questionnaire analysis

Indicators	Gi	Indicators	Gi
Typhoon hazard frequency	5.9362	House owner ratio	4.4965
Average typhoon precipitation	8.0291	Rainwater sewer construction ratio	7.2382
Average typhoon wind strength	6.7568	Pumping station ratio	7.2586
Injury and casualty number due to typhoon hazards	7.9936	Average infrastructure investment	6.5425
Number of collapsed or semi-collapsed houses	6.3548	Building level and height	6.1798
First-class industry loss	6.2641	Information broadcasting availability	7.0215
Public project loss	7.1315	Civilian disaster-prevention consciousness	7.8873
Potential flooded area ratio	7.0983	Prevention and rescue training	7.1856
Land development ratio	6.8193	Prevention rescue organization in public services	8.0569
Improper slope development ratio	7.3093	Firefighters, number of	7.1245
Drainage area ratio	5.5523	Fire trucks	6.9993
Population	7.2738	Medical institutes, number of	6.7273
Population density	7.9769	Temporary shelters, number of	7.3745
Female ratio	4.5117	Social assistance and welfare group	6.4441
Low income ratio	4.9658	Religious charity group	6.0043
Disabled ratio	4.7737	Community self-governing group	6.9661
Senior ratio (over 65 years old)	4.7139	Environment cleaning operation ability	6.8005
Children ratio (below 15 years old)	4.6447	Asylum	5.9895
Unemployment ratio	4.1633	Emergency rescue subsidy	6.729
Average household income	4.7109	Completeness of disaster-prevention rescue laws	6.9239
Dependence on single industry	4.7364		

$\mathrm{Gi} = \frac{\mathrm{CiM}+\mathrm{OiM}}{2}$

After the Fuzzy Delphi expert questionnaire, the indicators system for urban typhoon hazard vulnerability was established and adjusted. The assessment framework was subjected to the second-stage expert questionnaire. Expert knowledge and experience was utilized as the communication tool to reach decision-making consciousness. The questionnaire targeted experts in disaster prevention, urban planning, and hydraulic engineering. The proper number of experts was selected to cover public services and academia for the survey. Four people were chosen for each area, and 12 people formed the expert decision-making panel. Between March 30, 2008, and April 17, 2008, 12 expert questionnaires were sent out and recovered; the response rate was 100%. The study sought to integrate the analysis of dimension, indicators, and region. After expert questionnaire resilience, Super Decision 1.6.0 decision-making software was used for calculation. A few matrices that did not pass the test were adjusted by referring to the sequence of selected items in the original expert questionnaires to meet consistency requirement. The matrices for the index part are shown in Table 22.2.

Table 22.2 Indicator matrices for the decision-making group

	(A) Storm and rain effect	(B) Natural environment	(C) Social environment	(D) Software and hardware infrastructure	(E) Preparedness and emergency	(F) Resilience and adaptation
A1rainfall precipitation	0.1136	0.0000	0.0000	0.0000	0.0000	0.0000
A2 wind strength	0.0733	0.0000	0.0000	0.0000	0.0000	0.0000
A3injury and casualty	0.3871	0.0000	0.0000	0.0000	0.0000	0.0000
A4 flood height	0.0951	0.0000	0.0000	0.0000	0.0000	0.0000
A5 public loss	0.1290	0.0000	0.0000	0.0000	0.0000	0.0000
A6 wealth loss	0.2018	0.0000	0.0000	0.0000	0.0000	0.0000
B1 flood potential	0.0000	0.2779	0.0000	0.0000	0.0000	0.0000
B2 development	0.0000	0.4005	0.0000	0.0000	0.0000	0.0000
B3 land loss	0.0000	0.3216	0.0000	0.0000	0.0000	0.0000
C1 population	0.0000	0.0000	0.1949	0.0000	0.0000	0.0000
C2 density	0.0000	0.0000	0.5363	0.0000	0.0000	0.0000
C3Disadvantage Group disadvantaged	0.0000	0.0000	0.2689	0.0000	0.0000	0.0000
D1 watercourse	0.0000	0.0000	0.0000	0.2294	0.0000	0.0000
D2 flood prevention	0.0000	0.0000	0.0000	0.3507	0.0000	0.0000
D3 foundation	0.0000	0.0000	0.0000	0.1925	0.0000	0.0000
D4 information	0.0000	0.0000	0.0000	0.0694	0.0000	0.0000
D5 group	0.0000	0.0000	0.0000	0.0845	0.0000	0.0000
D6 community	0.0000	0.0000	0.0000	0.0736	0.0000	0.0000
E1 organization	0.0000	0.0000	0.0000	0.0000	0.3150	0.0000
E2 preparation	0.0000	0.0000	0.0000	0.0000	0.2187	0.0000
E3 medical	0.0000	0.0000	0.0000	0.0000	0.1238	0.0000
E4 shelter	0.0000	0.0000	0.0000	0.0000	0.0670	0.0000
E5 disaster training	0.0000	0.0000	0.0000	0.0000	0.2755	0.0000
F1 cleaning	0.0000	0.0000	0.0000	0.0000	0.0000	0.2216
F2 subsidiary	0.0000	0.0000	0.0000	0.0000	0.0000	0.1602
F3 laws	0.0000	0.0000	0.0000	0.0000	0.0000	0.2831
F4 project	0.0000	0.0000	0.0000	0.0000	0.0000	0.3351

Table 22.3 Eigenvector for each dimension in each decision-making area

	(A) Storm and rain effect	(B) Natural environment	(C) Social environment	(D) Software and hardware infrastructure	(E) Preparedness and emergency	(F) Resilience and adaptation
Disaster-prevention unit	0.1667	0.1714	0.0864	0.1311	0.2864	0.1581
Urban planning	0.1667	0.1329	0.0728	0.1263	0.2759	0.2255
Hydraulic engineering	0.1667	0.2285	0.0941	0.2024	0.1603	0.1480
All decision-making group	0.1667	0.1776	0.0844	0.1533	0.2408	0.1772

Taking the dimension of software and hardware facility as example, there is tremendous consensus in the distribution of the six indices. Experts demonstrated preference for the hardware aspects, such as flood-prevention facilities (0.3507), sewer construction ratio (0.2294), and infrastructure investment (0.1925). Meanwhile, the indices for software facilities, such as information broadcasting availability (0.0694), social assistance and welfare group (0.0845), and community self-governing group (0.0736), were clearly low. It is obvious that hardware construction is still the main factor in reducing typhoon hazard vulnerability.

The inter-dimensional network weight relationship is shown in Table 22.3. The super matrix from the limit operation makes the value convergent and obtains the final weight for all dimensions. It is very clear that the experts had a certain degree of consensus on the five intrinsic dimensions. In general, preparation and emergency responsiveness was considered the most important factor in vulnerability assessment, receiving a 30% importance rating. The experts in hydraulic engineering, however, stressed more software and hardware facilities than preparation and emergency responsiveness. Next, natural environment and resilience adaptation had similar weight value, and both were important assessment dimensions. The weight value of software and hardware facility condition was similar to the previous two. Social environment condition was commonly scored low, which was consistent with the previous expert consensus in the Delphi questionnaire. The main reason is that the overall urban social structure in Taiwan is close to average and with slight differences. In contrast, New Orleans (which was hit by hurricane Katrina) had mostly African-American and immigrant residents, and its social economic condition was much lower than that of other big cities with predominantly Caucasian populations. In New Orleans, therefore, the social and economic condition factor had higher weight.

The comparison and analysis of the index framework for urban typhoon hazard vulnerability is shown in Table 22.4.

Table 22.4 Relationship between metropolitan areas and dimensions

		(A) Storm and rain effect	(B) Natural environment	(C) Social environment	(D) Software and hardware infrastructure	(E) Preparedness and emergency	(F) Resilience and adaptation
		–	–	–	+	+	+
All decision-making group	G1 Taipei	0.2187	0.2228	0.2711	0.4382	0.3765	0.4507
	G2 Taoyuan	0.0748	0.1539	0.1122	0.1116	0.1240	0.0746
	G3 Taichung	0.2578	0.2690	0.2018	0.1387	0.1880	0.1924
	G4 Tainan	0.2303	0.2029	0.2106	0.0980	0.1074	0.0945
	G5 Kaohsiung	0.2183	0.1515	0.2044	0.2135	0.2041	0.1877
Disaster-prevention unit	G1 Taipei	0.1109	0.2650	0.2791	0.4853	0.2762	0.5356
	G2 Taoyuan	0.0358	0.1433	0.0832	0.1455	0.2700	0.0692
	G3 Taichung	0.2949	0.2455	0.1341	0.0732	0.0947	0.1266
	G4 Tainan	0.3621	0.2444	0.2767	0.0373	0.0817	0.0407
	G5 Kaohsiung	0.1964	0.1018	0.2269	0.2587	0.2774	0.2279
Urban planning	G1 Taipei	0.2198	0.1885	0.3545	0.5145	0.5154	0.5033
	G2Taoyuan	0.1105	0.2413	0.1292	0.0550	0.0472	0.0728
	G3 Taichung	0.2289	0.1987	0.1276	0.0995	0.1734	0.1822
	G4 Tainan	0.1614	0.2016	0.1156	0.0942	0.0701	0.0529
	G5 Kaohsiung	0.2793	0.1700	0.2731	0.2369	0.1939	0.1888
Hydraulic engineering	G1 Taipei	0.3255	0.2147	0.1798	0.3147	0.3379	0.3131
	G2Taoyuan	0.0781	0.0772	0.1241	0.1343	0.0549	0.0818
	G3 Taichung	0.2496	0.3627	0.3436	0.2433	0.2958	0.2685
	G4 Tainan	0.1675	0.1626	0.2393	0.1626	0.1704	0.1900
	G5 Kaohsiung	0.1793	0.1827	0.1131	0.1450	0.1410	0.1466

From the relationship between the regions and dimensions selected by decision-making group, it can be seen that the Taipei metropolitan area is positive in three dimensions: software and hardware facility, preparation and emergency responsiveness, and resilience and adaptation. The result is not surprising because Taipei is the capital city. The Taoyuan Chungli metropolitan area does not have a prominent dimension, and each eigenvector value is low, but it its negative dimensions, such as storm rain condition, natural environment condition, and social environment condition, were lower than those of other regions. Although it does not have artificial advantages, its inherent defects are fewer. Taichung is similar to Taoyuan in that it obtained average values for all dimensions. However, its eigenvector values in negative dimensions (storm rain condition, natural environment condition, and social environment condition) are a little higher, which translate into disadvantages. Meanwhile, Tainan has significantly higher eigenvector values in negative dimensions (storm rain condition, natural environment condition, and social environment condition) than positive dimensions (software and hardware facility, preparation and emergency responsiveness, and resilience and adaptation). Besides, the weight on negative dimensions is higher, so the overall assessment places the city at a disadvantage. This also greatly increases the vulnerability for Tainan metropolitan area. Finally, Kaohsiung has almost the same eigenvector values as Taichung, but it has better value in software and hardware facility and different weights in disaster prevention, urban planning, and hydraulic engineering.

After the analyses on index, dimension, and region, the study assessed the overall typhoon hazard vulnerability of each metropolitan area. With reference to Chang's (2009) BOCR model, the consideration of interest, cost, opportunity, and risk included different positive and negative dimensions. The selection was made according to the user's needs. The overall operation was conducted on eigenvector for positive and negative dimensions. The study based the assessment on the eigenvector values of six dimensions for each region. The first three dimensions (storm rain condition, natural environment condition, and social environment condition) were negative, while the next three (software and hardware facility condition, preparedness emergency responsiveness, and resilience and adaptation ability) were positive. Each had a different weight value. Thus, the study divided the analysis according to three major dimensions. The vector values of six dimensions for each region were multiplied by weight matrices and then all were added up to obtain the eigenvector value for the region; the smaller the value, the higher the region's vulnerability. The operation matrices are shown in Table 22.5.

The eigenvector from each dimension is not always the same. The study considered overall expert opinions. The typhoon hazard vulnerability is Tainan > Taichung > Taoyuan Chungli > Kaohsiung > Taipei. Tainan metropolitan area had quite high values for the first three negative dimensions because after a typhoon, the southwest stream usually brings heavy rain. The low land also makes the storm rain and nature dimensions generally disadvantageous. As a result, Tainan's vulnerability was the highest among all metropolitan areas. Taichung has also been subjected to typhoon hazards over years. Landslides and regional floods occur in Taichung every time there is a typhoon, for which hydraulic experts gave it

Table 22.5 Weight and performance of each metropolitan area

	Overall assessment	(A) Storm and rain effect (0.1667)	(B) Natural environment (0.1776)	(C) Social environment (0.0844)	(D) Software and hardware infrastructure (0.1533)	(E) Preparedness and emergency (0.2408)	(F) Resilience and adaptation (0.1772)
		–	–	–	+	+	+
G1 Taipei	0.1388	0.2187	0.2228	0.2711	0.4382	0.3765	0.4507
G2 Taoyuan	0.0109	0.0748	0.1539	0.1122	0.1116	0.1240	0.0746
G3 Taichung	−0.0072	0.2578	0.2690	0.2018	0.1387	0.1880	0.1924
G4 Tainan	−0.0346	0.2303	0.2029	0.2106	0.0980	0.1074	0.0945
G5 Kaohsiung	0.0346	0.2183	0.1515	0.2044	0.2135	0.2041	0.1877

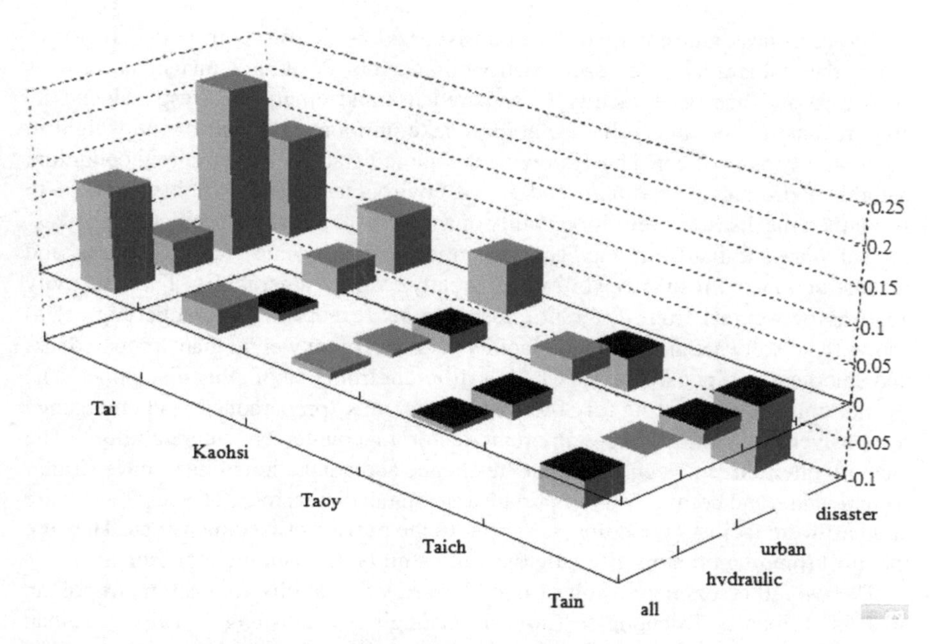

Fig. 22.3 Typhoon hazard vulnerability of each metropolitan area

very high vulnerability. Taoyuan Chungli is less affected by storms and heavy rain because of its mesa landscape, but it has poor software and hardware facility, emergency responsiveness, and resilience ability. Kaohsiung and Taipei are the two regions with the lowest vulnerability, as both have abundant capital, resources, and high-level software and hardware facility conditions, preparation and emergency responsiveness, and resilience and adaptation ability. Therefore, from the above analysis it was found that under the unpredictable storm rain condition, or natural environment condition and social environment condition, it is important to strengthen the disaster-prevention ability of a metropolitan area. In particular, increasing the capabilities in the areas of software and hardware facility, preparation and emergency responsiveness, and reconstruction and resilience will greatly reduce the vulnerability of metropolitan areas. This is summarized in Fig. 22.3.

22.5 Conclusions and Suggestions

This study established a flood disaster vulnerability indicators system for metropolitan areas and accommodated various assessment values for vulnerability and dimensions, including exposure, sensitivity, and resilience. Through the Fuzzy Delphi expert questionnaire, an indicators system was extracted, and ANP was utilized to adjust and revise the system and the assessment framework with network relationships, including six major dimensions and 27 indicators in Table 22.2.

"Preparedness and emergency responsiveness" is the most important factor to affect the vulnerability for each metropolitan area. Positive dimensions such as "software and hardware facility," "preparedness and emergency responsiveness," and "reconstruction and resilience ability" take up more than half of the weight in vulnerability assessment. Thus, knowing the unpredictability of storm rain condition, natural environment condition, and social environment condition, it is critical to strengthen the disaster-prevention ability of a metropolitan area. Specifically, increasing software and hardware facility, preparation and emergency responsiveness, and reconstruction and resilience ability will greatly reduce vulnerability. The sensitivity analysis of weights from different expert groups (except the hydraulic engineers) shows that "software and hardware facility" is given more weight than "preparedness and emergency responsiveness," which is different from that of other disciplines. The overall opinion, accounting for 30% importance, puts "preparedness and emergency responsiveness" as the most important factor to counteract "vulnerability." The "natural environment" condition and "resilience and adaptation ability" have similar weight value, and both are very important assessment dimensions. The next, "software and hardware facility" condition, is similar to the previous two dimensions. They are the most important factors affecting the vulnerability for each metropolitan area.

The overall assessment result of flood hazard vulnerability for each metropolitan area is Tainan > Taichung > Taoyuan Chungli > Kaohsiung > Taipei. Tainan metropolitan area had quite high values for the first three negative dimensions because usually, the southwest stream after typhoons brings heavy rains. The low land also makes overall storm rain and nature dimensions disadvantageous. As a result, its vulnerability was the highest among all metropolitan areas. Taichung has been subjected to typhoon hazards over the years; landslides and regional floods occur every time there is a typhoon. Therefore, hydraulic experts assigned very high vulnerability to Taichung. Taoyuan Chungli is less affected by storms and heavy rain because of its mesa landscape, but it has poor software and hardware facility, emergency responsiveness, and resilience ability. Kaohsiung and Taipei have the lowest vulnerability; both have abundant capital, resources, and high-level software and hardware facility conditions, preparation and emergency responsiveness, and resilience and adaptation ability.

The comparison of vulnerability in these metropolitan areas can be applied to the allocation of disaster-prevention resources. The framework of vulnerability assessment can be a useful tool for spatial planning and management; the potential applications are discussed in terms of the aspects of land use management, facility plan, risk communication, public finance and disaster insurance, and intergovernmental cooperation in the regional and local level. Further work in the formulation of planning guidance for each spatial administration level would be advisable. Several policy suggestions for reducing vulnerability are the following: (1) circulate hazard information and integrate and establish a complete disaster-prevention database; (2) strengthen preparation, emergency responsiveness, and resilience and adaptation ability; (3) balance the needs of disaster prevention among all regions; and (4) cooperate in preparation and disaster-prevention planning in different scales.

References

Bankoff G (2004) Mapping vulnerability: disasters, development, and people. James & James/ Earthscan, London

Birkmann J (2007) Risk and vulnerability indicators at different scales: applicability, usefulness and policy implications. Environ Hazards 7:20–31

Blaikie P, Cannon T, Davis I, Wisner B (1994) At risk: natural hazards, people's vulnerability, and disasters. Routledge, New York

Chang YH (1998) Transportation plan assessment and decision-making – study and application of fuzzy theory. Hwatai Culture, Taipei

Chang KF (2009) Super decisions software manual-breakthrough of AHP research limit by ANP. Maoding Publishing, Taipei

Chen CH (2001) Study of Asian Pacific port competitiveness and core strength index. Transportation 13(1):1–25

Chen LC (2003) Application of earthquake hazard risk- effect analysis on land use planning: HAZ-Taiwan System. Urban Plann 30(4):281–299

Chen LC, Chen HL (2007) Investigation of disaster-prone urban spatial development: formation of flood in Taipei basin as example. Urban Plann 34(3):293–315

Cheng TB (2001) Fuzzy assessment model for software organization to increase personnel ability maturity. Master thesis, Information Management, National Taiwan University of Science and Technology

Dalkey NC (1969) The Delphi Method: 9n experimental study of group opinion. RM-5888-PR. 48, The Rand Corporation, Santa Monica

Delbecq AL et al (1975) Group techniques for program planning: a guide to nominal group and Delphi processes. Scott, Foresman and Company, Glenview

Gallopin GC (2006) Linkage between vulnerability, resilience and adaptive capacity. Glob Environ Chang 16(3):235–316

Hsiao HC (2008) Establishment of flood vulnerability assessment model – Sijhih City as an example. PhD thesis, Department of Geology, Chinese Culture University

Lin KH (2004) Vulnerability and adaptation research method and methodology study under global climate change. Glob Climate Change Commun 43:33–38

Turner BL II et al (2003) A framework for vulnerability analysis in sustainability science. In: Proceedings of the National Academy of Sciences of the United States of America. www.pnas.org/cgi/reprint/100/14/8074.pdf

White GF (1945) Human adjustment to floods. Research Paper 29. Department of Geography, University of Chicago, Chicago

White GF, Haas JE (1975) Assessment of research on natural hazards. MIT Press, Cambridge

Chapter 23
The Post-Disaster Reconstruction and Socioeconomic Vulnerabilities in the Historical Site of an Island City: A Case Study of a Fire Incident in Nan-Gan Township, Lien-Chiang County, Taiwan

Chi-tung Hung, Wen-yen Lin, and Ju-yin Cheng

Abstract The case of Chun-You Street fire in downtown Nan-Gan Township revealed problems in Lien-Chiang County (also known as the Matsu Islands), such as conflicts in reconstructing post-disaster historical areas and the socioeconomic vulnerability to disasters. The changing urban form during previous decades generated politically oriented planning policies, tourism development policies, a unique population, and socioeconomic vulnerability. This study used a participatory observation methodology and in-depth interviews to investigate the following issues of the Chun-You Street fire: (1) the diversity of the county emergency management system during the response stage and the unique recovery process of a historical site that is based on the clan system, (2) the vulnerabilities caused by residents' experiences in lacking daily supplies and the custom to store great amounts of materials, and (3) victims' unclear property rights and land uses pertaining to past verbal agreements creating an obstacle to the recovery of a historical area.

Keywords Historical spatial development • Lien-Chiang • Disaster recovery • Socioeconomic vulnerability

Lien-Chiang County was a militarily administrated area until 1992. Compared with other counties in Taiwan, its historical background and the evolution of industrial patterns have made Lien-Chiang County unique. The Nan-Gan Township is the

C.-t. Hung (✉) • W.-y. Lin
Department of Urban Planning and Disaster Planning, Ming Chuan University,
5 De Ming Rd., Guishan Township, Taoyuan County 33348, Taiwan
e-mail: ct@mail.mcu.edu.tw; wylin01@mail.mcu.edu.tw

J.-y. Cheng
Department of Urban Development, Taipei City Government, 1, Shifu Rd.,
Xinyi District, Taipei City 11008, Taiwan
e-mail: mickey0988@uro.taipei.gov.tw

M. Kawakami et al. (eds.), *Spatial Planning and Sustainable Development: Approaches for Achieving Sustainable Urban Form in Asian Cities*, Strategies for Sustainability,
DOI 10.1007/978-94-007-5922-0_23,

political and economic center of Lien-Chiang County, and Chun-You Street is a central business district (CBD) and historical areas in downtown Nan-Gan. On January 20, 2008, the Chun-You Street fire burned 27 houses down in Nan-Gan Township (Lien-Chiang County Government 2008). This incident revealed that the local sociocultural environment, clan organizations, and urban regime have been ignored in disaster management and disaster planning research. For instance, the insecurity suffered by residents because of a lack of daily supplies during the militarily administrated period ultimately led to this disaster. Nevertheless, cooperation among residents and emergency management personnel was highly efficient during the disaster response stage; yet the traditional property rights system (wu-shi-mo-tau) in the clan society has resulted in a slow disaster recovery process, the loss of regional spatial culture, and urbanization.[1] In particular, this unique system also increased the socioeconomic vulnerabilities and substantially influenced the sustainability of a historical area. In general, collaboration between the local military forces and the clan society resulted in Nan-Gan Township being a superior community to other urbanized cities. These unique characteristics should accelerate the recovery process on Chun-You Street; however, the land ownership disputes revealed the quandary of this sociocultural phenomenon.

23.1 Introduction: The Socioeconomic Vulnerabilities, Military Culture, and the Clan Society in Lien-Chiang County

Increasing resilience can decrease vulnerability and economical losses during a disaster (Mileti 1999; Liu et al. 2006). Building a more highly resilient community and analyzing its vulnerability is a practical approach toward reducing physical damage and socioeconomic losses (Bruneau et al. 2003; Pelling 2003; Lin 2004). In this instance, the intensity of local social collaboration depends on the correlations among vulnerability, resilience, and spatial hazard reduction. Considering the Chun-You Street fire incident as an example, although the clan system and local military forces increased local resilience, vulnerability and the traditional property rights system resulted in higher reconstruction costs as well as the difficulty of applying zoning regulations (which require business owners to reduce the commercial zone and provide more lands for public use). Furthermore, the aforementioned circumstances are also resulted in higher rents.

Emergency Management in Australia (EMA) reported that networking among individuals in a community is critical for hazard resilience and recovery (Buckle et al. 2001). The United Nations International Strategy for Disaster Reduction (UNISDR) defines resilience as "the communities and individual's ability to bear

[1] Wu-shi-mo-tau is a local term meaning that a renter has the right (ownership) to his/her property as long as that renter's house still stands on the land in question; and the landlord has no right to reclaim this property unless the house is removed from the land.

disasters and to make the community functional and structural balanced." Similarly, this study defines resilience as "the systematic social abilities to self-adjust by their social organizations. Community resilience could actually enhance the disaster recovery process and decrease the vulnerability." This definition is based on two concepts: (1) spatial sustainability and resilience have become major concerns in disaster management, which relates to social, cultural, economical, political, and natural factors (Tobin 1999; Shaw and Goda 2004; Rozdilsky 2001; Flax et al. 2002) and (2) resilience is related to local development and economic growth, such as the tourism industry and supporting facilities, large-scale infrastructure construction, and commercial facility development.

This study considers resilience to be the positive capacity of people, communities, clan systems, organizations, and governments to cope with damages from disasters. The characteristics of resilience present primarily as (1) reduced strike probability, (2) reduced damage, and (3) quickened recovery. The Chun-You Street fire can be considered an example to explain the connections between vulnerability and resilience. The vulnerability revealed by this incident represents "status" aspects, such as the insecurity of possessing daily goods, isolated location (Lien-Chiang County is off the coast of the main island of Taiwan but is adjacent to a southeast economic region in China), environmental sensitivity, and land capacity concerns that result from the tourist industry. By contrast, the resilience of the community affected by the Chun-You Street fire is manifested in the "process" aspects, including the traditional property right system (wu-shi-mo-tau), and military/firefighting disaster response system.

23.2 Research Methods

The concept of socioeconomic vulnerability is often applied in hazard- and disaster-related research. Spatial vulnerability analysis can be used to identify the relationships between people and their living environments, including socioeconomic forces, government agencies, and cultural values (Hu 2008). Socioeconomic vulnerabilities can vary according to the type of disaster. Consequently, there are different aspects that need to be addressed in disaster management. Different levels of government also have different considerations on connections between the socioeconomical vulnerability and spatial planning.

To investigate socioeconomical vulnerability and planning further, in addition to surveying the Chun-You Street fire site and measuring the spatial locations, this study collected data by utilizing unstructured in-depth interviews and reviewing secondary literature. In-depth interviews are used to assist researchers in analyzing information provided by interviewees' own words and their understanding (Tayor and Bogdan 1984). Conducting face-to-face in-depth interviews could, therefore, assist this study in clarifying the facts of the Chun-You Street fire incident. The literature review method could facilitate recognizing the status quo interpretations of what occurred in the past; for example, by reviewing the local emergency plans and reports produced by "Empower the Matsu Government's Regional Hazard Reduction Plan Project" (Lee and Hung 2007; Hung 2008).

Interviews were conducted during 2007 and 2008 while the disaster reduction plan was being elaborated, and the interview subjects were emergency officials and public servants from the fire department, Civil Affairs Bureau, Construction Bureau, and military, along with scholars and researchers at National Taipei University and the National Science and Technology Center for Disaster Reduction. From March to May 2009, this study also conducted in-depth interviews with civilians such as the victims of the Chun-You Street fire, local residents, and hotel owners.

23.3 Results: Disaster Management in an Island City: Vulnerability, Safety, and Development

The environmental hazards of Lien-Chiang County have made fire a common problem. The affected area and socioeconomic loss of the Chun-You Street fire were tremendous. It was a result of increasing social vulnerability, the spatial development framework, traditional land use system, and traditional social connections (the clan system). The fire occurred on Chun-You Street in Jie-Shou Village, Nan-Gan Township (as shown in Fig. 23.1) on January 20, 2008. The fire started in a terraced house located at no.308, at the end of Chun-You Street. Based on the statement of the Lien-Chiang County government (2008), the cause of the fire was an electric short circuit, a common cause of fires in old housing units.

The fire destroyed 27 terraced houses along the Chun-You Street (the affected area was approximately 19,170 ft^2). The economic loss was approximately US$937,500. Two major factors contributed to the disaster: (1) because the fire occurred at night, the residents were not alert and were late in response, and (2) the strong northeastern monsoon winds of January caused the fire, which continued for nearly three and half hours, to spread rapidly. Chun-You Street is part of a historical area on the west side of Jie-Shou Village. Despite being developed later than the older streets on the east side of Jie-Shou, Chun-You Street suffered the common problems of old streets, such as being built without flameproof materials, no fire lanes between buildings, and highly dense partitions between living spaces. Eventually, firefighters used hydraulic excavators to destroy some buildings and created a fire line to prevent the fire from spreading. According to the analysis from interview results and the literature reviews, the following are the research findings of this street fire investigation.

(1) Traditional terraced houses sped up the fire's expansion
  Most of the building structures on Chun-You Street were terraced houses adjacent to each other, as shown in Fig. 23.2. Many of the houses were two or three stories (approximately 19–32 f. high), and most of their structures or interior decors were made of wood or other flammable materials. These terraced

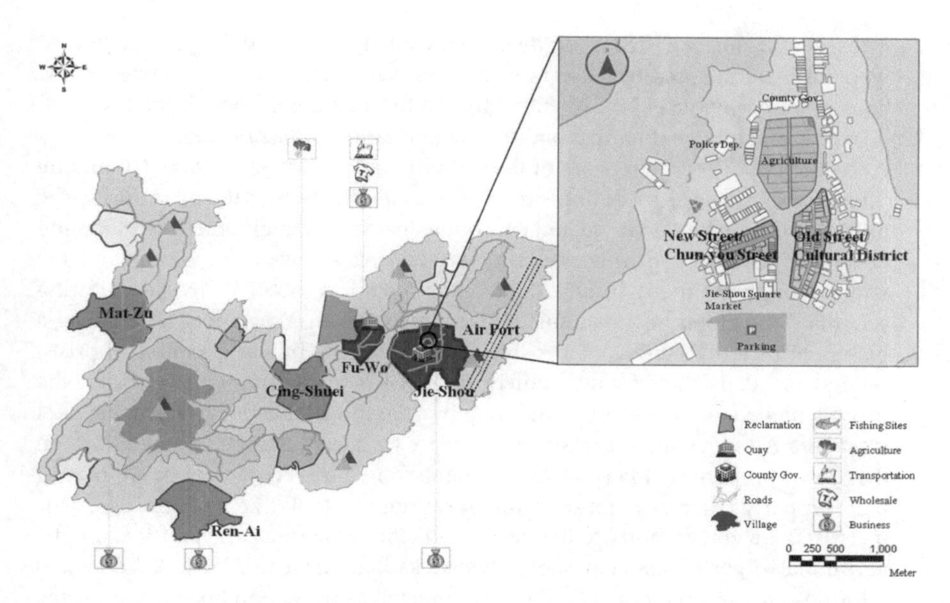

Fig. 23.1 Location of Chun-You Street

Fig. 23.2 Fire site on Chun-You Street

houses were built in the 1950s and had not been remodeled since then. According to the field surveys and interviews obtained from local residents (F_R2, M_R5, and F_R7), the following facts were observed:

[Eighty percent of the terraced houses belonged to a single owner. To expedite the construction process and reduce building costs, this owner used bricks to construct the buildings and wooden partition walls to divide the inner spaces into several units (approximately 533.75 sq. ft. each). The owner used wooden materials to construct the interior design as well.]

Because the interior decor of these terraced houses used large amounts of wooden materials and because no fire lanes were between the buildings, these houses were obviously highly vulnerable to fire that could spread quickly.

(2) Mixed use land development and the accumulation of hazard items

According to historical records of the militarily administrated period, Chun-You Street was famous as a "dining street." It was a reflection of the days during the military administrated period and part of the historical development of blooming economic activities. Several restaurants and snack bars were located on the street, along with many beauty salons, photo studios, book shops, and tutoring schools. This blooming economic activity made Chun-You Street a CBD and a historical area of Nan-Gan Township. Because Lien-Chiang County comprises several isolated islands with limited options for transportation, therefore, the inconvenience of acquiring daily required supplies made the residents feel insecure. As a result, the residents tended to accumulate their materials on their own properties. Many of the residents on Chun-You Street made part of their property (such as a house, shop, or restaurant) look like a warehouse. This increased the vulnerability to fire because (1) the restaurant owners, for example, accumulated gas tanks, disposable chopsticks, and dishes[2]; (2) book stores and tutoring schools stored many inflammables such as paper and books; and (3) the building structures were quite old and lacked fire lanes between them. In addition, because of limited indoor space, it was not uncommon for residents and restaurant owners to leave many flammable materials on the street for their own convenience, resulting, of course, in higher vulnerability to fire.

(3) Residents lacked disaster awareness

Lien-Chiang residents living in old buildings do not take housing safety seriously, unless obvious problems affect their lives. For the houses designed and built 50 years ago, electricity use is higher than the originally designed capacity, but few of the residents actually consider the capacities and conditions of electric wires. Such increased electricity use has caused short circuits, which resulted in small fire incident happening a lot in the past. Because the residents getting used to these events, they have resulted in lacking of fire disaster awareness and slower responding to the fire. According to the interviewee (F_R12):

> [Fires occurred frequently on this street. This is not the first time. . . It was always because of the old cable line and electric short circuit. Usually, the firefighters can put the fire out very quickly. So, we do not really pay much attention to fire incidents.]

In addition to the physical vulnerability (old cable line and electric short circuit), timing was a factor in exacerbating effects of the fire, which occurred from 8:00 to 9:00 p.m. This is the rush time for restaurants on Chun-You Street. On the

---

[2] In Taiwan, some old communities still use gas tanks to generate heat for kitchens and hot water. Regular households order only one gas tank from the gas store at a time; but households and restaurants in Lien-Chiang prefer to store two to four gas tanks to ensure a gas supply. In addition, restaurants in Taiwan usually provide disposable chopsticks and dishes to customers to avoid wasting manpower and time washing dishes.

night of the fire, residents believed that the electric short circuit occurring in one of the restaurants was a usual event and that the restaurant owners were accustomed to managing this type of problem by themselves. Such a lack of fire disaster awareness increased social vulnerability and resulted in a late response.

(4) Traditional social capital and the low rent rate

The building structures on Chun-You Street were quite old. The living conditions were also not ideal. Each terraced house unit had only 533.75 ft^2. Based on statements from the residents, at least two reasons explain why the business owners continued to operate on this street. First, Chun-You Street was popular CBD in Nan-Gan Township, creating greater business opportunities than other areas did. Second, the traditional social relationships and property rights system (wu-shi-mo-tau renting system) made the rent lower than in other places. By summarizing the interviews conducted with the residents (M_R13 and F_R12), this study found that the wu-shi-mo-tau renting system resulted in this vulnerable situation as follows: (1) the wu-shi-mo-tau renting system allows renters to have ownership of the terraced house for all matters, and the landlords have no right to reclaim their property. The rent was only US$125/year. The landlord can only terminate the lease if the building structure itself is removed from the property. (2) Landlord ceased continuing several wu-shi-mo-tau contracts after the 1990s. The rents for these new leases are three times that of the old wu-shi-mo-tau contracts. Furthermore, the existing wu-shi-mo-tau contracts and Chun-You Street's geographical location (close to a bus stop and Jie-Shou market square[3]) made this street one of the most popular business districts. Therefore, the wu-shi-mo-tau contracts attracted great numbers of business (restaurant) owners, and most of them have never left. The numerous restaurants increased the fire hazard vulnerability. In addition, the wu-shi-mo-tau contracts also resulted in a slow recovery process for this fire incident.

23.4 Discussion: The Difficulty of Post-Disaster Reconstruction Efforts and the Local High Resilience and Vulnerable Land Use System

According to our analysis, the Chun-You Street fire resulted from land use features such as the core business district, traditional terraced houses, accumulation of hazardous materials, and a highly vulnerable population (especially the elderly and children). Although, fortunately, no deaths or serious injuries resulted from this incident, the fire still revealed the vulnerable living conditions and difficulty of post-disaster reconstruction efforts in Lien-Chiang County. The geographical and

[3] The Jie-Shou market square is a place for housewives to shop for groceries. This type of market square provides fresh products every day. Some housewives maintain social connections in these markets square while shopping.

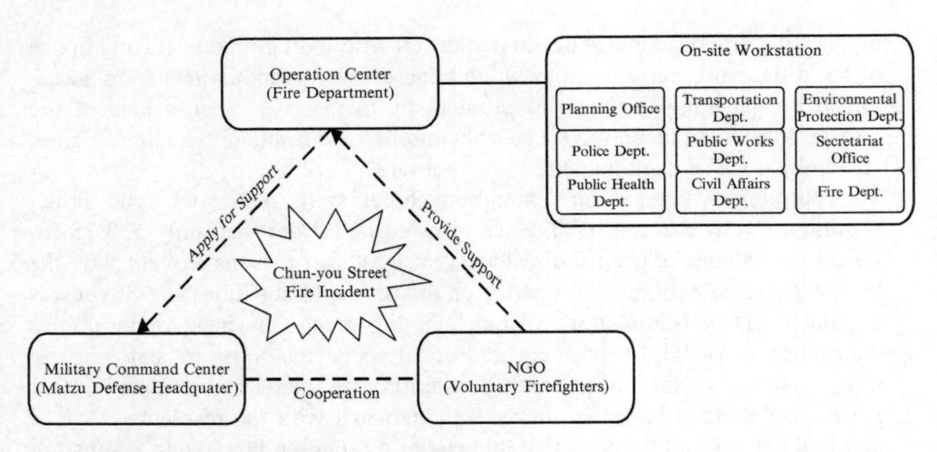

Fig. 23.3 Chun-You Street fire disaster response system

demographical characteristics make this a unique circumstance (Lien-Chiang County is a combination of islands, with many military bases on Nan-Gan Island). More importantly, the local culture made the disaster response and recovery systems different from those of most disaster management systems.

(1) Traditional clan system and high resilience
Most Lien-Chiang County residents are descendents of people from the coast of Fujian province in China, who immigrated long before the political separation between China and Taiwan. These immigrants came to Lien-Chiang County in the seventeenth century (at the end of Ming Dynasty and the beginning of the Qing Dynasty) and gradually evolved into clans with familial relationships (with Chen, Lin, Cao, Wang, and Liu as the most common family names). These social relationships still exist today and have become the traditional clan system. Most of the renters on Chun-You Street are related to each other in some way, and they are familiar with each other's families and houses. From a disaster management perspective, this clan system actually improved the rescue process of the Chun-You Street fire incident. For instance, (1) the residents helped firefighters locate victims and rescue them from the fire, thus making the rescue process more efficient and (2) the residents had higher incentive to work with or to provide useful resources to the firefighters because they were actually helping their families. Therefore, the immediate assistance demonstrated by the traditional clan society reduced the damages and casualties of this fire. Such a system can provide a neighborhood and community with self-defense security mechanisms.

(2) Local triangle disaster relief system
In the function of the Lien-Chiang local bureaucratic system, the fire department took charge of putting out the fire on Chun-You Street. The police department and military both assisted disaster relief logistics tasks. The rest of the government departments established an on-site emergency station to support victim's needs (as illustrated in Fig. 23.3).

Additional factors contributed to this disaster response system. First, regarding the disaster relief system: (1) the rescue operation was assisted by private organizations; (2) the military forces dispatched 40 soldiers to assist with the disaster response activities and established a military emergency command center to provide assistance in rescuing activities; (3) although the police, military, and the personnel in substitute military service provided necessary assistance,[4] they could not overcome the shortage of professional fire fighters (only 11) at the local fire station; therefore, the NGO (voluntary firefighters) was required to provide vital support to the firefighters[5]; and (4) the local disaster response system highlighted the cooperation among the military, civilians, fire stations, and other government departments.

Second, regarding the rescue process: (1) the local troops' self-organization reflected the regional military administration background of the past; (2) because of the reforming of the local government administration,[6] military units are no longer in control of political affairs; however, they can still provide vital assistance to local emergency planners; therefore, local emergency plan must incorporate the military as essential resource; the "standard operating procedure" including the roles of military bases, fire stations, and NGOs needs to be established; and (3) Chun-You Street fire incident also exposed the failure of the central government's funding standard. The entire funding is controlled and distributed by the central government, and the central government funds "the local disaster relief project" based on the local population and their territory. With the minor population and territory, it is obvious that Lien-Chiang County was not able to obtain enough grants.[7]

(3) High-quality shelters and immediate disaster aid

The local government established an emergency shelter at the Jie-Shou Village senior activity center immediately after the fire occurred and began accommodating victims. Because the major commercial activity in Lien-Chiang County is tourism, there are many hotels in Nan-Gan Township. The government officials began negotiating with the hotel owners and provided temporary shelter for the victims in hotel rooms with more comfortable environments. Compared to the victims who used tents or vehicles in public areas as temporary shelters for a

[4] The substitute military service is an alternative option for male adults who could not complete their military service physically or economically. In this instance, the substitute military service personnel work in the local fire station.

[5] The voluntary firefighters in Lien-Chiang County provide assistance to firefighters only and do not need to go to the incident scene to rescue victims.

[6] Lien-Chiang County is located between Taiwan and China and was militarily administrated for over 40 years. The Taiwanese government decided to relieve the military administration after 1998. However, military forces still play a vital role in emergency response.

[7] The island of Taiwan is approximately 13,866 miles^2. The total population was nearly 23 million in 2010. By contrast, the size of Lien-Chiang County is only 11.11 miles^2, and the population in 2010 approximately about 10,000 (Department of Household Registration 2009).

long period after the 1999 Ji-Ji earthquake,[8] the Lien-Chiang County government provided much higher quality "public shelters" to the victims of the Chun-You Street fire. In addition, the Lien-Chiang government allotted US$625 in disaster relief aid to every household on Chun-You Street. Compared to the Ji-Ji earthquake, the victims could only apply for financial aid under certain conditions such as serious injury, death, or a collapsed house. The Chun-You Street fire incident has altered the general conception that the sheltering conditions in remote islands are worse than the conditions on the main island of Taiwan. The smaller population and closer social connections between people make the sheltering and financial aid easier and more efficient.

23.5 Conclusion

The Chun-You Street fire incident exposed the socioeconomic vulnerabilities of the local government's planning for preserving historical buildings and culture. This incident revealed that the government did not properly manage and maintain well-worn infrastructures in the business and tourism areas that had transformed from military commercial districts. Nevertheless, as this chapter illustrates, the unique traditional property rights system (wu-shi-mo-tau) has been an obstacle to the rebuilding process, because of the limited rights caused by the small-divided lands, and building spaces squeezed by public facilities after recovery. Despite the local government's wishes to restore the traditional cultural landscape without further delay, the fire-consumed area has not changed much since the incident 3 years ago and has become an alternative parking area, creating hazards to the preservation of historical streets.

This chapter reports an empirical study of the Chun-You Street fire incident for discussion and analysis of post-disaster historical area restoration. Figure 23.4 illustrates the expectations of relationships among stakeholders: landlords (owners), renters (who varied in local customs and rents paid), and the local government (in developing local culture and industries). The findings can hopefully become useful references for other planning practices and research on similar topics.

(1) Community-culture-oriented restoration space agents: the role of stakeholders
Figure 23.4 illustrates the relationships of the most crucial stakeholders in this particular community reconstruction: landlords, renters, and the Lien-Chiang County government. Their roles are further discussed as follows:

1. The Landlords: The wu-shi-mo-tau contracts had limited the rent for the landlords for a long period. In the last decade, most of the original landlords

[8] The Ji-Ji earthquake is the costliest earthquake disaster in Taiwanese history.

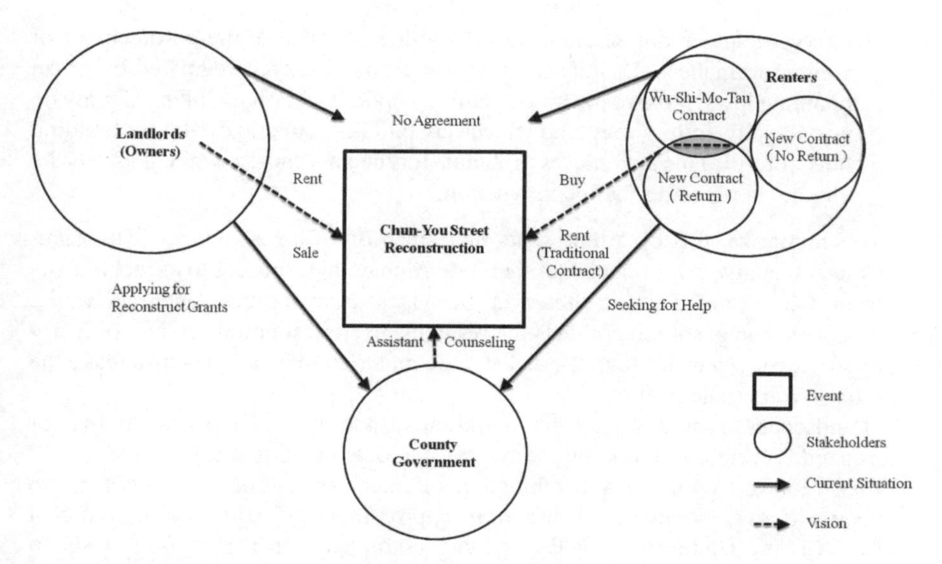

Fig. 23.4 Role of stakeholders in the Chun-You Street fire incident

have already passed the land ownerships to the younger generations. These young landlords have argued not to honor the traditional wu-shi-mo-tau contracts and have tried to raise the rents to match the market price level. In addition, they have wanted this street to become a pure commercial area, not a mixed land use area. Thus, they could increase their own profit; however, Chun-You Street would lose its Ming-Nan cultural capital forever, and the traditional spatial land use style would no longer exist.

2. The Renters: Renters have different perspectives regarding the reconstruction issue. There are three types of renters: renters with original wu-shi-mo-tau contracts, type-1 new contract renters (to reenter the Chun-You Street site), and type-2 new contract renters (to rent other properties and not return). The wu-shi-mo-tau contract renters and type-1 new contract renters want to return to the area and resume their businesses. They expect either to continue their old contracts or buy the land from the landlords. They do not believe that this fire incident should influence their property rights. By contrast, the type-2 new contract renters have begun new lives in other places and have no intention of returning to Chun-You Street. In fact, some of them feel that the wu-shi-mo-tau contracts had allowed them to take advantages of the landlords for a long time; therefore, it is time for them to move on and not come back. Thus, finding a common solution to satisfy everyone is difficult.
3. Lien-Chiang County Government (LCCG): Although the Lien-Chiang County government is trying to help residents return to their normal lives as soon as possible, the landlords and renters still hold different opinions about the reconstruction and property rights. Because the lands of the Chun-You Street site are private properties, the county government has no right to

intervene under this circumstance. Another obstacle is the enforcement of zoning ordinances. Despite the fact that entire county is regulated by urban planning laws, most of the houses built by civilians in Lien-Chiang County do not actually follow regulations such as building coverage ratio or building bulk control. This also causes problems for the government while it assists the victims in planning the reconstruction.

To summarize, this case represents the transition of an island city. The Lien-Chiang County government uses their environmental resources to attract tourists from China and Taiwan; however, the lack of resources (such as water, groceries, and gasoline) compels the residents to store supplies on their property to provide services for tourists, and storage materials such as gas tanks make the village vulnerable to fire.

(2) Landscapes transforming from a militarily administrated period to that of liberalized touring consumption: the role of the local government

Lien-Chiang County is a combination of many small islands; however, the county is over populated. There were approximately 50,000 troops living in Lien-Chiang County during the military administration period from 1949 to 1992 (Tam-Kang University 2003). After 1992, the number of troops has decreased, but the number of tourists has increased. A total of 66,427 tourists visited Lien-Chiang County in 2009, which was 6.6 times more than the total of permanent residents (Tourism Bureau 2010). The Lien-Chiang County government has tried to support this population of both troopers and tourists for many years. They have had to improve the transportation system, hotels, electricity systems, roads, and restaurants. Lien-Chiang County has become overloaded and has exceeded its carrying capacity. This situation and the Ming-Nan culture made Lien-Chiang County vulnerable to fire in various social, physical, and economic ways.

In the future, especially regarding the reconstruction of Chun-You Street, the local government should consider the sustainability of Lien-Chiang County and try to contend with the vulnerable situation. The county government should try to use their local environmental features to attract tourism, rather than simply throw government funds into reconstruction. Building an environmentally friendly street (or county) can make it sustainable and decrease its vulnerability in the disaster mitigation phase (Hung 2008). The Lien-Chiang County government, in managing an island city and reconstructing local historical streets, should consider the precious island tourism resources, limited capacity, and disaster compounding factors when planning for the impacts of extreme weather changes. Historical architecture styles such as "Feng-Huo-Shan-Qiang" (flame-shaped raised gables) can reduce the risk of spreading fire. In strengthening the resilience against disaster, future recovery efforts should also consider improving the physical environment and express it in historical contexts by integrating traditional architectural elements. Lien-Chiang, with local fishing village buildings of the Ming-Nan cultural style (which differs from Kinmen's South Pacific colonial architectures), should be carefully conserved. Public participation during recovery process and the local government's planning for island tourism and basic infrastructures to adapt to extreme weather changes both hold the keys to the success of the historical street reconstructions. The rebuilt cultural style of the future Chun-You Street will hopefully possess localized, hazard-mitigated, and creative characteristics of the tourism industry.

References

Bruneau M, Chang S, Eguchi R (2003) A framework to quantitatively assess and enhance seismic resilience of communities. Earthq Spectra 19(4):733–752 (in Chinese)

Buckle P, Graham M, Smale S (2001) Assessing resilience and vulnerability: principles, strategies and actions. Department of Defense Project 15/2000, Canberra

Department of Household Registration (2009) 2008 household statistics. http://www.ris.gov.tw/web_eng/eng_sta.html. Retrieved 20 Dec 2009

Flax LK, Jackson RW, Stein DN (2002) Community vulnerability assessment tool methodology. Nat Hazards Rev 3(4):163–176

Hu AG (2008) Reconstruction after the Wenchuan Earthquake: the Establishment of Public-Private Partnership Mode in China. The research reports of Reconstruction Planning Team of National Development and Reform Commission. Beijing: Tsinghua University (in Chinese)

Hung CT (2008) Revising Lien-Chiang County's Disaster Reduction Plan Project. Department of Urban Planning and Disaster Management, Ming-Chuan University, Taipei (in Chinese)

Lee C, Hung CT (2007) Empower the Matsu Government's Regional Hazard Reduction Plan Project, Taipei. Department of Urban Planning and Disaster Management, Ming-Chuan University, Taipei (in Chinese)

Lien-Chiang County Government (2008) The disaster response and recovery report of Chun-You Street Fire. http://www.lcfd.gov.tw/Web/%E6%96%B0% E8%81%9E/%E5%89%A A%E5% A0%B1/20080131/20080131-4.htm. Retrieved 20 Dec 2009 (in Chinese)

Lin KH (2004) Vulnerability issues under global change and adaptation research methods and methodology discussion. J Global Change Commun 43:33–38

Liu Z, Shi P, Ge Y (2006) Disaster resilience in the development process. Process Earth Sci 21 (2):211–218

Mileti D (1999) Disasters by design: a reassessment of natural hazards in the United State. Joseph Henry Press, Washington, DC

Pelling M (2003) The vulnerability of cities: natural disasters and social resilience. Earthscan Publications, London

Rozdilsky JL (2001) Second hazards assessment and sustainable hazards mitigation_disaster recovery on Montserrat. Nat Hazards Rev 2(2):64–71

Shaw R, Goda K (2004) From disaster to sustainable civil society: the Kobe experience. Disasters 28(1):16–40

Tam-Kang University (2003) History of Lien-Chiang County. Lien-Chiang County Government, Lien-Chiang (in Chinese)

Tayor SJ, Bogdan R (1984) Introduction to qualitative research methods, 2nd edn. Wiley, London

Tobin G (1999) Sustainability and community resilience: the holy grail of hazards planning. Global Environ Change B Environ Hazards 1(1):13–25

Tourism Bureau, M.O.T.C. Republic of China (Taiwan) (2010) Visitors to the principal scenic spots in Taiwan by Month, 2009. http://admin.taiwan.net.tw/statistics/File/200912/tourist_spots_2009.pdf. Retrieved 15 January 2010 (in Chinese)

Chapter 24
Sustainable Communities in Hilly, Mountainous and Heavy Snowfall Areas

Asako Yuhara and Kyung-rock Ye

Abstract Japan has large hilly and mountainous areas with heavy snowfall. Because of severe population decrease arising from the aging society, many communities in such areas have become quite vulnerable to natural disasters. In addition, public help in disaster reduction has been weakened because of difficulties in public finance. In order to consider sustainable communities in these circumstances, it is necessary to examine the power of both mutual and self-help. In this study, we examine some approaches to coping with severe snow disasters in several communities in hilly, mountainous, and heavy snowfall areas. First, problems arising from winter are considered through information from researching more than 300 local governments in heavy snowfall areas in Japan. Second, the outcomes of case studies in Semboku City in Akita Prefecture in northern Japan are discussed. The results show the importance of residents' associations and such local groups. In addition, by enhancing communication among the communities, these kinds of workshops are important for strengthening the power of mutual support.

Keywords Sustainable community • Hills • Mountains • Heavy snowfall area • Snow disaster • Mutual help • Self-help

A. Yuhara (✉)
National Institute for Land and Infrastructure Management,
1 Asahi, Tsukuba, Ibaraki 305-0804, Japan
e-mail: yuhara-a92ta@nilim.go.jp

K.-r. Ye
Graduate School of Frontier Sciences, The University of Tokyo,
Hongo 7-3-1, Bunkyo District, Tokyo, Japan
e-mail: ye.kr92@gmail.com

M. Kawakami et al. (eds.), *Spatial Planning and Sustainable Development: Approaches for Achieving Sustainable Urban Form in Asian Cities*, Strategies for Sustainability,
DOI 10.1007/978-94-007-5922-0_24,

24.1 Introduction

Japan has large hilly and mountainous areas with heavy snowfall (Shimizu and Abe 2001). The government specifies some places as "heavy snowfall areas" and "special heavy snowfall areas" by measuring several conditions, such as snow depth, under the Act on Special Measures Concerning Heavy Snowfall Areas, aiming to support the development of these areas. Although "heavy snowfall areas," including "special heavy snowfall areas," occupy approximately 50% of the country, only 16% of the population lives there. Population in these areas is declining: whereas total population in Japan increased 9% from 1985 to 2005, the population in heavy snowfall areas decreased 4%. Furthermore, in 2000 the Population Census showed that the share of population aged 65 and over in heavy snowfall areas was already 20%, while that of the total population was 17% (Ministry of Land Infrastructure, Transport and Tourism (MLIT) 2010).

Because of steady population decline in the aging society, many local communities in such areas have become quite vulnerable to heavy snowfall disasters. Kuriyama (1986) found that the elderly occupied high percentage of fatalities caused by snow disaster from the research on seven winters from 1978 to 1985 in heavy snowfall areas. And this trend was also found in the 2000s. Sixty-five percent of fatalities because of heavy snowfall disasters between 2005 and 2006 involved the elderly (Fire and Disaster Management Agency (FDMA) 2006). This situation prompted a revision of the basic plan for measures in heavy snowfall areas in order to emphasize the problems caused by depopulation and aging. However, a similar situation happened in the winter of 2010–2011 (FDMA 2011). In addition, public help for disaster reduction has been weakened because of recent financial difficulties (Hochrainer and Pflug 2009). Snow removal costs are an especially heavy burden for local governments in snowy areas.

MLIT (2009) reported that approximately 34,000 communities exist in the snowy regions, and population decline has been clearly seen in three-fourths of them in the last decade. Twenty percent of the communities are experiencing difficulty in maintaining themselves because of the severities of daily life in the winter.

In order to consider sustainable communities in these circumstances, it is necessary to examine how to manage the snow. The power of mutual and self-help in local communities is the particular focus of this chapter.

We here investigate the snow-related problems that local governments have attempted to solve. We also explore a collaborative approach to tackling such problems in local communities. Section 24.2 provides the methods used in this study. We describe the research results of questionnaires distributed by local governments in heavy snowfall areas and the results of case studies in Sect. 24.3, and we conclude in Sect. 24.4.

24.2 Methods

In order to examine what kinds of problems local governments have and how they address such problems, we conducted research by means of questionnaires directed at local governments, followed by case study research on approaches to solving snow-related problems through building a community of practice approach (Yamori 2009; Lave and Wenger 1991).

24.2.1 Questionnaire Research

The questionnaire survey for all the municipalities in heavy snowfall areas was aimed at examining the problems of and approaches to winter life with heavy snowfall. In 2008, 564 questionnaires were mailed, and 333 (61%) responses were received.

The questionnaires were designed as shown in Table 24.1. Each questionnaire consisted of three parts, and each part was designed for a specific section of the municipality. Consequently, three officers in each local government were required to answer the questionnaire. The three parts of the questionnaire covered general affairs, welfare, and road construction and maintenance.

24.2.2 Case Study Research

After the questionnaire survey was conducted, we began case study research in order to examine in more detail the problems posed by winter and how communities deal with them. This action research was carried out in Semboku City, which is one of the largest cities in a heavy snowfall area in Japan.

24.2.2.1 Location and Context of the Case Study Areas

Semboku City lies in the northern part of Japan. The area of the city grew to more than 1,000 km^2 after the amalgamation of Tazawako Town, Kakunodate Town, and Nishiki Village in 2005. Most parts of the city are categorized as hilly and mountainous areas between the Ou mountain range and the Dewa Mountains. Kakunodate is in a “heavy snowfall area,” whereas Tazawako and Nishiki are designated as “special heavy snowfall areas,” which have severer snow conditions.

Despite the large area, the population of the city is only 31,868. The percentage of population aged 65 and over was more than 30% in 2005, according to the Population Census. This rate is quite high even for heavy snowfall areas. The population has also been constantly decreasing (Semboku City 2011), and the share of aged population has become higher (Akita Prefecture 2010).

Table 24.1 Structure of the questionnaire

	Target section	Main questions
Part A	General affairs	Issues associated with snow and current approaches to those issues
Part B	Welfare	Issues in support of life in winter and approaches used
Part C	Road construction and maintenance	Issues concerning management of public spaces, including roads, during winter and methods used

Table 24.2 Types of areas

	Area type			
	Town		Farm and mountain village	
	City centre	Suburb	Along main roads	Cluster
With snow drainage	Kakunodate central	Obonai (Musashino)	Shiraiwa	Tazawa
Without snow drainage		Obonai (Osugisawa)	Kamihinokinai	

For the case study, we first chose six districts in the city because of their distinct types of infrastructure (see Table 24.2): Tazawa, Obonai (Musashino and Osugisawa), Shiraiwa, Kakunodate, and Kamihinokinai. Figure 24.1 shows the map of Semboku City and these areas. After detailed field research in each of these six areas, three—Tazawa, Obonai (Osugisawa), and Kakunodate—were selected for the program described in the next part. Tazawa (hereafter referred to as “Area 1”) is a farm and mountain village with snow drainage. Osugisawa in Obonai (“Area 2”) is a suburban town without snow drainage, and Kakunodate (“Area 3”) is a city center with snow drainage.

24.2.2.2 Approach to the Case Study

Figure 24.2 shows the plan for the program in this study, which promotes a kind of collaborative approach. The purpose of the program is to help the communities in these areas decide on their own approaches to heavy snowfall disasters. Furthermore, it aims at strengthening their mindset in favor of mutual help. The program consists of two workshops and one trial of a plan discussed in the workshops.

The schedule was as follows. First, a meeting was held in November 2007 with the heads of the three areas mentioned above to share the purpose of this program. Subsequently, the program, along with a questionnaire survey to participants, was conducted from November 2007 to February 2008. The follow-up meetings with leaders were also conducted late in February. To assess this program, a questionnaire survey was given to participants at the second workshop and the trial.

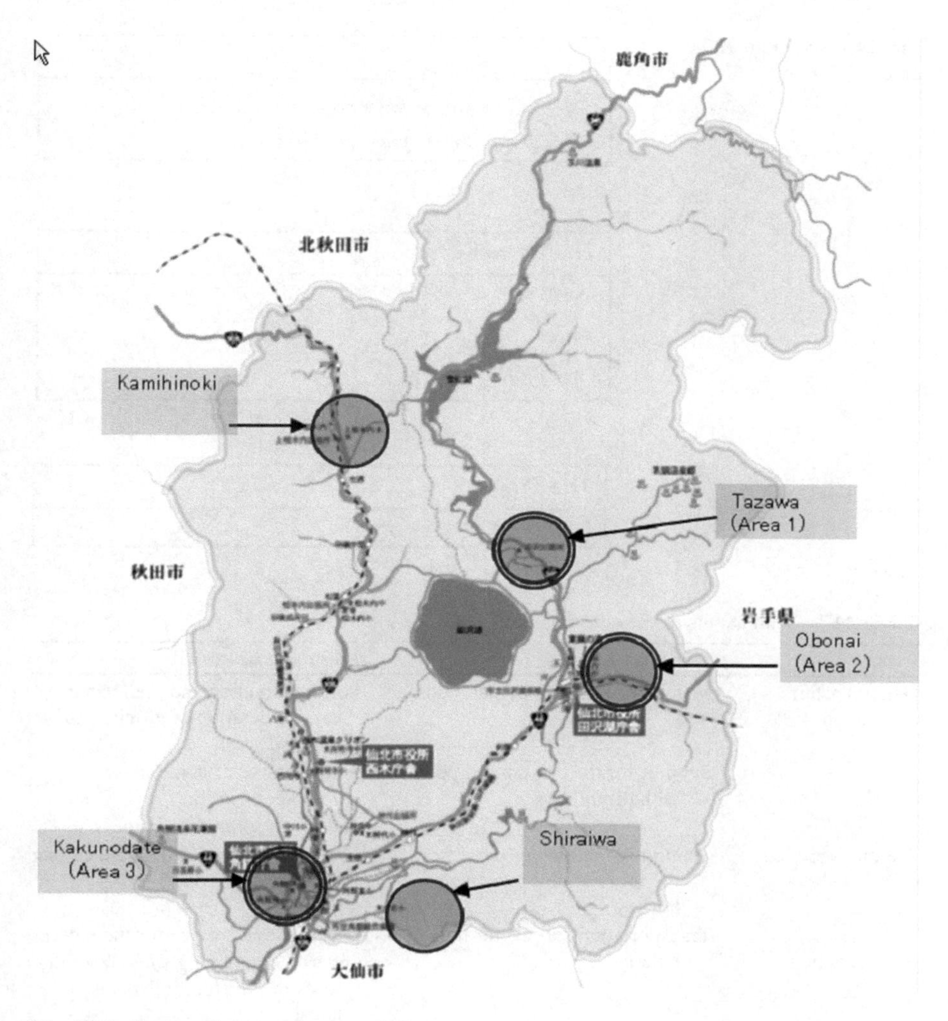

Fig. 24.1 Case study areas

24.3 Results

24.3.1 Issues in Heavy Snowfall Areas

24.3.1.1 The Results of Each Section

The main results from the questionnaires to local governments are shown in Table 24.3.

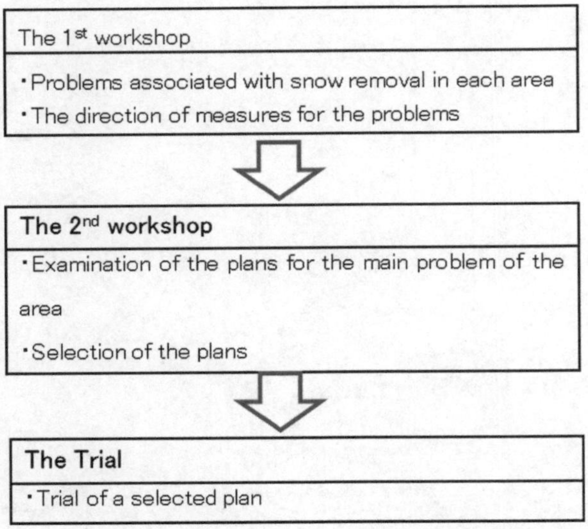

The 1st workshop
・Problems associated with snow removal in each area
・The direction of measures for the problems

The 2nd workshop
・Examination of the plans for the main problem of the area
・Selection of the plans

The Trial
・Trial of a selected plan

Fig. 24.2 Program in the case

Table 24.3 Main results of the questionnaire

	Current issues	Issues to be addressed
General affairs section	Maintenance of municipal roads	Road construction and maintenance in consideration of efficient snow removal
	Support for the elderly and people with disabilities	Reinforcement of community volunteer activities
		Public transportation service
Welfare section	Support for removing snow from the roofs of elderly persons' houses	Reinforcement of mutual help among neighbors
	Transportation service for the elderly	Transportation service for the elderly
		Support for removing snow from the roofs of the elderly persons' houses
Road construction and maintenance section	Support for removing snow from the entrances of elderly persons' houses	Reduction in snow removal costs
	Reduction in snow removal costs	Securing snow removal operators
	Snow removal around public buildings	Improvement in snow removal ability and manners of residents

(1) General Affairs Section

In the general affairs section, almost half of the responding municipalities answered that maintenance of municipal roads was the most important issue for them. Support for the living situations of the elderly and people with disabilities took second place.

Meanwhile, one of the most important issues for the future was road construction and maintenance that would allow for efficiency of snow removal. In second place was the importance of community volunteer activities.

(2) Welfare Section

The welfare officers indicated that support for the elderly was the main issue during winter. The top two issues were transportation and snow removal from elderly persons' houses.

With respect to future issues, transporting the elderly to shops and to hospitals was also recognized as important. They also answered that mutual help with neighbors was a significant consideration for the future. Helping the elderly with snow shoveling and encouraging community leaders were also considered important issues in the future.

(3) Road Construction and Maintenance Section

From the perspective of those responsible for road construction and maintenance, one of the most important issues was support for removing snow from the entrance of elderly persons' houses.

Issues for consideration in the future included, first, reduction in snow removal costs and, second, an adequate number of snow removal operators.

24.3.1.2 Issues to Be Addressed in Heavy Snowfall Areas

These results indicate that local governments in heavy snowfall areas attach much importance to hard infrastructure, such as roads. On the other hand, they are also aware of the importance of soft measures, such as encouragement to establish mutual support between neighbors.

Because of restricted finances, public funding is not adequate in most municipalities. The answers of local governments reflect this situation clearly. Therefore, we designed our case study with an approach for improving the power of mutual help among communities.

24.3.2 Collaborative Approaches in Local Areas

Our case study targeted three communities with differing environments and resident characteristics. In each community, workshops were held for discussing how to manage severe snow disasters. Approximately 20 residents participated in each workshop. On the administrative side, staff from Semboku City administration and staff members from the national highway branch office attended each session.

24.3.2.1 The Results of Workshops

The main results of the workshops are shown in Table 24.4. Here, we describe the result of the case from Area 1 in detail. In Area 1, the most important issues were,

Table 24.4 Main results of the workshops

	Area 1	Area 2	Area 3
Characteristics of the areas	Farm and mountain village with snow drainage	Suburb without snow drainage	City center with snow drainage
Snow-related problems	–Difficulty with emergency vehicles' access during heavy snowfall	–Lack of enough snow dumping places	–Difficulty in snow removal caused by aging in the historic preservation district
	–An increasing number of elderly households with difficulties in snow shoveling	–The increasing number of elderly households with difficulties in snow shoveling	–Use of channels and snow drainage
	–Blockage of snow drainage	–Road blockages from piled snow –Difficulty from an increasing water supply preventing the use of irrigation channels for snow drainage	
Plans to solve problems	–Snow removal from roads and spaces for ambulances and fire engines	–Making a map of snow dumping places	–Asking snowplows and trucks hired by local governments to remove snow
	–Snow shoveling by residents with the aid of a map of elderly households	–Organizing volunteer groups to support snow removal from elderly persons' houses	–Making a manual for use of the snow drainage
	–Making a manual for using the snow drainage		

first, how to secure the access of emergency vehicles such as ambulances and fire engines; second, the increasing number of elderly households needing support from others; and third, blockage of snow drainage as a result of too much clearing of snow by residents at the same time.

To solve these problems, participants designed the following plans. To promote easy access by emergency vehicles, participants planned to set emergency routes that residents would use for snow shoveling in case of a heavy snow disaster. Second, to cope with the increase in elderly households, they also examined a plan for making a map of elderly persons' houses to implement a mutual support system of snow shoveling. Finally, to deal with the blocking of snow drainage, they planned to make a manual showing the correct ways to use the drainage system.

According to Table 24.4, the main issue for all three of the communities was finding enough support for the elderly. On the other hand, each community had slightly different problems and approaches, depending on their situations, and their plans reflected the

Table 24.5 Plans in Area 1

Issues	Plans	Decision and consequences
Issue 1: Difficulty in emergency vehicles' access during heavy snowfall	–Set emergency routes and snow shoveling on the routes by cooperation among the residents	-Decided to try this plan
Issue 2: An increasing number of elderly households with difficulties in snow shoveling	–Make a map of elderly persons' houses	Needed to be examined –Tried by an NPO group
	–Snow shoveling by residents using the map of elderly households	Needed to be examined –Not tried (because of small amount of snow)
Issue 3: Blockage of snow drainage	–Making a manual for correct use of the snow drainage	Needed to be examined –Not tried
	–Organizing meetings for learning about snow drainage	Needed to be examined –Not tried

situation in each area. For example, the existence of snow drainage and snow dumping places seemed to be the most important problem in heavy snowfall areas.

24.3.2.2 The Results of the Trial

Although all of three communities made plans for snow-related problems, only one community in Area 1 could proceed with the trial experiment. On the day of the trial, almost 40 residents participated. The participants included people who had not attended the workshops.

Table 24.5 shows the decisions and consequences in the workshops. In sum, the main trial was the one involving emergency access. A map of elderly persons' houses was made by a residential NPO group in the area before the trial. However, snow shoveling at elderly persons' houses could not be carried out because of the small amount of snow on the day of trial. The plan for snow drainage was not attempted during this trial, either, because it had not yet been agreed upon at that time.

24.3.2.3 Findings from the Case Study

The reasons why only Area 1 was able to try the plan are inferred as follows. First, in Area 1, according to the results of the questionnaire survey ($N = 20$), 70% of the

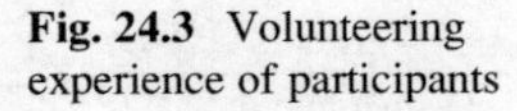

Fig. 24.3 Volunteering experience of participants

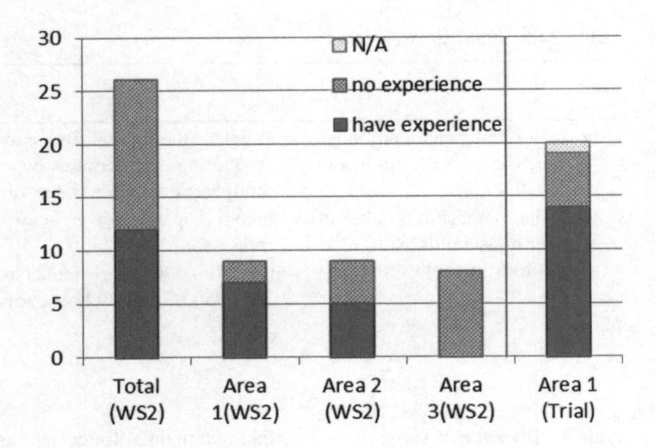

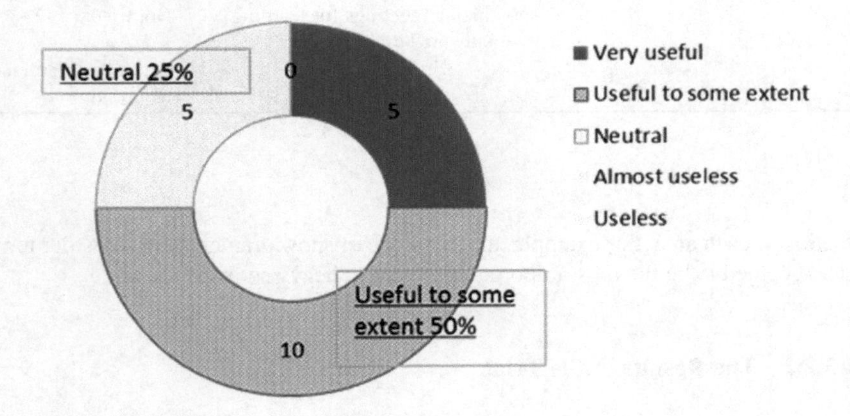

Fig. 24.4 Responses to this approach

participants had volunteer experience (see Fig. 24.3). Moreover, all of them answered that this kind of snow removal volunteer activity would be necessary in the future (see Fig. 24.4). The survey taken by participants in the second workshop ($N = 26$) also indicated that in comparison with other areas, more participants in Area 1 had volunteer experience.

In addition, the influence of strong leadership was found only in this area where a leader had just started a residential NPO group. The other two areas did not have such leaders and groups. This situation indicates that Areas 2 and 3 may need a different approach. For instance, more workshops are likely needed before any trials are attempted.

However, a common finding also emerged among the three areas, all of them demonstrating that this kind of workshop is useful for improving communication in communities: 25% answered that it was very useful, and 50% answered that it was useful to some extent.

24.4 Conclusions

From this research project, we realized that the severe conditions in heavy snowfall areas cause many problems. The result is that mutual help or volunteer activities have become important issues.

The results of the questionnaire research indicate that most local governments seem to think that maintaining roads and offering support to the elderly are important tasks. Maintaining municipal roads is an important concern for those who participate in general affairs, whereas those who construct and maintain roads are concerned with snow removal around the houses of the elderly. In addition, they seemed to understand the problems of the decreasing self-help power of residents in managing snowfall. All three sections found the issues regarding the elderly and community power to be highly relevant.

According to our results, residents acknowledged the increase of the elderly as the most important issue, just as the local governments did. Moreover, leadership and residents' associations seem to affect their collaborative activity. The type of workshop we offered also seemed to provide an opportunity for communication between residents and thereby to strengthen the power of mutual support.

Finally, even if a mutual and self-help system works adequately, more efficient snow removal systems are necessary in a severely aging society. Thus, for maximum efficiency, the arrangement of residences and roads should be considered. In summary, for sustainable communities in heavy snowfall areas, not only mutual support systems but also town planning for efficient snow removal may be essential.

References

Akita Prefecture (2010) The rate of ageing population in Akita Prefecture. Available from http://www.city.semboku.akita.jp/news_topics/population.php. Accessed 1 Mar 2011 (in Japanese)

Fire and Disaster Management Agency (FDMA) (2006) On heavy snow disaster during 2005–2006 winter season (Report No. 62). Available from http://www.fdma.go.jp/detail/649.html. Accessed 1 Mar 2011 (in Japanese)

Fire and Disaster Management Agency (FDMA) (2011) Damage by snow in this winter (Nov 2010–Mar 2011). Available from http://www.fdma.go.jp/bn/2011/detail/674.html. Accessed 29 July 2011 (in Japanese)

Hochrainer F, Pflug G (2009) Natural disaster risk bearing ability of governments: consequences of kinked utility. J Nat Disaster Sci 31(1):11–21

Kuriyama H (1986) The characteristics of the casualties due to snow in the snowy areas. Rep Natl Res Cent Disaster Prev 36:79–94

Lave J, Wenger E (1991) Situated learning: legitimate peripheral participation. Cambridge University Press, Cambridge

Ministry of Land, Infrastructure, Transport and Tourism (MLIT) (2009) Report of the ways to inherit lifestyle in snowy region, pp 4–6). Available from http://www.mlit.go.jp/kokudoseisaku/chisei/crd_chisei_tk_000038.html. Accessed 12 Aug 2011 (in Japanese)

Ministry of Land Infrastructure, Transport and Tourism (MLIT) (2010) Promoting supports for heavy snowfall areas. Available from http://www.mlit.go.jp/crd/crd_chisei_tk_000010.html. Accessed 1 Mar 2011 (in Japanese)
Semboku City (2011) The population of Semboku city. Available from http://www.city.semboku.akita.jp/news_topics/population.php. Accessed 1 Mar 2011 (in Japanese)
Shimizu M, Abe O (2001) Recent snow cover fluctuations in the mountainous areas of Japan. Ann Glaciol 32(1):97–101
Yamori K (2009) Action research on disaster reduction education: building a "community of practice" through a gaming approach. J Nat Disaster Sci 30(2):83–96

Chapter 25
A Vulnerability Study from Water Perspective on the Largest City of China

Guang-wei Huang and Zhen-jiang Shen

Abstract This chapter presents a preliminary study on the vulnerability of the Chongqing City to water-related hazards. The focus is on the factors that cause its vulnerability. The findings suggest that the root of its flood-related vulnerability may be the combination of its riparian development and the construction of the Three Gorges Dam downstream of the city.

Keywords Chongqing City • Vulnerability • Flood • Water quality

25.1 Introduction

The vulnerability of humans in relation to floods, drought, and water pollution has become a widespread concern. Over the past two decades, considerable conceptual and empirical research work has been carried out on vulnerability. The term "vulnerability" quietly appeared in the late 1980s, after researchers and practitioners expressed discomfort in defining marginality and poverty solely in economic terms. Since then, vulnerability has become a recognized conceptual framework and analytical approach in research and applications dealing with uncertainty and risks, especially in terms of livelihood security. To date, however, there is little consensus among researchers, planners, and disaster managers with regard to what is actually meant by vulnerability and vulnerable populations.

G.-w. Huang (✉)
Graduate School of Global Environmental Studies, Sophia University, 7-1 Kioicho, Chiyoda-ku, Tokyo 102-8554, Japan
e-mail: huang@genv.sophia.ac.jp

Z.-j. Shen
School of Environmental Design, Kanazawa University, Kakuma Machi, Kanazawa City 920-1192, Japan
e-mail: shenzhe@t.kanazawa-u.ac.jp

M. Kawakami et al. (eds.), *Spatial Planning and Sustainable Development: Approaches for Achieving Sustainable Urban Form in Asian Cities*, Strategies for Sustainability,
DOI 10.1007/978-94-007-5922-0_25,

The problem is that the concept addresses complex realities, which can hardly be classified with a few operational variables without loss of accuracy. Therefore, a vulnerability study depends on how we define it and what perspective we are interested in.

The pioneering definition by Chambers (1989) is the "exposure to contingencies and stress, and difficulty in coping with them. By this definition, vulnerability has thus two sides: an external side of risks, shocks, and stress to which an individual or household is subject; and an internal side which is defenselessness, meaning a lack of means to cope without damaging loss."

In climate change assessments, the Intergovernmental Panel on Climate Change (IPCC) advanced an alternative definition of vulnerability (www.ipcc.ch/pub/syrgloss.pdf): "The degree to which a system is susceptible to, or unable to cope with, adverse effects of climate change, including climate variability and extremes. Vulnerability is a function of the character, magnitude, and rate of climate variation to which a system is exposed, its sensitivity, and its adaptive capacity." By the definition of IPCC, vulnerability has three components: hazard exposure, consequences (impacts), and adaptive capacity. It has much in common with the concept of risk in the field of natural hazards study.

This study blended the concepts of Chambers and IPCC and came up with this definition: "Vulnerability has two sides: an external side of risks, shocks, and stress to which an individual or household is subject; and a social side which is used in a broad sense to describe all the factors which determine the outcome of a hazard event of a given nature and severity through interactions with hazard." This definition emphasizes the role of human systems in mediating the outcomes of hazard events. Like in a war situation, mediation could lead to either escalation or reduction. Sensitivity in the IPCC definition is the degree to which a system is affected by or responsive to climate stimuli. Nevertheless, the degree is a function of the system's resistance to external force. Besides, since this is a holistic study on the water-related vulnerability of Chongqing City, the grouping of sensitivity with the capacity to adjust (and recover and adapt) could be justified. For the past two decades, many vulnerability studies have focused on the least developed countries in Africa and Asia because many of them are situated in flood- or drought-prone regions. Meigh et al. (1999) undertook an assessment of water availability in eastern and southern Africa, including a demand/supply study. Cambula (1999) analyzed the impact of climate change on the water resources of Mozambique. On the other hand, a number of vulnerability studies have been conducted in relation to poverty. In the case of Bangladesh, Rayhan (2009) showed that poverty and idiosyncratic flood risks are positively correlated and that the possession of arable land and a small family can reduce poverty and flood risk. The Venezuela floods of 1999 left as many as 30,000 people dead and 150,000–200,000 homeless; many of the victims were poor. In general, groups who are already in weak economic and social standing tend to be more adversely affected by natural hazards (Mustafa 1998). In this chapter, this general perception is further discussed.

Chongqing City, the largest city of China and target of the present study, belongs to a very different category in terms of water availability, geographical context, topographical features, economic status, and social system. The main objective of this chapter is to contribute to this growing area of urban vulnerability analysis by identifying water-related issues in this city.

25.2 Materials and Methods

25.2.1 Study Site Description

Chongqing is located on the upper reaches of the Yangtze River in the mountainous areas of southwestern China. It became one of the four directly controlled municipalities in China in 1998 and the only such municipality in inland China (the other three are Beijing, Shanghai, and Tianjin). The central urban area of Chongqing is situated at the confluence of Yangtze and Jialing rivers. Figure 25.1 presents a geographical view of the location, spanning 470 km both from east to west and north to south. The history of Chongqing extends back at least 3,000 years. It now has jurisdiction over 40 districts and counties (autonomous counties and cities), sits on 8,240,000 km^2 of land, and has a population of 31,590,000. In sum, Chongqing is the largest city of China in terms of population and area; in fact, its name means "double happiness" in Chinese. Figure 25.2 outlines the evolution of Chongqing's urbanization over the past three decades. In the 1970s, the urbanized

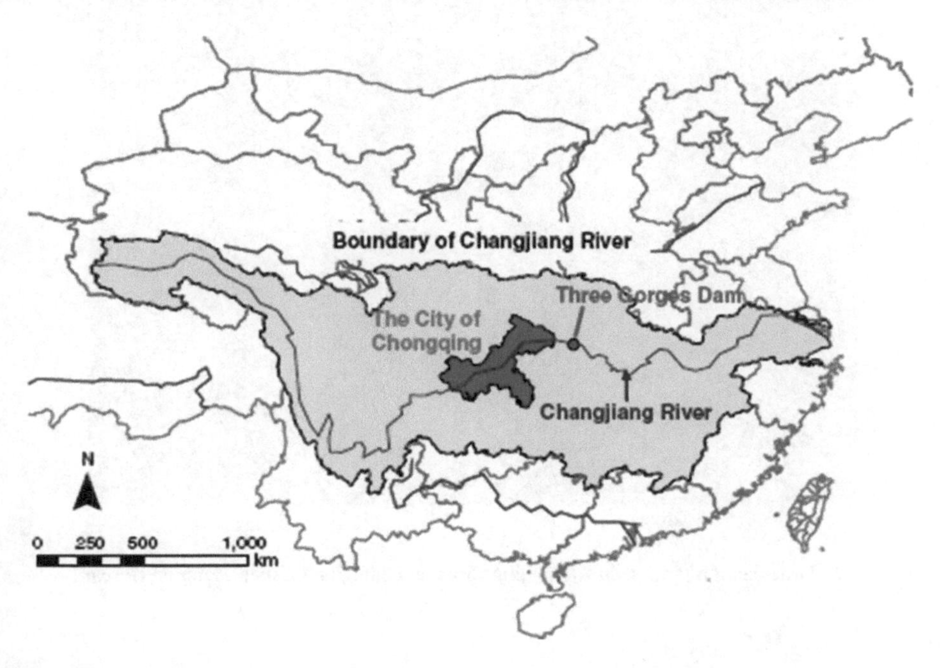

Fig. 25.1 Location of Chongqing

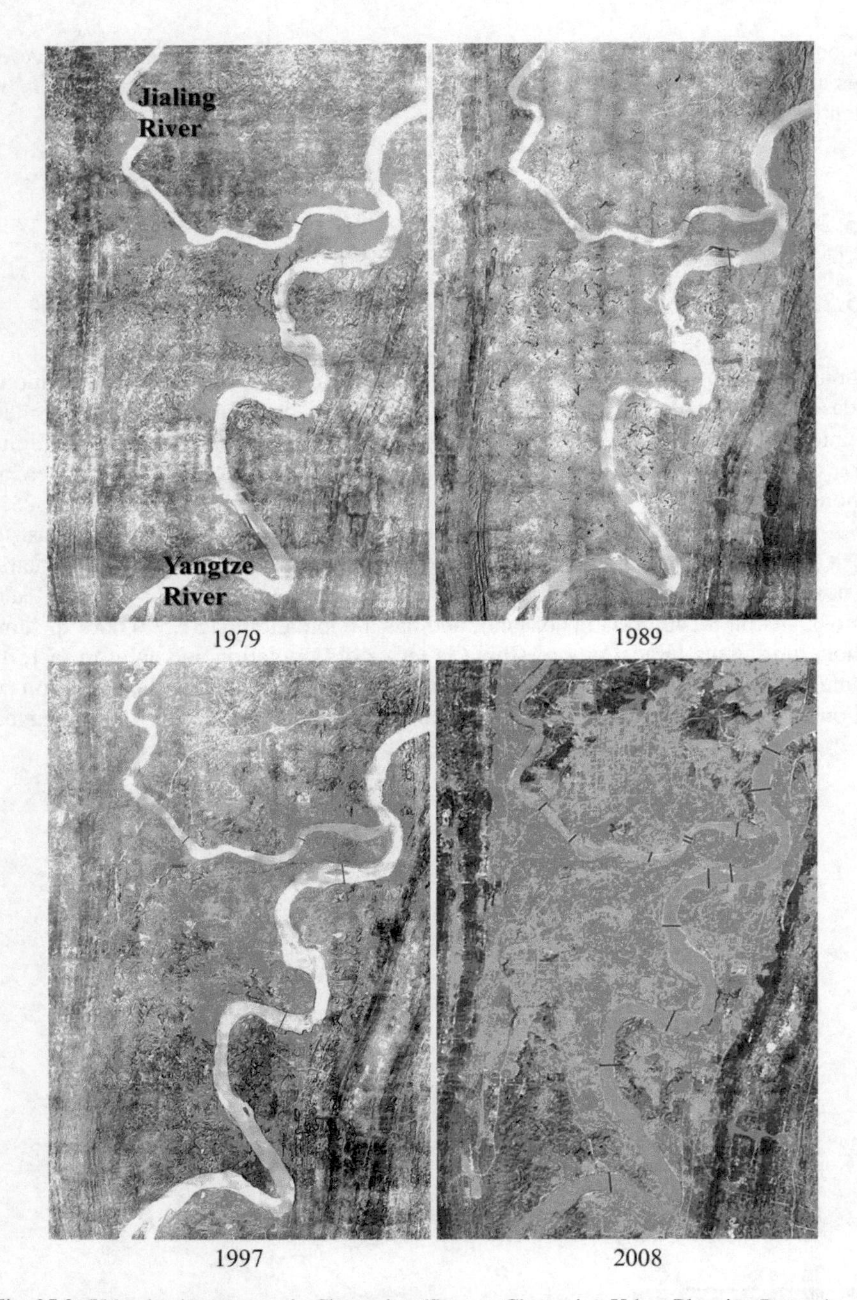

Fig. 25.2 Urbanization process in Chongqing (Source: Chongqing Urban Planning Bureau)

area was mainly confined to the Yuzhong District, which is on a narrow peninsula formed between the Yangtze and Jialing rivers. From 1989 to 1997, the urban area expanded but not significantly. However, during the 10-year period after Chongqing's administrative status was raised, urbanization progressed very rapidly toward the north and the west. Thus, it can be concluded that the central government's policy of raising Chongqing's status was the main driver of the rapid development of the city.

Chongqing has a complex geological conformation. Mountains and hills account for 75 and 18% of its area. Flat land is less than 10% of its area. By way of comparison, 75% of Japan is mountainous. Therefore, the situation in Chongqing is more severe than in Japan. The city is also characterized by abundant waterways. In addition to the Yangtze and Jialing rivers, there are 11 tributaries of the Yangtze River with a drainage area larger than 3,000 km^2 and another 40 tributaries with a drainage area of more than 500 km^2.

Chongqing's average population density of 377 people/km^2 is not high, but in the central part of the city, the density reaches is 27,228 people/km^2. The average density of all urbanized districts and counties is 1,005 people/km^2.

25.2.2 Methodology

Firstly, various data of the Chongqing City, including precipitation, flood disaster records, and urban development planning documents, were collected and analyzed to grasp the main characteristics of its natural environment and human interventions. In 2009, fieldwork was carried out to investigate the vulnerabilities of the city from various angles, including riparian conditions, water quality, transportation system, and urban slums. Since Yangtze River and Jialing River together form the backbone of the city, the field studies were conducted along the two river courses. A concept suggests looking into the basic reasons for the city's vulnerability in relation to river courses. Particular attention was placed on the impact of riverside road construction on flood risk. Meanwhile, the water quality investigation was centered on the concentration of ammonia and pH and on water temperature to detect potential toxicity problems.

Another issue in vulnerability analysis is the spatial scale. The vulnerability may be assessed at different scales: from the household, regional, and national to global. Many previous vulnerability studies were at the national level. Since China is a country with high heterogeneity in many aspects—from the landscape, water availability, and population density to lifestyle—the investigation of vulnerability from a regional or subnational perspective is more useful in formulating countermeasures. However, the vulnerability at the regional or subnational level is influenced by national policy. Therefore, the appropriate approach would be to assess vulnerability at a regional or subnational scale while considering the impact of national policy.

Based on the analysis of historical data, previous research results, and insights gained from the field survey, views on the vulnerability of Chongqing from the perspective of water were constructed.

25.3 Results

25.3.1 Flood Characterization

According to historical records, floods in Chongqing can be classified into three types. One is the overflow of floodwaters passing through Chongqing, which may be termed as transit floods. The second type is the waterlogging due to heavy rainfall, which may be termed as local floods. The third is the mixture of the transit and local floods. Due to the city's topography, transit floods are confined to the narrow belts along the two major rivers. However, the local and mixed floods may cause large areas to be inundated.

Figure 25.3 depicts the long-term mean monthly precipitation of Chongqing. June is the wettest month, and rainfall decreases significantly in July and August. Over the period 1960–2000, the wet years were 1965, 1968, 1987, and 1997. The extremely wet years were 1980, 1981, 1982, 1983, 1984, 1988, 1998, 2000, and 2010. Thus, it can be generally stated that the 1980s was a flood decade for Chongqing. In 1998 and 2010, however, transit floods that occur every 100 years and 115 years swept through the city. Chongqing's floods are also characterized by high spatial variability. The northeastern section has the highest incidence of flooding—about six to eight times every 10 years. The incidence in the west is also high: once every two years. The southwest suffers least from flooding, less than one every 5 years. The ratios of maximum to minimum flow discharge in the Yangtze River and the Jialing River are 37.8 and 183.6. As a result, the river water depth along the Chongqing reach of the Yangtze River can vary more than 30 m.

Due to the limited availability of inhabitable land, residential development in Chongqing has spread in the close proximity of the rivers and even into middle sandbars of the rivers. Riverside road construction has significantly reduced the width of the river channel. In the Caiyuanbar sub-reach, for example, the land reclaimed is 230 m wide, more than 20% of the original river channel width. Such a reduction in width affects the capacity of the river channel conveyance to pass floods. Besides, there is a large sandbar downstream of the Caiyuanbar sub-reach as shown in Fig. 25.4. The reduction of channel width may have an impact on the downstream flow pattern. To see how the land reclamation would affect the flood potential in the downstream reach, a 2-D flow simulation was conducted. It should be pointed out that the simulation was intended for quick diagnosis or potential problem identification only, not a detailed analysis of the flow structure in that reach. The simulation indicates that the flow direction is deflected toward the right

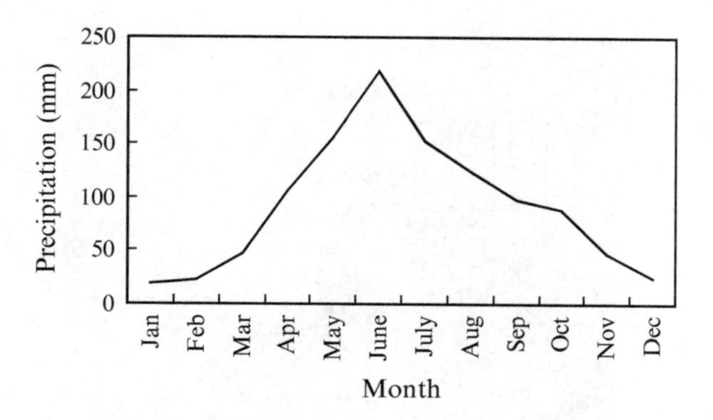

Fig. 25.3 Mean monthly precipitation of Chongqing

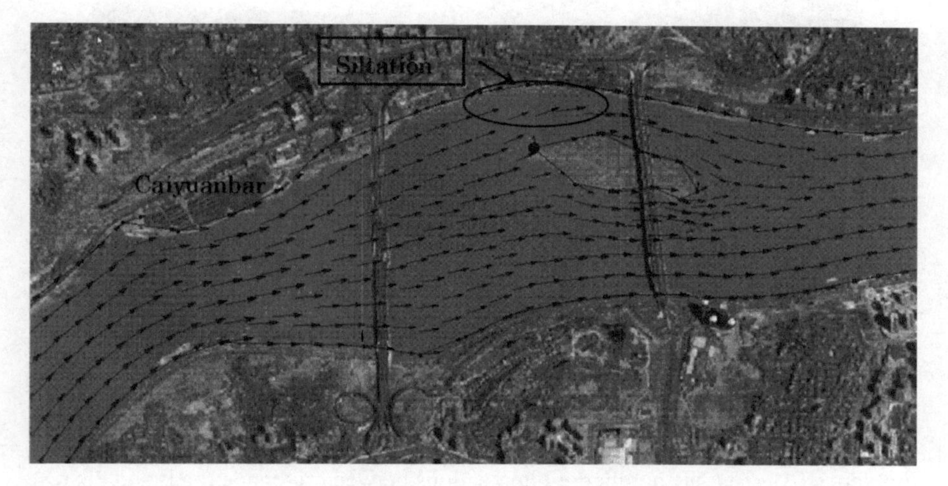

Fig. 25.4 Reduction of channel width due to riparian development and its impact on the downstream flow pattern

bank due to the land reclamation, and more water passes the sandbar via the right hand side of the sandbar (Fig. 25.4). Thus, the deposition can be expected on the left side of the sandbar (looking downstream). As the riverbed aggrades, it poses a higher flood risk on the adjacent residential area in the Yuzhong District. Field inspection found that the deposition is already in progress on the left side of the sandbar.

On July 20, 2010, a big flood passed through Chongqing. The maximum discharge at the Chutan hydraulic observation station was 70,000 m^3/s. Several portions of riverside roads were flooded, including the part that is downstream of the Caiyuanbar reach.

According to the municipal government report, there were eight lanes of riverside road along the Yangtze and Jialing rivers in Chongqing, totaling 46 km at the end of 2006. The construction had been accelerated since 1997. At present, the total length of the road, including some ongoing portions, is 140 km, and more riverside

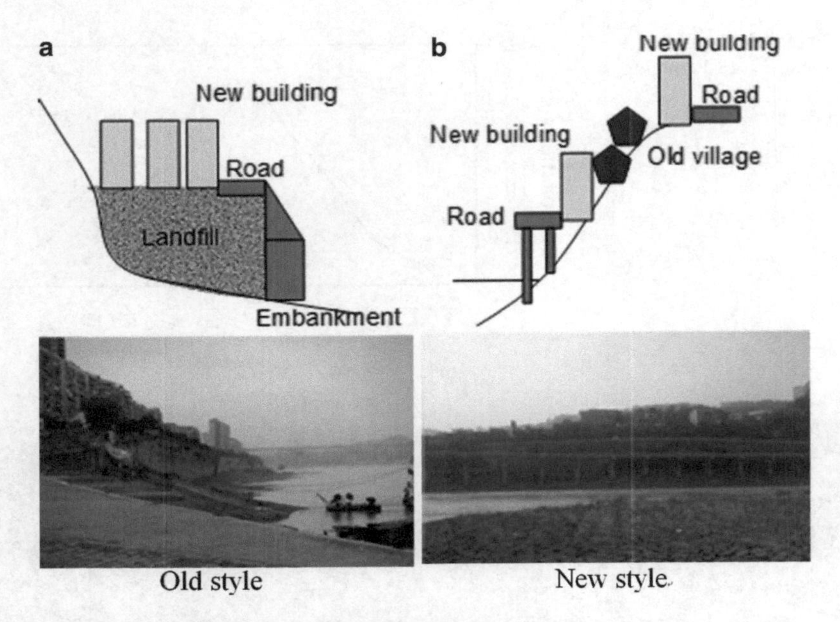

Fig. 25.5 Change in the development of riverside zones

roads are being planned. The author's field survey found, however, that there was a big positive difference between the old and new riverside roads. As shown in Fig. 25.5, the old riverside road was constructed by improperly filling the waterway, but the newly constructed one was supported by piers to minimize the impact on flood conveyance. Although the new construction method is a good step toward the integrated management of the river and riverside transportation system, another problem still exists: the number of intersections between riverside roads and the rest of the transportation network are rather few. Along a 5-km road segment we surveyed, there were just two intersections. The limited link of the riverside roads with rest of the city could impede evacuation and rescue in case of flood emergency.

In addition to riverside roads, the inhabitation of middle sandbars is also reducing flood conveyance in the region. Furthermore, there is even some illegal development on the riparian zones in Chongqing. For example, in Jiangjing City, a secondary city under the Chongqing municipality, and 50 km from central Chongqing, an illegal residential development intruded into the Yangtze River and reduced the river width by about 100 m. The problem was disclosed recently, and the solution to it remains undecided.

An issue that recently emerged is the backwater effect of the Three Gorges Dam. The Chongqing reach is part of the fluctuating backwater region of the Three Gorges Dam. The construction of the dam will lead to variations of the incoming water and sediment conditions and, in turn, changes in the deposition and erosion processes in the fluctuating backwater reach of the dam. Jiulongpo Harbor, the most important freight ship terminal in southwest China, is in Chongqing. However, the harbor could be affected or even blocked by sedimentation after the reservoir is impounded for 80 years (Wang 2005). As sediment deposition goes on, the flood water level at Chongqing may increase. Through computation analysis and model testing, it is estimated that after 100 years' operation of the reservoir, the floodwater level at Chongqing will increase from the present 194.3 to 199.2 m in case of a 100-year frequency flood (Tang 1992). Chongqing is a mountainous city; the elevation of most of the urban areas is over 200 m, and the center part, about 250 m. A higher floodwater level, therefore, would have little effect on it if the riparian zones were developed with more consideration to floods. Today, the riparian zones in Chongqing have become the engine of its economy, and a lot of residential development is also occurring there. Therefore, the inundation of riparian zones could cause unwelcome consequences.

During the 2010 flood, a number of riparian zones, such as Hechuan District and Ciqi Town, were submerged. The total death toll of the flood was 72, and 5.6 million people were affected. The direct economic loss reached 2.1 billion Yuan.

25.3.2 Water Quality Conditions

Rapid urbanization leads to water pollution in rivers and streams. There are too many examples to make a list. In Chongqing, there is something even bigger than urbanization: the Three Gorges Dam. The presence of the dam decreases the flow velocity in the reservoir substantially (Huang 2006), and as the flow slows down, so does the river's self-purification capacity. When the dam is operated at a water level of 175 m, the effect of the velocity reduction will reach Chongqing. Moreover, the pollutants generated or discharged into the river downstream of Chongqing could be brought up to Chongqing reach due to the fluctuating backwater effect. Under the combined influence of rapid urbanization and the completion of the world's largest hydro-dam, the water quality in Yangtze, Jialing, and many other rivers in the Chongqing administrative area will be a big concern to ordinary residents, researchers, and China's top leaders. Currently, there are more than 3,000 sources of pollution in Chongqing. The total industrial and domestic wastewater discharge is about 108 t/year and contains large amounts of N and P pollutants. Although numerous studies have been performed on the water quality of the reservoir (Feng and Zhu 2008; Liu and Zhang 2008; Zhang et al. 2008; Li and Huang 2006), uncertainties remained high.

According to the statistics of Chongqing Municipality Government (Chongqing EPA 2009), the water quality in the Yangtze, Jialing, and Wu rivers is stable. Among 23 monitoring stations, the water quality in 21 stations stood at Grade II and only two at Grade III. For secondary rivers, among 117 measuring stations, the percentages for water quality standard I, II, III, IV, V, and worse were 12.8, 25.6, 38.5, 13.7, 1.7, and 7.7%. Therefore, the general water quality was not too bad.

In 2009, we conducted our own water quality tests in the Yangtze, Jialing, and a few tributaries. The sampling points along the two major rivers are shown in Fig. 25.6.

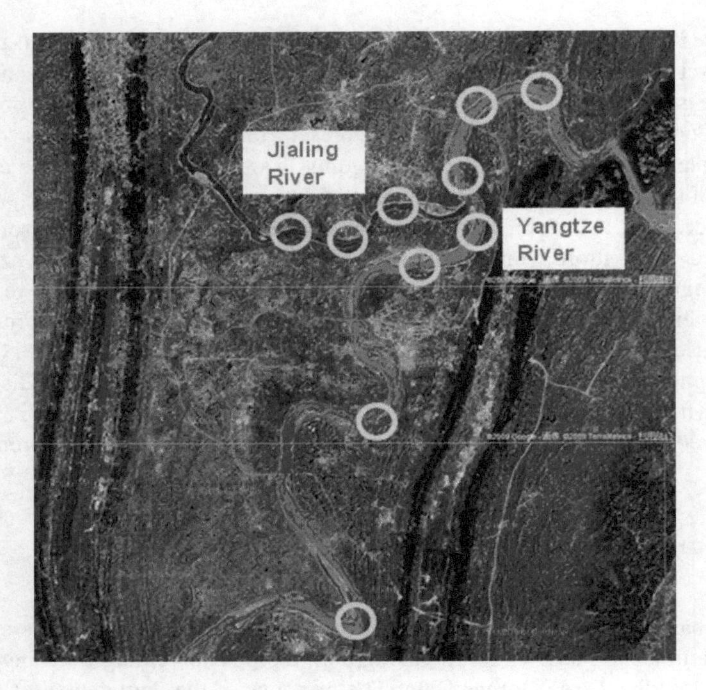

Fig. 25.6 Water quality survey locations

We found the ammonia concentrations to be in the range of 0.2 ~ 0.9 mg/L in the two major rivers. The maximum was recorded in the Banan County in summer. Ammonia exists in two forms in natural waters: ammonium ion (NH_4^+) and ammonia gas (NH_3). Whereas the former is innocuous at the levels encountered in most natural waters, the latter is an un-ionized form and toxic to fish. The NH_3 concentration depends on a number of factors in addition to the total ammonia concentration, the most important of which are pH and temperature. The concentration of NH_3 increases when either pH or temperature goes up. At high pH, ammonia exists principally in the toxic, un-ionized form. In our survey of the rivers, the pH ranged from 7 to 9.1. Therefore, the proportion of total ammonia in un-ionized form could reach as high as 30%. Besides, we also found that the nitrate concentration was 17 mg/L at one location in the Jialing River. A high concentration of nitrate can cause health problems. Moreover, the concentrations of chemical oxygen demand (COD) we measured near bank zones were all higher than 8 mg/L, with the highest being 12 mg/L. Although those sporadic measurements of water quality were not conclusive, they were sufficient for problem identification.

Fish is one of the most favored foods by Chongqing residents. However, fish catch has been falling dramatically in the region. In many small rivers, fish have disappeared completely due to pollution. In the Qijiang River, which is a tributary of the Yangtze River and 80 km south of central Chongqing, the ammonia level has reached 11 mg/L. Such a high concentration of ammonia may explain why fish no longer exist in this river. In a healthy cycle, fishery is a way to remove nutrients from water bodies. When fish is gone, this mechanism of purification stops.

Due to Chongqing's topography, it is difficult to channel river water to farmlands on hillsides. As a result, about 20% of the farmlands in Chongqing are irrigated by sewage water currently. At a number of sites, we spotted the use of nearly raw sewage water in irrigating vegetables. A report by the Chongqing Agricultural Monitoring Station revealed that more than 5% of vegetables produced in Chongqing contained one or more pollutants exceeding the national standards. The grain crops and vegetables produced with sewage water could become a long-term vulnerability of the city if sufficient countermeasures are not taken immediately.

Fortunately, the Chongqing municipality government has taken serious steps to combat pollution. Every year, the government discloses the names of companies that have failed to keep the pollutants they discharge below the permitted level. Besides, organizations for environmental protection are well established at different levels. The number of environmental monitoring staff is now more than 700, and a citizen volunteer monitoring program started recently. At present, investment in environmental protection accounts for 2.6% of the city's GDP.

25.4 Concluding Remarks

Rapid urban development in Chongqing was stimulated by the national policy to raise its status from a sub-provincial city to a city under the direct control of the central government. The construction of the Three Gorges Dam is also a national policy with multiple objectives. The two policies have had huge positive impacts on the region and the whole of China as well, but the combination of the two policies indirectly increases Chongqing's exposure to flood damage. Besides, the insufficient capacity to access or leave riparian zones may impede evacuation and rescue in case of an emergency. On the other hand, the design change of riparian roads from the landfill type to the pier-type reflects Chongqing's capability to self-adjust and mitigate.

As for the water quality, worrying signs were observed. Although the current situation is not critical in general, the whole society should pay serious attention to the issue of water pollution and continue to closely monitor any sign of change in water quality. The point is this: if you act only by the time you see bad conditions in statistics, it will be too late!

References

Cambula P (1999) Impacts of climate change on water resources of Mozambique. Republic of Mozambique. Final Report of the Mozambique/U.S. Country Study Program Project on Assessment of the Vulnerability of the Economy of Mozambique to Projected Climate Change, Maputo, Mozambique

Chambers R (1989) Vulnerability, coping and policy. IDS Bull 20:1–7
Chongqing EPA (2009) Environmental statistics 2008
Feng W, Zhu B (2008) Environmental hazard and ecological infrastructure construction of hilly city: a case of metropolitan area of Chongqing Municipality. Ecol Econ 3:19–27
Huang GW (2006) A diagnostic study on the environment of the three Gorges reservoir. J Environ Inf Sci 34(5):91–98
Li C, Huang ZL (2006) Study on the pollutant loads into Three Gorges reservoir—pollutant load prediction after impoundment. Resour Environ Yangtze Basin 3:99–108
Liu XA, Zhang MF (2008) Behaviour of evolution and growth of dominant algae in the Chongqing Urban Section along the Three-Gorges Valley. Environ Sci 29(7):1838–1843
Meigh JR, McKenzie AA, Sene KJ (1999) A grid-based approach to water scarcity estimates from Eastern and Southern Africa. Water Resour Manage 13:85–115
Mustafa D (1998) Structural causes of vulnerability to flood hazard in Pakistan. Econ Geogr 74 (3):289–305
Rayhan MI (2009) Assessing poverty, risk and vulnerability: a study on flooded households in rural Bangladesh. J Flood Risk Manage 3(4):18–24
Tang R (1992) Permanent use of the three Gorges Project Reservoir. J Yangtze River Sci Res Inst 1992(S1):41–45
Wang ZY (2005) Impacts of the TGP project on the Yangtze River ecology and management strategies. Int J River Basin Manage 3(4):237–246
Zhang S et al (2008) Trophic states and nutrient storage of reservoirs in Chongqing. Chin J Geochem 27:310–316

Index

A
Accessibility, 4, 6, 29, 38, 94, 102, 109, 112, 167, 171, 176, 199, 219, 249–265
Aging society, 9, 354, 432, 441
Agricultural labor force, 333, 334
Agricultural Monitoring Station, 453
Agricultural production, 7, 60, 61, 326, 327, 329–337, 339–351
Agricultural structural adjustment, 325–338
Agriculture, 7, 10, 60, 61, 151, 166, 169, 326–328, 333, 341–345, 347–351, 372, 375
Anguo City, 328, 334, 336, 337
Assessment, 7, 8, 31, 94, 96, 107, 185, 193, 306–309, 312–320, 354, 360–362, 369, 372, 401–414, 444
A-Team, 203–205, 207, 208, 211

B
Bangkok Metropolitan Region (BMR), 156, 158–167, 171
Beijing, 5–8, 10, 26, 173–186, 232, 235, 252, 311, 325–351, 445
Beijing Metropolis, 325–338
Bicycle industry, 190, 191, 193–197, 199, 200, 202–207, 210, 211
Biomass resources, 367
Biotope, 290, 302, 354
BMR. *See* Bangkok Metropolitan Region (BMR)
Bus availability, 6, 249–265

C
Carbon dioxide reduction, 272
Car dependence, 249–265
Care-requiring level, 77
Cave Dwelling, 5, 137–152
Chain production manufacturers, 204–205
Changhua county, 117–132
China, 2, 3, 5, 9, 10, 16, 17, 24–27, 92, 138, 190, 198, 203, 228, 249–265, 305–320, 325–338, 340, 341, 345, 347, 351, 419, 424, 425, 428, 443–453
Chiyoda Ward, Tokyo, 294
Chongqing city, 9, 444, 447, 448
Chungnam, 384
Chun-You Street fire, 418–420, 423–427
City-skirt area, 343–347
Climate change, 2, 92, 93, 95–97, 104, 110–113, 119, 120, 125, 253, 382, 402, 403, 444
CO_2 emission ratio, 8, 383, 384, 388, 392, 397
Collaborative approach, 432, 434, 437
Collaborative project, 119, 127, 131
Commercial crop, 328, 337
Community activity, 55–58, 61, 62
Commuting mode choice, 235, 244
Compact city, 2, 4, 9, 10, 229, 242
Conservation planning and design, 138, 140, 143, 148, 152
Continuity, 69–88, 161, 353–379
Cooling indoor temperature, 272
Cropland, 7, 340–343, 345–350
Cultural landscape, 119, 137–144, 147–149, 152, 426
Cultural resources, 47, 140, 141

D
Dachia bicycle valley, 5, 208
Daily living environments, 4, 71–73, 81–88
Dementia level, 76, 77

M. Kawakami et al. (eds.), *Spatial Planning and Sustainable Development: Approaches for Achieving Sustainable Urban Form in Asian Cities*, Strategies for Sustainability,
DOI 10.1007/978-94-007-5922-0,

Depopulation, 3, 4, 9, 54, 56, 61, 62, 65, 432
Detached house, 166
Disaster awareness, 422, 423
Disaster-prevention, 402, 407, 409, 410, 413, 414
Downtown decline, 34, 36–48
Drainage layer, 291, 293–303
Dwelling size, 175, 182–183

E

Early warning, 7, 309, 314, 315, 317–318
East Asian community, 16, 23
Eco-city, 2, 6, 7, 9, 10
Eco-environmental protection program, 306
Ecological planning, 305, 310, 311, 353–379
Ecological security, 7, 305–320
Ecological system, 2, 6–9, 308, 311–312, 320
Emergency response, 7, 314, 315, 317–318, 425
Employee, 36, 40–45, 53, 98, 99, 106, 208
Energy efficiency, 92, 108, 109
Environmental assessment, 354, 360, 362, 372
Environmental classification method, 355–362, 364, 367–369, 372–375
Environmental impact, 6, 7, 216, 227–245, 403
Environmental management, 315
Environmental performance, 95, 96, 110, 111, 271
Environmental planning, 4, 7, 8, 10
Environmental protection, 118, 272, 316, 349, 351, 424, 453

F

Family group, 180, 185, 186
Farm households, 326, 329, 331–334, 336–338
Farming and mountain areas, 60
Farm management, 58, 326, 327, 329–337
Flood disaster, 401–414, 447
Food crop production, 327, 328, 336, 340

G

Gaoxigou village, 5, 144–147, 149–152
Globalization, 121, 138, 190, 193, 194, 308, 326, 327, 337
Government resources, 206
Green curtains, 6, 7, 10, 271–285
Green design, 1, 6–7, 276
Greenhouse effect, 290, 382
Green roof system, 6, 7, 287–303
Green space, 6, 7, 10, 98, 107, 119, 120, 139, 289, 302, 340, 347–348
Green Station Greenway (GSG), 127–132
Group care facility, 71–74, 77–82, 84–88

H

Heat island phenomena, 272
Heavy snowfall area, 431–441
Historical site, 417–429
Hokuriku region, 52
Housing demand, 5, 10, 173–186
Housing demand structure, 5, 10, 173–186
Housing demand survey, 176–178
Housing preference, 174
Housing size demand model, 175
Humantope index, 353–379
Hypothetical city, 228, 230, 231, 233, 234, 237, 243

I

Image classification, 386, 389
Indicator framework, 93–96, 104, 112, 113
Indicators system, 404–407, 413
Industrial clustering, 190–192
Infrastructure and resource efficiency, 101, 108–110
Inner city, 93, 160, 163, 166–168, 171, 228, 229, 245
Inner-city ring, 346–347
Intangible culture, 139, 141–143, 146–148, 152
Intelligent transportation systems, 214
International statistical data, 24
i-REAL, 215, 217–222, 224, 225
i-Swing, 217

J

Japan, 3, 4, 6, 9, 10, 16, 17, 24–26, 52, 65, 70, 71, 93, 107, 109, 111, 194, 213–225, 272, 277, 282, 283, 288, 289, 292–295, 298, 302, 327, 354, 358, 369, 370, 372, 379, 432, 433, 447
Job location choice, 232–235, 239, 240, 244, 245

K

Kaohsiung, 205, 410–414
Knowledge-based economy, 190, 193
Korea, 3, 16, 17, 24, 25, 27, 33–48, 119, 383, 384, 390

Korea Land use Information System (KLIS), 390, 392
Kyushu, 27, 53, 355

L
Land cover analysis, 381–397
Land ownership, 353–379, 418, 427
Land reclamation, 448, 449
Landscape, 2, 4–9, 55, 60, 117–132, 137–152, 156, 236, 289, 290, 310–313, 318, 319, 340, 341, 343, 345–350, 353–379, 383, 392, 397, 413, 414, 426, 428, 447
Landscape Master Plan, 4, 118, 119, 123–125
Landscape pattern, 313, 318–319
Landscape planning, 119, 120, 125, 127, 354, 378
Land use change, 356, 360, 370
Land use continuity, 353–379
Land use/cover change (LUCC) model, 307, 308, 320
Land use patterns, 1, 5, 6, 228, 232, 233, 236, 244, 250, 354–356, 358–362, 364–367, 369, 370, 372, 373, 375–377
Leaf area density, 276–279
Lien-Chiang county, 417–429
Living-hub, 71–74, 76–78, 83–85, 87
Living space, 139, 140, 142–146, 148, 149, 152, 348, 420
Local council scale, 355, 356, 358, 360, 361
Low carbon society (LCS), 228

M
Marginal hamlet, 3–5, 51–66
Market economic systems, 328
McHarg method, 354, 377, 378
Medicinal crops, 326–330, 332–338
Metropolitan areas of Taiwan, 401–414
Minato Mirai 4, 21, 91–113
Modulation, 311–312
Monte Carlo approach, 244
Motor vehicle, 214, 215, 253
Multi-agent systems (MAS), 229, 230
Multi-layered area, 87
Multiple index layers, 354
Multispectral satellite imagery, 381–397
Mutual support systems, 438

N
National Spatial Plan, 52, 59, 62
Natural disaster, 8–10, 138, 318, 329, 402, 403
Natural ecological space, 141
Natural ecosystems, 7, 310, 312, 313, 316–317
Neo-traditional development, 1, 8, 9
New rural construction, 5

P
Personal mobility, 253
Personal mobility vehicle, 5, 10, 213–225
Physiotope indices, 354
Planning policies, 2, 5, 8, 230
Population change, 36, 40–42, 44, 45
Post-disaster reconstruction, 417–429
Production space, 139, 140, 142–144, 148, 149, 151, 152, 341
Public housing, 173, 179, 181, 186
Public transport, 2, 5, 6, 10, 92, 93, 100, 108, 130, 214, 215, 250, 252, 255, 257–262, 265, 436
Public transportation system, 2, 10, 93, 265

R
Rainfall simulator, 292
R&D activity, 199–203
Reform and opening policy, 326, 327, 337
Regional collaboration, 131
Regional self-sustaining development, 51–66
Regional sustention, 63
Regulation, 7, 156, 173, 180, 305–320
Residential commuting energy consumption, 6, 228–233, 235, 237, 241–245
Residential location, 156, 161, 167, 229, 232, 233, 235, 245, 367
Residential location choice, 229, 232, 233, 245
Roof garden, 289
Runoff experiments, 296–298
Runoff simulation, 298–301

S
Semboku city, 433, 434, 437
Seto Inland Sea, 74
Sewerage system, 288, 295, 296, 302
Shingu town, 355, 356, 362, 363
Small-and medium-size city, 3, 33–48, 194
Snow disaster, 432, 437, 438
Snow drainage, 434, 438, 439
Snow removal volunteer activity, 440
Social acceptability, 5
Socio-economic system, 7, 312, 317

Socio-economic vulnerability, 8, 402, 417–429
Solar radiation shading effect, 278–280
Solar shading effect, 273, 275–277, 279
South Korea, 3, 9, 17, 24
Spatial analysis, 3, 8, 10, 148
Spatial databases, 3, 8, 10
Spatial data infrastructure, 24, 31
Spatial development, 16, 17, 20, 21, 23–29, 31, 118, 312
Spatial index, 25
Spatial information, 17, 20, 24, 25, 303
Spatial unit, 24–25
Statistical unit, 20, 24, 31
Storm water runoff, 7, 287–303
Suburban, 4, 33, 36, 38–44, 46, 48, 160, 162–167, 169–171, 176, 250, 252, 343–345, 348–349, 351, 369, 383, 434
Suburban fringe, 160, 163–167, 169–171
Suburban plain zone, 344, 345
Suburban town, 176, 434
Sustainability indicators, 94, 95
Sustainable community, 8
Sustainable conservation, 139, 143, 148, 152
Sustainable landscape management, 8, 353–379
Sustainable urban form, 1–10, 229

T

Taichung, 4, 117–132, 194–198, 203, 205–206, 410–412, 414
Tainan, 194, 410–412, 414
Taipei, 194, 203, 410–414, 420
Taiwan, 4, 5, 8–10, 24, 27, 70, 117–132, 189–211, 252, 401–414, 417–429
Taoyuan, 194, 410–414
Terraced houses, 420–423
Thailand, 9, 155–172
Thermal environment, 6, 7, 10, 273, 277, 285
Three Gorges Dam, 450, 451, 453
Tourism functions, 7, 345
Townhouse development, 155–172
Townhouse settlements, 165
Toyota city, 213–225
Traditional clan system, 424
Traditional cultural landscape, 426
Traditional industry, 5, 190, 191, 200, 202, 210, 211
Traditional social connections, 420
Transportation and mobility, 97, 100, 104, 107–110
Transportation energy consumption, 6, 227–245
Transportation map, 25
Traveling behavior, 229

U

Urban and regional planning, 2
Urban and rural areas, 2, 3, 47, 52, 65, 347
Urban containment, 1, 3, 4, 9
Urban decline, 2–4, 9, 35, 38, 44, 48
Urban development phases, 39–40
Urban ecological regulation, 310, 311
Urban ecosystem, 96, 103, 108, 308, 311, 314–320, 340
Urban facilities, 4, 6, 162, 251, 252, 254–258, 260–262, 265
Urban form, 1–10, 93, 227–245, 250–253, 261, 265
Urban fringe, 160, 165–168, 171, 326, 337
Urban function, 34, 37, 46, 106
Urban growth, 2–4, 8, 9, 38, 39
Urbanization, 2–7, 9, 10, 39, 70, 118, 119, 138, 157, 166, 173, 305, 306, 308, 310, 337, 340, 402, 418, 445–447, 451
Urban natural ecosystem, 313, 316–317
Urban planning, 4, 7, 34, 46, 48, 113, 118, 160, 230, 251, 261, 265, 298, 305–320, 406, 409–411, 446
Urban planning and design, 4, 251
Urban policy, 37, 113, 249
Urban regeneration, 2–4, 91–113
Urban revitalization, 37, 38, 48
Urban–rural Landscape Improvement Movement, 118, 121, 122, 125, 290
Urban social and economic system, 313, 316, 317
Urban sprawl, 2–4, 310, 347
Urban thermal environment, 6, 10
Urban vulnerability, 2, 6, 8–10, 445

V

Vegetable production, 343–345
Ventilation, 7, 272–280, 282, 284, 285
Village cultural landscape, 5, 137–152
Village ecotourism, 138–143, 148–152
Volunteer activity, 436, 437, 440, 441

Vulnerability, 1, 2, 6, 8–10, 92, 93, 250, 306, 401–407, 409–414, 417–429, 443–453
Vulnerability assessment, 401–414
Vulnerability framework, 404

W
Walking accessibility, 6, 249–265
Water pollution, 305, 443, 451, 453
Water quality, 308, 447, 451–453
Winglet, 215–223, 225
Workshop, 332, 336, 434, 437–441

X
Xiamen, 6, 249–265

Y
Yangtze River, 445, 447, 448, 450, 452
Yokohama city, 4, 105, 106, 108, 110

Z
Zhengzhang village, 326, 328–337